Introduction to Optics

Advanced Texts in Physics

This program of advanced texts covers a broad spectrum of topics that are of current and emerging interest in physics. Each book provides a comprehensive and yet accessible introduction to a field at the forefront of modern research. As such, these texts are intended for senior undergraduate and graduate students at the MS and PhD levels; however, research scientists seeking an introduction to particular areas of physics will also benefit from the titles in this collection.

Germain Chartier

Introduction to Optics

With 440 Figures

 Springer

Germain Chartier
Institut National Polytechnique de Grenoble
38031 Grenoble, Cedex 1, France

Translated by Germain Chartier based on the French edition
(*Manuel d'optique*, Hermes, Paris, 1997).

Library of Congress Cataloging-in-Publication Data

Chartier, Germain.
    [Manuel d'optique. English]
    Introduction to optics / Germain Chartier.
       p.   cm.—(Advanced texts in physics)
    Includes bibliographical references and index.
    ISBN 0-387-40346-9 (acid-free paper)
    1. Optics. I. Title. II. Series.

    QC355.3.C4813 2005
    535—dc22

                                                                    2005042542

ISBN 0-387-40346-9   Printed on acid-free paper.

Printed in the United States of America.   (BS/EB)

9 8 7 6 5 4 3 2 1   SPIN 10936702

springeronline.com

To: Anne Marie
       Alain, Isabelle, Philippe
       Colin, Marie, Alice, Martin, Zoé

Preface

For those students who, like me, finished their studies toward the end of the 1960s, the advent of the laser was a magic new light illuminating a sector of science which had become somewhat moribund. In fact, physics seemed to be approaching an endpoint. Accelerators impelled particles at their targets where the resulting interactions did not seem to suggest any great difficulties for the perceived structure of elementary particle physics. The rapid advances in the progress of technology did not make itself generally felt; industry devoted more efforts to improving the function of automobile carburettors than seeking to harvest the fruits of research laboratories.

The history of lasers has its own fascination. Its starting point can be seen in the publication of an article in the *Zeitschrift für Physik*, in which a young physicist, Albert Einstein, compelled by the logic of his own reasoning, postulated a radically new form of interaction between radiation and matter—*stimulated emission of radiation*. That theory slept peacefully in the archives until the physicist Charles Townes, in 1956, showed that one could bring a microwave resonator into play as a basis for the realization of an ammonia "maser." Translating this concept to optical frequencies, using a Fabry-Perot resonator, Thomas Maiman succeeded in 1960 in constructing, for the first time, a coherent optical source—the ruby laser. One should understand that the advent of lasers stemmed directly from fundamental research, in that it was a discovery which owed nothing to any expectation of practical usage; it was the fruit of pure curiosity research.

Lasers appeared on the scene at just the right moment to revitalize laboratories devoted to optical instrumentation. Around 1960 the situation was as follows: geometric optics had been known for ever; wave optics almost as long; microscopes, telescopes, and spectroscopes—when adequately corrected for inherent aberrations—seemed wholly adequate. Moreover, optical physicists took quite a long time to embark on laser work; almost invariably it was in university Departments of Electrical Engineering, that the research on laser technology was initiated. This paradox stems no doubt from the fact that, in the period 1940–1960, the fantastically rapid development in radio telecommunications and in radar had led to a concentration, in these departments, of both human talent and physical resources.

The pure science origin of the laser had some unfortunate consequences in that there was a relatively long delay before applications surfaced. The fact that the discovery, which owed something to chance, had not emerged from a prolonged development during the course of which related technologies might have advanced, inhibited the rapid emergence of practical applications. Whilst the scientific community had appreciated the extraordinary potential of this new tool, the very real technological difficulties, together with the lack of experience of electronic engineers in the manipulation of optical beams, helps to explain why, twenty years after their birth, lasers were more apparent in science fiction than in factories. During the 1970s there was no shortage of witticisms at the expense of lasers: lasers are a solution in search of a problem; or, yet again, lasers are a solution for the future—destined to remain as such. . . .

Where are we now? Electronic engineers and optical physicists having largely overcome the technological difficulties set out to evolve a bewildering range of applications. These clearly are based on the coherence of the laser beams.

Spatial coherence allows beams to be focused into very small volumes (μm^3) with resulting energy densities of such a magnitude that one can envisage their use to initiate nuclear fusion reactions, or to create industrial cutting tools or again for use in surgical dentistry.

Temporal coherence of electromagnetic waves at frequencies so high (10^{14} Hz) enables a single laser beam to transmit the totality of all of the telephonic communications on the planet which are taking place at a given moment, that is, if one was able to deal with all the associated modulation and multiplexing problems. At this point we should mention another important technological development, namely the optical fibers which allow the propagation of optical signals with very low losses (fractions of a decibel per kilometer).

These applications are based on all the branches of classical optics, geometric optics, as well as diffraction effects, anisotropy, and interference phenomena. They have also brought into being the emergence of areas unknown before the appearance of lasers: nonlinear optics and the generation of ultrashort pulses. Picosecond pulses are now currently produced and the production of attosecond pulses has been reported.

The interaction of light with a material substance proceeds via the perturbed movement of the outer electrons of the atoms. One usually adopts a harmonic oscillator model to describe the movement that the optical wave imposes on the electrons; the restoring force is provided by the combined influence of the nucleus and of the other electrons. In the case of a low-intensity optical wave, the amplitude of the movement is sufficiently small so that the harmonic approximation remains valid. The electric field of a laser wave is not always negligible compared with the field experienced by the outer electrons, so that their movement can become anharmonic. Assuming that the laser beam excitation is purely sinusoidal at a frequency ν, the electrons will move periodically at the same frequency, but with an additional harmonic content at frequencies 2ν, 3ν. . . . If the waves are traveling in a suitably chosen medium, an infrared beam with a wavelength of $1.06\,\mu$m will generate a green beam with a wavelength of $0.53\,\mu$m. This new field of nonlinear optics was pioneered by Nicolas Blombergen in 1963. Frequency doubling is a specific example, but there are a host of other possibilities which bear on important practical applications such as, for example, optical beam modulation.

The aim of this book is to provide students as well as engineers with a simple account, covering both the traditional topics of optics as well as the most recent developments. Reading and comprehension of this text does not imply a significant prerequisite knowledge of the subject.

In the Spirit of Richard Feynman

Amongst all the authors of volumes intended for the teaching of physics, it is Richard Feynman who has contributed the most original ideas. The manner in which this Nobel Laureate of physics conveys the concept of "imagining" and "explaining" physics has had a major influence on generations of teachers who have sought to follow his approach.

In writing his books Richard Feynman thought about his students before considering the topics he was presenting. In his preface to *Lectures on Physics*, he explains this precisely:

The special problem we tried to get at with these lectures was to maintain the interest of the very enthusiastic and rather smart students coming out of the high schools and into Caltech.

Feynman's attitude with regard to the role of mathematics in the teaching of physics is well encompassed by G. Delacôte in his preface to the French edition of the Feynman course:

At no point is the machinery of mathematics allowed to detract from the comprehension of the physical phenomena. On the contrary, the mathematical tools emerge to respond to the problems defined by the physicist. The reader is thereby exposed both to the great problems of physics as well as to the mathematical tools which are needed to solve them in the simplest possible manner.

It is with the aim of following this *Feynmanist* philosophy of teaching physics that I have attempted in this volume on optics.

Chronology of the Discoveries of the Nature of Light

Undoubtedly in examining ancient Egyptian, Greek, Roman, Arabic, and Chinese texts one can sometimes find speculative ideas on the nature of light and how it is propagated. However, the real birth of modern optics begins in the seventeenth century.

- The law of refraction was formulated in 1621 by Snell in the United Kingdom and in 1637 by Descartes in France.
- 1657. Fermat makes a first theoretical attempt to explain the laws of reflection and refraction.
- 1665. Hooke advances the idea that light is a high-frequency self-propagating vibration.
- 1665. Newton proposes the corpuscular nature of light.
- 1669. The Dane, Bartholimus, discovers the phenomenon of birefringence.
- 1801. Young explains the phenomenon of interference by means of the wave theory of light.
- 1808. Malus discovers the polarization of light on reflection from a glass surface. From this Young and Fresnel deduce that the optical vibration must be vectorial and inherently transversal.
- 1818. In a celebrated lecture at the Academy of Sciences, Fresnel and Arago give an experimental demonstration which underlines the wave nature of light.
- 1849. Fizeau produced a highly accurate measurement of the velocity of light.
- 1876. Maxwell derives his famous equations showing that they embrace all the empirical knowledge of the behavior of light and that light is a special example of electromagnetic waves.
- 1879. Maxwell proposes an experimental system for measuring the velocity of light relative to a "hypothetical ether."
- 1881. Michelson carries out such an experiment and shows that light propagates at the same speed with respect to all references.
- 1888. Hertz verifies Maxwell's theory experimentally and shows that the generation of electromagnetic waves arises from the oscillatory movement of electric charges.
- At the beginning of the twentieth century optics advances rapidly and the nature of light is clarified.
- 1905. Einstein creates the theory of relativity according to which no material body nor any element of information can propagate faster than light in a vacuum.
- 1905. Einstein rehabilitates the corpuscular nature of light and introduces the concept of photons (photoelectric effect).
- 1920. De Broglie reconciles the two theories and introduces the concept of the dual nature of light, particle and wave, that in different situations one or the other aspect can manifest itself; he shows that electrons also have a wavelike attribute.

Lasers Made Their Appearance in 1960

Advances in telecommunications have always shown a trend toward ever higher frequencies. It was natural therefore to consider the possibility of using light waves (10^{15} Hz). Ordinary sources of light however have the unfortunate property that they emit wave packets at a high rate but in a random time sequence. Without going into detail this characteristic presents an insurmountable obstacle to the usage of such a light source for transmission of large bandwidth signals. Toward the end of the 1950s electronic engineers wondered why one might not seek to produce light waves by the same methods used to generate radio waves. The key element needed to produce a radio-frequency signal is an amplifier.

The requirement is then to discover a way of amplifying optical waves. There are not many physical processes which one could utilize for signal amplification—normally we are confined to just three:

- The control of the amplitude of an electron beam propagating in vacuum (vacuum tubes).
- The control of a current of electrons or holes in a semiconductor (the transistors).
- The process of stimulated emission which leads to the creation of lasers.

Most vacuum tubes are not able to operate at frequencies above 100 MHz; using very clever arrangements, such as gyrotrons, frequencies as high as 30 or 40 GHz can be reached; whilst transistors can be pushed to operate up to several tens of gigahertz. Around 1956 it had been remembered that Einstein, in the thermodynamic study of the interaction of an ensemble of atoms with electromagnetic radiation, had postulated a process of interaction called stimulated emission of radiation by the atoms. It was then appreciated that this phenomenon could lead to a process of coherent amplification of light. The generation of electromagnetic radiation by stimulated emission had in fact first been exploited at microwave frequencies in the realization of the ammonia MASER (1956) working at 23 GHz. Shortly thereafter it was the discovery of the ruby laser (1960) producing optical signals, in principle, well adapted to the transmission of information. The ruby laser generated pulses of coherent light, powerful but of very short duration; it was followed in 1961 by the helium-neon laser which provided a continuous output.

Since that time many new types of laser have been created and electronic engineers now share with optical physicists the privilege of amplifying coherent signals.

Acknowledgments

The publication of the French book *Manuel d'Optique*, and now of the English translation *Introduction to Optics*, would not have been possible without the friendly and efficient cooperation of many people.

First of all, AnneMarie, my beloved wife, for her patience and stimulating encouragements.

Sir Eric Ash, fellow of the British Royal Society, persuaded me that an English translation of my French book could be useful.

Many former PhD students, now well established as researchers or technical researchers, as well as engineers belonging to Teemphotonics, have agreed to read and amend this *Introduction to Optics*. Their names are given as footnotes at the beginning of the different chapters. Thank you to all of them.

Grenoble   Germain Chartier
April 2004

Contents

1

Orders of Magnitude in Optics

1.1. Main Applications of Electromagnetic Waves

1.1.1. *Electromagnetic Waves Can Carry Information*

During the twentieth century, technological development has been dominated by the use of radio waves for transporting signals over very long distances. This development is directly correlated to the property of electromagnetic (EM) waves to be propagated with no absorption in a vacuum, and with very low absorption in the atmosphere.

An electromagnetic wave gives some information about the source from which it has been emitted, often this information is not elaborated on at all, typically: *the source is ON, or the source is OFF*. Astronomical observation is very enlightening as an important indication of this; it just indicates that a star exists in a given direction and has emitted light having a certain color.

Engineers have been creative and have elaborated devices able to permanently emit electromagnetic waves, with well-controlled amplitudes and/or frequencies: it is said that a *carrier wave* has been produced, some of its characteristics being *modulated*. These same engineers have imagined and achieved devices able *to receive* the wave and *to extract* from it, this is said *to detect* the signal of modulation.

An electromagnetic wave is thus able to carry over great distances the information that is represented by the modulating signal. If, for this modulation, one uses the signal of a microphone in front of which an orator is speaking, it is seen that the speech can be broadcast. After *reception, detection,*

Sections 1.1 to 1.6 of this chapter have been reviewed by Andrew Benn who was, at that time, working for Teemphotonics and who is now with ATMEL. Sections 1.7 and 1.8 were reviewed by Dr. François Méot, Senior Physicist at the CEA (Commissariat à l'Energie Atomique).

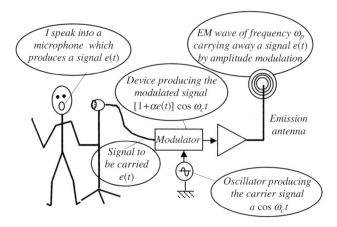

Figure 1.1. Production and emission of an amplitude modulated radio wave. A modulator is a device with two inputs and one output. Receiving $e(t)$ on one input and $\cos \omega_c t$ on the other, it elaborates an amplitude modulated signal $[1 + \alpha e(t)] \cos \omega_c t$.

and, eventually, *amplification*, the speech will be reproduced in a loud-speaker, see Figures 1.1 and 1.2.

The previous signal is rather simple, we may think of more elaborate signals, TV signals, for example, where the signal is said *to carry more information*. The more complex the signal is, the more necessary it is that the signal includes higher frequencies.

No need, at this stage, to go deeper into the details of *signal processing*; we will just consider it as evident that the frequency of the carrying wave has to be higher, by several orders of magnitude, than the frequencies involved in

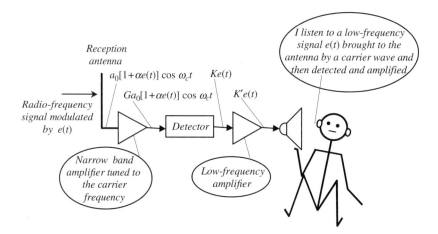

Figure 1.2. Reception and detection of a modulated wave. Receiving an amplitude modulated signal, a detector is able to extract the carried signal $e(t)$.

the signal to be carried. Then, higher and higher frequency carrying waves will prove to be necessary.

In Electronics, one knows how to produce oscillators working at frequencies ranging from zero (DC) up to tens of gigahertz (10^{10}). Optics had started to be really useful for Telecommunications only after the invention of *optical oscillators* exhibiting the same coherency properties as radio oscillators.

1.1.1.1. *In 1970, the Appearance of Optical Fibers*

Between emission and reception, a carrying wave, modulated by the signal to be carried, is propagated inside some medium. In the case of Earth radio telecommunications, this medium is just the atmosphere, in the case of Space telecommunications the medium of propagation is partly air (to reach the satellite) and partly cosmic vacuum.

Electromagnetic waves propagate with zero attenuation in a vacuum, whatever the frequency. Radio waves of course interact only slightly with the molecules that constitute the atmosphere, this is less true for light waves, and completely false if the weather is rainy or foggy. Hopefully it has been noticed that, on the one hand, amorphous silica, SiO_2, is remarkably transparent in the near infrared; and that, on the other hand, it can be given the shape of extremely long threads, inside of which light waves propagate with low attenuation ($0.16\,db/km - 3.6\%/km$, at a wavelength of $1.55\,\mu m$).

Optical fibers are able to carry light over amazing distances; oceans are now crossed by many undersea optical cables.

1.1.2. *Electromagnetic Waves Allow Material Investigation*

Any sample of material is a vast collection (think of the *Avogadro number*) of a great number of tiny objects. These objects are themselves made of even smaller particles (nuclei, atoms), that are electrically charged and may support electric and/or magnetic dipolar momentum. Electromagnetic propagation inside a material relies on the interaction of this collection of objects with the electric and magnetic fields of the electromagnetic wave.

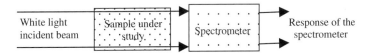

Figure 1.3. Absorption spectroscopy. White light is the superimposition of a continuum of many different lights with different colors. A spectrometer is a device which separates the various components of an incident optical signal and gives intensity variation versus frequency or wavelength. The response can be the darkening of a photographic plate. It can also be some electric signal delivered by a photodetector, in this case the signal will be displayed on a CRT screen or on a chart recorder, it can also be digitized and put into the memory of a computer.

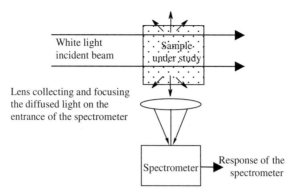

Figure 1.4. Emission spectroscopy. The incident light beam is totally, or partially, absorbed and reemitted in the 4π steradians. The spectral composition of the new light is not necessarily identical to the initial one, its analysis constitutes emission spectroscopy.

In the optical case, it is almost exclusively the interaction of the electric field with the electrically charged particles that is responsible for the *light/material interaction*. Since light interacts in a very intimate way with materials, it does constitute a powerful tool for investigating their properties. These interactions can be thought of in two different ways:

- *Either on the material side*: It is then studied how a material is modified in the presence of a light beam;

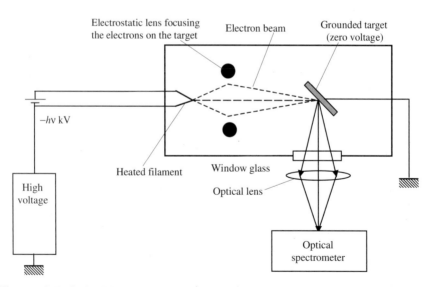

Figure 1.5. Cathodoluminescence emission spectroscopy. An electron beam of sufficient energy (usually 1–100 keV) is sent to a sample and this induces light emission, which is then analyzed by an optical spectrometer.

- *or on the wave side*: The properties of the wave are modified during propagation inside a material sample allowing some processing of optical signals.

Differential spectroscopy: An absorption spectrum is an intrinsic property of the sample under study, to obtain it with the previous method one has to be sure that the initial beam is *perfectly white*, which means that its intensity is fully independent of the frequency. This is never the case. This is why the experimental absorption spectrum should be compared with the spectrum that is obtained from the white light source. The real absorption spectrum will be obtained by taking the difference of the two spectra; so the real spectrum will be corrected to include the frequency variation of the source intensity, and also the spectral response of the spectroscope.

Emission spectroscopy: The material sample is placed under conditions where it will emit some light, this light then being analyzed in a spectroscope (see Figure 1.4). The two main methods for inducing light emission from the sample are *photoluminescence* and *cathodoluminescence* (see Figure 1.5).

Time-resolved spectroscopy: In emission spectroscopy, instead of using DC sources, pulsed sources are used. A piece of material is illuminated by short light pulses, and the analysis of the time variation of the spectral response gives useful information about the dynamics of the phenomena taking place inside the sample under study.

1.2. Wave-Particle Duality

Despite the fact that, in this book, we will almost exclusively refer to the wave aspect of light, it is not possible to start without mentioning the existence of photons.

The word *light* immediately evokes *something*, which is immaterial, and which propagates in space at very high speed. This *something* may receive two different descriptions about which we will make no attempt at any philosophical considerations. Light can be considered either as *energy grains*, called photons, or as a wave.

1.2.1. *Planck and De Broglie Relationships*

Planck and De Broglie relationships quantitatively connect the wave and particle aspects of light.

Photons are characterized by:	Waves are characterized by:
• *an individual energy W*;	• *a frequency ν*;
• *a momentum **p***.	• *a wave vector **k***.
Planck relationship: $W = h\nu$.	**De Broglie relationship: $p = \dfrac{h}{2\pi}k$.**

$(h = 6{,}626 \times 10^{-34}$ J s, *Planck's constant.)*

1.2.2. *Photons (Energy)—Light (Frequency)*

Moving from radio frequencies ($\nu \cong 10^6$ Hz = 1 MHz, $h\nu \cong 10^{-28}$ J) to nuclear physics ($h\nu \cong 10^{-10}$ J, GeV = giga electron volt, $\nu \cong 10^{24}$ Hz) the energies of photons and, consequently, the frequencies of the associated waves extend over many orders of magnitude.

> ***Optical photon energy is of the order of electron volts*** (eV).
> ***Optical frequencies are of the order of 10^{14} Hz.***

1.2.2.1. *Optical Photons Fit Very Well with Life on Earth*

Typical energies of most chemical reactions are of the order of 1 kJ/mol. If one divides this energy, first by the Avogadro number (6×10^{23}) to obtain the energy per molecule, and then by the elementary electric charge (1.6×10^{-19} C) to measure it in electron volts, the result is between 0.1 and 10 eV. As it is the peripheral electrons of atoms that are concerned with chemical reactions, it can be deduced that the bounding energies of peripheral electrons are also of the order of electron volts. This is the main reason why peripheral electrons efficiently interact with light. The former remark has important consequences about life on Earth:

- Photochemical reactions, the most important being chlorophyll synthesis.
- Production of nervous impulses in the cells of the retina of the eye.
- Photoelectric effect.

1.3. What Is a Wave?

Wave concept is closely related to two main notions:

- Time-varying phenomenon.
- Propagating phenomenon.

1.3.1. *Time-Varying Phenomenon, Necessity of a Source*

Let us suppose that, at some point $M_{(x,y,z)}$, a physical process produces a time variation of some physical parameter G, according to a law that will be written as $g_M(t)$. If we admit that the physical properties of space have to be continuous, it is reasonable to think that, at a point M' close to M, the parameter G will also vary versus time according to a law $g_M{}'(t)$ very similar to the law at M. The argument can be extended to a point M'' close to M'; which means that any variation produced anywhere will be felt later on in the surroundings.

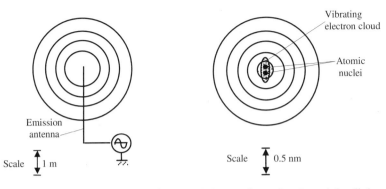

(a) Scheme of radio emission.   (b) Scheme of a molecule emitting light.

Figure 1.6. Vibrating electric charges generate electromagnetic waves. In the left-hand figure (a) an electric generator creates an AC voltage across a metallic conductor (antenna) and free electrons of the conductor are set into periodic motion with regard to the fixed positive charges of the nuclei. As a result, a roughly spherical wave is generated (the circles are supposed to suggest the shape of wave surfaces). The right-hand figure (b) represents a diatomic molecule: a cloud of peripheral electrons surrounds the two positive nuclei, keeping them at a roughly equal distance. When the molecule becomes excited, the cloud vibrates, generating an electromagnetic wave.

The main point of what has just been said, is the necessity, for a wave to be created, of a zone where a physical process will provide the energy required for generating the time variation of the parameter G. In the case of electromagnetic waves, the parameter G is an electromagnetic field.

Starting from radio frequency up to ultraviolet radiation, most electromagnetic sources originate from the *vibrations* of some electric dipole: *two electric charges with opposite signs being set into a periodic motion*. In the surrounding space *an electric field and a magnetic field* are simultaneously generated, synchronously vibrating.

1.3.2. *Electromagnetic Waves and Einstein's Relativity Principle*

The determination of electric and magnetic fields in the vicinity of moving electric charges is more complicated than would be supposed from a simple application of the Coulomb and Biot-Savart laws that are only valid for DC phenomena. *Maxwell's equations have to be used.* The main physical result is the following:

- There is no modification of the fields as long as the motion is rectilinear and uniform. Moving charges will produce some electromagnetic radiation if, and only if, they are *accelerated*.
- The effect of the acceleration is not felt immediately at a distant point. If the acceleration occurs at time $t = 0$, the radiated field will not be felt,

before a time $t = d/c$, by a physicist sitting at a distance d away from the charge (c is called the light speed in a vacuum): time must be left for the radiated field to reach the observer.

In the case of an oscillation, the acceleration is not permanently equal to zero: this is the reason why the antenna permanently radiates an electromagnetic field.

The above result is in agreement with the Einstein relativity principle, according to which no information should propagate faster than the light in a vacuum.

1.3.3. Description of a Propagating Phenomenon

1.3.3.1. The Example of Elastic Surface Waves

Physics is dominated by propagation phenomena. To give a taste of the main notions that are involved in propagation, we will start with a simple case, which has nothing to do with electromagnetic waves, and we will describe elastic surface waves propagating along the free surface of a liquid, the surface of a lake, for example.

Each of us has surely observed those waves that can be excited when stones are thrown into a lake. The small value (meter per second) of the speed of propagation makes it easier to feel what a propagation phenomenon is exactly.

The different points of the free surface of the liquid shown in Figure 1.7 are bound to one another because of the Van der Vaals forces that are responsible for capillarity.

Except in the immediate vicinity of the central point, the surface at time t_2 will have the same general shape that it had at time t_1: a circular ridge, the radius of which increases with time.

Permanent waves: Instead of dropping a stone, a float is placed on the surface and, in some way, is given a periodic motion: new circular ridges are

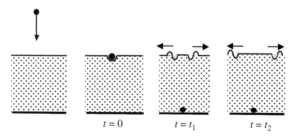

$t = 0$ $t = t_1$ $t = t_2$

Figure 1.7. Propagation of waves along the free surface of a liquid. A stone is dropped and reaches the surface at $t = 0$; the surface is distorted and, for the sake of continuity, the distortion is transmitted to the neighboring points: a circular ridge having a profile quite similar to the initial distortion propagates, starting from the point where the stone initially hits the surface.

permanently created at the level of the oscillating float. For example, permanent elastic surface waves can be excited by letting a needle slightly touch the surface, the needle being linked to an electrically excited tuning fork.

Waves with surface waves having any shape: Instead of only one, we can use many oscillating floats, each of them emitting a spherical wavelet similar to those that have just been described. The oscillation of a point of the liquid surface is the result of the actions of the different floats. Figure 1.8 gives an illustration of what happens if the oscillators are all synchronized and located along a straight line, it is easy to imagine that a planar wave can be generated. It should be easy to synchronize the children's motions by playing music.

More elaborate scenarios can be imagined. The edge of the swimming pool can be made circular, the envelope of the wavelets is now circular: a circular wave is then generated.

A rectilinear edge can still be used, but the children are now independently addressed by listening to the music through earphones that are individually excited, thanks to radio links, for example. The rhythm is the same for all the children, but the phase may be varied from one to the other. When all the children hit the surface "in phase," a rectilinear wave is generated parallel to the edge. If the phase repartition is varied linearly versus the distance from the child sitting in the middle, a rectilinear wave is still generated, but it propagates obliquely. By properly choosing the phase repartition, any shape may be given to the wave surfaces. It is left to the reader's imagination to determine which phase repartition is required for circular waves.

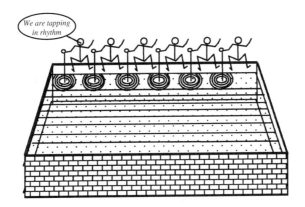

Figure 1.8. Generation of rectilinear waves from circular wavelets. Children are sitting along the rectilinear edge of a swimming pool. With rods, they hit the water surface generating circular waves. If the children hit at random, the surface takes on a chaotic appearance. Their motions can be synchronized, for example, by playing some rhythmic music. Close to the edge the surface has a fuzzy shape, further away many waves will interfere to give a rectilinear wave which is the envelope of the circular wavelets.

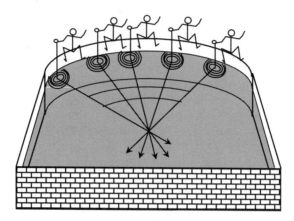

Figure 1.9. Generation of circular waves from circular wavelets. The swimming pool now has a circular edge; when the children periodically and synchronously hit the surface, the envelope of the wavelets takes the shape of circles having the same center as the circular edge.

1.3.3.2. *Mathematical Description of Propagation: $f(t \pm x/v)$*

Let us consider a function $f(u)$ of the variable u, and suppose that u is a linear combination of space x and of time t, according to a law that will be written as $u = (t \pm x/V)$, V is a parameter homogeneous to a velocity.

The function $f(u)$ then depends on x and t according to $g_{M(t,x)} = f(t \pm x/V)$.

Let 0, t_1, and t_2, respectively, be the initial time and two subsequent times, since $f_{(t=0,x=0)} = f_{(t_1,x_1=Vt_1)} = f_{(t_2,x_2=Vt_2)}$, it is seen that, if V is positive, the function $f(t - x/V)$ describes a forward propagation along the x axis, at a velocity equal to V, of the physical parameter attached to f. In the same way, the function $f(t + x/V)$ represents a backward propagation along the x axis.

Since y and z do not appear in the argument of the function $g_{M(t,x)} = f(t \pm x/V)$, g_M has the same value for all the points located in a given plane orthogonal to the x axis: surface waves are thus *parallel planes*, this kind of wave is said to be *planar*.

A function such as $g_{M(t,x)} = f(t \pm x/V)$ corresponds to propagation without any deformation of the initial time variation law: whatever the position, the physical parameter $g_M(t, x)$ varies according to the same law and with the same amplitude. As these variations represent some energy, it is foreseen that the previous description is ideal. A more realistic function could be $a(x)f(t \pm x/V)$, where $a(x)$ is a function which decreases with the modulus of the abscissa x. When propagating inside a material the wave will give energy to the atoms, a planar wave will not keep a constant amplitude. A formula, which is frequently encountered, is of the type:

$$e^{-ax}f(t - x/V), \quad \alpha \text{ is called the } absorption \ coefficient \text{ of the material.}$$

Spherical waves are described by $g_{M(t,r)} = a(r)f(t \pm r/V)$, where r is the distance to the center of the wave, which is often called the focus of the wave. The $\pm$ sign corresponds either to waves that diverge from the focus ($-$) or converge toward it ($+$).

If we admit that the power carried by a wave is proportional to the square of its amplitude, it is seen that $a(r)$ must be proportional to $1/r$, in order to keep constant the flux of energy across the different spheres centered at the focus.

For spherical waves propagating inside an absorbing material, the law $a(r)$ will take the shape $a(r) = e^{-\alpha r}/r$.

1.3.3.3. *Sine Waves (also Called Harmonic Waves)*

A wave is said to be harmonic when the time variation is sinusoidal. An harmonic planar wave will thus be written as $g_{M(u)} = A\cos u = A\cos[\omega(t \pm x/V)]$.

As harmonic planar waves are very useful, we will summarize the main parameters that are used to describe them.

Main parameters used to describe an harmonic planar wave.

Frequency, ν.

Pulsation ω (also called angular frequency), $\omega = 2\pi\nu$.

Period, $T = \dfrac{1}{\nu} = \dfrac{2\pi}{\omega}$.

Speed of propagation, V.

(Beware of typographical confusion between speed V and frequency ν.)

Wavelength *(space period)*, $\lambda = VT = \dfrac{V}{\nu} = \dfrac{2\pi}{k}$.

Wave vector module *(space pulsation)*, $k = \dfrac{2\pi}{\lambda} = \dfrac{\omega}{V}$.

Wave number *(space frequency)*, $N = \dfrac{1}{\lambda} = \dfrac{k}{2\pi}$.

The different forms of the argument of the propagation function.
 Since the propagation function $f(u)$ is sinusoidal and so is not linear, to cope easily with a change of units, its argument u should not have any dimension

$$u = [\omega(t - x/v)] = (\omega t - kx) = 2\pi(t/T - x/\lambda).$$

Wave vector of a planar harmonic wave: $\boldsymbol{k} = k\boldsymbol{x} = \boldsymbol{x}(2\pi/\lambda) = \boldsymbol{x}(\omega/V)$, where $\boldsymbol{x}$ is the unit vector of the direction of propagation.

The space pulsation $k = \omega/V = 2\pi/\lambda$ is a scalar number, which is often abusively called the *wave vector*.

1.3.4. *Schema of an Electromagnetic Propagation Experiment*

Figure 1.10 clearly shows the three main components of any electromagnetic propagation set-up: a source, a medium supporting the propagation, and, finally, a detection device that will reveal and possibly use the electromagnetic radiation.

In Optics, of course, the same elements are found: emission antennas are replaced by excited atoms. However, the physical mechanisms that are responsible for emission and detection are less intuitive.

The spectral composition of the electromagnetic field at any point is determined by the source, and only by the source. If the source is sinusoidal with a frequency ν, the time variation of the electromagnetic field at any point will also be a sine of the same frequency. Exceptions to the previous rule are very uncommon, and will occur under special conditions, when "some nonlinearity" appears in the interaction between the electromagnetic field and the atoms of the propagation medium; this situation will be considered in Chapter 10 devoted to Nonlinear Optics.

Frequency is thus an intrinsic parameter of the problem and is the same everywhere. This is not the case for the speed of propagation V that usually depends on the properties of the medium of propagation. It is the same for the different parameters that are linked to V, such as wavelength and wave vector.

In Electronics, as well as in Telecommunications, people almost exclusively make use of the frequency, or of the angular frequency, to describe the

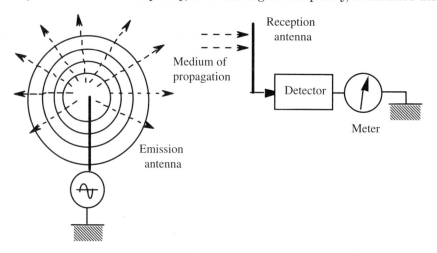

Figure 1.10. Propagation of a radio wave from the emission antenna to the reception antenna. An electric generator sets the free electrons of the antenna into motion, creating an electromagnetic field which propagates, starting from the antenna. This electromagnetic wave propagates in the vacuum (or in air) which constitutes the propagation medium. When reaching the reception antenna, the electric field of the wave makes the free electrons oscillate, thus generating an electric current which is detected by some measurement device.

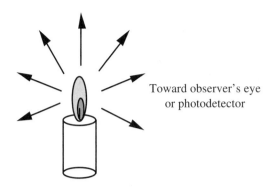

Figure 1.11. Production of light by a candle. The heat produced by the combustion of the wax provides energy to the air molecules and carbon atoms that are present in the flame. Part of this energy will excite the electrons, which vibrate with regard to the nucleus, at some frequency characteristic of the atom to which they belong. Electromagnetic radiation is emitted and, when arriving at a detector, it will excite its electrons producing a physiologic signal (eye) or an electric signal (photodetector).

signals. In Optics the wavelength is usually preferred; if not indicated, it is the wavelength in a vacuum. In Spectroscopy, the wave number (in cm^{-1}) is widely used, while in Chemistry the photon energy is frequently given in electron volts.

1.4. Electric Dipole Radiation

1.4.1. *Luminous Objects and Illuminated Objects*

Two kinds of visible objects have to be distinguished, on the one hand, light sources, and on the other hand, illuminated objects:

- Light sources receive from the outside some energy which is, totally or partially, converted into light.
- Illuminated objects receive light from some sources, then they reemit light waves having properties (frequency, coherence, polarization, . . .) in general very similar to that of the illuminating light.

Two kinds of light sources:

- *Classical sources*: These were the only ones before the invention of lasers. Most light sources that we use belong to this kind. Natural sources, such as the Sun or stars, as well as the sources that we make for lighting: candles, electric lamps where a wire is heated, or an electric discharge in a gas. These classical sources are said to be incoherent; they are never strictly monochromatic, the radiation that they produce always has a spectral broadness which may be quite large. We will come back later to the concept of incoherency. At this stage we will only say that interference phenomena cannot be observed by superimposing lights coming from two different classical light sources, as well as light coming from two different points of the same source. Strictly speaking, it's better that classical sources are not coherent, otherwise the world would appear as stripes with interference fringes.

- *Laser sources*: These are coherent and often very monochromatic.

For the two kinds of sources, the mechanism which is responsible for the emission of light is dipole electric radiation.

1.4.2. *Phenomenological Approach of the Motion of Electrons Inside an Atom*

1.4.2.1. *Phenomenological Equations*

A luminous object is always a piece of material, i.e., a collection of atoms. This collection receives energy and transforms it into electromagnetic energy. To start with we must imagine a model for the physical processes according to which an atom receives energy and emits electromagnetic waves.

Our approach will be phenomenological, that is to say that the phenomena will be described as we imagine them to be, *without any special attempt to be rigorous*. The only real justification of our equations is that they give good results in agreement with experimentation.

In an atom at rest we consider that electrons have equilibrium positions with regard to the nucleus, we then admit that, after having been released from those positions, electrons will go back to equilibrium by doing damped oscillations. The oscillation frequency is characteristic of an atom and of the special electron under consideration in an atom, its value is mainly determined by the electronic shell on which the electron is (K, L, M, ..., s, p, d, f, ...) and may correspond to very different spectral domains (X-ray, ultraviolet, visible, infrared, microwave, radio frequency).

Let x be the distance between an electron and the nucleus, and let x_0 be its value at rest; the distance to the equilibrium position is noted as $\xi = (x - x_0)$. If taken away from the equilibrium position, the electron is submitted to a restoring force proportional to ξ.

m and e being, respectively, the mass and the electric charge of the electron, k and γ' being the *phenomenological coefficients* and, finally, $f(t)$ being the value, at time t, of the electric field at the place where the electron is, the equation of motion of the electron can be written as

$$m\frac{d^2\xi}{dt^2} = -k\xi - \gamma'\frac{d\xi}{dt} + ef(t),$$

$$m\frac{d^2\xi}{dt^2} + \gamma'\frac{d\xi}{dt} + k\xi = ef(t).$$

Changing slightly our notations, we obtain the phenomenological equation (1.1),

---

Differential equation of the motion of an electron,

$$\frac{d^2\xi}{dt^2} + \gamma \frac{d\xi}{dt} + \omega_{\text{atom}}^2 \xi = \frac{e}{m} f(t). \tag{1.1}$$

Equation (1.1) is a second-order linear differential equation. It is well known that its solution is obtained by adding two terms:

- *Free regime*: This is the general solution of the differential equation obtained when the right-hand side of the equation is made equal to zero.
- *Forced regime*: This corresponds to a solution of the equation obtained when the right-hand side of the equation is made equal to $(e/m)f(t)$.

γ will be considered to be small enough, thus the free regime is damped and can be written as

$$\xi = \xi_0 e^{-\gamma t/2} \sin(\omega t - \varphi) \quad \text{with } \omega \cong \omega_{\text{atom}}, \tag{1.2}$$

where ξ_0 and φ are integration constants determined from initial conditions.

The damping coefficient γ does not correspond to any viscous friction, it is difficult to imagine in what kind of viscous medium the electron should move, γ is really a *phenomenological coefficient*. Its presence does represent a dissipation of energy. Instead of γ, a damping time $\tau = 1/\gamma$ is often introduced.

ω_{atom} is the eigenvalue of the angular frequency of the electron in the atom, from an experimental point of view it corresponds to an absorption band of the material. ω_{atom} may belong to the whole spectrum of electromagnetic radiation, between X-ray and radio frequency.

1.4.2.2. *Shape of the Solution in the Case of Light Sources*

To be able to emit dipole electric radiation, an atom should have previously received some energy from the surrounding medium. In most cases the energy is provided to the atom in a random way, during atomic collisions. The collisions may occur with other atoms, or with thermal phonons of a crystalline lattice, or with electrons of a gas discharge. . . .

Things can be thought of as follows: at random times θ_i (i is just an index of numeration), the oscillator describing the electron receives *very short bursts of energy*.

We now come back to the phenomenological equation (1.1), the right-hand side is almost always equal to zero, except during very short intervals very close to θ_i. The average value of the interval, $(\theta_{i+1} - \theta_i)$, between two con-

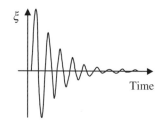

Figure 1.12. Motion of an electron going back to equilibrium. The damping time constant τ is far longer than the pseudo-period $T_{atom} = 2\pi/\omega_{atom}$, so the diagram is just an illustration, since it is impossible to draw with the same scale the exponential and the sine.

secutive collisions is longer than the damping time constant $\tau = 1/\gamma$; under such conditions, the motion of an electron is a succession of damped oscillations that are reinitiated from time to time. Because of the large difference of the orders of magnitude between $T_{atom} = 2\pi/\omega_{atom}$ and $\tau = 1/\gamma$, there is time enough for many oscillations to be produced before complete damping.

It will be considered that the amplitude of the electromagnetic field emitted by the oscillating dipole follows the same time variation law as the motion of the electron. The electromagnetic field is just a succession of what we will call "*wave packets.*" The initial phases of the different wave packets are not correlated to one another.

One can be tempted to assimilate a wave train and a photon, this is completely false, and *each wave train does correspond to many photons.*

For a coherent beam—emitted by a perfect laser—the graph of the variation of the electromagnetic field amplitude would be an *everlasting sine*, i.e., a single wave packet of infinite damping time.

It is not at all obvious, and certainly not true, that all the wave trains of Figure 1.13 should have the same amplitude; however, for the sake of sim-

Figure 1.13. Representation of the motion of an electron of an incoherent source and of the variation of the electromagnetic field produced by the oscillatory motion of the electron. The motion is reinitiated from time to time. The initial phases of the different damped sinusoidal wave trains are not correlated to one another.

plicity, we will admit that this is the case. The time variation of the emitted electromagnetic field will be written as

$$E(t) = E_0 \sum_p H_{(t-\theta_p)} e^{-(t-\theta_p)/\tau} \cos(\omega_{\text{atom}} t - \varphi_p). \tag{1.3}$$

$H(u)$ is a Heaviside step, equal to zero when its argument u is negative and equal to one for any other value.

1.4.2.3. Object Illuminated by a DC Coherent Light Source

The electric field of the light coming from a DC coherent source follows a sinusoidal law of constant amplitude E_0 and angular frequency ω.

The motion of an electron belonging to an object illuminated by a coherent light source is described by the following equation:

$$\frac{d^2\xi}{dt^2} + \gamma \frac{d\xi}{dt} + \omega_{\text{atom}}^2 \xi = \frac{e}{m} E_0 \cos \omega t. \tag{1.4}$$

The best way to solve equation (1.4) is to use complex numbers, the solution then has the following shape:

$$\xi = |\xi| \cos(\omega t + \phi) = \text{Re}\{|\xi| e^{j\phi} e^{j\omega t}\}.$$

It can be shown that the squared modulus of the variation is given by

$$|\xi|^2 = \frac{e^2}{m^2} E_0^2 \frac{1}{(\omega_{\text{atom}}^2 - \omega^2)^2 + \gamma^2 \omega^2}. \tag{1.5}$$

The amplitude of the motion is maximum at resonance, when the light frequency is equal to the eigenfrequency of the electron of the illuminated object.

1.4.2.4. Object Illuminated by an Incoherent Source

Let us now consider the case of an object illuminated by an incoherent light source. The electric field which must be written on the right-hand side of the differential equation (1.1) is given by (1.3):

Most of the time:

$$\frac{d^2\xi}{dt^2} + \gamma \frac{d\xi}{dt} + \omega_{\text{atom}}^2 \xi = 0. \tag{1.6}$$

When a collision occurs $(t \cong \theta_i)$:

$$\frac{d^2\xi}{dt^2} + \gamma \frac{d\xi}{dt} + \omega_{\text{atom}}^2 \xi = \frac{e}{m} E_0 e^{-(t-\theta_i)/\tau} \cos \omega t. \tag{1.6'}$$

There are two kinds of parameters in equation (1.6):

- γ and ω_{atom} are characteristic of the atoms of the illuminated object.
- E_0, ω, and τ are characteristic of the illuminating source.

For the sake of simplicity we will consider the motion of electrons belonging to the illuminated object as far more damped than the motion of the electron of the light source. The free regime of the illuminated electrons rapidly goes to zero and will not be considered, keeping only the forced regime of equation (1.6′).

The right-hand side of equation (1.6′) can be written as

$$\frac{e}{2m} E_0 e^{-(t-\theta_i)/\tau} \left[e^{j\omega t} + e^{-j\omega t} \right] = \frac{e}{2m} E_0 e^{(\theta_i/\tau)} \left[e^{(-1/\tau+j\omega)t} + e^{(-1/\tau-j\omega)t} \right].$$

It is easy to see that the forced regime of equation (1.6′) is a combination of terms such as $(e/2m)E_0 e^{(\theta_i/\tau)} e^{(-1/\tau \pm j\omega)t}$. A rather tedious calculation shows that the motion of the electron looks very much like the wave packets of the incoherent light source. The amplitude of the motion of the electron is proportional to the amplitude of the wave packet, and the proportionality coefficient is higher as the wave packet frequency is nearer to the eigenfrequency of the electron in its atom.

1.4.2.5. *Colored Appearance of Objects*

Objects have the same color as their absorption bands.

Let us now consider what happens when some object is illuminated by a white classical source. A white classical source emits incoherent wave packets having any frequency in the visible spectrum. Suppose that the object is absorbing in the red, which means that its atoms have an eigenfrequency ω_{atom} in the red. The different spectral components of the white light will be reemitted with an intensity proportional to the incident intensity, as the proportionality coefficient is larger for red light, the object will be of a red appearance.

In fact, a given object always has several absorption bands, their spectral widths are never perfectly thin and they are characterized by a *profile*, i.e., a law of variation of the proportionality coefficient versus frequency, or wavelength. Illuminated by a white source, such objects will take a color which is characteristic of the profiles of the different absorption bands. Of course, the spectral composition will also play an important role in determining the aspect of the object. For example, whatever its color under a white light, when illuminated by a monochromatic blue light, an object can only reemit blue light.

A white object is an object that has no absorption band in the visible, and thus rediffuses with the same efficiency the various components of a white light. A black object absorbs, with an efficiency that is higher as the object is darker, all the visible components of the spectrum.

Trichromatism: The color attributed to an object is the result of the appreciation, by an observer, of the superposition of the various frequencies that have been reemitted under white illumination. Experiment shows that the observer may be accurately given the impression of any color, by mixing the lights of three sources having three different colors. The sources should have narrow enough spectra, or either be monochromatic. The color is tuned by changing the proportions of the three components.

1.4.3. Radiation Emitted by an Oscillating Electric Dipole

1.4.3.1. Physical Significance of the Electric Field and of the Magnetic Field

This part of the book is very favorable in getting students to think about *long distance actions* and *radiated fields*. The notion of *field* has two aspects:

- First, it is some physical property, electric, magnetic, acoustic, ..., that is described as a mathematical object, such as a scalar, a vector, a tensor, ...
- Second, it is a portion of space inside of which this physical property has been modified, thanks to long distance actions.

The creation of a field, whatever its nature, always needs energy. This energy is stored at the place where the field exists, with a density that is often proportional to the squared modulus of the field.

In the case of Electromagnetism, *Coulomb interaction* is the long distance action responsible for the existence of an electric field, the *Biot-Savart interaction* being responsible for the magnetic field.

The Coulomb Law	**The Biot-Savart Law**
Placed at some point O, an electric charge e will modify the properties of the surrounding space by creating an electric field at any point M, $$E = \frac{1}{4\pi\varepsilon_0} e \frac{OM}{OM^3}.$$ An electrostatic energy is accumulated in space with a density $$\varpi_{\text{electric}} = \tfrac{1}{2}\varepsilon_0 E^2.$$	A moving electric charge e, passing point O at a velocity V, modifies the properties of the surrounding space by creating a magnetic field at any point $$B = \frac{\mu_0}{4\pi} e \frac{(V \wedge OM)}{OM^3} \quad \text{and} \quad H = B/\mu_0.$$ A magnetic energy is accumulated in space with a density $$\varpi_{\text{magnetic}} = \frac{1}{2\mu_0} B^2 = \frac{\mu_0 H^2}{2}.$$

1.4.3.2. Weaknesses of the Coulomb and Biot-Savart Formulas in Describing Waves

An oscillating dipole is, at the same time, *two charges* and an *electric current*, this is the reason why it generates an electric and a magnetic field. Unfortunately, the calculation of the fields is more complicated than one would suppose, just thinking of the Coulomb and Biot-Savart laws, since they would imply that any modification occurring somewhere would be immediately felt all over the space, which is incompatible with the Einstein relativity principle.

Oscillating dipole radiation is well described, in many Electromagnetism textbooks, as the topic *Hertz dipole*. Calculations are developped starting from Maxwell's equations, the results are quite intuitive for today's physicists.

1.4.3.3. *Field Radiated from an Oscillating Dipole*

Let us consider a dipole made of an electron and a positively charged nucleus, its electric dipole moment μ is the product of the absolute value of the elementary electric charge e by the vector d joining the nucleus to the electron $\mu = ed$. We now imagine that d is imposed as a sinusoidal time variation, so that the electric moment will follow the same kind of time variation.

At a distance OM which is large as compared with the wavelength, field expressions become rather simple expressions:

E is orthogonal to OM and inside the plane Π defined by OM and μ (this is very different from what would be obtained from Coulomb's law), see Figure 1.14.

H is orthogonal to OM and also orthogonal to the plane Π,

$$E = E_0 \frac{1}{r} \sin[\omega_{\text{atom}}(t - r/c)], \quad H = H_0 \frac{1}{r} \sin[\omega_{\text{atom}}(t - r/c)].$$

1.4.3.4. *Remarks*

- Because of the $1/r$ law for the field, and the $1/r^2$ law for the radiated power, the amplitude and intensity decay are relatively slow. This is a typical characteristic of such *electric interaction*, in opposition to nuclear interactions, the range of which doesn't extend much further than the radius nucleus.
- The amazing range of electromagnetic waves has enormous practical consequences. For example, the possibility for light to propagate along astro-

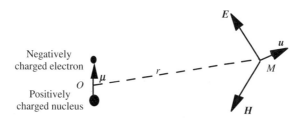

Figure 1.14. Field radiated from an oscillating dipole. The expressions of the electric and magnetic fields are quite complicated in close vicinity to the dipole, but they become considerably simpler if the distance OM is large, as compared with atomic size and with the distance covered by light during one period (wavelength).

nomic distances, which allows us to receive light rays from the extreme edges of the universe. This is responsible for the ability of radio waves to propagate over large distances.

1.5. Light Detectors

1.5.1. *Definition of Detection*

An electromagnetic wave is an alteration of space, a modification which is due to the action of a source. This becomes real and exists only after it *has been detected*, which means after it has interacted with a *detector*, that is to say, a device which delivers a signal that can be measured.

An electromagnetic wave represents information concerning the emitting source, this information is carried to the detector. The kind of information that is carried may be as simple as: the source is on, or the source is off; however, it can also be very elaborate, due to sophisticated modulation procedures.

To detect is to destroy: As long as the signal has not been detected, it keeps transporting its information. The signal stops existing as soon as it has been detected. Truly speaking, a detector usually receives a small part of the wave, and so, only a small fraction of the signal will be destroyed by detection.

The information of an electromagnetic wave may be carried by one of the four basic parameters that are: amplitude, frequency, phase, and polarization.

Detection is the operation which allows us to obtain again the information carried by the wave, and which brings it to the attention of an observer who can then make use of it.

Detection has a mathematical aspect, as well as a technological one. The reader will usefully think of the following examples: astronomy, radar, broadcasting and, more generally, telecommunications, spectroscopy, ...

We have already met the notion of detection in this chapter, this was on the occasion of the transportation of information, using a carrying wave. We then considered a wave of the following form, which is called *amplitude modulation*,

$$y(t) = [1 + \alpha e(t) \cos \omega_{\mathrm{car}} t].$$

The information is contained in the function $e(t)$, the frequency spectrum of which should only have components having frequencies far lower than the frequency ω_{car} of the carrying wave. The role of the carrier wave is just to *bear* the signal. In this case detection is made in several steps:

(i) Reception by the antenna of the electromagnetic wave and generation of an electric signal $z(t)$ proportional to $y(t)$.

(ii) Starting from $z(t)$, elaboration of another electric signal proportional to the initial signal $e(t)$.

(iii) Giving the previous electric signal a more useful shape for utilization (an acoustic signal, for example).

In radio telecommunications it is often considered that detection only corresponds to topic (ii). In Optics, because of the extremely high value of light wave frequencies, the two operations (i) and (ii) are made at the same time by the *photodetector*.

From the previous examples we will be reminded that, if modulation and detection are, in essence, signal processing operations (i.e., mathematical operations) to be performed, they will always involve physical and technological processes.

1.5.2. *Measuring a Power in Decibels*

The power carried by a signal is obviously an important parameter: the more powerful a signal, the easier it is to detect. The exact definition of the power transported by a wave will be given in the next chapter, for the moment, it will be considered as intuitive.

A signal processing set-up is always made of several successive elements, each of which is characterized by the ratio of the transmitted power to the received power (transmission coefficient):

$$\rho_i = \frac{\text{Power}_{\text{transmit}}}{\text{Power}_{\text{received}}}.$$

The output power is equal to the product of the initial incident power multiplied by the transmission coefficient ρ_i of the different elements. It is because the human brain prefers performing summations rather than multiplications, that the *decibel has been introduced*. Given two physical parameters of the same physical kind, G and G', and their usual ratio, G/G', we will call *a ratio in decibels* (db) by the following expression:

$$\left(\frac{G}{G'}\right)_{\text{db}} = 10 \, \log_{10}\left(\frac{G}{G'}\right). \tag{1.7}$$

Decibels fit very well with the way physicists usually speak by order of magnitude. It's not certain that the factor of 10 makes things easier, tradition has not decided so . . . , if this were not the case *bel* would be used instead of *decibel*.

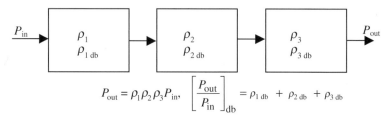

$$P_{\text{out}} = \rho_1\rho_2\rho_3 P_{\text{in}}, \quad \left[\frac{P_{\text{out}}}{P_{\text{in}}}\right]_{\text{db}} = \rho_{1\,\text{db}} + \rho_{2\,\text{db}} + \rho_{3\,\text{db}}$$

Figure 1.15. Cascade of several signal processing components. Each component is characterized by a transmission coefficient.

Scientists from Optics, or from Electronics, have a slightly different understanding of decibels. Because of the great value of the light frequency, in Optics we have no direct experimental access to the instant field, $E(t) = \hat{E}\cos(\omega t - \varphi)$ ($\hat{E}$ is the amplitude of the field), and usually we deal with its averaged squared value. On the contrary, in Electronics, the fields are easily handled and the time variations can be displayed on the screen of a cathode ray tube (CRT). So, in Electronics, the temptation was high to extend the notion of decibels to the field's amplitudes.

First remark that since a logarithm is only defined for positive numbers, decibels can only be used for the modulus of a signal. The power of an AC signal being proportional to the square of the absolute value of its amplitude, if we want the expressions to be comparable for both amplitude and power, a factor of 20 (instead of 10) must be introduced in the definition of decibels in the amplitude case,

$$\left(\frac{\hat{E}_{\text{out}}}{\hat{E}_{\text{in}}}\right)_{\text{db}} = 20\log_{10}\left(\frac{\hat{E}_{\text{out}}}{\hat{E}_{\text{in}}}\right).$$

The use of decibels may be troublesome for beginners, and also for others. It must always be kept in mind that decibels have been introduced to describe the ratio between two powers.

Decibel milliwatt (dbmW)

By definition only ratios can be expressed in decibels. It's nonsense to express a power in decibels, nevertheless, this is very often the case. When a power is given in decibels, a reference power has, more or less, been implicitly introduced. As in optical telecommunications the powers are of the order of milliwatts, the reference power is 1 mW. The number giving the power will be followed by "dbmW" (decibel milliwatt),

$$P_{\text{dbmW}} = 10\log_{10}\left(\frac{P}{P_0 = 1\,\text{mW}}\right) \quad \rightarrow \quad P_{\text{mW}} = \left[10^{\frac{P_{\text{dbmW}}}{10}}\right],$$

$0\,\text{dbmW} \rightarrow 1\,\text{mW}, \quad 10\,\text{dbmW} \rightarrow 10\,\text{mW}, \quad 20\,\text{dbmW} \rightarrow 100\,\text{mW}.$

1.5.3. *Physical Considerations About Photodetectors*

The first point is to understand the physical processes that permit a photodetector to elaborate a signal from the energy that is brought to it by a beam of light. The second point is to find the best way to make this signal useful.

1.5.3.1. *Photochemical Reactions*

These are used in photographic plates, the most frequently used being the chemical reduction of silver salts in the presence of light:

$$Ag^+ + electron + photon \rightarrow Ag.$$

The silver ion belongs to a silver halide, a water insoluble salt that is immobilized inside a thin film of gelatin. As long as it remains in darkness, the silver halide stays white. The metallic silver atoms are black: the darkening of the film is controlled by the light exposure.

1.5.3.2. *Production of a Nerve Impulse*

When retina cells receive visible photons, a type of photochemical reaction occurs, the energy of the reaction induces a physiological signal: a nerve impulse is produced and driven to the brain, which is well equipped for processing and interpreting the signal.

Sensitivity: The human eye is remarkably sensitive; it can detect a *few photons per second*. Of course, it is the spectral sensitivity of the human eye that defines what is called the *visible domain*, which ranges from the red ($\lambda = 0.76\,\mu m \cong 0.8\,\mu m$) to the blue ($0.4\,\mu m$).

Response time: The eye is not a fast detector; its response time is of about $0.1\,s$. This slowness is compensated for by the enormous amounts ($\sim 10^8$) of eye cells that are equivalent to photodetectors working in parallel. The fascinating power of the human vision process really comes from the efficiency of the way in which our brains can process in parallel all the data coming from the retina cells.

Resolving power: The human retina has a surface area a little smaller than $1\,cm^2$; it is made of cells having a diameter d of a few micrometers. For two points to be seen separately, their images should be focused on two different retina cells. If Δ is the diameter of the eye sphere, the angular resolving power ε is equal to the ratio $\varepsilon = d/\Delta$.

From a practical point of view the human eye angular resolving power is 1 min (3×10^{-4} rad).

1.5.3.3. *Classification of Photoelectric Detectors*

The two previous detectors were of a very special kind, in all other cases the light signal is transformed into an electric signal (current or voltage): *photons create charge carriers which then flow in electrical conductors*.

At this stage of the book it cannot be omitted that it is on the occasion of the discovery of the photoelectric effect that Einstein got the idea of introducing *photons*.

Main Properties of a Photodetector

- *Spectral response*: Mainly fixed by the work function of the photocathode material or by the energy gap in a semiconductor. In most cases it corresponds to wavelengths shorter than $1\,\mu m$.
- *Sensitivity*: This gives the electrical intensity given by the photodetector versus optical power, usually expressed in milliamps per watt.
- *Quantum efficiency*: The probability of a photon impinging on a photocathode to generate a photoelectron is called "quantum efficiency." Of course, this smaller than one, the quantum efficiency of good detectors is between 0.1 and 0.5.
- *Darkness current*: Even with no incident light a photodetector gives a small current which is called the darkness current. This phenomenon, of course, limits the possibility of exhibiting low signals. For each detector, it is defined as a *least detectable light flux* ϕ_{min}, the flux that produces a current equal to the darkness current.
- *Dynamic range*: A very powerful incident beam will generate an intense current and probably damage the detector. A detector is characterized by a maximum allowed incident power ϕ_{Max}. The detector dynamic range is the difference between ϕ_{Max} and ϕ_{min}.
- *Zone of linear response*: The most favorable conditions are those where the photoelectric current is proportional to the light power, for this to be the case, it's necessary that the incident power should not be too near ϕ_{Max} or ϕ_{min}.
- *Response time*: Two different processes, a priori, determine the response time. First, the mechanism which is responsible for photocarrier generation and, second, by the time delay necessary for the charges to flow inside the electric circuit. The first delay is always negligible with regard to the second, which is fixed by some RC time constant of the circuit. Because of parasitic capacitors the shortest response times are measured in nanoseconds (10^{-9}), a few microseconds being a typical value.

1.5.3.4. Vacuum Photodiode

When arriving on a conducting material, photons give free electrons enough energy for them to escape outside the conductor. Their individual energy, $h\nu$, must be higher than some threshold energy W_{thresh} that is characteristic of the material called its work function. This phenomenon is *photoemission*. A frequency ν_{thresh} is associated to the work function by the formula $W_{thresh} = h\nu_{thresh}$.

After the departure of photoelectrons an isolated conductor becomes electrically positively charged and, because of Coulomb interaction, a restoring force then prevents the electrons going very far away from the surface. On the contrary, if a second positively polarized conductor (anode) has been placed in front of the first one, an electric field will pull the photoelectrons to

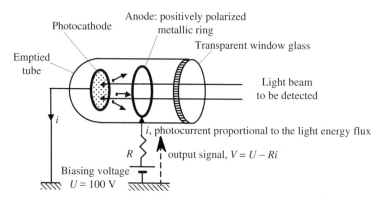

Figure 1.16. Vacuum photodiode. A high vacuum has been obtained inside the tube, so that the electrons are very unlikely to collide with residual gas molecules and have a mean free path larger than the tube size. Thanks to the photoelectric effect, electrons are ejected outside the cathode. The anode, a ring through which the light can easily go and reach the cathode, is positively polarized to attract the electrons. An electric current, proportional to the light flux, flows in the circuit. The energy of each photon must be larger than the work function of the cathode material.

the anode: a *photocurrent* is generated, its intensity is proportional to the number arriving each second on the photocathode of the photons.

1.5.3.5. Semiconductor Photodetectors

The energy difference between electrons and holes in semiconductors being of the order of an electron volt, it is easy to make semiconductor photodetectors having interesting optical properties. The two most important are: (i) photoresistive cells; and (ii) cells using PN junctions.

Photoresistive cells: When a semiconductor sample receives photons having an energy $h\nu$ larger than its energy gap, electron-hole pairs are created; as a consequence the *electrical resistance* will be varied. If the sample is connected to some electric circuit, the electric intensity is determined by the light flux.

PN junction photocells are the most common. When a PN junction is backward polarized, the junction area is depleted from free carriers, electrons as well as holes. In the presence of suitable illumination, electron-hole pairs are generated. The polarizing voltage sweeps the electrons in one direction and the holes in the other; the resulting current is fixed by the light intensity.

The main advantages of PN photocells are their sensitivity and their short rise time. It must be added that microelectronic technologies allow for easy mass production at low cost. Detectors can be as small as a few micrometers, or even less; they can be very close to one another. Arrays or matrices of detectors are easily made. The remaining difficulty is then to be able to read,

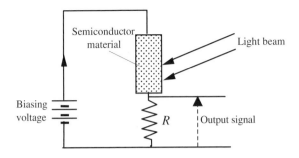

Figure 1.17. Photoresistive cells. An incident light flux generates electrons and holes that are free to move inside a semiconducting sample. These electric charges are swept by an electric field due to a polarizing electric generator. The electrical resistance gets smaller with increasing light intensities.

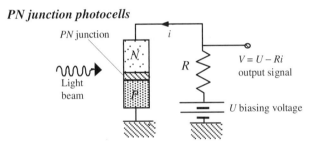

Figure 1.18. The *PN* junction is backward polarized; the junction area is depleted from free carriers. In full darkness the electric current is in principle equal to zero, except for a usually very small "dark current." Photon energy must be larger than the gap in the semiconductor. When the light intensity remains inside a suitable range of values, the photocurrent is proportional.

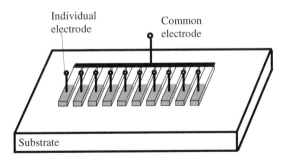

Figure 1.19. Ten arrayed *PN* junctions. The width and the periodicity are of the order of 5–10 μm. The number of cells is usually a power of 2 (1024 is often met). Two-by-two matrices are quite common, typically 1024/1024.

successively or simultaneously, an enormous number of individual detectors (a few hundreds to several millions); these difficulties have found a solution thanks to computers.

1.5.3.6. *In Optics it Is Possible to Count the Photons*

At radio frequencies, the energy of the photons is so weak compared to the thermal agitation energy $kT/2$, that it's completely out of the question to characterize one, or even a few, of them. On the other side of the spectrum, nuclear physicists often deal with events where only one γ photon is involved.

Optics is the domain where photons start to be countable. On this occasion it should be emphasized that, from an historical point of view, it is with Optics that the concept of photons was brought up for the first time.

Avalanche Devices

Detector sensitivity, using either photoemission or *PN* junctions, may be considerably increased by using *charge multiplying processes*. Once it has been emitted, a photoelectron will induce the *production of an avalanche of other electrons*. The quantum efficiency of avalanche devices is much larger than one.

Avalanche Photodiode

The *PN* junction is polarized so that, with regard to the *Zener effect*, the conditions are slightly less critical. In darkness the current is extremely small (dark current), when photons arrive in the junction area the generated photoelectrons switch on a *Zener discharge*, giving high photocurrents.

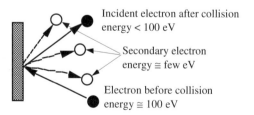

Figure 1.20. An incident electron collides with a metallic target, thus initiating the creation of n other electrons, this is called secondary emission, n is several units. Secondary electrons have energies of a few electron volts, they can be collected by an anode and driven to earth through a load resistance.

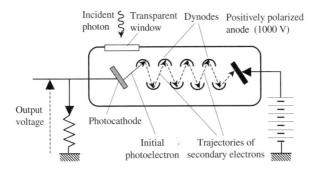

Figure 1.21. Principle of a photomultiplier. Between the photocathode and the anode, special electrodes, called dynodes, have been placed, there are p of them. The voltage between the two following dynodes is 100 V. Each photoelectron creates a "burst" of n secondary electrons on each of the different dynodes. To each initial photoelectron will then correspond n^p secondary electrons ultimately reaching the anode.

Photomultipliers

The physical process responsible for the avalanche is secondary emission, it occurs when an electron collides with a metallic target after having been accelerated to an energy of the order of 1 keV.

1.5.4. *Detection of Light Waves. Response Times of Photodetectors*

In the case of radio waves it has already been said that we have first a reception in an antenna which gives a signal, very similar, apart from a proportionality coefficient and also from some phase difference, to the signal that had left the emission antenna. Detection then occurs, before detection, as well as after, the signals can be displayed on the screen of a cathode ray tube (CRT).

Optical frequencies are far too high to be able to excite an electric circuit, or for the carrier signal to be displayed on a CRT.

The photodetection of a signal can be put in the following way: What is the response $r(t)$ of a detector receiving a signal such as

$$E(t) = A(t) \cos \omega_{car} t.$$

The notation ω_{car} has been used to suggest a carrier wave.

When the carrier wave is a light wave, the time variation of the modulation $A(t)$ is very slow compared to the oscillations of $\cos \omega_{car} t$. More accurately, we can say that the frequencies of the Fourier components of $A(t)$ are much lower than ω_{car}. Let ω_{mod} be an order of magnitude of the highest component of the modulation spectrum, the response of the detector will be quite

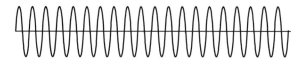

Figure 1.22. CRT representation of a wave with a constant amplitude (no modulation). After detection such a wave gives a constant electric signal proportional to the square of the amplitude of the initial wave.

different if the response time τ_{resp} is small or, on the contrary, large as compared to $1/\omega_{\text{mod}}$.

Light Intensity

The power is proportional to the square of the amplitude. The proportionality coefficient is called *wave impedance*, its accurate definition and its expression depend on the kind of wave that is considered (electromagnetic, acoustic, mechanical, ...), an exact definition will be given in the next chapter in the case of electromagnetic waves. In many cases the light intensity will very simply be assimilated to the square of the wave amplitude

$$I(t) = A^2(t).$$

Fast Photodetectors

A photodetector is considered to be fast if $\tau_{\text{resp}} \ll 1/\omega_{\text{mod}}$. A fast photodetector is able follow the variations of $A(t)$. In this case, it's enough to consider that the response $r(t)$ is proportional to the mean value $\langle E^2(t) \rangle$ of the square

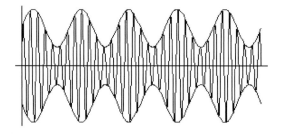

Figure 1.23. Wave having its amplitude modulated by a sinusoidal signal. If the frequency ω_{signal} of the modulating signal is low enough, a photodetector will be able to follow the time variations of the amplitude and will deliver a linear representation (same period, amplitude proportional to the initial amplitude). If that's not the case, the detector cannot follow the modulation and will just give a constant output proportional to the average value.

of the electromagnetic signal, the mean value being evaluated during a time equal to the response time,

$$r(t) = K\frac{1}{\tau_{\text{resp}}}\int_{t}^{(t+\tau_{\text{resp}})} E^2(u)\,du = K\frac{1}{\tau_{\text{resp}}}\int_{t}^{(t+\tau_{\text{resp}})} A^2(u)\cos^2\omega_c u\,du,$$

$$r(t) = KA^2(t)\frac{1}{\tau_{\text{resp}}}\int_{t}^{(t+\tau_{\text{resp}})} \cos^2\omega_c u\,du = \frac{KA^2(t)}{2}.$$

It can be considered that, during the interval $(t, t + \tau_{\text{resp}})$, the function $A^2(t)$ remains roughly constant and can be taken out of the integral. The average value of a squared cosine depends on the interval θ during which the evaluation is made, it is exactly equal to 1/2 if θ is a whole number of periods, and is very near to 1/2 when θ is very long, but not exactly equal to a whole number of periods. Finally, the photodetector response is proportional to $A^2(t)$ and to the light intensity.

Slow Photodetectors

When the detector is not saturated, it can be described by a differential equation and, to remain as simple as possible, we will consider that it is a first-order linear equation

$$\frac{dr}{dt} + \frac{1}{\tau_{\text{resp}}}r = KI(t), \tag{1.8}$$

where K and τ_{resp} are two proportionality coefficients characteristic of the detector.

τ_{resp} is called the rise time and often originates from capacitive effects. For most detectors it is of the order of microseconds. Even for the best detectors, τ_{resp} will never be shorter than a few tens of picoseconds (10^{-12} s).

Response to a Wave of Constant Amplitude

$$I = I_0 = \text{constant}.$$

The solution to equation (1.8) is very simple: $r(t) = K\tau_{\text{resp}}I_0$.

Response to a Slowly Varying Signal

Slow means that the variations of $I(t)$ are considered to be unimportant during an interval of the order of τ_{resp}, meaning that

$$I(t+\tau_{\text{resp}}) \approx I(t) \quad \text{as} \quad I(t+\tau_{\text{resp}}) \approx I(t) + \tau_{\text{resp}}\frac{dI}{dt},$$

$$\tau_{\text{resp}}\frac{dI}{dt} \ll I(t) \quad \rightarrow \quad \tau_{\text{resp}}\frac{dI}{dt} \ll I(t) \quad \rightarrow \quad r = K\tau_{\text{resp}}I(t).$$

For slow phenomena, the time variation of the response exactly follows the variation of the intensity.

Response to Rectangular Signal

The input signal is described in Figure 1.24. It remains equal to zero as long as t is negative or is greater than τ and keeps a constant value within the interval $0 \leq t \leq \tau$.

The left-hand diagrams of Figure 1.24 correspond to the case when the pulse duration is much longer than the response time. The response is given by

$$r(t) = I_0 K \tau_{\text{resp}} (1 - e^{-t/\tau_{\text{resp}}}) \quad \text{for } 0 \leq t \leq \tau,$$

$$r(t) = I_0 K \tau_{\text{resp}} e^{-(t-\tau)/\tau_{\text{resp}}} \quad \text{for } t > \tau.$$

For the right-hand diagrams of Figure 1.24, the light pulse is shorter than the response time. The response is a pulse which lasts for about $\theta = \tau + \tau_{\text{resp}}$, its height is all the smaller as τ becomes shorter, exact values of θ and I_{Max} can easily be obtained from previous equations.

A photodetector cannot discriminate two pulses which are separated by an interval shorter than its response time.

Response of a Photodetector to a Sinusoidal Excitation

$I(t)$ is given by $I(t) = I_0 \cos \Omega t$, Ω is far below the light frequency.

Let us start with the case when $\Omega \ll 1/\tau_{\text{resp}}$, the excitation is a slowly varying function, the response is proportional to the light intensity and is a sine in phase with $I(t)$. In the most general case, the response is still a sine

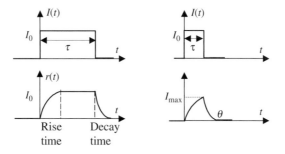

Figure 1.24. The upper diagrams (left) describe a long pulse and (right) a short pulse. The lower diagrams represent the respective response of a detector. In most cases the rise time and decay time are equal; examples may be found where this is not the case.

of frequency Ω, but in this case the amplitude decreases with the frequency. The photodetector behaves like a filter. The response is given by

$$r(t) = K\tau_{\text{resp}} \, \text{Re}\left\{\frac{e^{j\Omega t}}{1+j\Omega\tau}\right\}.$$

Response of a Photodetector to the Light of a Classical Source

The time variation of the light intensity of a classical source has been given in Section 1.4.2, it's a succession of random light pulses, each of them lasting for a time which is much shorter than the response time of a photodetector.

When a classical light source is said to have a constant intensity, it doesn't mean that it emits a permanent and constant electromagnetic field. It can be considered that it works under random stationary conditions and emits light pulses that can be described as follows:

- The pulses are all identical.
- They are randomly emitted.
- The number of pulses emitted during a given time θ is proportional to θ, with the condition that θ should be large with respect to the duration of a pulse.
- The light intensity is proportional to the average number of pulses emitted per second.

Each light pulse is very short, the photodetector gives a short pulse having a duration just about equal to the response time. As the time between individual pulses is shorter than the response time, the many electric pulses become superimposed giving a constant electrical intensity proportional to the light intensity.

1.6. Interference, Diffraction

1.6.1. *The Paradox of Interference*

An interference experiment consists of superimposing, on the same photodetector, the electromagnetic fields of several different waves. For the sake of simplification, we will just consider the interference between only two waves. The detector is supposed to give a response proportional to the square of the electromagnetic field.

Receiving alternatively the signals $E_1(t)$, then $E_2(t)$ and, ultimately, $[E_1(t) + E_2(t)]$, a photodetector would, respectively, give the following responses:

$$\langle E_1^2(t)\rangle, \ \langle E_2^2(t)\rangle, \ \text{and} \ \langle E_1^2(t)+E_2^2(t)+2E_1(t)E_2(t)\rangle.$$

It may happen that $E_1(t) = -E_2(t)$ and under such conditions it is seen that

$$\left\langle\left[E_1(t)+E_2(t)\right]^2\right\rangle = 0. \tag{1.9}$$

Interference is sometimes given the following paradoxical description:

Light + light → darkness.

However, usual observation shows that most often the situation is

Light + light → twice as much light,

When a phenomenon of interference occurs:

The intensity resulting from the superimposition of two beams is not equal to the addition of their individual intensities.

Interference is a manifestation of what is called, by mathematicians, the Schwartz inequality

$$(a-b)^2 \le (a^2 + b^2) \le (a+b)^2.$$

1.6.2. *No Interference if no Detection*

Interference patterns can only be observed if a photodetector receives simultaneously several interferring beams. It is by time-averaging the square of the sum of the amplitudes that an interference phenomenon is revealed. From this point of view, the possibility of observing interference is a direct consequence of the fact that detection is a *nonlinear interaction* between the electromagnetic field and the detector.

When two light beams cross somewhere, nothing special is to be observed at the place where they intersect, even if they are coherent. At the place where the two beams are superimposed, the electromagnetic field, which is the sum of the electromagnetic field's vectors of the two incident beams is however spatially modulated. After they have crossed, the two beams don't keep any memory at all of the fact that they have just crossed.

If, in the arrangements described in Figures 1.25 and 1.26, a photoplate is disposed at the place where the two beams are intersecting, interference fringes can be recorded, under the condition that the two beams should fulfill *interference conditions* that we will define below in Section 1.6.3.

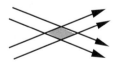

Figure 1.25. Two light beams cross inside a transparent material. The electrons of the atoms located in the shaded area are simultaneously submitted to the electromagnetic fields of the two beams. As the medium is perfectly transparent and homogeneous no light is diffused outside: no interference pattern can be seen.

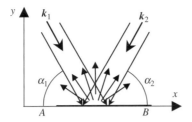

Figure 1.26. Two light beams illuminate a sheet of paper. The electrons of the paper behave in the same way as those of the transparent medium of Figure 1.25, but they reemit light above the screen; this light then goes to a photodetector, the eyes of some observer, for example. It is when the detector averages the square of the light amplitude that a nonlinear interaction occurs and when the fringes are created. In fact the fringes only exist on the surface of the retina of the eye.

1.6.3. *Conditions for Two Light Beams to Produce Interference*

1.6.3.1. *The Two Frequencies Must Be Equal*

Let us suppose that on the sheet of paper of Figure 1.26 the two light beams have the same amplitudes and are written in the following form:

$$y_1 = a \cos(\omega_1 t + \varphi_1) \quad \text{and} \quad y_2 = a \cos(\omega_2 t + \varphi_2).$$

The phases φ_1 and φ_2 depend on which point M is chosen between A and B. The detector response is proportional to the average value of $(y_1 + y_2)^2$:

$$\left\langle a^2 [\cos(\omega_1 t + \varphi_1) + \cos(\omega_2 t + \varphi_2)]^2 \right\rangle$$

$$= \left\langle a^2 \left\{ 1 + \cos[(\omega_1 - \omega_2)t + \varphi_1 - \varphi_2)] + \cdots \right. \right.$$

$$\left. \left. + \frac{\cos(2\omega_1 t + 2\varphi_1) + \cos(2\omega_2 t + 2\varphi_2) + 2\cos(\omega_1 + \omega_2 t + \varphi_1 + \varphi_2)}{2} \right\} \right\rangle.$$

As ω_1 and ω_2 are optical frequencies (10^{15} Hz), the terms of the frequencies equal to $2\omega_1$, $2\omega_2$, and $(\omega_1 + \omega_2)$ will give components having amplitudes equal to zero in the response of the detector. An exception is possible for the term $(\omega_1 - \omega_2)$ which can contribute to the detected signal under the condition that $2\pi/(\omega_1 - \omega_2)$ should be of the order of the detector response time τ_{resp}. This last condition corresponds to a rather severe limitation, and to be able to interfere the two interfering frequencies must be very close: for example, if

$$\tau_{resp} = 10^{-6}\,\text{s} \quad \rightarrow \quad \frac{\Delta\omega}{\omega} = \frac{(\omega_1 - \omega_2)}{\omega_1} \le 10^{-9}.$$

If this condition is not satisfied, the photodetector response will be proportional to the sum of incident intensities. This is why it is said that light

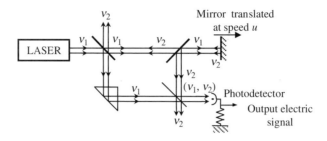

Figure 1.27. Experiment showing the Doppler shift in Optics. When reflected back onto the moving mirror, the frequency of the light is shifted because of the Doppler effect: the output signal contains a beat at the frequency difference $(\nu_1 - \nu_2)$.

sources must have strictly identical frequencies to be able to produce interference.

Exercise: Evaluate the frequency of the term $(\omega_1 - \omega_2)$ in the case of the two components of a sodium doublet, with the following wavelengths: $\lambda_1 = 589.0$ nm and $\lambda_2 = 589.6$ nm.

Figure 1.27 represents an experimental set-up where two light beams with very close frequencies interfere at a photodetector. A mirror is moving at a speed u and receives a very monochromatic light beam of frequency ν_1, because of the Doppler effect the frequency of the reflected beam is shifted according to the formula $\Delta\nu = (\nu_2 - \nu_1) = \nu_1(u/c)$. If $\Delta\nu$ is well inside the bandwidth of the photodetector, an electric signal will be produced at this frequency, allowing a measurement of the mirror speed.

1.6.3.2. *The Two Light Sources Must Be Coherent*

Coherence is a key point in studying interference, its study needs an appropriate approach to the statistical properties of light. It is out of the question to do this in a first introductory chapter; it will be enough to introduce qualitatively the required notions.

The diffusing screen in Figure 1.26 is illuminated by two planar optical waves, they have the same frequency ω and two different wave vectors $\boldsymbol{k}_1$ and $\boldsymbol{k}_2$. At any point $M(x, y)$ the amplitudes of the electromagnetic fields are, respectively, written as

$$E_1 = a\cos(\omega t - \boldsymbol{k}_1 \boldsymbol{OM} + \varphi_1) \quad \text{and} \quad E_2 = a\cos(\omega t - \boldsymbol{k}_2 \boldsymbol{OM} + \varphi_2),$$

with

$$\boldsymbol{k}_1 = \frac{2\pi}{\lambda}(\boldsymbol{x}\cos\alpha_1 - \boldsymbol{y}\sin\alpha_1) \quad \text{and} \quad \boldsymbol{k}_2 = \frac{2\pi}{\lambda}(-\boldsymbol{x}\cos\alpha_2 - \boldsymbol{y}\sin\alpha_2).$$

Let us suppose that the origin O has been chosen inside the plane Π of the sheet. $\boldsymbol{x}$ is a unit vector of a direction lying along Π, $\boldsymbol{y}$ is a unit vector orthogonal to Π. If M belongs to Π, then $\boldsymbol{OM} = x\boldsymbol{x}$ and the electromagnetic field at point M can be written as

$$E_1 + E_2 = 2a \cos\left[\frac{\pi x}{\lambda}(\cos\alpha_1 + \cos\alpha_2) + \frac{\varphi_1 - \varphi_2}{2}\right] \times \cdots$$

$$\times \cos\left[\omega t - \frac{\pi x}{\gamma}(\cos\alpha_1 - \cos\alpha_2) + \frac{\varphi_1 + \varphi_2}{2}\right].$$

The light intensity I is obtained by taking the time-averaged value of $(E_1 + E_2)^2$:

$$I = \left\langle (E_1 + E_2)^2 \right\rangle = 4a^2 \cos^2\left[\frac{\pi x}{\lambda}(\cos\alpha_1 + \cos\alpha_2) + \frac{\varphi_1 - \varphi_2}{2}\right]. \qquad (1.10)$$

The two sources are coherent: This means that their phase difference doesn't vary at random, for the sake of simplicity we will just consider the phase difference to be constant. On the plane Π, fringes can be observed with a periodic repartition governed by the term $(\pi x/\lambda)(\cos\alpha_1 + \cos\alpha_2)$, the fringe separation is then $\lambda/(\cos\alpha_1 + \cos\alpha_2)$.

The two sources are not coherent: The phase difference $(\varphi_1 - \varphi_2)$ cannot at all be considered as constant, but varies randomly with time. At a given time, $(\varphi_1 - \varphi_2)$ has some value to which can be associated an interference fringe pattern; as time advances, different patterns follow one another at a pace that cannot be resolved by any photodetectors: the fringes are not clear anymore.

When the two sources are incoherent, the light intensity is the same whatever the point on the screen, and is obtained by averaging (1.10), $(\varphi_1 - \varphi_2)$ being considered as a function of time,

$$\left\langle 2a^2 \cos^2\left[\frac{\pi x}{\lambda}(\cos\alpha_1 + \cos\alpha_2) - \frac{\varphi_1 - \varphi_2}{2}\right]\right\rangle$$

$$= a^2\left\langle 1 + \cos\left[2\frac{\pi x}{\lambda}(\cos\alpha_1 + \cos\alpha_2) - (\varphi_1 - \varphi_2)\right]\right\rangle = a^2.$$

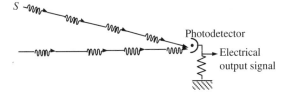

Figure 1.28. Interference between waves coming from two different sources. The two sources emit wave packets of the same frequency but having no phase coherence. No interference can be seen by the detector for two reasons: first, it is very unlikely that two wave packets should arrive simultaneously and, second, even if this were the case, the phase difference would randomly vary from one coincidence to the next.

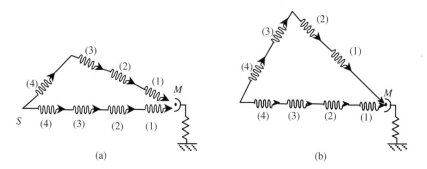

Figure 1.29. Superposition of two waves coming from the same source after they have traveled independently. In (a) the two light paths are only slightly different, the two wave packets arriving at the detector come from the same initial one: the phase difference is not at random but is determined by the lengths of the two optical trajectories: interference is possible. In (b) the length difference is larger than the length of coherence of the source: even if two different wave packets meet on the detector they cannot interfere.

The intensity is just the addition of the intensities of each beam and no interference can be observed.

1.6.3.3. *Thermal Sources, Length of Coherence*

We will now deal with two thermal sources of the kind described in Section 1.4.2.4, they emit damped wave packets described by $E(t) = e^{-(t-\theta_i)/\tau} \cos(\omega t - \varphi_i)$: the instant θ_i and the phase φ_i randomly vary with i; τ is a time characteristic of the source and is called its coherency duration; the distance $d = c\tau$ covered by light during a time equal to τ is called the coherence length.

1.6.4. *The Validity of Geometrical Optics Is Limited by Diffraction*

Development of geometrical optics relies on the notion of light rays. In a homogeneous transparent medium, rays are straight lines along which light energy is propagated. The notion of a ray is suggested by normal observation. Who has never witnessed light rays filtering through the tiles of a roof and revealed by the dust floating in the air? Or, who has never contemplated, during a night by the seashore, the rays of a lighthouse illuminating the haze?

Accurate definition of a light ray meets serious theoretical difficulties, as well as its accurate experimental observation. One can of course think of limiting the cross section of a beam, using diaphragms with smaller and smaller diameters, but then diffraction will occur. Diffraction is very general and is observed for any kind of propagating phenomenon. Acoustic waves, as well as *De Broglie* waves associated to particle beams, substantially diffract every

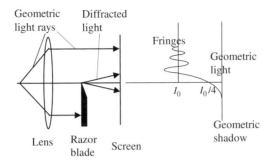

Figure 1.30. Diffraction by a sharp edge. A beam of parallel rays is partially hindered by an obstacle limited by a straight and very sharp boundary having a tiny radius of curvature (a razor blade). Some light is diffracted by the edge and interferes with the undiffracted part of the light: fringes can be seen on the observation screen.

time they go through obstacles of the same order of magnitude as their wavelength.

On the observation screen of Figure 1.30, instead of having a clear separation between shadow and light, it seems that some light penetrates inside the shadow area. On the other hand, diffraction fringes exist in the clear area. Let I_0 be the illumination of the observation screen in the absence of the razor blade:

- The illumination is equal to $I_0/4$ at the boundary between shadow and light, and goes smoothly to zero far enough into the shadow.
- In the clear part of the screen, the illumination has well-contrasted oscillations and goes to I_0 for points far from the edge.

Diffraction appears as a limitation to the rectilinear propagation of light. To describe simply difficult phenomena, we will say that when a light beam collides with an obstacle, after the collision, part of the light will travel in many directions: the angle with the initial direction being all the larger as the obstacle is smaller.

Diffraction is always followed by interference between the initial beam and the diffracted beams; this is the reason why diffraction produces *fringes*, usually called *diffraction fringes*.

Diffraction is enhanced when the diffracting obstacles have a periodic repetition and it becomes quite spectacular if the periodicity has the same order of magnitude as the wavelength.

1.6.5. *Diffraction of Electrons*

Let us consider a beam of electrons that have been accelerated by a voltage $V = 150$ V and evaluate the associated wavelength. We will use the De Broglie

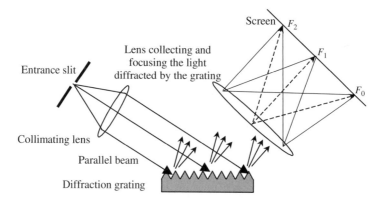

Figure 1.31. Diffraction of a light beam by a diffraction grating. It is possible to engrave, or to imprint or to etch, parallel and equidistant linear grooves on planar substrates having noticeable dimensions (10×10 cm). Such arrangements are called diffraction gratings. Each groove efficiently diffracts light. If a grating is illuminated by a parallel light beam, there are only a few directions, determined by the wavelength and by the geometry, for which the waves diffracted by the various grooves interfere positively. A lens then focuses the different diffracted beams and each focal point associated with a direction of positive interference is brightly illuminated.

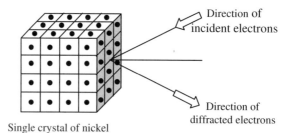

Figure 1.32. Davisson and Germer electron diffraction experiment. A monocinetic electron beam is sent onto the cleaved surface of a single crystal of nickel. Electrons are diffracted by the atoms of the crystal lattice along directions that are in good agreement with De Broglie wavelength and with the crystal cell size. Conversely, such an experiment is now used to measure the parameters of a crystal cell.

formula $h = p\lambda$, where p is the momentum of the accelerated electrons. For such weak energies it's not necessary to use a relativistic theory and it's readily obtained that

$$\lambda = \frac{h}{\sqrt{2emV}} \cong 10^{-10} \text{ m} = 1 \text{ Å}, \quad \text{if } V \approx 150 \text{ V}.$$

For diffraction of the previous De Broglie waves to be observed, the periodicity of a regular arrangement of the diffracting obstacle should be of a few

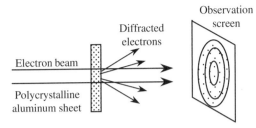

Figure 1.33. Thomson's experiment. Any sheet of metal is an arrangement of microcrystals randomly oriented. Because of this random orientation in the sample, the problem has cylindrical symmetry around the axis of the beam. The diffraction pattern is made of concentric rings, the radii of which are determined by De Broglie wavelength and the size of the crystal cell.

angstroms. This has been done first by Davisson and Germer in 1928 (the diffracting elements were simply the atoms of a single crystal of nickel) and then by Thomson (using the atoms of a polycrystalline aluminum sheet).

Electron diffraction experiments must be performed in a high vacuum, so that the electrons can propagate with a low probability of collision with a molecule of some residual gas. Observation screens are painted with a material which emits light when it receives electrons. Photographic plates can also be used, since electrons are able to chemically reduce Ag^+ ions to metallic silver atoms.

The same kind of experiments can be done using X-rays having the same wavelength as the electrons. Similar diffraction patterns are obtained. In the same way, diffraction can also be observed using neutron beams.

1.7. Photometry

Electromagnetic radiation carries energy. The purpose of *photometry* is to clarify this notion as far as light is concerned and to introduce the physical parameters allowing us to quantify the visual observation of light sources and illuminated objects.

1.7.1. *Physical Parameters in Relation to Energy*

Planar Monochromatic Wave

The problem is rather simple when we are concerned with planar waves. It's then enough to consider the amount of energy ϕ (energy flux) which crosses (per second) a surface S orthogonal to the wave vector

$$\boldsymbol{E} = E_0 \boldsymbol{x} \cos(\omega t - \boldsymbol{k} \boldsymbol{r}), \quad \text{electric field of the planar light wave.} \quad (1.11)$$

Using the Poynting theorem we can calculate what is called, by definition, the *luminous intensity I*:

$$I = \frac{\varphi}{S} = \frac{1}{2} \frac{E_0^2}{Z} \quad (\text{W/m}^2), \tag{1.12}$$

where Z is the wave impedance; the 1/2 factor corresponds to time-averaging over a great number of periods.

Planar Nonmonochromatic Wave

A nonmonochromatic wave is the superimposition of an infinite number of monochromatic waves. To calculate the flow of energy through a surface S, the spectral domain is divided into narrow bands with wavelengths between λ and $(\lambda + d\lambda)$, in formula (1.12) the light intensity has to be replaced by an elementary luminous intensity dI defined by

$$dI = I_\lambda \, d\lambda \quad \text{and} \quad I = \int_0^\infty I_\lambda \, d\lambda \quad (\text{W/m}^2), \tag{1.13}$$

where I_λ is the spectral density of the luminous intensity; its law of variation versus wavelength is determined by the physical properties of the emitting source.

Monochromatic Point Source

The luminous intensity of a point source is defined for a *given direction* Oz: this is the ratio of the elementary luminous flux $d\phi$ to the elementary solid angle $d\Omega$ of the elementary cone inside of which the flux is emitted, see Figure 1.34(a).

$$I = \frac{d\phi}{d\Omega}. \tag{1.14}$$

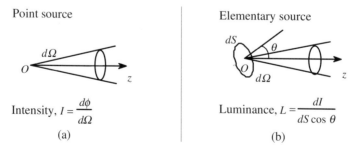

Figure 1.34. Definition of the luminous intensity.

Case of a Surface with a Finite Size

The surface is divided into elementary surfaces of area dS, each of them then being considered as a point source with the elementary luminous intensity dI. A new parameter is then introduced, the *luminance L* of the emitting surface at the point under consideration and in a direction that makes the angle θ with the normal to the surface (see Figure 1.34(b)).

$$L = \frac{dI}{dS\cos\theta}.\qquad(1.15)$$

The elementary flow of energy sent inside an elementary cone $d\Omega$, centered around Oz, is given by

$$d^2\phi = L\,d\Omega\,dS\,\cos\theta.\qquad(1.16)$$

Source Emittance

The *emittance M*, also called *exitance*, of a source at one of its points is a characteristic of the total energy sent into all the surrounding space by an elementary surface drawn around this point

$$M = \frac{d\phi}{dS} = \int L\cos\theta\,d\Omega.\qquad(1.17)$$

If the sources, point sources or more general sources, are not monochromatic, the notion of spectral density should be extended to luminance and to emittance, using formulas similar to (1.13).

Irradiance of a Screen

Let us consider a point P of a screen *receiving* light coming from one or several light sources. An elementary area dS drawn around P receives per second a total energy equal to $d\varphi$. The *irradiance E* at point P is, by definition,

$$E = \frac{d\phi}{dS}\quad \text{(beware of confusion with the electric field which is also noted } E\text{)}.$$

An illuminated screen can also be considered as a light source, its emittance is smaller than its irradiance. In the case of negligible losses, irradiance and emittance are equal.

Geometrical Width of a Light Pencil

Let us consider a light pencil emitted by some elementary surface dS inside some elementary cone $d\Omega$ centered around the Oz axis, and let θ be the angle between Oz and the normal to the surface; by definition we will call the

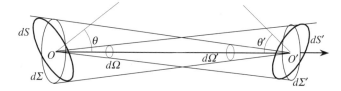

Figure 1.35. Geometrical width of a pencil.

elementary geometrical width d^2G of the beam the quantity defined by the following formula:

$$d^2G = d\Omega \, dS \cos\theta = d\Omega \, d\Sigma. \tag{1.18}$$

In Figure 1.35 have been drawn two elementary surfaces dS and dS' disposed with regard to one another. Their respective centers are O and O', θ and θ' are the angles of OO' with the normal to the two surfaces. We consider two pencils, respectively, issued from O and O' and based on dS and dS', a simple application of the definition of a solid angle shows that they have *equal geometrical widths*:

$$d\Omega = \frac{dS' \cos\theta'}{OO'^2}, \quad d\Omega' = \frac{dS \cos\theta}{OO'^2} \quad \rightarrow \quad d\Omega \, dS \cos\theta = d\Omega' \, dS' \cos\theta'.$$

Expression of the Energy Carried by a Pencil

Referring to formula (1.16), it is seen that the energy carried by a pencil is proportional to its geometrical width

$$d^2\phi = L d^2G \quad \rightarrow \quad L = \frac{d^2\phi}{d^2G}. \tag{1.19}$$

Conservation of the Geometrical Width by Refraction

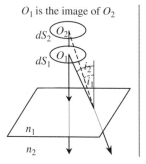

O_1 is the image of O_2

Magnification is equal to unity $\rightarrow dS_1 = dS_2$.
Snell-Descartes law of refraction: $n_1 \sin i_1 = n_2 \sin i_2$.
Geometrical width of the incident pencil

$$d^2G_1 = dS_1 \cos i_1 \, d\Omega_1 = 2\pi \sin i_1 \cos i_1 \, dS_1 \, di_1.$$

Geometrical width of the refracted pencil

$$d^2G_2 = dS_2 \cos i_2 \, d\Omega_2 = 2\pi \sin i_2 \cos i_2 \, dS_1 \, di_2,$$

$$n_1^2 d^2G_1 = n_2^2 d^2G_2. \tag{1.20}$$

Figure 1.36. Conservation of the geometrical width of a pencil after refraction.

Formula (1.20) is a relationship between the geometrical widths of an incident pencil and of the corresponding refracted pencil. The product of the geometrical width by the square of the refractive index remains constant when a pencil is refracted. Some authors include the square of the refractive index in the definition of the geometrical width.

Conservation of the Geometrical Width in a Centered System

Formula (1.20) is readily extended to centered systems, since the propagation of rays through a centered system is nothing other than a succession of refractions on various interfaces. Nevertheless, we are going to show that the conservation of $n^2 d^2 G$ is also a consequence of the Lagrange-Helmholtz relationship.

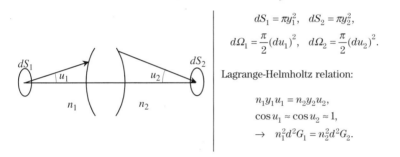

$$dS_1 = \pi y_1^2, \quad dS_2 = \pi y_2^2,$$
$$d\Omega_1 = \frac{\pi}{2}(du_1)^2, \quad d\Omega_2 = \frac{\pi}{2}(du_2)^2.$$

Lagrange-Helmholtz relation:

$$n_1 y_1 u_1 = n_2 y_2 u_2,$$
$$\cos u_1 \approx \cos u_2 \approx 1,$$
$$\rightarrow \quad n_1^2 d^2 G_1 = n_2^2 d^2 G_2.$$

Figure 1.37. Conservation of the geometrical width in a centered system.

Lambert's Law

Lambert's law introduces an ideal source called a *Lambertian source*, which is very much like a real heated body source. The radiation emitted by the *blackbody* introduced in Thermodynamics does follow Lambert's law.

The radiation emitted by a Lambertian source is the same for any direction and, if the source is extended, it is the same for any of its points.

Point Lambertian Source

The flux of energy (flow of energy per second) emitted by a point Lambertian source having an intensity I, in the 4π steradians of the whole space, is equal to

$$\phi = I \int d\Omega = 4\pi I.$$

The flux emitted inside a cone having a summit half-angle α is given by

$$\phi = I\int d\Omega = I\int_0^\alpha 2\pi\sin\theta\,d\theta = 2\pi I(1-\cos\alpha), \qquad (1.21.\text{a})$$

$$\phi = \pi a^2 I \quad \text{(if the angle } \alpha \text{ is small enough).} \qquad (1.21.\text{b})$$

Because of the cosine in formula (1.21.a), Lambert's law is often called Lambert's cosine law.

Extended Lambertian Source

Let us consider an extended Lambertian surface and determine the relationship between its luminance L and its emittance M.

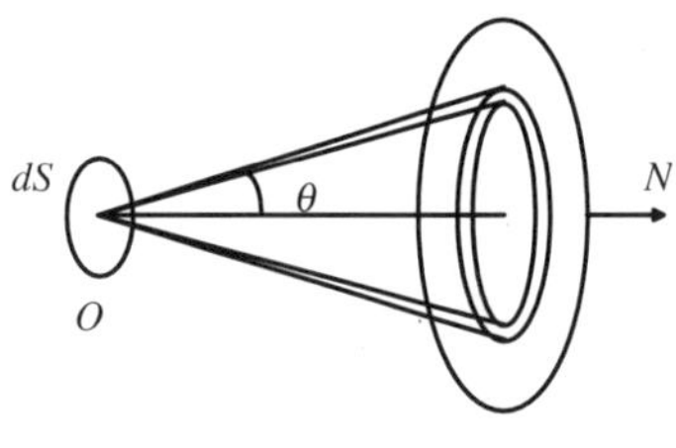

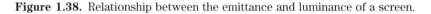

Flux sent inside the conical shell,

$$d^2\phi = L\,dS\,d\Omega\cos\theta = 2\pi L\,dS\sin\theta\cos\theta\,d\theta.$$

Total flux sent in the half-space in front of dS,

$$d\phi = \pi L\,dS\int_0^{\pi/2} d(\sin\theta)^2 = \pi L\,dS,$$

$$M = \frac{d\phi}{dS} = \pi L. \qquad (1.22)$$

Figure 1.38. Relationship between the emittance and luminance of a screen.

1.7.2. Physiologic Parameters

1.7.2.1. Introduction to Physiologic Parameters

In the first part of this chapter, *joules* and *watts* have been used in order to characterize the light intensity and the radiated light flux, corresponding parameters are said to be *energetic parameters*. This approach is not well adapted when the photodetector is the human eye; new parameters, called *physiologic parameters*, need to be introduced. Visual perception is a complex phenomenon; it varies from one person to another and is different during daylight or at night. After statistical studies concerning thousands of people, an *average human eye* has been introduced. Two light beams, with two different wavelengths λ and λ' and two different intensities, illuminate alternately the same sheet of white paper, an observer is then asked to say whether the illumination is the same in both cases. It is then possible to associate an elementary *physiologic flux* $d\varphi_\lambda$ with an *elementary energetic flux* $d\phi_\lambda$:

$$d\phi_\lambda = k_\lambda\,d\varphi_\lambda, \qquad (1.23)$$

where k_λ is a proportionality coefficient that depends on the color (wavelength) and on the light intensity. The graph of $k_\lambda(\lambda)$ versus wavelength is

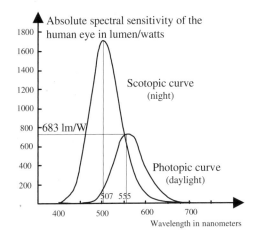

Figure 1.39. Spectral sensitivity curves of the human eye. Two curves are usually given: the photopic curve corresponding to a comfortable lighting (the sensitivity is maximum at $0.555\,\mu m$) and the scotopic curve which is obtained for a low level of illumination, the maximum of sensitivity is shifted toward the blue ($0.507\,\mu m$).

given in Figure 1.39, it's a kind of bell-shaped curve; of course, the curve goes down to zero at the boundaries of the visible spectrum (roughly 0.4–$0.8\,\mu m$ and, more accurately, 0.390–$0.790\,\mu m$). For daylight observation the maximum is at $0.555\,\mu m$, which means that the sensitivity of the human eye is maximum in the yellow. Table 1.1 gives the relative values of k_λ for several wavelengths.

1.7.2.2. *Units*

Candela: By definition a candela is the intensity, in a given direction (see formula (1.14)), of a source emitting a monochromatic radiation of wavelength $0.555\,\mu m$ and having an energetic intensity of $1/683 = 0.00146$ W/sr.

  Lumen: This is the physiologic flux associated to the candela, it corresponds to the flux sent inside a solid angle of one steradian by a source having an intensity of one candela.

  Lux: This is the physiologic irradiance associated to the lumen, it corresponds to the irradiance of a surface receiving a physiologic flux of one lumen per square meter. The lumen is also the irradiance unit.

Table 1.1. Human eye sensitivity versus wavelength for daylight observation.

λ (μm)	0.4	0.45	0.50	0.555	0.60	0.65	0.70	0.75
$k_\lambda/k_{0.555}$	4×10^{-4}	0.038	0.032	1	0.63	0.107	4×10^{-3}	10^{-4}

1.7.2.3. *Order of Magnitude*

To establish the link between physical and physiological parameters, two kinds of experimental conditions must be considered: first, the spectral variations of the eye sensitivity and, second, the spectral composition of the light source (Table 1.2). In fact the engineers and physiologists who have elaborated these concepts have been very pragmatic and obtained quite concrete proposals; since most observers agree when they compare their impressions when watching color movies, either on TV or in the cinema, we can say that they have been successful.

Table 1.2. Orders of magnitude concerning various light sources.

	Physiological units	Physiological units	Physical units	Efficiency
	Luminance (cd/m^2)	Illumination (lx)	Luminance $(W/m^2 sr)$	(lm/W)
Sun before atmosphere	2×10^9	1.4×10^5	2×10^7	100
Sun on Earth's surface	1.3×10^9	10^5	1.3×10^7	90
White paper in the Sun	2×10^4			
Tungsten lamp 2700 K	6.5×10^6			
Dazzling threshold	10^4			
Daylight vision	> 10			
Night	< 0.01			
Visibility threshold	10^{-6}			
Comfortable lighting		400 to 1000		
Full moon	2.5×10^3	0.2		
Brilliant star	10^{-5}			

When a flow of energy is calculated, an integration is performed over the whole spectral range; in the case of a "physiological flow," the different wavelengths will be attributed to a pondering coefficient which can be obtained from Table 1.1. Let us take the case of a 100 W electric bulb, for which the maker indicates 1000 lm. 100 W is the electrical power obtained from the electrical network, if the conduction and convection losses are evaluated to 25%, 75 W are radiated in the 4π sr, which makes 6 W/sr; 1000 lm at the wavelength of maximum sensitivity represent 1.46 W/sr, the difference corresponds to the energy which is emitted outside the visible domain.

1.7.3. *Thermodynamics and Conservation of the Geometrical Width*

1.7.3.1. *Luminance of a Blackbody*

As illustrated in Figure 1.40, a *blackbody* is a closed box, the walls of which are perfectly reflecting for electromagnetic waves, whatever the frequency, and are *adiabatically* isolated from the outside. Initially the blackbody has been brought into contact with a heat sink raised to the temperature T K. Inside the blackbody, electromagnetic waves, with frequencies between zero and infinity, endlessly reflected on the walls, follow zigzag paths having all possible directions.

Let us consider a point M inside the blackbody and an elementary area $d\sigma$ drawn around M. An elementary cone, solid angle $d\sigma$, having its summit in M and centered on the normal to $d\Omega$, carries an elementary energy flux given by

$$d\phi = L\,d\sigma\,d\Omega, \tag{1.24.a}$$

$$d\phi = \int_\lambda d\phi_\lambda \quad \text{with} \quad d\phi_\lambda = L_\lambda\,d\sigma\,d\Omega\,d\lambda, \tag{1.24.b}$$

where L_λ is called the spectral luminance.

We are going to demonstrate that the spectral luminance L_λ is the same for all points inside the blackbody and doesn't depend on the orientation of the elementary surface $d\sigma$. Let us consider two different blackbodies C_1 and C_2 having the same temperature T K, but differing in their shape and by the various objects that have been put inside. Holes of respective areas S_1 and S_2 have been drilled in each of them, and it is assumed that they are small enough so that the thermal equilibrium between C_1 and C_2 is not affected by

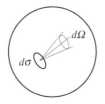

Figure 1.40. Schematic illustration of a blackbody.

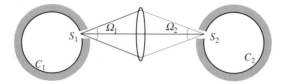

Figure 1.41. A lens conjugates two holes drilled in two different blackbodies having identical temperatures. Since, because of the second principle, the exchange of energy between them should be balanced, it can be shown that the two holes have the same spectral luminance and it can then be deduced that the electromagnetic energy density $u_\nu(v,\,T)$ must be the same inside the two blackbodies. The final result is that the law of variation of $u_\nu(v,\,T)$ is a universal function, whatever the blackbody.

the losses of the electromagnetic radiation escaping through the hole. A perfectly transparent lens images S_1 on S_2, and conversely. Finally, a spectral filter with a narrow band $\Delta\lambda$ is placed on each hole. Ω_1 and Ω_2 being the solid angles of the two cones having S_1 on S_2 as summits and leaning on the lenses' contours, and σ_1 and σ_2 being their respective projections on planes normal to the lens axis, the flow of energy $\Delta\phi_1$ radiated each second from C_1 to C_2 is equal to

$$\Delta\phi_1 = L_{1,\lambda}\sigma_1\Omega_1\Delta\lambda.$$

In the same way the energy flow radiated from C_2 to C_1 is equal to

$$\Delta\phi_2 = L_{2,\lambda}\sigma_2\Omega_2\Delta\lambda.$$

$\Delta\phi_2$ and $\Delta\phi_1$ must be equal, if not, one blackbody would receive more energy than it dissipates which would raise its temperature, which is in contradiction to the thermodynamic second principle. On the other hand, we have $\sigma_1\Omega_1 = \sigma_2\Omega_2$, because of the conservation of the geometrical width, and we finally obtain

$$L_{1,\lambda} = L_{2,\lambda}.$$

In conclusion, the spectral luminance in a blackbody at thermal equilibrium doesn't depend on the particular blackbody under consideration. Changing the direction of the lens axis, it is seen that the spectral luminance is also independent of the direction of the light rays: which means that a hole drilled in a blackbody is a *Lambertian source*.

1.7.3.2. *Electromagnetic Energy Density Inside a Blackbody*

u being the amount of electromagnetic energy per unit of volume (radiant energy volume density) inside a blackbody, and u_λ the corresponding spectral density, we seek a relationship between u_λ and L_λ. The electromagnetic energy is the energy of electromagnetic waves going back and forth inside the blackbody at the speed of light c. We consider an elementary surface dS inside the

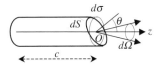

Figure 1.42. The energy inside the cone was previously inside the cylinder.

blackbody; the total amount of energy $d\phi$ carried per second inside an elementary cone (summit on dS, elementary solid angle $d\Omega$) was, before reflection, inside a cylinder parallel to the cone axis and based on dS and having a length equal to the distance covered by the light during one second, see Figure 1.42. If $d\sigma$ is the projection of dS on a plane normal to the cone axis, we can write $d\phi = c\,du\,ds$ and, according to formula (1.24.a), $d\phi = L\,d\Omega\,d\sigma$. By integration over all the half-spaces located on the same side of dS, and using the fact that the luminance doesn't depend on the direction, we obtain

$$u = \frac{2\pi}{c}\,L. \tag{1.25.a}$$

The same treatment can be used using spectral densities

$$u_\lambda = \frac{2\pi}{c}\,L_\lambda. \tag{1.25.b}$$

We have thus demonstrated that the electromagnetic energy spectral density is independent on the kind of blackbody that is considered, hence it is a *universal function of temperature and wavelength*. The determination of this universal function, which needs more Physics and more Mathematics, can be found in Section 9.2 (see formula (9.18)).

du_ν being the amount of electromagnetic energy corresponding to waves having their frequencies between ν and $(\nu + d\nu)$, it can be shown that

$$du_\nu = u_\nu\,d\nu \quad \text{with} \quad u_\nu = \frac{8\pi h\nu^3}{c^3}\,\frac{1}{e^{h\nu/kT} - 1}. \tag{1.26.a}$$

If $du_\lambda = u_\lambda\,d\lambda$ is the amount of electromagnetic energy corresponding to waves having their wavelengths between λ and $(\lambda + d\lambda)$, we can write

$$\nu = \frac{c}{\lambda} \quad \rightarrow \quad d\nu = -\frac{c}{\lambda^2}\,d\lambda \quad \rightarrow \quad u_\lambda = \frac{8\pi h}{\lambda^5}\,\frac{1}{e^{hc/\lambda kT} - 1}. \tag{1.26.b}$$

1.7.3.3. *Luminance of a Heated Object*

We consider an object that is not perfectly transparent and can be characterized by an absorption coefficient a_λ depending on the color; we are looking for its spectral emittance e_λ when raised to some temperature $T\,K$. This object is put inside a blackbody at temperature T K, each elementary area dS of its surface receives per second an energy given by

$$\iiint_{S\,\Omega\,\lambda} \alpha_\lambda L_\lambda\, dS\, d\Omega\, d\lambda, \quad L_\lambda \text{ is the luminance of the blackbody.}$$

The power reemitted by dS is given by

$$\iiint_{S\,\Omega\,\lambda} e_\lambda\, dS\, d\Omega\, d\lambda.$$

Equating these two powers, we obtain an important relation

$$e_\lambda = a_\lambda L_\lambda. \tag{1.27}$$

The ratio of the spectral emittance to the spectral absorption of some object is a constant which doesn't depend on the kind of object under consideration and which is equal to the luminance of a blackbody raised to the same temperature.

For a blackbody $a_\lambda = 1$ ($\forall\ \lambda$), a blackbody is a perfect absorber, whatever the wavelength. At a given temperature an absorbing object radiates more energy than a transparent one. A piece of iron heated to 800° C seems quite red and radiates a lot; a piece of quartz, raised to the same temperature, is just a little brighter than at room temperature and one must be cautious and not take it with bare hands.

1.8. Perception and Reproduction of Colors

Although photography and perhaps, even more, color TV appear as nice jewels of modern technology, the analysis of the mechanisms of color vision and the famous trichromatic system had been analyzed two centuries ago. As early as 1802 the British physicist Thomas Young said that it was impossible that any point of our retina should have an infinity of cells able to vibrate in unison with all the monochromatic waves of the visible spectrum; he estimated that only three colors would be enough (red, yellow, blue) and admitted the existence of nerve threads bringing the required information to our brain.

Given the enormous market concerned, a considerable effort of standardization has been necessary. Problems of normalization, and especially the experimental protocols to be followed in the experimentation suggested in Figure 1.43 are very tricky and have needed many international agreements. The quality of color pictures as well as the reproducibility of most visualization devices witnesses in favor of the quality of the work that has been produced.

Monochromatic Light

The color of a light beam is directly connected to the spectral composition of the corresponding electromagnetic radiation. The case of a monochromatic

Table 1.3. Wavelength and color correspondence in the monochromatic case.

λ (nm)	400		500		600		700		800
		390	455 492	577	597	622			760
Color	ultraviolet	violet	blue	green	yellow	orange	red		infrared

light is very simple, and there is a univocal relation between the wavelength and the color that is described by some observer. Table 1.3 describes this correspondence.

Light of Arbitrary Spectral Composition

The light coming from most colored objects is not monochromatic; moreover, it's the diversity of the different spectra which make the painter's palette so rich. The automatic reproduction of color (photo, cinema, television, display, ...) has required an analysis of the reasons that makes us attribute such color to such light. Lights of quite different spectral composition may produce identical impressions. The basic experiment consists in illuminating a white screen (i.e., a screen diffusing the light with the same efficiency whatever the color) with a superposition of different colored lights and to ask an observer, or rather many observers, to say what color they attribute to the mixture.

Two mixtures are said to be "metameric" if they appear to be identical although they don't have identical spectral compositions.

The principle of the *additive trichromatic system* relies on the fact that it is possible to reproduce most of the colors by mixing, in suitable proportions, three light beams having different colors. The scheme of Figure 1.43 shows the principle of an experimental set-up allowing the determination of

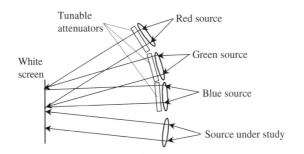

Figure 1.43. Principle of the additive trichromatic system. The light of a source under study is put side by side with the superposition of the lights coming from three different colored sources.

the "suitable proportions." The three colored sources can be extremely mono-chromatic (laser) or only be reasonably monochromatic (white source + colored filter). The light flux that each source sends toward the screen is varied with calibrated tunable attenuators. The three sources are called *primary sources*; in fact there is an infinity of possible triads, from which any color can be reconstructed. A very common triad is blue-green-red.

The Space of Colors Follows a Linear Algebra

We now come back to the experiment described in Figure 1.43 and we con-sider that the three colored sources are very stable, let [B], [G], and [R] be the luminance that they, respectively, produce on the screen with no attenu-ation. Within experimental error, it is found that there is only one setting of the three attenuations for which the observer has the impression that the superposition of the three colored lights reproduces exactly the color of the source under study; let B, G, and R be the respective transmission coefficients of the attenuators. In some way the triad (B, G, R) is a kind of coordinate of the source on the basis of the three chosen primary sources.

The experiment also shows that if the triad $(\alpha B, \alpha R, \alpha P)$ is used the observer will say that he sees the same color, except that the global bright-ness is higher if $\alpha > 1$, and lower if $\alpha < 1$. The following quantity will be called *stimulus*:

$$[S] = B[B] + G[G] + R[R].$$

Given two stimuli $[S_1] = B_1[B] + G_1[G] + R_1[R]$ and $[S_2] = B_2[B] + G_2[G] + R_2[R]$, we have the following relations:

$$\text{If } [S_1] = [S_2] \quad \text{and} \quad \text{if } [S_1] = [S_3], \quad \text{then} \quad [S_3] = [S_2],$$

$$[S_1] = [S_2] \quad \leftrightarrow \quad \{B_1 = B_2 \text{ and } G_1 = G_2 \text{ and } R_1 = R_2\},$$

$$[S_1] + [S_2] = (B_1 + B_2)[B] + (G_1 + G_2)[G] + (R_1 + R_2)[R].$$

As soon as the stimuli follow the rules of linear algebra, their manipula-tion becomes easier, especially the change of primary colors. To go from the reference triad (B, G, R) to another one (*X, Y, Z*), we must at first give the expression of the primary colors of one system on the basis of the primary colors of the other system, and then apply the usual rules of matrix calculation:

$$\begin{bmatrix} X \\ Y \\ Z \end{bmatrix} = T \begin{bmatrix} R \\ G \\ B \end{bmatrix} \quad \text{with} \quad T = \begin{bmatrix} X_R & X_G & X_B \\ Y_R & Y_G & Y_B \\ Z_R & Z_G & Z_B \end{bmatrix} \quad \text{and} \quad \begin{bmatrix} R \\ G \\ B \end{bmatrix} = T^{-1} \begin{bmatrix} X \\ Y \\ Z \end{bmatrix}.$$

Chromatic Coordinates

Chromatic coordinates are the following normalized parameters:

$$r = \frac{(R)}{(R)+(G)+(B)}, \quad v = \frac{(G)}{(R)+(G)+(B)}, \quad b = \frac{(B)}{(R)+(G)+(B)},$$

$$r+g+b=1.$$

Colorimetric Coefficients

We come back again to Figure 1.43, in the position of the source under study; we place a monochromatic source (wavelength λ) having a *calibrated intensity* (1 μW/sr, for example). The chromatic coordinates $\overline{b}_{(\lambda)}$, $\overline{g}_{(\lambda)}$, $\overline{r}_{(\lambda)}$ of the calibrated source are then measured on the basis of the three chosen primary sources; they are called the colorimetric coefficients of the apparatus.

Chromatic Coordinates of a Metamer

A spectral density $S_\lambda(\lambda)$ characterizes the light flux emitted by a non-monochromatic source S. The triad of coordinates (r, g, b) associated to a given source is unique and fixed by $S_\lambda(\lambda)$; on the contrary it is not possible, starting from the triad, to go back to $S_\lambda(\lambda)$ just by definition of a metamer. (r, g, b) is given by

$$r = \iint_\lambda S_\lambda(\lambda)\overline{r}(\lambda)\,d\lambda, \quad b = \iint_\lambda S_\lambda(\lambda)\overline{b}(\lambda)\,d\lambda, \quad g = \iint_\lambda S_\lambda(\lambda)\overline{v}(\lambda)\,d\lambda.$$

2

Electromagnetic Waves

2.1. Mathematical Formulation of Electromagnetism

2.1.1. *E, D, H, B*

The description of an electromagnetic (EM) field in any material, including vacuum, requires four basic vectors. Everybody uses the following notation and we call them E, D, H, B; everybody, of course, agrees about their physical interpretation, however there are some disagreements about their names. E is always designated as the electric field. When introduced for the first time by Maxwell, D had been called the "electric displacement vector," some people now prefer the expression "electric induction vector." The same ambiguity is found for B and H which are often, respectively, called the "magnetic induction vector" and the "magnetic field vector." In a more recent trend B is the "magnetic vector" and H the "magnetic excitation vector." The author has no clear-cut opinion, but he considers that E and D are attached to electric properties, while H and B correspond to magnetic properties.

Since optical materials, by definition, are transparent, they don't usually have any magnetic properties; so H and B are strictly proportional and are thus collinear, the proportionality coefficient is called the vacuum permeability $\mu_0 = 4\pi \times 10^{-7}$ (SI units, henry/meter (H/m)), $B = \mu_0 H$.

For electric vectors, the situation is more complicated. Except in the very special case of nonlinear optics, the relationship between E and D is linear. For isotropic media, E and D are simply proportional, $D = \varepsilon E = \varepsilon_0 \varepsilon_r E$, which implies collinearity of the two vectors. The proportionality coefficient ε is the permittivity; in a vacuum, its value is $\varepsilon_0 = 1/(36\pi \times 10^9)$ (SI units, farad/meter (F/m)). In the case of an isotropic dielectric material ε is proportional to ε_0, and the proportionality coefficient ε_r is the relative permittivity. For a given

Chapter 2 has been reviewed by Dr. François Méot, Senior Physicist at the CEA (Commissariat à l'Energie Atomique).

material ε_r varies with the color (i.e., frequency) of the waves (dispersion). ε_r is a dimensionless parameter, with typical values between 1 and 10.

For anisotropic media the relationship between E and D remains linear but is now described by a tensor, so that the two vectors are no longer collinear, the coefficients of the matrices associated to the tensor vary with color and of course with the medium under consideration.

2.1.2. The Electric Field Vector

From an electromagnetic point of view a system is characterized by:

- A medium, or different media, which can be:
  - ◦ absorbing or transparent;
  - ◦ isotropic or anisotropic;
  - ◦ homogeneous or inhomogeneous.
- Discontinuities across the surfaces separating the previous media:
  - ◦ separation between two dielectric materials (vacuum being considered as a dielectric material);
  - ◦ separation between a dielectric material and a metal (a metal being considered as a dielectric material for which ε is a complex number).
- A time variation law imposed on the electromagnetic field at some points of the system, these points are called *sources of radiation*.

Even in the simple case of an isotropic transparent medium, Electromagnetism is a very mathematical game which is played in a six-dimensional space. For a given system we must find, at any time and at any point, a six-component vector $EM_{(x,y,z,t)}$ which is called the "electromagnetic vector." The representation, in the three-dimensional usual geometrical space of a six-component vector, is usually a difficult exercise; hopefully in this case the six components may be associated in two individual sets of three components; the electric vector $E_{(x,y,z,t)}$ on the one hand and the magnetic vector $H_{(x,y,z,t)}$ on the other hand,

$$EM_{(x,y,z,t)} = \{E_{(x,y,z,t)} \cap H_{(x,y,z,t)}\}.$$

2.1.3. General Rules for the Determination of an Electromagnetic Field

To fulfill the laws of Electromagnetism $E_{(x,y,z,t)}$ and $H_{(x,y,z,t)}$ must satisfy:

- Four basic equations, called Maxwell's equations.
- Continuity (or discontinuity) conditions when crossing surfaces separating different media.
- Boundary conditions in the vicinity of the points where the fields have imposed values corresponding to the sources of radiation.

We will admit that existence and uniqueness theorems can be established, so that, following the previous conditions, the problem has a solution and that *this solution is unique*. In many cases this unique solution will be obtained thanks to intuitive considerations. In spite of its lack of purity the method will be considered as satisfactory. An objective of this book is to develop the physical insight of the reader allowing him to guess the response of a system to a given electromagnetic excitation.

- Any given field $EM_{(x,y,z,t)}$ will not, in general, fulfill Maxwell's equations. However, there is an infinite number of fields obeying the famous equations.
- Among this infinity of possible solutions, the continuity/discontinuity conditions will select solutions that fit the special system under consideration.
- Eventually the boundary conditions, that's to say, the excitation conditions, will determine, among the possible fields, which one really exists.

Because of the linearity of Maxwell's equations, any possible field in a given system may be written as

$$EM_{(x,y,z,t)} = \sum_i \alpha_i EM_{i(x,y,z)}.$$

The set of vectors $EM_{i(x,y,z)}$ constitutes a complete, although not unique, basis which is characteristic of the system and allows the representation of any possible fields inside the system.

The sources of radiation play a key role in the problem, eventually they fix the field which really exists. The presence of an electromagnetic field corresponds to some electromagnetic energy stored inside the system, it is at the places where the sources are located that this energy is transferred, this is the reason why the sources should always be connected to an external source of energy.

Let (x_S, y_S, z_S) be the coordinates of a radiation source located at point S, the field $EM_{(x_S,y_S,z_S)}$ at point S cannot follow any arbitrary law. It must be possible to use the set of basis vectors $EM_{i(x,y,z)}$ to obtain an expression of the electromagnetic field at point S. The following formula, which also defines the $\alpha_{i(t)}$ coefficients are finally obtained from

$$EM_{(x_S,y_S,z_S,t)} = \sum_i \alpha_i EM_{i(x_S,y_S,z_S,t)}.$$

Remarks

- The previous method, very attractive because it is very general, is currently used to study microwave problems, less often in Optics, except in the case of optical waveguides.
- The previous method is a magnificent example of the formulation of so-called *modern mathematics*. In fact Electromagnetism, and especially Optics, has probably played an important role in establishing the formalism of the set theory and in raising the notion of *vectorial spaces*.

- An acoustic analogy can be made to illustrate previous considerations: the acoustic fields that can be generated by a musical instrument are mainly determined by mechanical considerations. A violin is able to generate an infinity of acoustic fields, each of them obeying continuity conditions at the extremities of a string; the music which is produced is mainly determined by the way the player will excite the strings and, in lesser proportion, by the shape of the concert room.

2.1.4. *Example of a Planar Sine Wave*

The conditions for exciting a planar sine wave is a typical example of the previous method. An electromagnetic field $EM_{(x,y,z,t)}$ is said to be a planar sine wave propagating in the direction of a wave vector k, when it can be written as

$$EM_{(x,y,z,t)} = EM_0 \cos(\omega t - kOM).$$

It can be shown that, provided that the frequency ω and the wave vector modulus k should be related by a specific law called the dispersion law of the medium of propagation, such a time-space variation law does fulfill Maxwell's equations. To excite such a wave we just have to achieve a situation where, at any point of a plane (P) orthogonal to the vector k, the electromagnetic field should have a sine variation, with the same phase.

Figure 2.1 shows a situation where, at points spread over a plane (P), the electromagnetic field is forced to follow a sinusoidal and in-phase variation law $EM_{(x_s,y_s,z_s,t)} = EM_0 \cos \omega t$; (x_S, y_S, z_S) are the coordinates of any point located on (P). The plane (P) is the only discontinuity in the problem. The law $EM_{(x,y,z,t)} = EM_0 \cos(\omega t - kOM)$ fulfills Maxwell's equations and is in good agreement with the law of variation at any point of (P). Thus it constitutes the unique solution of the problem. Let us now consider the family of planes (P') that are parallel to (P), a point O belonging to (P) and H is its projection on (P'), at any point M of a plane (P'), the scalar product $kOM = kOH$ keeps a constant value. This means that the vibrations of the electromagnetic field have the same phase all over (P'), this is the reason why such a wave is called a planar wave. (P') are called wave planes.

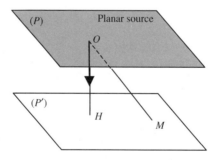

Figure 2.1. Generation of a planar wave by a planar source.

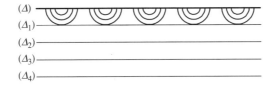

Figure 2.2. Illustration of a planar wave with elastic surface waves. Along Δ, Δ_1, Δ_2, and Δ_3 the vibrations are in-phase (modulo 2π).

Figure 2.2 is just a top-view of the experiment suggested in Figure 1.8 where children were periodically hitting the free surface of a liquid. At points located along a straight line (Δ), a sinusoidal motion is given to the molecules of water. A rectilinear wave is generated with a direction of propagation orthogonal to (Δ), all the points belonging to the same line (Δ_1) parallel to (Δ) vibrate with the same phase. Of course, the phases are not the same along two different parallel lines (Δ_2) or (Δ_3): there is a propagation phenomenon; the distance between two lines for which the out-phasing is 2π is the wavelength. For obvious reasons it's only possible to achieve a discrete repartition of sources, rather than a continuous one; this is not dramatic, since it can be demonstrated that if the distance between a source and its nearest neighbor is small enough, everything occurs as if the repartition was continuous. The Shannon theorem (also called the sampling theorem) indicates that the previous condition is not difficult to fulfill, the distance between two sources should be smaller than one wavelength.

2.2. The Different Kinds of Waves

2.2.1. *Wave Equation*

Analytic Formulation of a Propagation Phenomenon

Physics is largely dominated by the notion of waves and of propagating phenomena. Some physical entity G is said to propagate when its value, at point $M_{(x,y,z)}$ and at time t, is given by a relationship of the following kind:

$$G = G_0 f(\eta) \quad \text{with} \quad \eta = (t \pm \boldsymbol{u}\boldsymbol{OM}/V), \tag{2.1}$$

where $\boldsymbol{O}$ is the origin of coordinates, $\boldsymbol{u}$ is a unit vector defining the direction of propagation, and V is a parameter homogeneous to speed and specific to the medium in which the propagation occurs. For the sake of simplification and without any loss of generality, we will consider that the vector $\boldsymbol{u}$ which defines the propagation direction is along the Ox axis and we keep only the − sign in the argument of the function $f(\eta)$, the relationship is then written as

$$G = G_0 f(t - x/V). \tag{2.2}$$

Such a formula describes a propagation phenomenon, since it implies that G takes, at point x_2 and at time t_2, the same value that it had at the previous time t_1 at the point having an abscissa x_1 such that $(x_2 - x_1) = V(t_2 - t_1)$. The physical meaning of the $\pm$ sign becomes clearer: the $-$ sign corresponds to a forward propagation (direction of increasing x), while the $+$ sign is for a backward propagation (direction of decreasing x).

Necessary Condition for a Function to Describe a Planar Wave

First we are going to establish that a function $f(x, y, z, t)$ as defined by formula (2.1) necessarily obeys a very specific partial differential equation, called the wave equation. Such equations are very often met in Physics (acoustics, mechanics, electromagnetism, quantum mechanics, . . .), they are always associated to some propagation effect.

$Oxyz$ being a Cartesian system of reference, let (x, y, z) be the coordinates of some point $M(x, y, z)$ and (u_x, u_u, u_z) the three components of some unit vector $\boldsymbol{u}$, the scalar product of $\boldsymbol{u}$ by the vector $\boldsymbol{OM}$ is equal to $\boldsymbol{OMu} = xu_x + yu_y + zu_z$, from formula (2.1) we obtain

$$G = G_0 f(\eta) = G_0 f\left(t \pm \frac{xu_x + yu_y + zu_z}{V} \right) \quad \text{with} \quad \eta = t \pm \frac{xu_x + yu_y + zu_z}{V}.$$

Since $\boldsymbol{u}$ is a unit vector, we have $u_x^2 + u_y^2 + u_z^2 = 1$, it is then easy to evaluate the time and space partial derivatives of η and G:

$$\frac{\partial \eta}{\partial t} = 1 \quad \rightarrow \quad \frac{\partial^2 \eta}{\partial t^2} = 0, \quad \frac{\partial \eta}{\partial x} = \pm \frac{u_x}{V} \quad \rightarrow \quad \frac{\partial^2 \eta}{\partial x^2} = 0,$$

$$\frac{\partial G}{\partial t} = \frac{\partial G}{\partial \eta} \quad \text{and} \quad \frac{\partial G}{\partial x} = \frac{\partial G}{\partial \eta}\frac{\partial \eta}{\partial x} = \pm \frac{u_x}{V}\frac{\partial G}{\partial \eta},$$

$$\frac{\partial^2 G}{\partial x^2} = \frac{\partial}{\partial x}\left[\frac{\partial \eta}{\partial x}\left(\frac{\partial G}{\partial \eta} \right) \right] = \frac{\partial^2 G}{\partial \eta^2}\left(\frac{\partial \eta}{\partial x} \right)^2 + \frac{\partial G}{\partial \eta}\frac{\partial^2 \eta}{\partial x^2} = \left(\pm \frac{u_x}{V} \right)^2 \frac{\partial^2 G}{\partial \eta^2},$$

$$\frac{\partial^2 G}{\partial y^2} = \frac{\partial}{\partial y}\left[\frac{\partial \eta}{\partial y}\left(\frac{\partial G}{\partial \eta} \right) \right] = \frac{\partial^2 G}{\partial \eta^2}\left(\frac{\partial \eta}{\partial y} \right)^2 + \frac{\partial G}{\partial \eta}\frac{\partial^2 \eta}{\partial y^2} = \left(\pm \frac{u_y}{V} \right)^2 \frac{\partial^2 G}{\partial \eta^2},$$

$$\frac{\partial^2 G}{\partial z^2} = \frac{\partial}{\partial z}\left[\frac{\partial \eta}{\partial z}\left(\frac{\partial G}{\partial \eta} \right) \right] = \frac{\partial^2 G}{\partial \eta^2}\left(\frac{\partial \eta}{\partial z} \right)^2 + \frac{\partial G}{\partial \eta}\frac{\partial^2 \eta}{\partial z^2} = \left(\pm \frac{u_z}{V} \right)^2 \frac{\partial^2 G}{\partial \eta^2}.$$

Finally G is a solution of the following partial differential equation:

$$\frac{\partial^2 G}{\partial x^2} + \frac{\partial^2 G}{\partial y^2} + \frac{\partial^2 G}{\partial z^2} - \frac{1}{V^2}\frac{\partial^2 G}{\partial t^2} = 0. \tag{2.3}$$

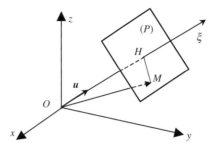

Figure 2.3. The scalar product, $OMu = ru = OH$, keeps a constant value for all points M of plane (P). At a given time, $f(\eta) = f(t - ur/V)$ has the same value anywhere on (P): $f(\eta)$ describes a planar wave orthogonal to the unit vector u.

2.2.2. General Considerations About the Planar Wave Solutions of a Wave Equation

Planar waves are a solution of the wave equation (2.3), they are not the only possible solution, however they are an important family of solutions.

Let us first consider a direction associated to some unit vector u, by definition a planar wave is a function $f(x, y, z, t)$ which, at a given time t, keeps the same value at any point M belonging to a plane (P) orthogonal to u. If we set $\zeta = ur = u_x x + u_y y + u_z z$, then $f(x, y, z, t)$ only depends on t and ζ,

$$\frac{\partial \xi}{\partial x} = u_x \quad \rightarrow \quad \frac{\partial^2 \xi}{\partial x^2} = 0, \quad \frac{\partial}{\partial x} = u_x \frac{\partial}{\partial \xi}, \quad \frac{\partial^2}{\partial x^2} = \frac{\partial}{\partial x}\left(\frac{\partial}{\partial x}\right) = u_x^2 \frac{\partial^2}{\partial \xi^2},$$

$$\frac{\partial \xi}{\partial y} = u_y \quad \rightarrow \quad \frac{\partial^2 \xi}{\partial y^2} = 0, \quad \frac{\partial}{\partial y} = u_y \frac{\partial}{\partial \xi}, \quad \frac{\partial^2}{\partial y^2} = \frac{\partial}{\partial y}\left(\frac{\partial}{\partial y}\right) = u_y^2 \frac{\partial^2}{\partial \xi^2},$$

$$\frac{\partial \xi}{\partial z} = u_z \quad \rightarrow \quad \frac{\partial^2 \xi}{\partial z^2} = 0, \quad \frac{\partial}{\partial z} = u_z \frac{\partial}{\partial \xi}, \quad \frac{\partial^2}{\partial z^2} = \frac{\partial}{\partial z}\left(\frac{\partial}{\partial z}\right) = u_z^2 \frac{\partial^2}{\partial \xi^2},$$

$$\rightarrow \quad \frac{\partial^2}{\partial x^2} + \frac{\partial^2}{\partial y^2} + \frac{\partial^2}{\partial z^2} = (u_x^2 + u_y^2 + u_z^2)\frac{\partial^2}{\partial \xi^2} = \frac{\partial^2}{\partial \xi^2}.$$

If we restrict ourselves to the planar solutions of the wave equation, we obtain

$$\frac{\partial^2 f}{\partial \xi^2} - \frac{1}{V^2}\frac{\partial^2 f}{\partial t^2} = 0 \quad \rightarrow \quad \left\{\left(\frac{\partial}{\partial \xi} - \frac{1}{V}\frac{\partial}{\partial t}\right)\left(\frac{\partial}{\partial \xi} + \frac{1}{V}\frac{\partial}{\partial t}\right)\right\}f = 0.$$

We introduce new variables, p and q, and evaluate the new partial derivatives

$$p = \left(t - \frac{\xi}{V}\right) \quad \text{and} \quad q = \left(t + \frac{\xi}{V}\right) \quad \rightarrow \quad t = \left(\frac{q+p}{2}\right) \quad \text{and} \quad \xi = V\left(\frac{q-p}{2}\right),$$

$$\frac{\partial}{\partial p} = \frac{1}{2}\left(\frac{\partial}{\partial t} - V\frac{\partial}{\partial \xi}\right) \quad \text{and} \quad \frac{\partial}{\partial q} = \frac{1}{2}\left(\frac{\partial}{\partial t} + V\frac{\partial}{\partial \xi}\right).$$

The wave equation can then be written as

$\partial^2 f/\partial p\ \partial q = 0$, which implies that $\partial f/\partial p$ is independent of p and only depends on q.

Two successive integrations lead to

$$\frac{\partial f}{\partial q} = g(q) \quad \text{and} \quad f(p,q) = \int g(q)\ dq = f_1(q) + \text{constant.}$$

The constant of integration doesn't depend on q, but it can depend on p. In other words, the general planar wave solution $f(p, q)$ is the sum of a function $f_1(q)$ of q only, and of a function $f_2(p)$ of p only,

$$f(\xi, t) = f_1\left(t - \frac{\xi}{V} \right) + f_2\left(t - \frac{\xi}{V} \right). \tag{2.4}$$

We don't know, a priori, anything about the functions $f_1(t - \zeta/V)$ and $f_2(t + \zeta/V)$, they are obtained from the boundary and excitation conditions.

2.2.3. General Considerations About the Spherical Wave Solutions of a Wave Equation

We are now looking at solutions of the wave equation that would keep a constant value at any point on a sphere (centered at point O, radius equal to r). Such waves are called spherical waves. Because of the symmetry of the problem it is convenient to use spherical coordinates. We will present the following formulas:

$$\frac{\partial^2 f}{\partial x^2} + \frac{\partial^2 f}{\partial y^2} + \frac{\partial^2 f}{\partial z^2} = \frac{1}{r}\frac{\partial^2}{\partial r^2}(rf) \quad \rightarrow \quad \frac{1}{r}\frac{\partial^2}{\partial r^2}(rf) - \frac{1}{V^2}\frac{\partial^2}{\partial t^2}(f) = 0.$$

Multiplying both sides of the last equation by r and taking advantage of the fact that time and space derivatives commute, we obtain

$$\frac{\partial^2}{\partial r^2}(rf) - \frac{1}{V^2}\frac{\partial^2}{\partial t^2}(rf) = 0.$$

Considering the product (rf) as an auxiliary function, we can use the results obtained for planar waves. It is seen that (rf) is the sum of two functions of the variables $(r \pm Vt)$. One of these functions represents a wave diverging from O, the second one is a wave converging toward O,

$$rf = g_1(t - r/v) + g_2(t + r/v) \quad \rightarrow \quad f(r, t) = \frac{1}{r}[g_1(t - r/v) + g_2(t + r/v)]. \tag{2.5}$$

Some Remarks About Spherical Waves

The presence of the $1/r$ factor in (2.5) is not at all surprising and only corresponds to the conservation of the flux of energy across spherical surfaces centered at point O. Point O plays a special role for a spherical wave and is often called its *focus*, as well as its *center*. Unfortunately, when performing the calculation, we had to multiply the two sides of the equation by r, we have no right to do so when r is equal to zero: thus the solution described by formula (2.5) is not valid in the immediate proximity of the focus.

2.3. Solutions of Maxwell's Equations for Harmonic Planar Waves

2.3.1. *Maxwell's Equations*

2.3.1.1. *Linear Operators Involved in Solving Maxwell's Equations*

Although Electromagnetism is exhaustively described by Maxwell's equations, we have waited until the third section of this chapter before writing them. This was deliberate, since we wanted the reader to become accustomed to waves and to the associated physical parameters.

In fact, the resolution of many optical problems doesn't explicitly require going back to Maxwell's equations. We can also emphasize the fact that very important developments in Optics (geometric optics, diffraction, interference, polarization of light, ...) had been extensively studied between the end of the seventeenth century and the end of the nineteenth century, at a time when Maxwell's equations had not yet been formulated. Still, Maxwell's equations constitute an extremely powerful tool. As the teaching of Physics developed, it is on the occasion of Electrostatics, of Magnetism, and of Electromagnetism, that Maxwell's equations are progressively introduced. From that point of view it can be considered that they are *experimentally rooted*.

Conversely, it is possible, a priori, to write Maxwell's equations and then deduce all the laws that govern Electrostatics, Magnetism, and Electromagnetism (Coulomb, Laplace, Biot-Savart, Faraday, ... laws).

Main Operators Used in Electromagnetism.

$V(x, y, z)$ is a scalar.

$\boldsymbol{E}_{(x,y,z)} = E_{x(x,y,z)}\boldsymbol{x} + E_{y(x,y,z)}\boldsymbol{y} + E_{z(x,y,z)}\boldsymbol{z}$ is a vector.

$(\boldsymbol{x}, \boldsymbol{y}, \boldsymbol{z})$ is an orthogonal and normalized trihedral.

Nabla operator: $\nabla = \left[\boldsymbol{x}\dfrac{\partial}{\partial x} + \boldsymbol{y}\dfrac{\partial}{\partial y} \; \boldsymbol{z}\dfrac{\partial}{\partial z} \right]$.

Gradient: $\text{grad}(V_{(x,y,z)}) = \left[\boldsymbol{x}\dfrac{\partial V}{\partial x} + \boldsymbol{y}\dfrac{\partial V}{\partial y} + \boldsymbol{z}\dfrac{\partial V}{\partial z} \right] = \nabla V$ vector.

Divergence: $\text{div}(\boldsymbol{E}) = \left[\dfrac{\partial E_x}{\partial x} + \dfrac{\partial E_y}{\partial y} + \dfrac{\partial E_z}{\partial z} \right] = \nabla \boldsymbol{E}$ scalar.

Scalar Laplacian: $\Delta V = \left[\dfrac{\partial^2 V}{\partial x^2} + \dfrac{\partial^2 V}{\partial y^2} + \dfrac{\partial^2 V}{\partial z^2} \right] = \text{div}[\text{grad}(V)] = \nabla(\nabla V) = \nabla^2 V$.

curl: $\text{curl}(\boldsymbol{E}) = \boldsymbol{x}\left(\dfrac{\partial E_y}{\partial z} - \dfrac{\partial E_z}{\partial y} \right) + \boldsymbol{y}\left(\dfrac{\partial E_z}{\partial x} - \dfrac{\partial E_x}{\partial z} \right) + \boldsymbol{z}\left(\dfrac{\partial E_x}{\partial y} - \dfrac{\partial E_y}{\partial x} \right) = \nabla \wedge \boldsymbol{E}$.

$\text{curl}(\boldsymbol{E})$ is a vector.

Vector Laplacian: $\Delta(\boldsymbol{E}) = \boldsymbol{x}\Delta(E_x) + \boldsymbol{y}\Delta(E_y) + \boldsymbol{z}\Delta(E_z) = \Delta^2 \boldsymbol{E}$ vector.

The following important identity can be established:

$$\text{curl}[\text{curl}(\boldsymbol{E})] = \text{grad}(\text{div } \boldsymbol{E}) - \Delta \boldsymbol{E} = \nabla \wedge (\nabla \wedge \boldsymbol{E}) = \nabla(\nabla \boldsymbol{E}) - \nabla^2 \boldsymbol{E}.$$

From a formal point of view, Maxwell's equations are linear relations between the first time and space partial derivatives of the electric and magnetic field components. Some combinations of the space partial derivatives play a very important role, and are so commonly met that they have been given special names: gradient, curl, divergence, scalar, and vector Laplacian; they are mathematical operators operating in the space of the functions of x, y, z.

Historically, Maxwell's equations were first formulated using partial derivatives, then using operators. Finally, a more general operator, called Nabla (∇), has been introduced. Each of the previous operators can be expressed using this Nabla operator which allows a very elegant and powerful presentation of Maxwell's equations.

2.3.1.2. *Writing Maxwell's Equations Using Operators*

Maxwell's equations will not be written for the most general case, but only for conditions corresponding to the propagation of the usual light waves: the propagation will occur in *nonmagnetic transparent dielectric materials*, inside of which the electric charge density is equal to zero and where the only electric currents will be the displacement currents.

The value of the magnetic permeability will be taken equal to the vacuum permeability ($\mu_0 = 4\pi \times 10^{-7}$ (MKS units, H/m)). Except in the case of an inhomogeneous media, the dielectric constant ε will be considered to have the same value at any point.

ε is submitted to dispersion: its value varies with the frequency. This corresponds to serious theoretical difficulties in writing Maxwell's equations in the more general case, for example, when the time variation laws of the fields are not sinusoidal. When the Fourier components of the signals occupy a broad frequency domain, as soon as ε depends on the frequency, a question must immediately be answered: Which value should be used for ε? The problem is not that serious, since in many practical situations ε will keep an almost constant value over all the frequency bands of the signals. Formulas (2.6) are strictly valid in a vacuum which is definitely a nondispersive medium ($\varepsilon = \varepsilon_0$), they are quite acceptable as long as the frequencies remain outside the absorption bands of the material where the waves are propagated.

Maxwell's equations	
$\text{curl}(\boldsymbol{E}) = -\mu_0 \dfrac{\partial \boldsymbol{H}}{\partial t}$,	$\nabla \wedge \boldsymbol{E} = -\mu_0 \dfrac{\partial \boldsymbol{H}}{\partial t}$,
$\text{curl}(\boldsymbol{H}) = +\dfrac{\partial \boldsymbol{D}}{\partial t}$,	$\nabla \wedge \boldsymbol{H} = +\dfrac{\partial \boldsymbol{D}}{\partial t}$,
$\text{div}(\boldsymbol{D}) = 0$,	$\nabla \boldsymbol{D} = 0$,
$\text{div}(\boldsymbol{H}) = 0$,	$\nabla \boldsymbol{H} = 0$,
$\boldsymbol{D} = \varepsilon \boldsymbol{E}$.	$\boldsymbol{D} = \varepsilon \boldsymbol{E}$.
(2.6.a)	(2.6.b)

2.3.2. Deducing the Electromagnetic Wave Equation from Maxwell's Equations

Using the fact that time derivation commutes with space derivation, and taking advantage of vector identities previously written, we have

$$\nabla^2 \boldsymbol{E} - \varepsilon \mu_0 \frac{\partial^2 \boldsymbol{E}}{\partial t^2} = 0 \quad \text{and} \quad \nabla^2 \boldsymbol{H} - \varepsilon \mu_0 \frac{\partial^2 \boldsymbol{H}}{\partial t^2} = 0. \tag{2.7}$$

Apart from the fact that they are concerned with vectors, equations (2.7) are identical to the wave equation (2.3). Any component EM_i of the electromagnetic field will follow equation (2.8),

$$\frac{\partial^2 EM_i}{\partial x^2} + \frac{\partial^2 EM_i}{\partial y^2} + \frac{\partial^2 EM_i}{\partial z^2} - \varepsilon \mu_0 \frac{\partial^2 EM_i}{\partial t^2} = 0 \frac{\partial^2 EM_i}{\partial x^2}, \tag{2.8}$$

with $i \in \{x, y, z\}$ and $EM_i \in \{E_x, E_y, E_z, H_x, H_y, H_z\}$.

Therefore a wave equation is obtained from Maxwell's equations, an important result being that we obtain an expression of the speed of light V versus ε and μ_0,

$$\varepsilon\mu_0 V^2 = 1 \quad \rightarrow \quad V = \frac{1}{\sqrt{\varepsilon\mu_0}}. \tag{2.9}$$

2.3.3. Maxwell's Equations for Harmonic Planar Waves

A harmonic planar wave is a wave for which both time *and* space variation laws are sinusoidal. If we restrict ourselves to this category of waves, Maxwell's equations take a far simpler form: they become *algebraic* equations instead of *partial derivative* equations (or equations between operators).

Of course we will use the complex notation for the analytic expression of the electromagnetic field, any of its components EM_i can thus be written as

$$EM_i = A_i e^{j\omega(t-\mathbf{ur}/V)} = A_i e^{j\omega t} e^{-j\omega \mathbf{ur}/V} = A_i e^{j\omega t} e^{-j\mathbf{kr}},$$

$$\mathbf{k} = \mathbf{u}\frac{\omega}{V} \quad \text{is the wave vector.}$$

Within the restriction to complex exponential functions, the derivative operators are replaced by the products:

$$\frac{\partial}{\partial t}(EM_i) = j\omega EM_i, \qquad\qquad \frac{\partial^2}{\partial t^2}(EM_i) = -\omega^2 EM_i,$$

$$\frac{\partial}{\partial x}(EM_i) = -jk_x EM_i, \qquad\qquad \frac{\partial^2}{\partial x^2}(EM_i) = -k_x^2 EM_i,$$

$$\text{(2.10.a)} \qquad\qquad\qquad\qquad \text{(2.10.b)}$$

$$\frac{\partial}{\partial y}(EM_i) = -jk_y EM_i, \qquad\qquad \frac{\partial^2}{\partial y^2}(EM_i) = -k_y^2 EM_i,$$

$$\frac{\partial}{\partial z}(EM_i) = -jk_z EM_i, \qquad\qquad \frac{\partial^2}{\partial z^2}(EM_i) = -k_z^2 EM_i,$$

$$\nabla EM = -j\mathbf{k}EM, \quad \text{(2.10.c)} \qquad\qquad \nabla^2 EM = -k^2 EM. \quad \text{(2.10.d)}$$

- deriving once versus x (or y, or z) → multiplying by $-jk_x$ (or $-jk_y$, or $-jk_z$);
- deriving twice versus x (or y, or z) → multiplying by $-k_x^2$ (or $-k_y^2$, or $-k_z^2$);
- taking the divergence of a vector → scalar multiplication of the vector by $-j\mathbf{k}$;
- taking the Laplacian of a vector → multiplication by the number $-k^2$;
- taking the curl of a vector → vector multiplication by the vector $-j\mathbf{k}$.

The formulation of Maxwell's equations is considerably simplified by the following equivalence rules:

$$\frac{\partial}{\partial t} \to j\omega, \quad \nabla \to -j\boldsymbol{k} \quad \text{and} \quad \nabla^2 \to -k^2.$$

Maxwell's equations (general case)	Maxwell's equations (planar waves)
$\nabla \wedge \boldsymbol{E} = -\mu_0 \dfrac{\partial \boldsymbol{H}}{\partial t},$ $\nabla \wedge \boldsymbol{H} = \varepsilon \dfrac{\partial \boldsymbol{E}}{\partial t},$ $\nabla \varepsilon \boldsymbol{E} = 0,$ $\nabla \boldsymbol{H} = 0.$	$\boldsymbol{k} \wedge \boldsymbol{E} = \omega\mu_0 \boldsymbol{H},$ $\boldsymbol{k} \wedge \boldsymbol{H} = -\omega\varepsilon \boldsymbol{E},$ $\boldsymbol{k}\boldsymbol{E} = 0,$ $\boldsymbol{k}\boldsymbol{H} = 0.$ $\qquad(2.11)$

2.3.4. *Decomposition of a Wave in Planar Harmonic Waves*

The previous operators are all linear since we have

$$\frac{\partial}{\partial t}(A+B) = \frac{\partial}{\partial t}(A) + \frac{\partial}{\partial t}(B) \quad \text{and} \quad \nabla(A+B) = \nabla(A) + \nabla(B).$$

The linearity of the time derivative operator is very often met in Physics, and is responsible for the possibility of Fourier developments: a time function $f(t)$ can always be considered as the superposition of an infinite number of sine functions lasting for ever and having a suitably chosen repartition of frequencies

$$f(t) = \alpha \int\limits_{-\infty}^{+\infty} e^{-j\omega t} a(\omega)\, d\omega,$$

where α is a normalizing coefficient and $a(\omega)$ is a complex number that varies with the frequency. The time variation of $f(t)$ is the result of the interference of all the different sine components.

Planar waves play the same role for functions of space variables $(x,\ y,\ z)$, as the sine functions for functions of time. To use a slightly emphatic, but sometimes useful language, it can be said that planar waves constitute a set of orthogonal functions and allow a representation of the space of the functions of $(x,\ y,\ z)$. In other words, a function $f(x,\ y,\ z)$ can always be considered as a superposition of an infinite number of planar waves, each of them having a different wave vector,

$$f(\boldsymbol{r}) = \alpha \iiint e^{-j\boldsymbol{k}\boldsymbol{r}} a_{(k_x,k_y,k_z)}\, dk_x\, dk_y\, dk_z,$$

with $r = x\boldsymbol{x} + y\boldsymbol{y} + y\boldsymbol{y}$ and $\boldsymbol{k} = k_x\boldsymbol{x} + k_x\boldsymbol{y} + k_x\boldsymbol{z}$.

2.4. Structure of an Electromagnetic Planar Wave

2.4.1. *General Topics*

Let us consider:

- A propagation medium having an infinite extension, filled with a dielectric material (ε and μ_0).
- An angular frequency ω and a wave vector $\boldsymbol{k}$.
- An electromagnetic field $\boldsymbol{EM}$, defined by the union of an electric field $\boldsymbol{E}_0$ and of a magnetic field $\boldsymbol{H}_0$.

Let us ask the following question: Which conditions should obey the previous parameters, if we want the planar waves defined by formulas (2.12) to satisfy Maxwell's equations?

$$\boldsymbol{E} = \boldsymbol{E}_0 e^{j\omega t} e^{-jkr},$$
$$\boldsymbol{H} = \boldsymbol{H}_0 e^{j\omega t} e^{-jkr}. \tag{2.12}$$

After a few elementary manipulations of vector algebra, and making good use of the double vector product formula and of its equivalent formulation using the Nabla operator, equations (2.12) become

$$\boldsymbol{E} = -\frac{1}{\omega\varepsilon} \boldsymbol{k} \wedge \boldsymbol{H}, \quad \boldsymbol{k} \wedge (\boldsymbol{k} \wedge \boldsymbol{E}) = +\omega\mu_0 \boldsymbol{k} \wedge \boldsymbol{H},$$

$$\boldsymbol{H} = +\frac{1}{\omega\mu_0} \boldsymbol{k} \wedge \boldsymbol{E}, \quad \boldsymbol{k} \wedge (\boldsymbol{k} \wedge \boldsymbol{H}) = -\omega\varepsilon \boldsymbol{k} \wedge \boldsymbol{H}. \tag{2.13}$$

2.4.2. *The Geometry of a Planar Electromagnetic Wave*

Because of formulas (2.13), $\boldsymbol{E}_0$ and $\boldsymbol{H}_0$ are mutually orthogonal; they are also orthogonal to the wave vector and thus belong to a plane orthogonal to $\boldsymbol{k}$, such a plane is called a *wave plane*. The trihedral ($\boldsymbol{E}_0$, $\boldsymbol{H}_0$, $\boldsymbol{k}$) is orthogonal and direct.

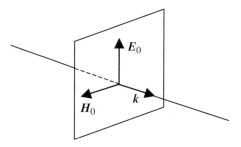

Figure 2.4. Respective positions of vectors $\boldsymbol{E}_0$ and $\boldsymbol{H}_0$ lying in the wave plane.

Dispersion Law of the Medium Supporting the Waves

Starting from formulas (2.7) it is readily obtained that

$$(k^2 - \varepsilon\mu_0\omega^2)\boldsymbol{E} = 0, \quad (k^2 - \varepsilon\mu_0\omega^2)\boldsymbol{H} = 0 \quad \rightarrow \quad k^2 - \varepsilon\mu_0\omega^2 = 0. \quad (2.14)$$

The frequency, on the one hand, and the modulus of the wave vector, on the other, cannot be given arbitrary values, they must obey a relation involving ε and μ_0; this relation is called the *dispersion law of the medium* supporting the wave,

$$k^2 = \varepsilon\mu_0\omega^2 \quad \rightarrow \quad k = \omega\sqrt{\varepsilon\mu_0} \quad \text{dispersion law.} \quad (2.15)$$

From (2.15) we can deduce the value of what is called the *phase velocity V* of the wave,

$$k = \omega/V \quad \rightarrow \quad \varepsilon\mu_0 V^2 = 1 \quad \rightarrow \quad V = \frac{1}{\sqrt{\varepsilon\mu_0}}. \quad (2.16)$$

If ε doesn't depend on the frequency, the modulus of the wave vector is proportional to the frequency. The dispersion law is then said to be *linear* and the material is said to have no dispersion; in this case, the phase velocity doesn't depend on the frequency.

Vacuum Is the Only Medium to Have No Dispersion

The speed of propagation in a vacuum is called the *celerity of light* and is usually designated by the letter c,

$$\varepsilon\mu_0 c^2 = 1 \quad \rightarrow \quad c = \frac{1}{\sqrt{\varepsilon_0\mu_0}} = 299{,}792.458 \text{ km/s} \approx 3 \times 10^8 \text{ m/s.}$$

Any material shows dispersion: The dispersion law is never linear over the whole spectral range (frequencies going from zero to infinity). This dependence of ε is rooted very deep in the physical processes governing light/material interactions, it is connected to the *causality principle* by means of the *Kramers-Krönig* formulas.

Wave Impedance

Relations (2.13) imply that the modulus of the electric and magnetic fields are proportional. It can be shown that the proportionality coefficient is measured in ohms, this is the reason why it is called the *wave impedance Z*, in a vacuum its value Z_0 is equal to 120π (about 400 Ω),

$$E_0 = \frac{k}{\omega\varepsilon} H_0 = \sqrt{\frac{\mu_0}{\varepsilon}} H_0,$$

$$H_0 = \frac{k}{\omega\mu_0} E_0 = \sqrt{\frac{\varepsilon}{\mu_0}} E_0, \quad Z = \frac{E_0}{H_0} = \sqrt{\frac{\mu_0}{\varepsilon}} = \mu_0 V = \frac{1}{\varepsilon V}. \quad (2.17)$$

Index of Refraction

For a given material, the ratio between the celerity of light and the phase velocity in the material is the index of refraction, n. The index of refraction can be simply related to the relative dielectric permeability of the material, $\varepsilon_r = \varepsilon/\varepsilon_0$,

$$n = \frac{c}{V} = \frac{\sqrt{\varepsilon\mu_0}}{\sqrt{\varepsilon_0\mu_0}} = \sqrt{\frac{\varepsilon}{\varepsilon_0}} \quad \rightarrow \quad n^2 = \varepsilon_r. \tag{2.18}$$

Formula (2.18) is only valid for nonmagnetic transparent materials, otherwise the relative magnetic permeability should be introduced.

2.4.3. Energy Transportation by an Electromagnetic Wave

Having felt the heat of the Sun, everybody is surely convinced that light waves carry some energy. When trying to evaluate the amount of energy transported by a planar wave, a slight difficulty is immediately met, since such a wave has an infinite extension in any direction orthogonal to the wave vector. We can easily get round this difficulty by considering the energy flux across a given closed loop; of course, we will find it more convenient to choose a planar loop orthogonal to the direction of propagation.

To calculate the amount of energy, ϕ, that is sent across a given area, S, during one second, a new vector is introduced, this is called the Poynting vector and is equal to the vector product of the electric field by the magnetic field,

$$\boldsymbol{\Pi} = \boldsymbol{E} \wedge \boldsymbol{H} \quad \text{Poynting vector.} \tag{2.19}$$

ϕ is simply equal to the flux of the Poynting vector across the loop under consideration. We will now establish this proposition and we recommend the reader to refer to a textbook about Electromagnetism where a more rigorous demonstration will be found. To obtain ϕ from the parameters that characterize a planar wave, we will develop a less powerful and less general theory. This method is however quite meaningful from a physical point of view and

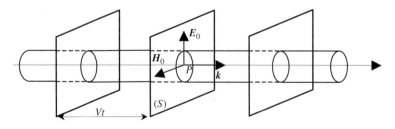

Figure 2.5. Energy transportation by an electromagnetic wave.

emphasizes the fact that some electromagnetic energy is stored at places where an electromagnetic field has been created.

If an electromagnetic field (E, H) exists at a point P and around it, it should be considered that some energy is stored with a density u expressed in joules per unit volume and given by

$$u = \frac{\varepsilon E^2}{2} + \frac{\mu_0 H^2}{2}. \tag{2.20}$$

Let us evaluate the electromagnetic energy density at a point where the fields are given by $E = E_0 i \cos(\omega t - \varphi)$ and $H = H_0 j \cos(\omega t - \varphi)$ (i and j are two orthogonal unit vectors), the instantaneous value $u(t)$ is equal to

$$u(t) = \frac{\varepsilon E^2 + \mu_0 H^2}{2} \cos^2(\omega t - \varphi).$$

We consider that the stored energy is the time-averaged value of $u(t)$, since $\langle \cos^2 \omega t \rangle = \frac{1}{2}$ we obtain

$$u_{\text{stored}} = \langle u(t) \rangle = \frac{\varepsilon E_0^2 + \mu_0 H_0^2}{4} = \frac{\varepsilon E_0^2}{2} = \frac{\mu_0 H_0^2}{2}. \tag{2.21}$$

The amount of energy which crosses the surface S of Figure 2.5 during time t was previously stored inside a cylinder having a cross section equal to S and length equal to Vt (V is the speed of the propagation of light). If the area S is taken equal to one area unit and the time equal to one second, the corresponding amount of energy is called the light intensity I of the wave,

$$I = \frac{1}{S} \frac{1}{t} u_{\text{stored}} SVt = u_{\text{stored}} V = V \frac{\varepsilon E_0^2}{2} = V \frac{\mu_0 H_0^2}{2}. \tag{2.22}$$

Using the wave impedance Z that was introduced in formula (2.17), formula (2.22) can be made more elegant,

$$I = u_{\text{stored}} V = \frac{1}{2} \frac{E_0^2}{Z} = \frac{1}{2} Z H_0^2. \tag{2.23}$$

The above formula reminds us of the expression of the electric power, $P = RI^2/2 = V^2/2R$, in a resistor R when the current and the voltage are, respectively, $I \cos \omega t$ and $V \cos \omega t$.

2.5. General Harmonic Waves

2.5.1. Helmholtz's Equation

Planar waves are not the only possible solutions of the wave equation. They are the simplest and correspond to many practical situations.

We now consider waves for which the time variation is still sinusoidal (harmonic waves), but for which the space variation is of a more general kind. The field components will be taken as

$$EM_{i(r,t)} = a_{i(r)} \cos(\omega t - g_{i(r)}).$$

We are now looking at conditions which the two functions $a_{i(r)}$ and $g_{i(r)}$ should obey for $EM_{i(r,t)}$ to be an acceptable solution of Maxwell's equations and, more specifically, of the wave equation

$$\frac{\partial^2 EM_i}{\partial x^2} + \frac{\partial^2 EM_i}{\partial y^2} + \frac{\partial^2 EM_i}{\partial z^2} - \frac{1}{V^2}\frac{\partial^2 EM_i}{\partial t^2} = 0.$$

We introduce the complex function $U_{(x,y,z)} = U_{(r)}$ of the real variables x, y, z using the following relations:

$$U_{(r)} = a_{(r)}e^{-jg(r)} \quad \rightarrow \quad EM_{i(r,t)} = \mathrm{Re}\{U_{(r)}e^{j\omega t}\}.$$

Using the equivalence $\partial^2/\partial t^2 \rightarrow -\omega^2$ we obtain, for $U_{(x,y,z)}$, a partial derivative equation that is called the *Helmholtz equation*:

$$\frac{\partial^2 U}{\partial x^2} + \frac{\partial^2 U}{\partial y^2} + \frac{\partial^2 U}{\partial z^2} + \frac{\omega^2}{V^2}U = 0,$$

$$\Delta U + \frac{\omega^2}{V^2}U = 0 \quad \rightarrow \quad \nabla^2 U + \frac{\omega^2}{V^2}U = 0, \quad \text{Helmholtz equation.}$$

$$(2.24)$$

Many problems of Electromagnetism consist in finding a solution of the Helmholtz equation, which can fit with the special boundary conditions of the system under consideration. The functions $U_{(x,y,z)} = e^{-jkr}$ (planar waves) and $U_{(r)} = e^{-jkr}$ (spherical waves) are solutions of the Helmholtz equation.

It should be remembered that, after having found a possible solution, we must solve the problem of the excitation: Along which curve will the children of Figure 1.8 be asked to sit around the swimming pool to generate the right function $U_{(x,y,z)}$?

2.5.2. Helmholtz's Equation for Slowly Varying Amplitudes

There are many practical situations where the propagation mostly occurs along one direction that will be chosen as the Oz axis. It is then convenient to separate the function $U_{(r)} = a_{(r)}e^{-jg(r)}$ into two parts: one part varies periodically and *rapidly* with z, while the second part has smoother variations:

$$U_{(x,y,z)} = \psi_{(x,y,z)}e^{-jkz} \quad \text{with} \quad k = \frac{2\pi}{\lambda}, \quad (2.25)$$

where e^{-jkz} is a periodic function, which takes again the same value when z is increased by one wavelength. In Optics λ is quite short, so the variation is very fast. On the contrary, $\psi_{(x,y,z)}$ varies slowly with z, this will be expressed in the following way:

$$\psi_{(x,y,z+\lambda)} \approx \psi_{(x,y,z)},$$

$$\psi_{(x,y,z+\lambda)} = \psi_{(x,y,z)} + \lambda \frac{\partial \psi}{\partial z} + \frac{\lambda^2}{2} \frac{\partial^2 \psi}{\partial z^2} + \cdots,$$ (2.26.a)

$$\psi_{(x,y,z+\lambda)} = \psi_{(x,y,z)} + \frac{2\pi}{k} \frac{\partial \psi}{\partial z} + \frac{2\pi^2}{k^2} \frac{\partial^2 \psi}{\partial z^2} + \cdots.$$

The condition of *slow variations* can then be written in the following way:

$$\frac{\partial^2 \psi}{\partial z^2} << k \frac{\partial \psi}{\partial z} << k^2 \psi.$$ (2.26.b)

To use relation (2.24), we must first calculate the derivatives of the function $U_{(x,y,z)}$:

$$\frac{\partial U}{\partial x} = \frac{\partial \psi}{\partial x} e^{-jkz}, \quad \frac{\partial U}{\partial y} = \frac{\partial \psi}{\partial y} e^{-jkz}, \quad \frac{\partial U}{\partial z} = \left(\frac{\partial \psi}{\partial z} - jk\psi \right) e^{-jkz},$$

$$\frac{\partial^2 U}{\partial x^2} = \frac{\partial^2 \psi}{\partial x^2} e^{-jkz}, \quad \frac{\partial^2 U}{\partial y^2} = \frac{\partial^2 \psi}{\partial y^2} e^{-jkz},$$

$$\frac{\partial^2 U}{\partial z^2} = \left(\frac{\partial^2 \psi}{\partial z^2} - 2jk \frac{\partial \psi}{\partial z} - k^2 \psi \right) e^{-jkz}.$$

The Helmholtz equation then becomes

$$\left(\frac{\partial^2 \psi}{\partial x^2} + \frac{\partial^2 \psi}{\partial y^2} + \frac{\partial^2 \psi}{\partial z^2} - 2jk \frac{\partial \psi}{\partial z} - k^2 \psi + k^2 \psi \right) e^{-jkz} = 0,$$

$$\Delta \psi - 2jk \frac{\partial \psi}{\partial z} = 0.$$

As the coordinate z plays a special role, it seems convenient to divide the Laplace operator into two parts:

- A longitudinal part $\partial^2/\partial z^2$.
- A transverse part $\Delta_t = \partial^2/\partial x^2 + \partial^2/\partial y^2 = \nabla_t^2$.

For functions that slowly vary with z, the longitudinal part of the Laplace operator is negligible with regard to its transverse part, the Helmholtz equation can be written as

$$\Delta_t \psi - 2jk \frac{\partial \psi}{\partial z} = 0.$$ (2.27)

2.5.3. *Light Rays and Propagation Speed for Inhomogeneous Waves*

When the functions $a_{(r)}$ and $g_{(r)}$ have no special properties and, for example, don't correspond to planar or spherical waves, the corresponding wave is said to be *inhomogeneous*. For such waves two kinds of surfaces are usually introduced:

- equiamplitude surfaces along which $a_{(r)}$ keeps a constant value; and
- wave surfaces along which it is the phase $g_{(r)}$ which remains constant.

For planar waves and spherical waves, equiamplitude and wave surfaces coincide.

Surface waves, sets of points all vibrating in phase, play an important role in studying the properties of inhomogeneous waves. It is probably because the amplitude of vibration varies along a surface wave that they are said to be inhomogeneous.

Let us imagine an observer moving along some trajectory T, see Figure 2.6(b), and ask the question: How should he move if he wants to see an instant phase $(\omega t - g_{(r)})$ that remains constant? At time t, and at point M, the phase is $[\omega t - g_{(r)}]$; at time $t + dt$, and at point M', the phase is $[\omega_{(t+dt)} - g_{(r+dr)}]$, if the phases are the same at points M and M' we obtain

$$\omega \, dt = g_{(r+dr)} - g_{(r)} = dr \, \mathrm{grad}(g_{(r)}).$$

Let q be a unit vector of the tangent to T at point M, and let $s(t)$ be the abscissa of the observer, measured along T. We have $dr = q \, ds$ and the speed of the observer on his trajectory is thus defined by

$$\frac{ds}{dt} = \frac{\omega}{q \, \mathrm{grad}(g_{(r)})}.$$

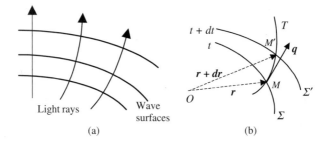

Light rays   Wave surfaces

(a)   (b)

Figure 2.6. (a) shows the surface waves and light rays of inhomogeneous waves. (b) indicates how to evaluate the speed of propagation of an inhomogeneous wave.

This speed is minimum when the scalar product $\boldsymbol{q}\,\mathrm{grad}(g_{(r)})$ is maximum, this occurs when the two vectors $\boldsymbol{q}$ and $\mathrm{grad}(g_{(r)})$ are parallel. Trajectories for which the speed is minimum are orthogonal to the wave surfaces: they constitute the light rays, and the corresponding phase velocity is given by

$$V = \frac{\omega}{\|\mathrm{grad}(g_{(r)})\|}.$$

Let us use the above formula to again find the dispersion law of a planar wave

$$\boldsymbol{r} = x\boldsymbol{x} + y\boldsymbol{y} + z\boldsymbol{z}, \quad \boldsymbol{k} = k_x\boldsymbol{x} + k_y\boldsymbol{y} + k_z\boldsymbol{z},$$

for a planar wave

$$g_{(r)} = \boldsymbol{kr} = k_x\boldsymbol{x} + k_y\boldsymbol{y} + k_z\boldsymbol{z} \quad \rightarrow \quad \mathrm{grad}(g_{(r)}) = k_x\boldsymbol{x} + k_y\boldsymbol{y} + k_z\boldsymbol{z},$$

$$\|\mathrm{grad}(g_{(r)})\| = (k_x^2 + k_y^2 + k_z^2)^{1/2} = k \quad \rightarrow \quad V = \frac{\omega}{\|\mathrm{grad}(g_{(r)})\|} = \frac{\omega}{k} = \frac{c}{n}.$$

2.6. Spherical Waves

2.6.1. *Physical Difficulties Associated with Spherical Waves*

An important part of Optics consists in obtaining the image of an object. Ideally a light point source emits spherical waves, diverging waves if the object is real and converging waves if virtual. The main role of optical instruments is to transform an initial spherical wave into another spherical wave. Although spherical waves are easily conceivable, it is impossible to obtain them rigorously. From a mathematical point of view this comes from the fact that, when integrating the wave equation for spherical waves, we had to exclude the focus where r is equal to zero. From a physical point of view, it can be said that if the wave surfaces remain strictly spherical up to the focus, all the rays would intersect at the same point where the electromagnetic field would be infinite.

The modification of the shape of the surface waves near the focal point is a manifestation of diffraction: it is not possible to focus a wave in a spot having a size much smaller than the wavelength.

2.6.2. *π Out-Phasing of a Spherical Wave when Crossing its Focus*

A spherical wave is always limited by some diaphragm and never occupies the 4π steradian of the whole space; it is usually limited to a cone of half-angle θ (see Figure 2.7). When crossing its focus, the wave first converges, and

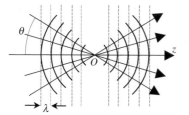

Figure 2.7. Representation of an ideal spherical wave near focus.

then diverges; if r is arithmetic and thus positive, the representation of the wave is:

- before focus $(1/r)\cos(\omega t + kr)$;
- after focus $(1/r)\cos(\omega t - kr)$.

In Figure 2.7 we have drawn, as solid lines, a family of wave surfaces, i.e., the points where the vibrations have the same value, φ or $\varphi + 2p\pi$ (p is an integer); a new surface is obtained each time p is increased by one. Using dotted lines we have drawn planes that are tangent to the wave surfaces at the points where they intersect the axis Oz of the cone. These planes are parallel, far away from the focus they are equidistant from and separated by one wavelength, then they coincide with the wave planes of a planar wave that would propagate parallel to the cone axis.

For a planar wave, the phase smoothly varies when z varies from $-\infty$ to $+\infty$; for spherical waves, there is an abrupt discontinuity or, more likely, a rapid variation of π in the close vicinity of the focus.

If will be shown (see Section 2.6.5) that the rectilinear converging rays of a spherical wave are replaced by hyperbolas. Farther away the rays are the asymptotes; near the focus the spherical wave merely looks like a planar wave (!). The region where the spherical wave can be assimilated to a planar one is called the *Rayleigh zone*. At the focus most of the energy is inside a circle having a diameter W, so the field amplitude doesn't go to infinity. The half-angle θ between the asymptotes of the hyperbola are related to the wavelength λ and to W by $\theta = K\lambda/W$ (K is a dimensionless coefficient of the order of unity). This formula reminds us of the expression of the resolving power of a microscope ($1.22\lambda/d$).

2.6.3. *Series Development of a Collimated Spherical Wave*

It will be considered that we are far enough from the center of a spherical wave so that the wave surfaces are again quite spherical; in fact, the limitation is not very severe and it's sufficient to be separated from the focus by several wavelengths. We are now going to introduce a new approximation that is traditionally called Gaussian approximation.

In practice, spherical optical waves are always limited by a diaphragm: the aperture angle θ of the cone which contains the energy of the wave is usually

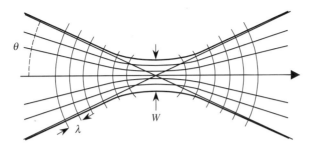

Figure 2.8. A more realistic representation of a spherical wave crossing its focus.

small enough, so that the higher-order terms can be omitted in the development of trigonometric functions ($\cos\theta \cong 1$, $\sin\theta \cong \tan\theta \cong \theta$). Our purpose is now to give an approximate expression for the mathematical description of a spherical wave: calculations using Cartesian or cylindrical coordinates will then be quite simple.

Let us consider a spherical electromagnetic wave and its center O, the various components of the field can be written as

$$EM_{(x,y,z,t)} = \text{Re}\left\{\frac{a}{OM}e^{j\omega t}e^{\pm jk_0 OM}\right\} = \text{Re}\{U_{(x,y,z)}e^{j\omega t}\},$$

with

$$U_{(x,y,z)} = \frac{a}{\sqrt{x^2+y^2+z^2}}e^{\pm jk_0\sqrt{x^2+y^2+z^2}}.$$

The $-$ sign is associated with a wave that diverges from point O, while the $+$ sign corresponds to a wave that converges toward O, and diverges after. Attention must be paid to the fact that in the expression $e^{-jk_0 OM}$, we don't have the scalar product of two vectors, but rather the product of the length of the wave vector k_0 by the length of the vector $r = OM$, both lengths being positive numbers.

In Figure 2.9, (Σ) represents a given surface wave of a spherical wave centered at point O and propagating in the vicinity of the Oz axis, with a small numerical aperture. PP' is the trace of the plane which is tangent to (Σ) at point $H(0, 0, z)$ where PP' intersects the Oz axis; Ox and Oy are two orthogonal axes on (Π). To any point $M_{(x,y,z)}$ belonging to (Σ), we associate the point $M'_{(x',y',z')}$ where OM intersects pp'. Far enough away from the center, the values of the coordinates (x, y), as well as (x', y'), are very small as compared to the value of z, an approximate expression of the field components is readily obtained using the following development of OM:

$$r = OM = (x^2+y^2+z^2)^{1/2} = \bar{z}\left(1+\frac{x^2+y^2}{z^2}\right)^{1/2} \approx \bar{z} + \frac{x^2+y^2}{2\bar{z}},$$

$$U_{(x,y,z)} \approx \frac{a}{\bar{z}}e^{\pm jk_0\frac{x^2+y^2}{2\bar{z}}}e^{-jk_0\bar{z}}, \quad \bar{z} \text{ is the absolute value of } z.$$

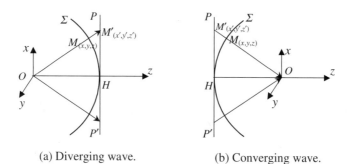

(a) Diverging wave.   (b) Converging wave.

Figure 2.9. Spherical waves. The points M and M' are considered to be almost at the same place.

The main interest of the previous expression of $U_{(x,y,z)}$ comes from the fact that, in the argument of the exponential, the square root is replaced by a polynomial expression.

The variable z is found at three different places in the expression of $U_{(x,y,z)}$:

(i)   In the denominator of the real term, a/z decreases slowly with z and can be considered to keep a *constant* value inside rather long distances.

(ii) In the argument of the complex exponential $e^{\pm jk(x^2+y^2)/2z}$: this term is responsible for the fact that the phase is not the same for all the points of a plane perpendicular to the Oz axis.

(iii) In the argument of the complex exponential $e^{\pm jkz}$: this term represents a very fast variation, since its sign changes every time z increases by one half-wavelength.

Let us now consider the two waves illustrated in Figures 2.9(a) and (b); since they both propagate in the positive direction of the Oz axis, we will choose the exponential term e^{-jkz} and describe the amplitude repartition over a plane (P) orthogonal to Oz by the following expression:

$$U_{(x,y)} = Ke^{ja(x^2+y^2)}, \quad K \text{ is just a proportionality coefficient.} \quad (2.28.a)$$

Such a wave can be considered to be a spherical wave having its center at the origin O of the Oz axis; the radius of curvature of the wave surface that is tangent to the plane PP' is related to the modulus k_0 of the wave vector or to the wavelength λ by relations (2.28.b):

$$U_{(x,y)} = Ke^{ja(x^2+y^2)}, \quad \rho = \frac{k_0}{2\alpha} = \frac{\pi}{\lambda\alpha}, \quad (2.28.b)$$

where $\alpha > 0$ corresponds to a wave converging toward point O; and
$\alpha < 0$ corresponds to a wave diverging from point O.

2.6.4. *Using an Optical Device to Transform a Planar Wave into a Spherical Wave*

To illustrate the interest of the previous development, we are going to show how this allows understanding as to how a lens, or a spherical mirror, transforms a planar wave into a spherical one.

2.6.4.1. *Reflection of a Wave on a Spherical Mirror*

When an electromagnetic wave hits the mirror surface, its electric field sets into vibration the electrons which then become emitting sources: a reflected wave is thus generated. For the sake of simplicity it will be considered that the phase of the motion of an electron is equal to the phase of the incident wave at the place where the electron is. The complex amplitude of the sources is determined by the incident wave and by the shape of the mirror. In the case of an incident planar wave and a spherical mirror, it will be shown that the law of variation of the complex amplitudes of the reflected wave is just the same as for a spherical wave.

Let us consider, Figure 2.10, a planar wave arriving at a concave spherical mirror after having propagated parallel to one of its diameters. The mirror is defined by the following parameters: its center Ω and its radius of curvature $\Omega S = \Omega N = R$. We will compare the complex amplitudes of the incident and reflected waves on the plane (Π) tangent to the mirror at point S. If the mirror is a perfect reflector, the moduli of these amplitudes are equal. However, there is a phase difference associated to a distance of the light equal to $2NM$, N is on the mirror, M is on plane (Π). Using $Sxyz$ as the axis of coordinates, the equation of the mirror surface is $x^2 + y^2 + (R - z)^2 = R^2$, in the vicinity of point S it can be simplified as $x^2 + y^2 - 2Rz = 0$.

For the incident planar wave the phase would be constant along (Π), we use the conditional since the incident wave doesn't reach (Π). On the contrary, the phase of the reflected beam varies along (Π), the phase variation being given by $e^{-jk_0 MN}$.

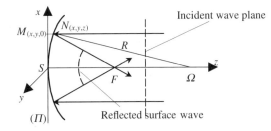

Figure 2.10. Reflection of a planar wave on a spherical mirror. Ω and R are, respectively, the center and radius of the mirror.

MN is easily obtained from the simplified equation of the mirror: $MN = r = (x^2 + y^2)/2R$.

Finally, the complex amplitude of the reflected wave at point M is given by

$$Ke^{jk_0\left(\frac{x^2+y^2}{2R/2}\right)} = Ke^{jk_0\left(\frac{x^2+y^2}{2f}\right)} \quad \text{with } f = R/2.$$

The above expression coincides with the first term of the development of a spherical wave centered at point F in Figure 2.10; the focal length of a spherical mirror is found to be equal to half of its radius of curvature. It is left as an exercise to consider the case of an incident spherical wave and establish the conjugation law for spherical mirrors.

2.6.4.2. *Focusing a Parallel Beam with a Spherical Lens*

To simplify the presentation and Figure 2.11 we will consider the case of a planoconvex lens; the method is however quite general and the reader is advised to treat the case of a biconvex lens, either by adapting the previous method, or by considering a biconvex lens as the assembly of two planoconvex lenses.

The lens is a piece of glass (index of refraction n) limited on one side by a plane (SH) and on the other side by a sphere (center Ω, radius $\Omega S' = R$). Using the coordinate axis $(S'xyz)$, the sphere equation can be approximated as

$$x^2 + y^2 - 2Rz = 0.$$

All the points of the plane (SH) are set in motion with the same phase by the incident wave. Inside the lens, the light propagates at speed $V = c/n$, outside the lens the speed is equal to c. The light takes a longer time to cross the lens at the level of SS' than it does at the level of HH'.

If we now consider the vibrations at points located on the plane $S'H'$, they are not in phase: the phase retardation is maximum at the center and diminishes as the distance to the axis $S'z$ increases. $t_{SS'}$ and $t_{HH'}$ being, respectively, the times for going from S to S' and from H to H', we have

$$t_{SS'} = \frac{SS'}{V} = n\frac{SS'}{c} \quad \text{and} \quad t_{HH'} = \frac{HK}{V} + \frac{KH'}{c} = \frac{(nHK + KH')}{c} = \frac{(nSS' + KH')}{c}.$$

From the simplified equation of the sphere it is seen that $KH' = (x^2 + y^2)/2R$, the definition of K is given in Figure 2.11. The phase difference between S' and H' is given by the following expression, where ω and k_0 are, respectively, the angular frequency and the vacuum wave vector of the light,

$$\varphi_{SS'} - \varphi_{HH'} = \omega(t_{SS'} - t_{HH'}) = k_0(n-1)KH' = k_0(n-1)\frac{(x^2 + y^2)}{2R}.$$

The above phase repartition is quadratic in x and y and corresponds to a spherical wave; at the level of S' the radius of curvature of the surface wave is the focal length $f = R/(n-1)$.

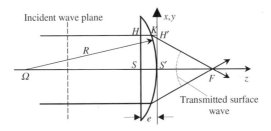

Figure 2.11. Focusing a parallel light beam with a lens. Ω and R are the center and radius of the spherical interface. H and H' are considered to be at the same place.

This example shows the mathematical meaning of Gauss approximation: stigmatism is obtained only within the limit of validity of the development of formula (2.26.b). Aberrations correspond to the omitted terms of the development.

2.6.5. Gaussian Beams

Gaussian beams are a very important type of light beam. They have been made very popular by lasers, since such devices usually emit such beams. There are two reasons for using the word Gaussian, first, the transverse variations of the amplitude is governed by a Gaussian function and, second, Gaussian beams provide a very convenient mathematical description of the propagation of light along centered optical systems using paraxial rays (Gauss approximation).

2.6.5.1. Helmholtz's Equation Solution for Beams with Slow Amplitude Variations

We will only consider problems having a radial symmetry about the propagation Oz axis. The Gaussian beams that are obtained are said to be zero-order Gaussian beams, they are surely not the most general, however they are the most commonly met. To introduce the Gaussian beams we will use a rather formal approach and solve equation (2.27), the mathematical development is rather tedious and can be bypassed going directly to the final result in equations (2.30) and (2.31).

Because of the radial symmetry, the transverse Laplace operator is much simpler and reduces to

$$\nabla_t^2 = \frac{\partial^2}{\partial x^2} + \frac{\partial^2}{\partial y^2} = \frac{\partial^2}{\partial r^2} + \frac{1}{r}\frac{\partial}{\partial r},$$

the function $\psi(x)$ follows the following equation:

$$\frac{\partial^2 \psi}{\partial r^2} + \frac{1}{r}\frac{\partial \psi}{\partial r} - 2jk\frac{\partial \psi}{\partial z} = 0. \tag{2.29}$$

To find a solution of equation (2.29), we draw our inspiration from a description already used for spherical waves fulfilling Gauss approximation. Let us introduce two auxiliary functions $P(z)$ and $a(z)$ and write the function $\psi(r, z)$ as

$$\psi(r,z) = \psi_0 e^{-j\left[P(z)+\frac{1}{2}kr^2 a(z)\right]}$$

$$\rightarrow \quad \frac{\partial \psi}{\partial z} = -j\left(\frac{dP}{dz} + \frac{kr^2}{2}\frac{da}{dz}\right)\psi, \quad \frac{\partial \psi}{\partial r} = -jkar\psi, \quad \frac{\partial^2 \psi}{\partial r^2} = -jka(1 - jkar^2)\psi,$$

$$\rightarrow \quad k^2 r^2\left(a^2 + \frac{da}{dz}\right) + 2k\left(\frac{dP}{dz} + ja\right) = 0. \tag{2.29.a}$$

As equation (2.29.a) should be satisfied for any value of r, it is deduced that the two terms of (2.29.a) are both equal to zero:

$$\left(a^2 + \frac{da}{dz}\right) = 0 \quad \rightarrow \quad a = \frac{1}{(z+\alpha)},$$

where α is a constant of integration that will be taken as $\alpha = jz_0$,

$$\left(\frac{dP}{dz} + ja\right) = 0 \quad \rightarrow \quad \frac{dP}{dz} = -ja = -\frac{j}{z+jz_0},$$

$$\rightarrow \quad P = -j\,\mathrm{Log}\left(1+\frac{z}{\alpha}\right) + \text{constant} = -j\,\mathrm{Log}\left(1 - j\frac{z}{z_0}\right) + \beta.$$

The integration constant β corresponds to a phase shift that can be put in the complex amplitude ψ_0, thus it will be made equal to zero, $\psi_{(x,y,z)}$ is given by

$$\psi_{(x,y,z)} = \psi_{(r,z)} = \psi_0 e^{-j\left[-j\mathrm{Log}\left(1-j\frac{z}{z_0}\right)+\frac{kr^2}{2}\frac{1}{(z+jz_0)}\right]}.$$

Using the identity

$$\mathrm{Log}(a+jb) = \frac{1}{2}\mathrm{Log}\sqrt{a^2+b^2} + j\,\mathrm{Arctan}\left(\frac{b}{a}\right),$$

we obtain

$$e^{-j\left[-j\mathrm{Log}\left(1-j\frac{z}{z_0}\right)\right]} = e^{-\mathrm{Log}\left(1-j\frac{z}{z_0}\right)} = \frac{1}{\sqrt{1+z^2/z_0^2}}e^{j\mathrm{Arctan}\frac{z}{z_0}}.$$

Going back now to equation (2.25), the propagating field is finally obtained as

$$U_{(x,y,z)} = U_0 \frac{w_0}{w(z)} e^{-r^2/w_{(z)}^2} e^{-jkr^2/2R(z)} e^{j\text{Arctan}(z/z_0)} e^{-jkz}, \tag{2.30}$$

with the following notations:

$$w_0^2 = \frac{2z_0}{k} = \frac{z_0 \lambda}{\pi}, \quad w(z) = w_0 \sqrt{\left(1 + \frac{z^2}{z_0^2}\right)}, \quad R(z) = z\left(1 + \frac{z_0^2}{z^2}\right). \tag{2.30.a}$$

Equation (2.30) is made more convenient by setting $q(z) = (z - jz_0)$:

$$U_{(x,y,z)} = \frac{U_1}{q_{(z)}} e^{-jkr^2/2q_{(z)}}, \tag{2.31}$$

where U_1 is a new constant that could be related to U_0. It can be shown that

$$\frac{1}{q_{(z)}} = \frac{1}{R_{(z)}} - j\frac{2}{kw_{(z)}^2}.$$

Rayleigh zone: The function $\text{Arctan}(\zeta)$ remains constant over large variations of its argument ζ, except in the immediate vicinity of zero. It can be roughly considered that the domain of variation can be restricted to ± 1. Going back to equation (2.30), is seen that the term $\text{Arctan}(z/z_0)$ is only important when $z_0 < z < +z_0$. The corresponding region is called the Rayleigh zone, z_0 is known as the Rayleigh range.

Physical interpretation of the various terms of equations (2.30) and (2.31).	
$e^{-(r/w_{(z)})^2}$	In a plane orthogonal to Oz, the amplitude of vibration decreases according to a Gaussian law, as we get away from the axis. The width of the bell-shaped Gaussian curve is minimum when $z = 0$, its value is then equal to w_0 and is called the beam waist.
$\dfrac{w_0}{w_{(z)}} = 1/\left(1 + z^2/z_0^2\right)^{1/2}$ $\dfrac{w_0}{w_{(z)}} \approx 1/z \quad \text{if } z \to \infty$	This term describes a diminution of the amplitude when moving away from the origin. Further away a $1/z$ law is obtained, just as in the case of a spherical wave.

(continued)

e^{-jkz}	This term simply describes the propagation along Oz. Further away from the origin, it is the only complex term of equation (2.30): a Gaussian wave, as well as a spherical wave, coincides with a planar wave.
$e^{-jkr^2/2R_{(z)}} \to 1$ if $z \to \infty$	Because of this term, the phase varies inside a plane orthogonal to Oz. The curvature of the wave surfaces is introduced by a quadratic variation of the phase versus distance. Further away from the origin there are no more transverse phase variations, as for a planar wave.
$R_{(z)} = z(1 + z_0^2/z^2)$ $R_{(z)} \approx z$ if $z \to \infty$ $R_{(z)} \to \infty$ if $z \to 0$	$R(z)$ is the radius of curvature of the wave surface at abscissa z. Further away the radius is equal to the distance from the origin. In the vicinity of the origin, a Gaussian wave is very similar to a planar wave.
$e^{j\,\mathrm{Arctan}(z/z_0)}$	This term remains almost constant when z is varied, except for a fast variation from $-\pi/2$ to $+\pi/2$ when crossing the origin (see Section 2.6.2).

If we consider the expression $R_{(z)} = z(1 + z_0^2/z)$, it can be considered that:

- Outside the Rayleigh zone $R_{(z)} \approx z$, the transverse phase repartition, as well as the amplitude variation law versus distance to origin, are very similar for Gaussian and spherical waves.
- Inside the Rayleigh zone:
  - $R_{(z)} \approx z_0^2/z$, the phase repartition and the amplitude variations are quite different for Gaussian and spherical waves.
  - $\mathrm{Arctan}(z/z_0) \cong z/z_0$, the phase variation is linear versus z: the Gaussian wave behaves more like a planar wave, the phase velocity is however different from the speed of light in the same medium. (It is left as an exercise to evaluate the phase velocity.)

Diffraction and Gaussian beams: Out of the Rayleigh zone, a Gaussian wave is nothing but a spherical wave limited by some circular aperture of radius $w(z)$. Since it is a solution of the wave equation the effect of diffraction is immediately taken into account.

Let us consider the variations of the modulus of the field amplitude, $U_{(x,y,z)}$, inside a plane orthogonal to Oz at abscissa z; this amplitude is maximum on the axis and decreases as the absolute value of x and/or y increases; let

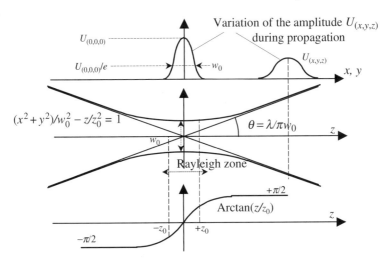

Figure 2.12. Illustration of a Gaussian beam. Amplitude variations $U_{(x,y,z)}$ versus distance to the axis follow a Gaussian law, the width of which increases with z, while its maximum value, $U_{(0,0,z)}$, decreases. The locus of the points where $U_{(x,y,z)}/U_{(0,0,z)} = 1/e$ is a hyperbola.

us consider a point $M_{(x,y,z)}$ where the modulus is equal to the maximum divided by e (Neperian logarithm basis)

$$U_{(x,y,z)} = \frac{U_{(0,0,z)}}{e} \quad \rightarrow \quad \frac{x^2+y^2}{w_0^2} - \frac{z^2}{z_0^2} = 1.$$

$M_{(x,y,z)}$ belongs to a hyperboloid of revolution about Oz, the half-angle θ of the asymptotic cone is obtained by taking the limit of r/z as z goes to infinity:

$$\frac{r^2}{z^2} = \frac{x^2+y^2}{z^2} = \frac{w_0^2}{z_0^2} + \frac{w_0^2}{z^2}, \quad \text{hence} \quad \frac{r}{z} \xrightarrow{z\to\infty} \frac{r}{z} = \frac{\lambda}{\pi w_0} = \theta. \quad (2.32)$$

We consider the plane (Ox, Oy), it is located at the center of the Rayleigh zone; the energy of the field is almost entirely contained inside a circle having a radius equal to the waist w_0. The divergence of the beam can be approximated by $\theta = \lambda/\pi w_0$, this result is sensible and should be compared to the diffraction of a planar wave by a circular aperture of radius w_0, the light is then diffracted inside a cone having an angle equal to $1.22\lambda/2w_0$.

The energy of a Gaussian beam remains constant during propagation. The demonstration of this point, which is a physical necessity, is left as an exercise. It will be shown that the surface integral $\iint_{(P)}\|U_{(x,y,z)}\|^2\,ds$ doesn't depend on z, (P) is a plane of abscissa z and orthogonal to the z axis, $U_{(x,y,z)}$ is given by equation (2.30).

2.6.5.2. *Parameters Required to Characterize a Gaussian Beam*

Given a frequency and a direction of propagation, to fully define a planar wave it's necessary to know its complex amplitude at some point. In the case of a spherical wave a few more indications are needed, since in addition to the complex amplitude at a given point, the center of the wave must be defined; the extension of the beam is always limited and we must know the axis in the vicinity of which the propagation is made and the size of the aperture.

For a Gaussian beam, the following parameters should have been defined: frequency, direction of propagation, location, and size w_0 of the waist; instead of w_0, the Rayleigh range z_0 can be used (the two quantities are connected by equation (2.30.a)).

How to Find the Characteristics of a Given Gaussian Beam?

We are looking at the properties (place and size of the waist) of a Gaussian beam, knowing the complex amplitude repartition along a given plane. Thanks to some uniqueness and existence theorems concerning the solutions of Maxwell's equations, we can be assured that this Gaussian beam is unique.

The complex amplitude repartition along a given plane (P) is completely determined from a complex number $q = z + jz_0$ which is then introduced in equation (2.31). The real part of q defines the place of the waist, it indicates where to place the origin of coordinates so that (P) will be located at abscissa z. If z is positive, the waist is located before (P); if z is negative, the waist is after (P). The imaginary part, z_0, should be a positive number, otherwise the amplitude and, consequently, the energy of the beam would go to infinity as x or y increases indefinitely. Finding a negative value for z_0 at the end of some calculation cannot be anything other than an error.

Let us give some practical examples to show how the parameter $q_{(z)}$ should be manipulated. The wavelength being equal to $\lambda = 0.5\,\mu$m, we want to characterize two Gaussian beams for which the q parameters, along a plane (P) are, respectively, equal to $q = (3 + j4)\,$cm and $q = (-2 + j0.5)\,$cm, see Figure 2.13.

In the first case, the real part is positive, which means that the waist is in front of the plane (P), at a distance equal to $z = 3\,$cm, (P) is inside the Rayleigh zone. The waist radius is $w_0 = \sqrt{z_0\lambda/\pi} \approx 80\,\mu$m.

In the second case, the waist is $2\,$cm after (P), the beam converges toward the waist and then diverges, the radius of the waist is equal to $w_0 = 28.2\,\mu$m.

2.6.5.3. *Transformation of a Gaussian Beam by a Spherical Lens*

A thin spherical lens (focal length f) receives a Gaussian beam which is characterized by $q = (z + jz_0)$ along the input face L^- just in front of the lens. Starting from the complex amplitude repartition along $U(x, y, L^-)$, we have to calculate the complex amplitude along the output face L^+, just after the lens.

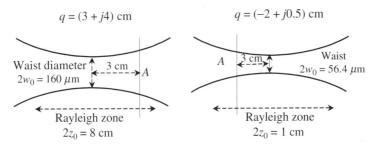

$q = (3 + j4)$ cm

$q = (-2 + j0.5)$ cm

Waist diameter
$2w_0 = 160 \ \mu m$

3 cm

A

A

3 cm

Waist
$2w_0 = 56.4 \ \mu m$

Rayleigh zone
$2z_0 = 8$ cm

Rayleigh zone
$2z_0 = 1$ cm

Figure 2.13. Examples of Rayleigh zones.

$U_{(x,y,L^+)}$ is obtained by a simple multiplication by a term which makes a transversal correction of the phase, just as we did in Sections 2.4.6.1 and 2.4.6.2.

We are going to show that the new repartition is that of a Gaussian beam: a lens transforms an initial Gaussian beam, $q = (z + jz_0)$, into a new Gaussian beam, $q' = (z' + jz'_0)$,

$$U_{(x,y,L^+)} = U_{(x,y,L^-)} e^{jk\frac{x^2+y^2}{2f}} = U_1 e^{jk\frac{x^2+y^2}{2(z+jz_0)}} e^{jk\frac{x^2+y^2}{2f}} = U_1' e^{jk\frac{x^2+y^2}{2(z'+jz'_0)}},$$
(2.33)

$$\frac{1}{z' + jz'_0} = \frac{1}{z + jz_0} - \frac{1}{f} \rightarrow \frac{1}{q'} = \frac{1}{q} - \frac{1}{f}.$$

Equation (2.33) is the lens formula for Gaussian beams.

Identification of the real and imaginary parts of equation (2.33) allows the expression of z' and z'_0 versus z and z'. If the lens is largely outside the Rayleigh zone ($z \gg z_0$), equation (2.33) becomes very similar to the usual lens formula; this is quite normal since a spherical wave can be considered as the limit of a Gaussian wave, if the Rayleigh range goes to zero. However, when comparing equation (2.33) with the usual lens formula, care should be taken of the fact that z is algebraically measured from the object to the lens, while in Geometric Optics the distances are usually measured from the lens to the object.

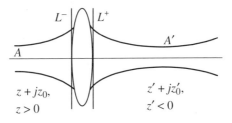

L^- L^+

A

A'

$z + jz_0,$
$z > 0$

$z' + jz'_0,$
$z' < 0$

Figure 2.14. Image-forming with Gaussian beams.

Rather tedious calculations using complex numbers give the following, and sometimes useful, expressions in which the following parameter has been introduced:

$$M = \frac{|f/(z-f)|}{\sqrt{1 + z_0^2/(z-f)^2}}.$$

Waist radius	$w_0' = M^2 w_0$	(2.34.a)
Position of the waist	$(z' - f) = M^2(z - f)$	(2.34.b)
Rayleigh range	$z_0' = M^2 z_0$	(2.34.c)
Divergence	$\theta_0' = \theta_0/M$	(2.34.d)

Image of a Waist Placed at the Object Focal Point of a Lens

From equation (2.34.b), it is seen that the output waist is at the image focal point. The incident and transmitted beams are however very different, since $M = f/z_0 \rightarrow w_0' = w_0 f^2/z_0^2$ and $\theta_0' = \theta_0 f/z_0$.

Focusing a Beam with a Lens Disposed at the Center of the Rayleigh Zone

$$z = 0, \quad w_0' = \frac{w_0}{(1 + z_0^2/f^2)}, \quad z' = \frac{w_0}{(1 + f^2/z_0^2)}.$$

If the focal length is much larger than the Rayleigh zone of the incident beam, the image almost coincides with the focal point; if this condition is not satisfied, the distance between the image and the lens focal point can no longer be neglected.

3

Geometrical Optics

3.1. Geometrical Propagation of Light

3.1.1. *Light Rays*

Geometrical optics is that part of Optics where the formation of an image starting from an object is mostly studied. The notion of a light ray is used extensively, allowing an intuitive and efficient understanding of the way optical instruments are working. In any optical experiment some light energy propagates from a source toward a detector, or toward infinity if the medium of propagation is perfectly transparent and if no detector is present.

From a mathematical point of view a light ray is just a *curve*, which means a vectorial space having only one dimension. From a physical point of view, a light ray is a trajectory followed by the electromagnetic (EM) energy. The impossibility of isolating a light ray considerably weakens any physical interpretation, a ray will be considered as nothing other than a useful mathematical tool.

3.1.2. *Medium Supporting the Propagation of Light*

In this chapter we will only deal with isotropic mediums of propagation, having identical properties, whatever the direction of propagation. These mediums will be considered as perfectly transparent, the light intensity, as defined in formula (2.23), remains constant along a given ray.

Chapter 3 has been reviewed by Dr. Olivier Delléa from Teemphotonics.

Homogeneous Medium—Inhomogeneous Medium

The mediums of propagation can be either homogeneous or inhomogeneous: their optical properties will be, or will not be, the same whatever the point under consideration. Here we propose a classification of the different transparent mediums.

	Homogeneous medium. Same properties at any point. The speed of light and the refractive index have, respectively, the same value everywhere. To go from one point to another the light follows the path taking the shortest time, i.e., a straight line.
	Discontinuously inhomogeneous medium. Succession of different mediums, each of them being homogeneous. Two consecutive mediums are in contact along "surfaces of discontinuity." Inside a given medium the light follows a straight line, the light path is made of several rectilinear segments intersecting on discontinuity surfaces.
	Continuously inhomogeneous medium. Properties are not the same at the different points, however, their variations are continuous functions of the coordinates. The path actually followed by the light when going from a point A to a point B is no longer a straight line but a curve along which the transit time t_{AB} is the shortest possible one. $V_{(x,y,z)}$ being the local value of the light speed and s the abscissa along the curve t_{AB} is given by the integral: $t_{AB} = \int_A^B \dfrac{ds}{V_{(x,y,z)}}$.
	Example of discontinuity: the plane mirror. The discontinuity occurs along a perfectly polished surface (irregularities are smaller than the wavelength). The path going from A to B is made of two rectilinear segments joining at the mirror.

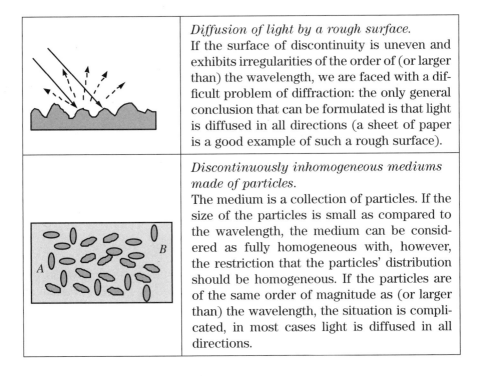

	Diffusion of light by a rough surface. If the surface of discontinuity is uneven and exhibits irregularities of the order of (or larger than) the wavelength, we are faced with a difficult problem of diffraction: the only general conclusion that can be formulated is that light is diffused in all directions (a sheet of paper is a good example of such a rough surface).
	Discontinuously inhomogeneous mediums made of particles. The medium is a collection of particles. If the size of the particles is small as compared to the wavelength, the medium can be considered as fully homogeneous with, however, the restriction that the particles' distribution should be homogeneous. If the particles are of the same order of magnitude as (or larger than) the wavelength, the situation is complicated, in most cases light is diffused in all directions.

3.1.3. *Speed of Propagation—Index of Refraction*

In this chapter no difficulty will be raised concerning the definition of the speed of propagation, only one speed will be introduced and it will be assimilated to the phase velocity.

As far as geometrical optics is concerned, a transparent medium is exhaustively characterized if the value of the propagation speed of the light V and its law of variation with the color (that's to say, with the frequency) are known.

The speed V is enormous and measured in hundreds of thousands of kilometers per second. It's the reason why a normalized speed is introduced thanks to a division by the speed c of the light in vacuum. Very often we will have to use $1/V$, this parameter is a very small number. The *index of refraction* n is the inverse of the normalized speed.

The index of refraction of a transparent medium is given by

$$n = \frac{c}{V} = \frac{\text{light speed in a vacuum}}{\text{light speed in the medium}}.$$

Dispersion

The index of refraction is a very basic parameter in Optics, it can be measured with an accuracy which can be extremely high and reach 10^{-8} and which is easily equal to 10^{-4}. The law of dispersion of a given material is the law of variation of the index of refraction versus wavelength (or frequency). The values of refraction indices of transparent mediums may be found in Optical Handbooks for almost any spectral lines (sodium doublet, mercury lines, main laser frequencies, ...). Dispersion laws are also given, with good precision, by semitheoretic/semiempiric analytical formulas.

3.2. Fermat's Principle

3.2.1. *Different Ways of Introducing the Fermat Principle*

From an historical point of view, the principle of Fermat, also known as the principle of the shortest optical path, was introduced by Pierre de Fermat as early as the seventeenth century. It's a very powerful formulation of geometrical optics. The Snell-Descartes laws of reflection and refraction can be mathematically demonstrated from the principle of Fermat. On the other hand, it is also possible to start from the Snell-Descartes laws and then demonstrate what will now be the *theorem of Fermat*; as reflection and refraction laws can be experimentally verified, the Fermat principle can thus be considered as having experimental roots. Finally, it should be noticed that the principle of Fermat, as well as the Snell-Descartes laws, are easily obtained from Maxwell's equations.

3.2.2. *Formulation of the Fermat Principle*

The path followed by light going from point A to point B is such that the transit time is stationary.

By *stationary* we mean that the transit time along the special path followed by the light is either minimum or maximum with regard to the times that would be taken along other paths having the same extremities and remaining very close to the light trajectory. In most cases the time taken will be minimum; an example is given in Figure 3.10(c) where the time is maximum, of course it's not an absolute maximum, it's only a minimum relative to the neighboring paths.

Discontinuously inhomogeneous medium

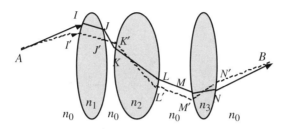

Above has been drawn an optical system using three lenses, with indices, respectively, equal to n_1, n_2, and n_3, and they are immersed in a medium with index n_0. Using a solid line we have drawn the path that the light follows *AIJKLMNB*, and using a dotted line we have drawn another path very close to the first one *AI′J′K′L′M′N′B*.

The transit times along *AIJKLMNB* and *AI′J′K′L′M′N′B* at the speed of light in the various mediums requires, respectively, the following times:

$$t = (n_0 AI + n_1 IJ + n_0 JK + n_2 KL + n_0 LM + n_3 MN + n_0 NB)/c,$$

$$t' = (n_0 AI' + n_1 I'J' + n_0 J'K' + n_2 K'L' + n_0 L'M' + n_3 M'N' + n_0 N'B)/c,$$

if the different points I', J', K', ... get nearer and nearer the points I, J, K, ..., the limit of t' is equal to t. Among the different times, associated to the various paths $I'J'K'$, ..., t is stationary and usually the shortest.

Continuously inhomogeneous medium

$n_{(x,y,z)}$ is the law of variation of the index of refraction. The single arrowed curve is the path that is taken by the light going from A to B, the double arrowed curve is a neighboring path. The times to be compared are now given by integrals. The following integral is stationary when taken along the trajectory followed by the light:

$$t = \frac{1}{c} \int_A^B n_{(x,y,z)} \, ds, \quad s \text{ is the abscissa measured along the curve.}$$

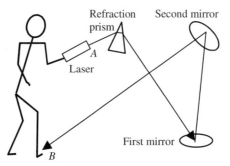

Figure 3.1. Illustration of Fermat's principle; a child plays the following game: using a well-collimated laser source he wants to light up his foot, but he also wants that the light should travel across a glass prism and be reflected from two mirrors. He immediately finds that only one path fulfils the imposed conditions and goes from his hand to his foot.

If we consider an optical system and two points, A and B, in general there is one, and only one, light ray joining A to B; this is a consequence of the fact that the light path corresponds to an extreme value (minimum or maximum) and that, by definition, an extreme value is unique. This property is illustrated in Figure 3.1.

It may happen that many different paths take *exactly* the same time for the light to go from A to B, each of these paths represents a possible path for the light: in such a case, there are many rays going from A to B, and it is then said that B is the image of A.

3.2.3. *Principle of Reversibility*

At a given point of a trajectory the speed of light is the same, whatever the direction. An important consequence is that the curve followed by the light going from A to B is the same as the curve followed by going from B to A, since the transit times will be the same in both cases. This property is known as the principle of reversibility.

3.2.4. *Demonstration of the Snell-Descartes Laws*

The Snell-Descartes laws allow the calculation of the trajectory of a light beam after it has arrived at an interface between two different transparent mediums (respective indices of refraction, n_1 and n_2). The situation is that of a discontinuously inhomogeneous medium. The experiment shows that the trajectory of the light is made of straight lines joining at the interface: an incident ray generates a reflected ray and a refracted ray. Laws of refraction and reflection indicate in which directions the two new rays will propagate.

Figure 3.2. Fermat's principle in the case of reflection: the shortest path from S to R is obtained when the angle of incidence is equal to the angle of reflection i'.

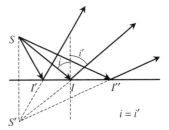

$i = i'$

Law of Reflection

Let us consider Figure 3.2, to go from point S to point R, the light will follow a broken line made of two rectilinear segments joining at the point of I where the incident beam hits the interface. The length of the path SIR should be as short as possible. Let S' be the symmetrical point of S with regard to the planar interface, it's easy to see that $SIR = S'IR$, $SI'R = S'I'R$, $SIR = S'IR$, the minimum path is obtained when $S'IR$ is a straight line.

Law of Refraction

We refer to Figure 3.3 where i and r are called, respectively, the angles of incidence and of refraction. The first medium is supposed to be faster than the second and the speeds of propagation are, respectively, equal to $V_1 = c/n_1$ and $V_2 = c/n_2$. We will first evaluate the time taken by the light to go from A to B versus various geometrical parameters, the path really followed by the light is then obtained by cancelling the derivative of t versus x,

$$t = \frac{1}{c}\left(n_1\sqrt{x^2+a^2} + n_2\sqrt{(b'-x)^2+b^2}\right) \quad \rightarrow \quad \frac{dt}{dx} = \frac{1}{c}\left(\frac{n_1 x}{\sqrt{x^2+a^2}} - \frac{n_2(b'-x)}{\sqrt{(b'-x)^2+b^2}}\right),$$

$$\frac{dt}{dx} = 0 \quad \rightarrow \quad \frac{n_1 x}{\sqrt{x^2+a^2}} = \frac{n_2(b'-x)}{\sqrt{(b'-x)^2+b^2}},$$

Figure 3.3. Fermat's principle for refraction: the time from A to B is minimum when $n_1\sin i = n_2\sin r$, which corresponds to the path AIB.

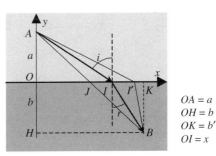

$OA = a$
$OH = b$
$OK = b'$
$OI = x$

since

$$\sin i = \frac{x}{\sqrt{x^2 + a^2}} \quad \text{and} \quad \sin r = \frac{(b' - x)}{\sqrt{(b' - x)^2 + b^2}},$$

we obtain $n_1 \sin i = n_2 \sin r$.

This is the famous Snell-Descartes law, also known as the "sine law"; this demonstration gives only the directions of the reflected and refracted beams, but it doesn't give any information about the percentages of energy carried away by each beam. Another demonstration, starting from Maxwell's equations, will be given in Section 4.3.

3.2.5. *Fermat's Principle for Stigmatic Optical Systems*

An important role of Optics is to obtain images. In this section we intend to use the formalism of Fermat to give an interpretation of the mechanism of imaging and to examine the condition for obtaining high-quality images.

3.2.5.1. *Stigmatism and the Principle of Fermat*

It may happen, although it's not generally the case, that all the rays proceeding from a point source A, after they have been transmitted by an optical system, will converge toward the same point B (Figure 3.4); it is then said:

- That the system is stigmatic for points A and B.
- That the points A and B are conjugate across the optical system.

As emphasized in Figure 3.1, among all the beams emitted by the point source A, one, and only one, will go to a given point B. The case when A and B are conjugate is very special; since all the paths from A to B take the same time, the notion of minimum (or maximum) vanishes: there is a kind of degeneracy and every path is a possible trajectory for the light.

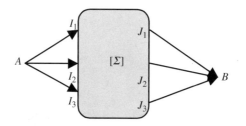

Figure 3.4. [Σ] is the optical system limited by two interfaces. Input and output mediums are homogeneous, before and after [Σ] light rays are rectilinear. [Σ] is made of a succession of either continuously or discontinuously inhomogeneous mediums. All the paths from A to B take the same time.

3.2.5.2. *Perfect Imaging*

Stigmatism as it has been introduced is also called perfect imaging and also *sharp* imaging. Systems where all light beams issued from the same object point then intersect at the same image point are really rare. First, it should be noticed that stigmatism is a notion that is restricted within the domain of geometrical optics, since we know that diffraction forbids any electromagnetic wave to be focused inside a spot smaller than the wavelength.

Even if we forget the preceding remark about diffraction, there are very few strictly stigmatic systems, that's to say, systems for which the rules of geometrical optics lead to an exact convergence at point *B* of light beams issued from point *A*. What will often happen is the following: the emerging beams will run very close to point *B*, in such a case the system will be said to be *approximately* stigmatic. We will now describe systems that are strictly stigmatic.

A Planar Mirror Is Stigmatic for any Point

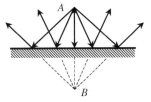

Figure 3.5. A planar mirror is strictly stigmatic. Any incident beam issued from *A* gives a reflected beam passing through point *B* symmetric to *A* with regard to the plane of the mirror. *B* is the image of *A*.

An Elliptic Mirror Is Stigmatic for its Focus

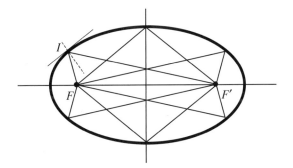

Figure 3.6. An ellipse is a set of points for which the sum of the distances to two given points, called focus, is constant. Any beam issued from one focus is reflected toward the other focus. It's a well-known property of an ellipse that the normal to the ellipse surface at point *I* is a bisector of the angle *FIF'*.

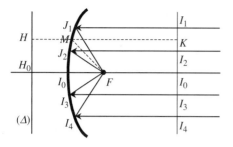

Figure 3.7. A parabolic mirror is stigmatic for its focuses F, the other focus is at infinity. A parabola is a set of points located at equal distance from a point called the focus and a line (Δ) called the directrix. $MF = MH \rightarrow$ the paths I_1J_1F, I_2J_2F, I_3J_3F, I_4J_4F are all equal to I_0H_0.

A Parabolic Mirror Is Stigmatic for its Focus

Hyperbolic and parabolic mirrors are also stigmatic for their focuses. From the point of view of applications the most important case is that of parabolic mirrors which are used for making telescopes for astronomy (radio as well as optical telescopes).

Stigmatic Points of a Spherical Interface

We consider a sphere (center O, radius $OC = R$, see Figure 3.8), filled with a transparent material of refractive index n and surrounded by a second transparent medium of index n'. A point object is immersed inside the sphere and emits light rays in all directions. We are looking for conditions so that all the beams that are refracted on the spherical interface will pass point A'. For the sake of symmetry A and A' should belong to the same diameter BC of the sphere. Let us choose A and A' so that we have $CA'/CA = -BA'/BA = n/n'$. For any point I of the sphere we also have $IA'/IA = n/n'$. We now draw a sphere with a radius equal to L and centered at A', then we consider a light ray AI and the corresponding refracted ray IM (M belonging to the sphere).

The time taken by the light going from A to M is equal to $t = (nAI + n'IM)/c$ $= (nAI - n'IA + n'L)/c = n'L/c$, t is the same for all rays emitted by A, according to Fermat's principle, the previous result implies that A' is a sharp image of A. A and A' are called the stigmatic points of the spherical interface, they are known as points of Weierstraß in Germany and points of Young in the United Kingdom. They are often used to make high-resolution microscope objectives.

From the theory of diffraction it will be shown that a high resolving power needs an important numerical aperture (NA) for the objective of the microscope. The numerical aperture is the product of the sine of the half-angle of the cone of light rays arriving at the first lens of the objective, by the index of refraction of the medium where incident rays propagate. To improve the resolution rays making an important angle with the axis are required, so Gauss conditions are not satisfied. In the objective of Figure 3.9, the first lens takes advantage of the stigmatic points of a sphere, the light rays emerging from

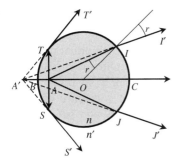

For any point of the sphere: $\dfrac{IA'}{IA} = \dfrac{n}{n'}$,

$$\dfrac{CA'}{CA} = -\dfrac{BA'}{BA} = \dfrac{n}{n'},$$

$$OA = \dfrac{n}{n'}OB = -\dfrac{n'}{n}OC,$$

$$OA' = \dfrac{n}{n'}OB = -\dfrac{n}{n'}OC.$$

Figure 3.8. Points of Young-Weierstrass. A' is a sharp image of A. A' is at the intersection of the refracted rays TT', II', JJ', and SS'.

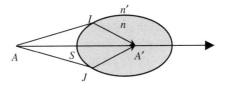

Figure 3.9. Schematic arrangement of a microscope objective using stigmatic points of a spherical interface to obtain a high numerical aperture. The first lens is a truncated sphere, the object is immersed in a liquid having the same index as the sphere. The second and third lenses work in the conditions of Gauss.

the sphere make a small angle with the axis, so the second part of the objective rays propagate in Gauss conditions.

Attempt of Perfect Imaging Using Refraction at an Interface

We refer to Figure 3.10, light rays emitted by some point source A arrive at a boundary between two transparent mediums (respective indices of refraction n and n'), is it possible to obtain a perfect image A'? According to the principle of Fermat, we should have $nIA + n'IA' = nSA + n'SA' =$ constant. The previous formula defines a surface of revolution about the axis AA', this surface is called a Cartesian oval which is of course rather different from a sphere.

Figure 3.10. Cartesian oval: $nIA + n'IA' = nSA + n'SA' =$ constant.

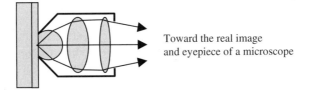

(a) 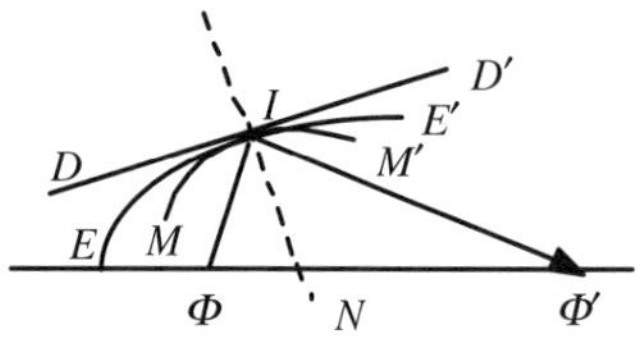 	Here *DD'*, *EE'*, and *MM'* are three mirrors having the same tangent at point *I'*. *DD'* is a planar mirror and *EE'* is an elliptical mirror of focuses ϕ and ϕ'. The radius of curvature of the surface of mirror *MM'* at point *I* is smaller than the radius of curvature of the ellipse. *ΦI* is an incident ray and *IΦ'* is the associated reflected ray.
(b) 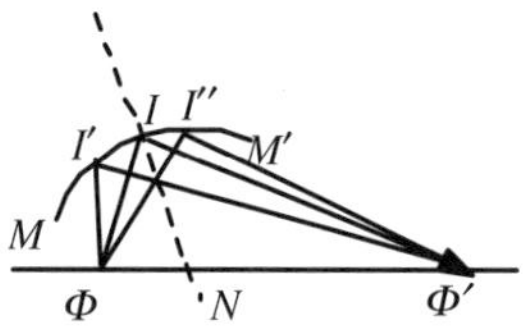 	In the case of the elliptical mirror, the three paths *ΦIΦ'*, *ΦI'Φ'*, and *ΦI''Φ'* are equal and correspond to three possible paths for the light.
(c) 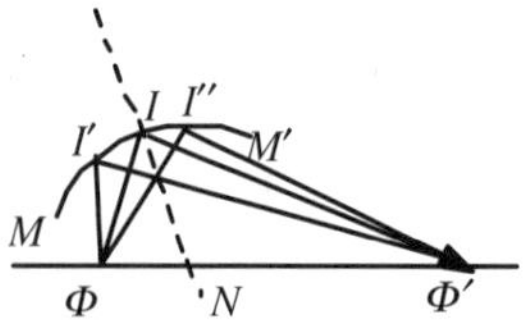 	The radius of curvature of mirror *MM'* is smaller than the radius of the ellipse. The path *ΦIΦ'* followed by the light is larger than *ΦI'Φ'* and *ΦI''Φ'*: this is a case where the light path is maximum relative to the adjacent paths.
(d) 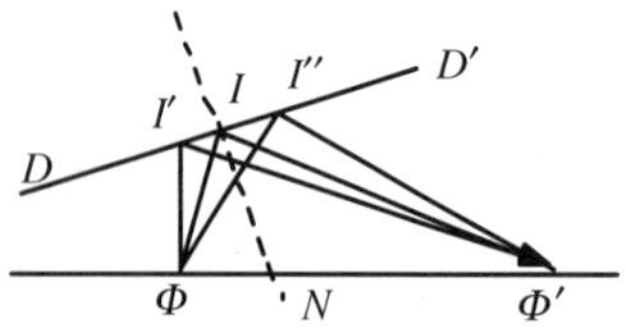 	In the case of a planar mirror, *ΦI'Φ'* and *ΦI''Φ'* are longer than the light path *ΦIΦ'* which is then minimum. The same result would be obtained for a mirror having a radius of curvature larger than the ellipse radius of curvature.

Figure 3.11. Examples of where the transit times are minimum or maximum.

3.3. Formation of Images

3.3.1. *Real or Virtual Objects and Images*

An image-forming instrument is a device which, receiving light rays coming from point sources, gives emerging beams which converge toward point sources. Such a device is always limited by some input interface and some output interface, according to the fact that the point source and the point image are before or after those interfaces, the object and image will be *real* or *virtual*.

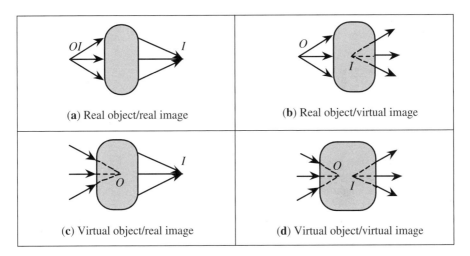

(a) Real object/real image	(b) Real object/virtual image
(c) Virtual object/real image	(d) Virtual object/virtual image

Figure 3.12. The different kinds of objects and images.

- A real object is located before the input interface (outside the instrument), and sends on it a diverging beam.
- A virtual object is located after the input interface (inside the instrument), this one receives a converging beam.
- A real image is located after the output interface (outside the instrument), the emerging beam is convergent.
- A virtual image is located before the output interface (inside the instrument), the emerging beam is divergent.

3.3.2. Perfect Imaging—Approximate Imaging

The situation where all the light rays initially coming from a given point are exactly focused at the same point image is an ideal, and thus very rare, situation. We then speak of *perfect* or *sharp imaging*. Some examples have already been described.

Approximate Imaging

In many cases all the rays issuing from a given point source will not be strictly focused at the same point, however, it will often happen that the emerging rays will be very near to a point which will then be considered as an image.

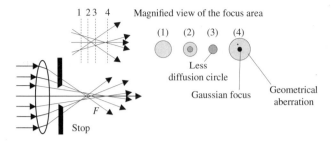

Figure 3.13. A beam of parallel rays is transformed by a lens into a beam of rays, which almost converge to the same point. An incident ray is all the more bent toward the axis, that it propagated further from the axis of the lens. Unfortunately the marginal rays are bent too much; the role of the stop is to avoid them contributing to the formation of the image.

A Lens Is too Converging at its Periphery, as Compared to its Center

When a screen is moved behind a converging lens illuminated by a beam of light rays propagating parallel to its axis, it can be observed that the focusing of the beam is not perfect. Referring to Figure 3.13 it is seen that:

- For position (4) of the screen a very bright spot is observed, the size of which is determined by diffraction phenomena and is equal to $1.22f\lambda/d$ (f is the focal length, λ is the wavelength, and d is the diameter of the lens). The spot is surrounded by a less luminous circular halo which constitutes the geometrical aberrations.
- Position (3) of the screen corresponds to what is called the *circle of less diffusion*; in the absence of stop, the screen should be placed there.

If a stop is used in order to keep only the rays that propagate close enough to the axis, geometrical aberrations are avoided: the system is then working under the conditions of Gauss.

Acceptable Images May be Obtained in Spite of Aberrations

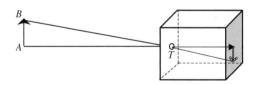

Figure 3.14(a). Pinhole camera: a small circular hole T has been drilled in one side of a box, the opposite side is made of a translucent paper. Some object AB is placed in front of the drilled face, each of its points sends to the inside of the box a conic pencil of light rays limited by the edge of the hole. Each pencil will produce on the translucent paper a small illuminated spot: if the size of the spot is small enough, the set of spots has an appearance which really looks like the object AB.

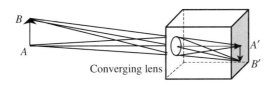

Figure 3.14(b). Elementary camera. The set-up is the same as in Figure 3.13 except that a converging lens has been placed just in front of the hole. The lens conjugates the bottom of the box and the plane of the object AB. The image $A'B'$ is sharper. The reader will ask why, in the second case, the image is brighter?

3.4. Thin Lenses

3.4.1. *Definition*

The more general spherical lens is made of a transparent material limited by two spherical interfaces (respective centers O_1 and O_2, respective radii of curvature R_1 and R_2). O_1O_2 is called the axis of the lens; the points S_1 and S_2 where the axis meets the interfaces are the summits of the lenses. According to the signs of the radius of curvature, a spherical lens may have one of the different shapes indicated in Figure 3.16.

Very often the thickness of the lens is far smaller than the radius of curvature, and the points S_1 and S_2 are assimilated to only one point S which is called the *center of the lens*.

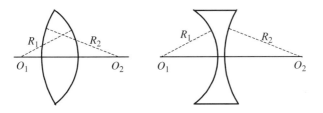

Figure 3.15. Cross section of spherical lenses.

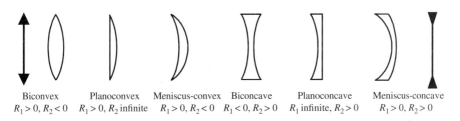

Biconvex	Planoconvex	Meniscus-convex	Biconcave	Planoconcave	Meniscus-concave
$R_1>0, R_2<0$	$R_1>0, R_2$ infinite	$R_1>0, R_2<0$	$R_1<0, R_2>0$	R_1 infinite, $R_2>0$	$R_1>0, R_2>0$

Figure 3.16. Different kinds of spherical lenses. Since the lenses are often thin, a symbolic representation is often used as shown on the left for converging lenses and on the right for diverging lenses. The radii of curvature are algebraic quantities, the signs correspond to an orientation from left to right.

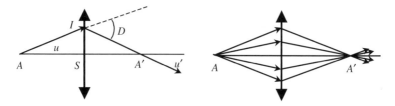

Figure 3.17. The imaging process in a thin spherical lens: since the angle of deviation D is proportional to SI and since this angle is small, all the rays coming from A converge to the same point A'.

We refer to Figure 3.17, the incident ray AI gives an emerging ray IA' which is all the more bent toward the axis that the point of incidence I is further from the center S of the lens. It will be shown in Annex 3.A that the angle of deviation D is proportional to the distance SI: $D = KSI$, K is a specific constant of the lens under consideration. K is homogeneous to the inverse of a length: $f = 1/K$ is called the *focal length* of the lens. K is related to the index of refraction and to the radius of curvature and given by the following formula:

$$K = \frac{1}{f} = (n - 1)\left(\frac{1}{R_1} - \frac{1}{R_2}\right).$$

The axis being oriented by the direction of propagation of the light and R_1 and R_2 are algebraic quantities.

3.4.2. Ray Tracing in a Thin Spherical Lens

From a mathematical point of view, the following operation is achieved by a lens: to each point A of the axis is associated another point A'. The correspondence between A and A' is a bijection: there is one, and only one, point A' associated to the point A, and reciprocally. The terminology of the set theory is well adapted to this problem; this is not at all surprising, since the association "object space $\leftrightarrow$ image space" is the first example of correspondence between items of two sets that have been met by both physicists and mathematicians. The set theory jargon is nothing but a generalization of the object $\leftrightarrow$ image correspondence in Optics.

Focal Points: A Focal Point Is Conjugated with
a Point that Is at Infinity

If the point is rejected at infinity in the direction of the axis of the lens, the associated point is a *principal focus*. If the point is rejected in another direction, the associated point is a *secondary focal point associated with the direction under consideration.*

If the point at infinity is an object, the associated point is an *image focal point*. If the point at infinity is an image, the associated point is an *object focal point*.

Principal image focal point = image of a point source at infinity in the direction of the axis.

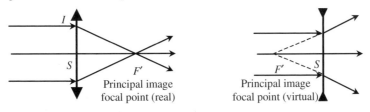

Figure 3.18. Principal image focal point.

Principal object focal point = point having its image at infinity in the direction of the axis.

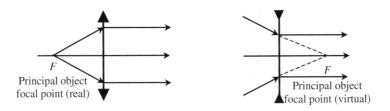

Figure 3.19. Principal object focal point.

Existence and Property of the Optical Center of a Thin Lens

The two summits of the spherical interfaces of a *thin* spherical lens are considered to coincide at one point *s* that is called the *optical center* of the lens. An important property of the optical center is that any incident beam directed toward the optical center will not be bent. The existence of an optical center is special to thin lenses and will not be generalized to centered optical systems.

A ray which is directed toward the center of a thin lens is not bent.

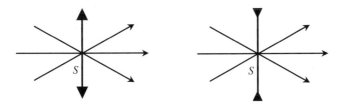

Figure 3.20. Optical center of a thin lens. A ray directed toward the optical center is not bent and keeps straight on.

Focal Planes—Secondary Focal Points

> *Secondary image focal point Φ′ = image of a point at infinity in the direction of the line Φ′S, joining the center and the focal point.*

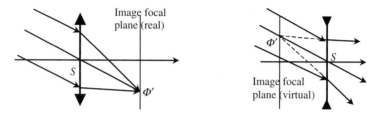

Figure 3.21. All the secondary focal points ϕ′ belong to the image focal plane.

The object focal plane is the plane that contains the principal object focal point and is orthogonal to the axis. All the secondary focal points belong to the object focal plane.

The image focal plane contains the principal image focal point and is orthogonal to the axis. All the secondary image focal points belong to the image focal point.

> *Secondary object focal point Φ: its image is at infinity in the direction of the line ΦS, joining the center and the focal point.*

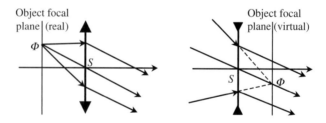

Figure 3.22. All the secondary object focal points belong to the object focal plane.

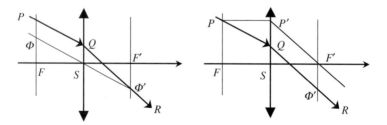

(a) The emerging ray *QR* should pass by the image focal point *Φ′* associated to the direction of the incident ray *PQ*.

(b) *P* belongs to the object focal plane. Considered as an incident ray *PP′* gives an emerging ray *P′F′*: *QR* should be parallel to *P′F′*.

Figure 3.23. Construction of the emerging ray *QR* associated to the incident ray *PQ*.

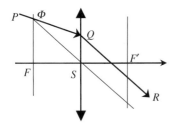
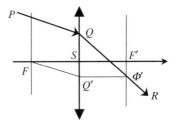

(a) *PQ* intersects the object focal plane at Φ. Considered as an incident ray, ΦS is not bent by the lens. *QR* is parallel to ΦS.

(b) *FQ'* is drawn parallel to *PQ*, considered as an incident ray it gives an emergent ray $Q'\Phi'$ parallel to the axis. *QR* intersects the image focal plane at Φ'.

Figure 3.24. Other construction of the emerging ray *QR* associated to the incident ray *PQ*.

Construction of the Image of a Point Object A Belonging to the Axis

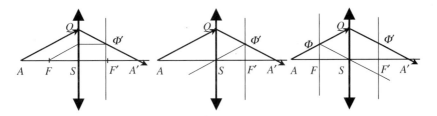

Figure 3.25. The image *A'* is at the intersection of the axis with an emergent ray associated to an incident ray *AQ* coming from *A*. Each of the methods of Figures 3.23 and 3.24 may be used.

Construction of the Image of a Point Object Out of the Axis

It can be proved from elementary geometrical considerations that the image formed by a lens of a small object belonging to a plane *P* normal to the axis will belong to another plane *P* also orthogonal to the axis. The two planes *P* and *P'* are said *to be conjugated* by the lens.

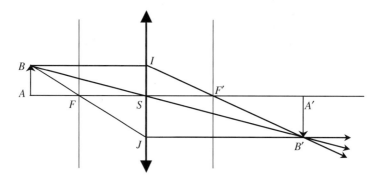

Figure 3.26. Construction of the image B' of a point B taken out of the axis. We consider two incident rays coming from B, B' is at the intersection of the associated transmitted rays. The more commonly used incident beams are those of Figures 3.23 and 3.24.

Some Useful Arrangements

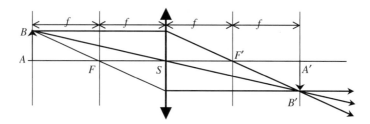

Figure 3.27. The 4-f arrangement. The object is placed at a distance equal to twice the focal length in front of a converging lens. The image is real and located at the same distance behind the lens, it has the same size as the object, but is inverted. In this disposition the object-image separation is the shortest.

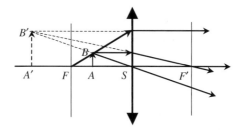

Figure 3.28. Virtual image of a real object. A real object placed between the lens and its focal plane gives a virtual image.

Figure 3.29. The image focal plane plays no special role in the object space. Here a virtual object *AB* is in the image focal plane; the image *A'B'* is real, erected and located in the middle of *SF'*.

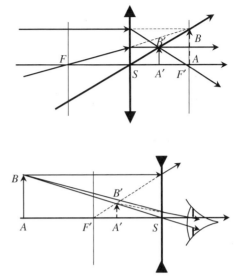

Figure 3.30. Lens for a short-sighted eye. The real object *AB* is replaced by the virtual image *A'B'* located nearer to the eye.

In Figure 3.30 have been drawn the paths followed by light rays issuing from the real point object *B* and the pencil of rays which finally penetrate the eye of an observer. The real object *AB* is replaced by a virtual image *A'B'* which in fact plays the role of a *real object* for the observer. Such a lens is used to help a short-sighted eye to accommodate an object located far away. It's left as an exercise to see how a converging lens can be used to correct the vision of a long-sighted eye.

3.4.3. *Thin Lenses Equations*

We consider a point object *A* which is imaged at *A'* by some lens, *A'* can as well be considered as an object and *A* as its image. *A* and *A'* are said to be conjugated by the lens, the expression *conjugate points* is often used. The lenses equations are formulas relating, on the one hand, the abscissas of two conjugated points and, on the other hand, the ratio of the object to image size. There are different kinds of equations, they differ in the choice of the origins used to represent, respectively, the object and image spaces. Two are more commonly used:

* *Descartes' formula*: The summit *S* of the lens is taken as a common origin for both the object and image space.
* *Newton's formula*: Two different origins are chosen, namely the object focal point *F* for objects, and the image focal point *F'* for the image.

Conjugation equations are readily obtained from simple geometric considerations in Figure 3.26, such as expressing the similarity of several triangles (*SAB/SA'B'*, *FAB/FSJ*, *F'IS/F'A'B'*).

Descartes' equations	Newton's equations
$$\dfrac{1}{SA'}-\dfrac{1}{SA}=\dfrac{1}{SF'}=-\dfrac{1}{SF},$$	$$FAF'A' = SFSF' = -f^2. \quad (3.1.b)$$
setting $SA = p$, $SA' = p'$, and $SF' = f$,	$$\gamma = \dfrac{A'B'}{AB} = \dfrac{FS}{FA} = \dfrac{F'A'}{F'S}. \quad (3.1.c)$$
$$\dfrac{1}{p'}-\dfrac{1}{p}=\dfrac{1}{f'}, \quad \gamma = \dfrac{A'B'}{AB} = \dfrac{SA'}{SA} = \dfrac{p'}{p}. \,(3.1.a)$$	(All quantities are algebraic.)

3.5. Centered Systems Under Gauss Conditions

3.5.1. *Definition of a Centered System*

3.5.1.1. *Spherical Interface*

It has been shown that perfect imaging cannot be obtained by refraction at a spherical interface. However, if we use only paraxial rays (i.e., rays making a small angle with the axis), acceptable, although nonperfect, images are obtained. In such conditions the Snell-Descartes law, $n\sin i = n'\sin i'$, is replaced by the Kepler law, $ni = n'i'$, which is *linear*.

We refer to Figure 3.31, on the left-hand side of the figure the interface has been drawn as a sphere and, on the right-hand side, the interface has schematically been replaced by a plane, which is orthogonal to the axis. Whatever the angle of incidence, the following relations are rigorously valid, even the angles are not small:

$$\Omega = (i - u) = (i' + u'), \qquad (3.2.a)$$

$$\frac{IA}{\sin \Omega} = \frac{IC}{\sin u} = \frac{CA}{\sin i} \quad \text{and} \quad \frac{IA'}{\sin \Omega} = \frac{IC'}{\sin u'} = \frac{CA'}{\sin i'},$$

$$n\frac{CA}{IA} = n'\frac{CA'}{IA'}. \qquad (3.2.b)$$

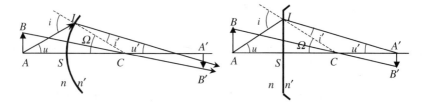

Figure 3.31. Image construction at a spherical interface. When the angles u and u' are small enough the point of incidence I can be considered as belonging to the plane tangent to the sphere at point S.

A given incident ray AI gives a refracted ray that intersects the axis at a point A'. According to equation (3.2.b), it is clear that the position of A' on the axis will depend on the incident ray: the spherical interface is not stigmatic.

Conjugation Equation (Small Angles)

If the angles are small enough, the sine can be assimilated to the angle

$$\Omega = \frac{SI}{SC}, \quad u \cong \frac{SI}{SA}, \quad u' \cong \frac{SI}{SA'}.$$

If we place the origin at point S, after some algebraic manipulations we obtain the equation of conjugation for a spherical interface

$$\frac{n}{SA} - \frac{n'}{SA'} = \frac{n-n'}{SC}, \tag{3.3}$$

$$\gamma = \frac{A'B'}{AB} = \frac{CA'}{CA} = \frac{n}{n'} \frac{SA'}{SA}. \tag{3.4}$$

According to equation (3.3), A' is at the same place, whatever the incident ray: if the angles are small, a spherical interface is stigmatic for points located on the axis of revolution. For a given spherical interface, SC, n and n' are constant, the correspondence between SA and SA', as expressed by equation (3.3), is a *homographic transformation*.

Aplanetism

The stigmatism is also obtained for points that are not on the axis. A point B belonging to the plane (P), that contains A and is perpendicular to the axis, has its image B' in the plane (P') that contains the image A' of A and is perpendicular to the axis. The previous property is known as *aplanetism*. The two planes (P) and (P') are said to be conjugated. Under the condition that the angles should remain small, the ratio $\gamma = A'B'/AB$ keeps the same value, whatever the size of AB. γ is called the magnification associated to the two conjugate planes, it's a real number which can be positive or negative, and larger or smaller than one.

The Lagrange-Helmholtz Equation

We refer again to Figure 3.31, a light ray emitted by AI gives a refracted beam IA', let u and u', respectively, be the angles of the two rays with the axis, we have the important relationship, called the Lagrange-Helmholtz formula,

$$nABu = n'A'B'u'. \tag{3.5}$$

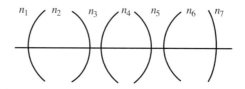

Figure 3.32. General arrangement of a centered system. Several spherical interfaces with their centers aligned along a common diameter, called the axis of the system. The first and last interfaces are, respectively, called the input and output interfaces. The refractive indices n_1 and n_7 are often equal and correspond to the transparent medium inside of which the system is immersed.

3.5.1.2. *Centered Systems*

A centered system is a succession of spherical interfaces having their centers aligned along a unique diameter called the axis of the system. Very often the interfaces will go two by two, and a centered system is nothing other than an assembly of spherical (thin or thick) optical lenses.

A real, or virtual, point source A is placed in front of the system, the first interface gives an image A_1 that acts as an object for the second interface, hence a second image A_2 is formed, and this acts as an object for the third interface, and so on. . . .

3.5.1.3. *Basic Properties, Centered Systems*

The properties that have just been established for a spherical interface can be generalized by transitivity to the centered system:

- *Approximate stigmatism under Gauss conditions*: All the rays emitted by a point object are focused at the same point image.
- *Aplanetism*: A plane perpendicular to the axis is called a *front plane*. All the points of a given front plane (P) have their images in another front plane (P'). (P) and (P') are said to be conjugated by the system.
- *The object $\leftrightarrow$ image transformation is a homographic transformation*: It's known in mathematics that the product of two homographic transformations is also a homographic transformation. The position of some point object A being referred to as origin S, and the position of its image being referred to as origin S' (S and S' can be either identical or distinct), SA and SA' obey a relationship of the following kind:

$$\alpha SASA' + \beta SA + \chi SA' + 1 = 0, \tag{3.6}$$

where α, β, and χ are constant parameters that are only determined by the optogeometric characteristics of the system (radius of curvature, positions

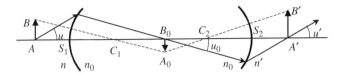

Figure 3.33. Successive images in a centered system with two interfaces. The first object AB, the intermediate image A_0B_0, and the final image $A'B'$ are, respectively, immersed in mediums of refractive indices n, n_0, and n'.

of the centers, indices of refraction) and by the positions of the two origins S and S'.

- *Linear magnification*: The image to object ratio, $\gamma = A'B'/AB$, doesn't depend on the positions of A and B in a front plane, its value is characteristic of a given pair of conjugate planes. To a given value of γ is associated one, and only one, pair of conjugate planes.
- *Angular magnification*: An incident light ray, intersecting the axis at some point A and making the angle u with the axis, gives an emerging ray that intersects the axis at the conjugate point A' and makes the angle u' with the axis. The ratio $g = u'/u$ is called the *angular magnification*; g takes the same value for any pair of rays, respectively, going through A and A'. To a given value of g is associated one, and only one, pair of conjugate points.
- *The Lagrange-Helmholtz invariant* (see Figure 3.33): If equation (3.5) is successively used for the different interfaces of a centered system, it is seen that the product $nABu$ keeps a constant value from interface to interface, so it is called the Lagrange-Helmholtz invariant,

$$nABu = n_0 A_0 B_0 u_0 = n'A'B'u'. \tag{3.7}$$

3.5.2. Cardinal Elements of Centered Systems

A conjugating equation is an equation between the positions of two conjugate points A and A'. Equation (3.6) is a conjugating equation, the coefficients of which depend on the two points of reference S and S'; it becomes simpler by a clever choice of the origins: if S and S' are conjugate points, SA and $S'A'$ cancel simultaneously, equation (3.6) becomes

$$SAS'A' + \beta SA + \chi S'A' = 0 \quad \text{or} \quad \frac{\beta}{S'A'} + \frac{\chi}{SA} + 1 = 0. \tag{3.8}$$

Cardinal elements (points and planes) are special conjugate elements playing an essential role for a centered system. The most important cardinal elements are: focuses and focal planes on the one hand, and principal points and principal planes on the other.

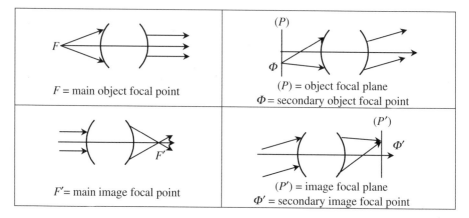

Figure 3.34. Focal points and focal planes.

Focus and Focal Planes

A focus, or focal point, is conjugated with a point at infinity. A main focus is the conjugate of a point at infinity in the direction of the axis; a secondary focus is conjugate with a point rejected at infinity in any other direction. All the different focuses are located in the same front plane, called a focal plane.

Principal Points and Principal Planes

The correspondence between conjugate objects and image front planes is a one-to-one correspondence; in the same way, the correspondence between the values of the linear magnification γ and a pair of conjugate front planes is also a one-to-one correspondence.

> *Principal planes are the conjugate planes for which the linear magnification is equal to $\gamma = +1$.*

The points H and H' of intersection of the principal planes with the axis are called *principal points*.

In the case of a thin lens, there is a kind of degeneracy, the two principal planes are not separate and coincide with the plane of the lens.

Once the principal points of a system are known it's easy to draw the emerging ray associated to some incident ray. Let I be the intersection point of the incident ray with the object principal point, the emerging ray will intersect the image principal point at a point I' having the same ordinates as I: $HI = H'I'$.

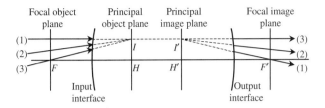

Figure 3.35. Object and image principal planes. In this case the two principal planes are virtual, since they are inside the system. I' is the image of I.

Object focal length: This is the algebraic distance, $HF = f$, between the principal object point and the object focal point.

Image focal length: This is the algebraic distance, $H'F' = f'$, between the image principal point and the image focal point. In the absence of other indications, the distance focal of a system is its image focal length.

There are two focal lengths, f and f', starting from the Lagrange-Helmholtz equation and it can be seen that the two focal lengths fulfill the following equation:

$$\frac{f}{n} = -\frac{f'}{n'}. \tag{3.9}$$

If the input and output mediums have the same index of refraction, the image and object focal lengths are equal: $f = -f'$.

3.5.3. *Image Construction in a Centered System*

3.5.3.1. *Image Construction Using Cardinal Elements*

To obtain the position of the image from the position of the object, we could use the basic homographic relationship (3.8); the three parameters α, β, and γ can be obtained from the optogeometrical characteristics of the different spherical interfaces, although very tedious this method can be used with computers. Very often it will proceed in another way, three couples of conjugate points will first be directly determined; then by writing equation (3.8) with the coordinates of those three couples of points, three equations are obtained, from which α, β, and γ may be calculated. From a practical point of view, things are even simpler, the three couples of conjugate points are:

> Point object at infinity $\leftrightarrow$ image focal point.
> Object focal point $\leftrightarrow$ point image at infinity.
> Principal object point $\leftrightarrow$ principal image point.

It will not be necessary to determine α, β, and γ since geometrical constructions will be proposed to obtain the image of an object. For constructing the image of some point object B located out of the axis, see

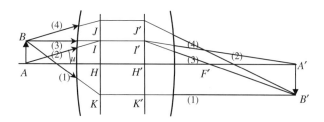

Figure 3.36. Ray tracing using principal planes and focal points. In the absence of further indications about the system, only the lines supporting the rays can be drawn. This is the reason why, inside the system, the rays are represented by dotted lines, since it cannot be known which parts are real or virtual.

Figure 3.36, the two following rays are considered:

- Incident ray parallel to the axis → emerging ray going from point I' belonging to the principal image plane and having the same ordinate as B to the image focal point F'.
- Incident ray going from B to the object focal point F and intersecting the object principal plane at some point J → emerging ray parallel to the axis and at the same distance from the axis as J.

3.5.3.2. Newton's Equations

Two different origins are chosen: the image focal point F' for the image and the object focal point F for the object. Expressing the similarity of the triangles ABF and HJF on one hand and of $H'I'F'$ and $A'B'F'$ on the other, and noticing that $AB = IH = I'H'$ and $A'B' = H'J' = HJ$, we obtain

$$\frac{AF}{HF} = \frac{H'F'}{A'F'} \quad \rightarrow \quad FAF'A' = FHF'H' = ff'. \tag{3.10}$$

3.5.3.3. Descartes' Equations

Two different origins are also used: the principal object and image points H and H'. Starting from the Newton equations, using the equalities $FA = FH + HA$, $F'A' = F'H' + H'A'$, and $f/n = -f'/n'$, we finally obtain

$$\frac{f'}{p'} + \frac{f}{p} = 1 \quad \rightarrow \quad \frac{n'}{p'} - \frac{n}{p} = \frac{n'}{f'} = -\frac{n}{f} \quad \text{with} \quad HA = p \text{ and } H'A' = p'. \tag{3.11}$$

3.5.3.4. Magnification

Linear magnification: This is the ratio of the size of the image to the size of the object,

$$G_{\text{linear}} = \frac{A'B'}{AB} = \frac{f}{FA} = \frac{F'A'}{f'} \quad \rightarrow \quad G_{\text{linear}} = -\frac{f}{f'} \frac{p'}{p} = \frac{p'/n'}{p/n} = \frac{n}{n'} \frac{p'}{p}. \tag{3.12}$$

Angular magnification: Let u and u', respectively, be the angles with the axis of an incident ray and of the associate emerging ray, the angular magnification is the ratio $G_{\text{angular}} = u'/u$. Thanks to the Lagrange-Helmholtz equation (3.7), the two different magnifications are easily related to one another:

$$G_{\text{angular}} G_{\text{linear}} = \frac{n}{n'} \quad \rightarrow \quad G_{\text{angular}} = \frac{p}{p'}. \tag{3.13}$$

As illustrated by rays AK and $K'A'$ the linear and angular magnifications have the same signs.

Axial magnification: This third magnification is not as useful as the other two. Given two object points A_1 and A_2 belonging to the axis, let B_1 and B_2 be their respective images, the axial amplification is the ratio $G_{\text{axial}} = B_1B_2/A_1A_2$, its value can be obtained by differentiating Descartes' equation (3.11):

$$G_{\text{axial}} = \frac{dp'}{dp} = \frac{n}{n'} \left(\frac{p'}{p} \right)^2 = \frac{n'}{n} G_{\text{linear}}^2. \tag{3.14}$$

The axial magnification is always positive, which means that an object and its image always move in the same direction along the axis.

3.5.4. *Matrix Methods for Centered Systems*

The problem is to follow a light ray as it propagates across the system and is refracted from the different interfaces. In order to reference a light ray we must have an indication about its position, which needs the two coordinates (x, y) of the point of intersection with a given front plane and we must also have an indication about its direction, which needs the two angles (u, v) that makes the axis with the two projections on two orthogonal planes (Oxz and Oyz). The situation is illustrated in Figure 3.37.

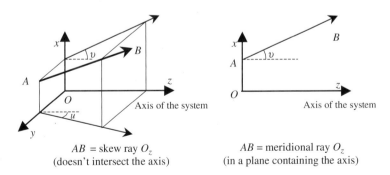

AB = skew ray O_z AB = meridional ray O_z
(doesn't intersect the axis)   (in a plane containing the axis)

Figure 3.37. To reference a ray, a point and a direction are needed. In the most general case, two coordinates and two angles are required. The case of a meridional ray is simpler, one coordinate and one angle are sufficient.

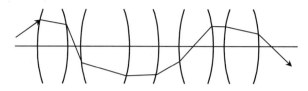

Figure 3.38. Trajectory of a light ray propagating across a centered system. The path of the light is made of segments joined along the different interfaces. If the initial ray is in a plane containing the axis, because of symmetry all the refracted rays will remain inside this plane.

3.5.4.1. *Ray's Equations Are Linear and Homogeneous Versus Angles and Positions*

For the sake of simplicity we will only consider rays that belong to a meridional plane. Figure 3.39 indicates the notations we use, (P_1) and (P_2) are two standard front planes and, in general, they are not conjugate through the system. We start from the input parameters (u_1, r_1) and we would like to obtain the output associate parameters (u_2, r_2). The angles being small (Gauss conditions), the sine law is replaced by the Kepler law. The relations between u and r are thus linear, furthermore, if u_1 and r_1 are equal to zero, u_2 and r_2 are also equal to zero: the relations must be homogeneous. Under such conditions the use of matrices is very appropriate.

The matrix elements are determined by two considerations: on one hand, by the special centered system under consideration and, on the other

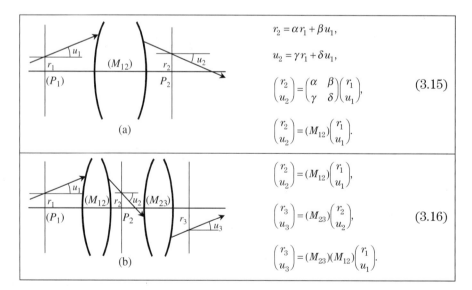

$$r_2 = \alpha r_1 + \beta u_1,$$

$$u_2 = \gamma r_1 + \delta u_1,$$

$$\begin{pmatrix} r_2 \\ u_2 \end{pmatrix} = \begin{pmatrix} \alpha & \beta \\ \gamma & \delta \end{pmatrix} \begin{pmatrix} r_1 \\ u_1 \end{pmatrix}, \qquad (3.15)$$

$$\begin{pmatrix} r_2 \\ u_2 \end{pmatrix} = (M_{12}) \begin{pmatrix} r_1 \\ u_1 \end{pmatrix}.$$

$$\begin{pmatrix} r_2 \\ u_2 \end{pmatrix} = (M_{12}) \begin{pmatrix} r_1 \\ u_1 \end{pmatrix},$$

$$\begin{pmatrix} r_3 \\ u_3 \end{pmatrix} = (M_{23}) \begin{pmatrix} r_2 \\ u_2 \end{pmatrix}, \qquad (3.16)$$

$$\begin{pmatrix} r_3 \\ u_3 \end{pmatrix} = (M_{23})(M_{12}) \begin{pmatrix} r_1 \\ u_1 \end{pmatrix}.$$

Figure 3.39. Definition of a matrix connecting two front planes (P_1) and (P_2). The matrix of cascaded systems is the product of the matrices of the different elements.

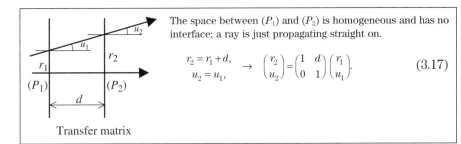

The space between (P_1) and (P_2) is homogeneous and has no interface; a ray is just propagating straight on.

$$r_2 = r_1 + d, \quad \rightarrow \quad \begin{pmatrix} r_2 \\ u_2 \end{pmatrix} = \begin{pmatrix} 1 & d \\ 0 & 1 \end{pmatrix} \begin{pmatrix} r_1 \\ u_1 \end{pmatrix}. \tag{3.17}$$
$$u_2 = u_1,$$

Transfer matrix

Figure 3.40. Matrix of transfer.

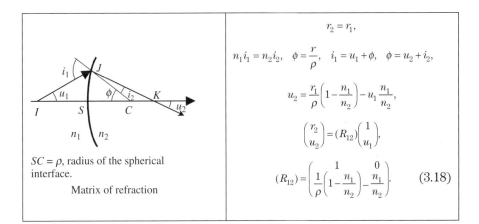

$$r_2 = r_1,$$

$$n_1 i_1 = n_2 i_2, \quad \phi = \frac{r}{\rho}, \quad i_1 = u_1 + \phi, \quad \phi = u_2 + i_2,$$

$$u_2 = \frac{r_1}{\rho}\left(1 - \frac{n_1}{n_2}\right) - u_1 \frac{n_1}{n_2},$$

$$\begin{pmatrix} r_2 \\ u_2 \end{pmatrix} = (R_{12})\begin{pmatrix} 1 \\ u_1 \end{pmatrix},$$

$$(R_{12}) = \begin{pmatrix} 1 & 0 \\ \frac{1}{\rho}\left(1 - \frac{n_1}{n_2}\right) & \frac{n_1}{n_2} \end{pmatrix}. \tag{3.18}$$

$SC = \rho$, radius of the spherical interface.

Matrix of refraction

Figure 3.41. Matrix of refraction.

hand, by the positions of the two planes (P_1) and (P_2), their values are determined by the laws of geometrical optics.

Refraction produces a discontinuity for the directions of the rays, but not for their positions. The two planes (P_1) and (P_2) coincide with the interface: (P_1) can be considered to be "just" before, and (P_2) "just" after the interface. (R_{12}) is called the matrix of refraction.

3.5.4.2. Matrices for Skew Rays

The result given in Figure 3.42 for a transfer matrix is quite general: the two sets of variables (x, u) and (y, v) follow matrix equations that are identical to the equations obtained for meridional rays. The two sets can be treated simultaneously by introducing the following complex variables:

- *Complex coordinate:* $r^* = (x + jy)$. $\qquad$ (3.19.a)
- *Complex angle:* $\theta^* = (u + jv)$. $\qquad$ (3.19.b)

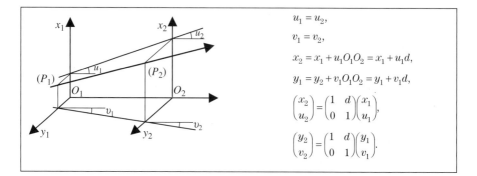

Figure 3.42. A skew ray is projected onto the two planes of coordinates, the set of variables (x, u) and (y, v) describing the two projections obey the same matrix law as a ray belonging to a meridional plane.

Of course we have the following equations:

$$x = \mathrm{Re}[r^*], \quad j = \mathrm{Im}[r^*], \quad u = \mathrm{Re}[\theta^*], \quad v = \mathrm{Im}[\theta^*].$$

Optical Angles

Optical angles are introduced to simplify the different matrices. The matrices are all unitary matrices, i.e., matrices having a determinant equal to unity. The main interest is that the product of unitary matrices is also a unitary matrix. For a given ray, an optical angle is equal to the usual angle multiplied by the index of refraction of the transparent medium inside of which the ray propagates.

- *Usual angles:* u and v.
- *Optical angles:* nu and nv. (3.20.a)
- *Complex optical angles:* $n\theta^* = n(u + jv)$. (3.20.b)

3.5.4.3. *Matrix of an Association of Centered Systems*

The propagation of a light ray across the four interfaces of the system of Figure 3.43 is obtained by the product of transfer T_{ij} and refraction R_{ij} matrices,

$$\begin{bmatrix} r_6^* \\ n'u_6^* \end{bmatrix} = T_{56} R_{45} T_{45} R_{34} T_{34} R_{23} T_{23} R_{12} T_{12} \begin{bmatrix} r_1^* \\ nu_1^* \end{bmatrix} = M \begin{bmatrix} r_1^* \\ nu_1^* \end{bmatrix}, \qquad (3.21.a)$$

$$M = T_{56} R_{45} T_{45} R_{34} T_{34} R_{23} T_{23} R_{12} T_{12}. \qquad (3.21.b)$$

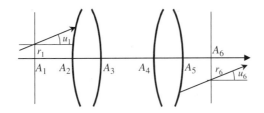

Figure 3.43. Optical centered system made of four spherical interfaces. The respective refractive indices of the first and last transparent mediums are equal to n and n'.

3.5.4.4. *Matrices of a Centered System*

We consider a centered system where the two extreme mediums have the same index of refraction. The two focal lengths have equal absolute values. The image focal length will be called $H'F = f$. The elements of the matrix connecting the object to the image principal plane are evaluated in Figure 3.44.

In Figure 3.45 are evaluated the elements of a matrix connecting two planes having no special positions and, respectively, located at distances equal to d_1 and d_2 from the object and image principal planes.

It should be noticed that all the previous matrices are unitary. In this type of calculation great attention should be paid to the homogeneity of the expressions: some elements are homogeneous to a length, while others have dimensions of the inverse of a length.

3.5.4.5. *Vergence—Matrix of Conjugation*

Given a centered system, any pair of planes (A) and (A') is associated to one, and only one, matrix which can be written as $T_{AA'} = \begin{pmatrix} T_{11} & T_{12} \\ T_{21} & T_{22} \end{pmatrix}$.

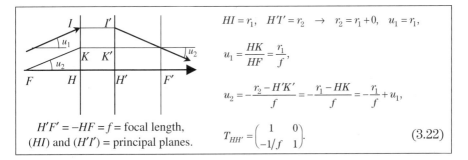

$HI = r_1, \quad H'I' = r_2 \quad \rightarrow \quad r_2 = r_1 + 0, \quad u_1 = r_1,$

$u_1 = \dfrac{HK}{HF} = \dfrac{r_1}{f},$

$u_2 = -\dfrac{r_2 - H'K'}{f} = -\dfrac{r_1 - HK}{f} = -\dfrac{r_1}{f} + u_1,$

$H'F' = -HF = f = $ focal length,

(HI) and $(H'I')$ = principal planes.

$T_{HH'} = \begin{pmatrix} 1 & 0 \\ -1/f & 1 \end{pmatrix}.$ (3.22)

Figure 3.44. Matrix connecting the two principal planes of a centered system.

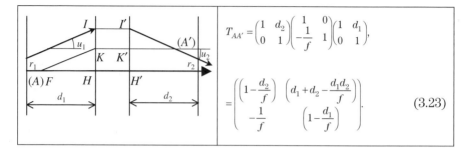

$$T_{AA'} = \begin{pmatrix} 1 & d_2 \\ 0 & 1 \end{pmatrix} \begin{pmatrix} 1 & 0 \\ -\dfrac{1}{f} & 1 \end{pmatrix} \begin{pmatrix} 1 & d_1 \\ 0 & 1 \end{pmatrix},$$

$$= \begin{pmatrix} \left(1 - \dfrac{d_2}{f}\right) & \left(d_1 + d_2 - \dfrac{d_1 d_2}{f}\right) \\ -\dfrac{1}{f} & \left(1 - \dfrac{d_1}{f}\right) \end{pmatrix}. \tag{3.23}$$

Figure 3.45. The matrix connecting two planes is equal to the product of the matrix between the two principal planes by two transfer matrices.

It can be seen that the element T_{21} has the same expression, $T_{21} = -1/f$, *for* the two matrices $T_{AA'}$ and $T_{HH'}$ (see formulas (3.23) and (3.24)). This is a general property that will be admitted: whatever the two planes (A) and (A'), the matrix element T_{21} is an *intrinsic parameter* of a centered system.

By definition, the *vergence* of a centered system is equal to $V = -T_{21}$.

In the general case where the input and output indices, respectively, n and n', are different, it can be established that V is related to the focal lengths

$$V = -\frac{n}{f_{\text{object}}} = \frac{n'}{f_{\text{image}}}. \tag{3.24}$$

The vergence is homogeneous to the inverse of length and is measured in m^{-1}; opticians have given a special name, *diopter*, to this unit.

Matrix of Conjugation

When two planes (A) and (A') are conjugate planes, the matrix $C_{AA'}$ is called the *matrix of conjugation*, the matrix elements then have special physical meanings that are given in Figure 3.46. Let us consider an object AB and its image $A'B'$, and let (r, u) and (r', u') be the two couples of parameters that, respectively, describe an incident beam and the associate emerging beam,

$$\begin{pmatrix} r' \\ u' \end{pmatrix} = \begin{pmatrix} C_{11(AA')} & C_{12(AA')} \\ C_{21(AA')} & C_{22(AA')} \end{pmatrix} \begin{pmatrix} r \\ u \end{pmatrix} = (C_{(AA')}) \begin{pmatrix} r \\ u \end{pmatrix},$$

$r' = C_{11(AA')}r + C_{12(AA')}u$:

- Since B' is the image of B the value of r' cannot depend on the orientation of the ray, thus the matrix element $C_{12}(AA')$ is equal to zero.
- $C_{11}(AA') = r'/r$ is nothing other than the linear magnification G_{linear} associated to the couple of conjugate planes (A) and (A').

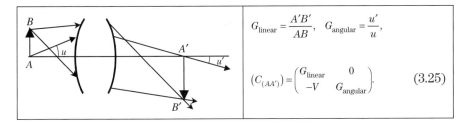

$$G_{\text{linear}} = \frac{A'B'}{AB}, \quad G_{\text{angular}} = \frac{u'}{u},$$

$$(C_{(AA')}) = \begin{pmatrix} G_{\text{linear}} & 0 \\ -V & G_{\text{angular}} \end{pmatrix}. \qquad (3.25)$$

Figure 3.46. Matrix of conjugation.

$u' = C_{21(AA')}r + C_{22(AA')}u = -Vr + G_{\text{angular}}u$:

- We have seen that C_{21} is equal to the vergence.
- To have an interpretation of C_{22}, we consider the case where point A is on the axis: $r = 0 \rightarrow C_{22} = u'/u$. $C_{22}(AA') = u'/u$ is seen to be equal to the angular magnification, G_{angular}, associated to the two conjugated planes.

Annex 3.A

Thin Lenses

3.A.1. Lens Considered as a Prism Having a Variable Angle

Let C_1, $C_1S_2 = R_1$, C_2, and $C_2S_1 = R_2$, respectively, be the centers and the radii of curvature of the two interfaces. B and B' are two conjugate points. We use the following notations: $I_1J_1 = I_2J_2 = r$, $OB = p$, and $OB' = p'$, as well as the geometric indications given in Figure 3.A.1. The lens can be considered as a prism with an angle $\hat{A}$, which would vary proportionally to the distance r to the axis. A ray is all the more deviated so that it hits the input interface at a point located farther from the axis. The angles being small, we use the law of Kepler to evaluate the deviation of the ray $B_1I_1 \to \hat{D} = (n-1)\hat{A}$, we obtain

$$\frac{1}{p'} - \frac{1}{p} = (n-1)\left(\frac{1}{R_1} + \frac{1}{R_2}\right) = \frac{1}{f}. \tag{3.A.1}$$

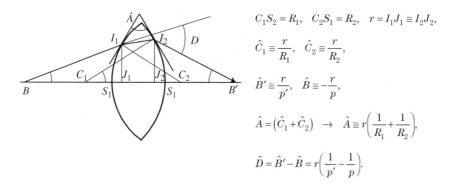

$$C_1S_2 = R_1, \quad C_2S_1 = R_2, \quad r = I_1J_1 \cong I_2J_2,$$

$$\hat{C}_1 \cong \frac{r}{R_1}, \quad \hat{C}_2 \cong \frac{r}{R_2},$$

$$\hat{B}' \cong \frac{r}{p'}, \quad \hat{B} \cong -\frac{r}{p},$$

$$\hat{A} = (\hat{C}_1 + \hat{C}_2) \quad \to \quad \hat{A} \cong r\left(\frac{1}{R_1} + \frac{1}{R_2}\right),$$

$$\hat{D} = \hat{B}' - \hat{B} = r\left(\frac{1}{p'} - \frac{1}{p}\right).$$

Figure 3.A.1. For the light ray $B_1I_1I_2B_2$ the lens is equivalent to a refracting prism of the same index and having an angle A equal to the angle of the two planes that are tangent to the interfaces at points I_1 and I_2. As the lens is considered to be thin, the points S_1, J_1, J_2, and S_2 coincide.

Figure 3.A.2. An incident beam propagates parallel to the axis and is focused by a plano-convex lens. The part of the optical light path which is inside the glass, is longer if the light propagates nearer to the axis: the vibration at point C is in advance with regard to the vibration at point H.

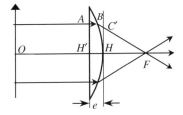

In formula (3.A.1) the radii of curvature are positive for convex interfaces, and negative for concave interfaces. The distances to the origin O are algebraically measured along the axis that is oriented in the direction of propagation of the light. In the example of Figure 3.A.1, the focal length $f = OF'$ is positive, corresponding to a converging lens.

3.A.2. Lens Considered as a Phase Correcting Device

The lens of Figure 3.A.2 receives a planar wave having its wave planes orthogonal to the axis. A light beam propagating parallel to the axis, respectively, intersects the input and output interfaces at points A and B. (H) is a plane orthogonal to the axis at the point H of intersection with the spherical interface, C is the point where the emerging beam intersects the plane (H).

The input planar interface coincides with a wave plane of the incident beam. Along the plane (H) the phase of the vibration is not constant since the delay is not the same at point C or at point H: the emerging wave is no longer a planar wave. The distance between A, B, and C is of the order of a few micrometers, if we just think of the positions or of the directions of the rays, only a small error is made by assuming that the three points are at the same place. On the contrary, if we want to compare the phases of the vibrations, we must be more careful, since in Optics, several micrometers represent several wavelengths and a phase difference of several times 2π. This problem was studied in Section 2.6.4.2 and formula (3.A.1) was already established.

3.A.3. Matrices for the Association of Thin Lenses

To study a system made of a succession of cascaded thin lenses it is convenient to have the expression of the matrix $T_{L^-L^+}$ connecting the plane (H^-) located "just before" the lens, with the plane (H^+) located "just after." A thin lens is nothing other than a centered system in which the two principal planes

coincide, the matrix $T_{L^-L^+}$ has already been calculated and is given by formula
(3.22) of *Section 3.5.4.4*

$$T_{L^-L^+} = \begin{pmatrix} 1 & 0 \\ -\dfrac{1}{f} & 1 \end{pmatrix}.$$

Association of Two Lenses

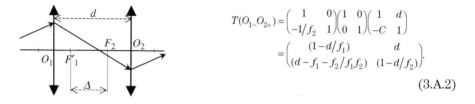

$$T(O_{1^-}O_{2^+}) = \begin{pmatrix} 1 & 0 \\ -1/f_2 & 1 \end{pmatrix}\begin{pmatrix} 1 & 0 \\ 0 & 1 \end{pmatrix}\begin{pmatrix} 1 & d \\ -C & 1 \end{pmatrix}$$

$$= \begin{pmatrix} (1-d/f_1) & d \\ (d-f_1-f_2/f_1f_2) & (1-d/f_2) \end{pmatrix}.$$

$$(3.A.2)$$

Figure 3.A.3. Matrix connecting the plane (O_{1^-}) located "just before" the first lens, with the plane (O_{2^+}) located "just after" the second lens. A matrix of transfer is inserted between the matrices of the two lenses.

The Gullstrand Formula

The Gullstrand formula gives the focal length f of a doublet made of two thin lenses. Recalling that the matrix element T_{21} is an intrinsic parameter of a centered system that is equal to its vergence, we may write

$$T_{21} = -\frac{1}{V} = -\frac{1}{f_{\text{image}}} \quad\rightarrow\quad \frac{1}{f} = \frac{1}{f_1} + \frac{1}{f_1} - \frac{d}{f_1f_2} \quad \text{and} \quad f = -\frac{f_1f_2}{\Delta}, \quad (3.A.3)$$

where d is the separation between the two lenses. $\Delta = F_1'F_2$ is the distance between the image focal point of the first lens and the object focal point of the second.

Periscopic Stability

In submarines, periscopes are (or were) long (several meters) pipes with an objective at one end and an eyepiece at the other. If no special attention is paid, even with low diverging beams (Gauss conditions), the diameter of the beam can easily reach 0.5 m . . . , which would require enormous lenses. The solution had been to place a regular succession of converging lenses, as indicated in Figure 3.A.4. The next problem is to ask if the light rays will remain confined in the vicinity of the axis (the system will then be said to be *stable*) or, on the contrary, if they will escape (*unstable* system). Thanks to TV cameras, periscopic stability is now quite obsolete for submarines, however it remains up to date for particle accelerators, where the accelerated charges

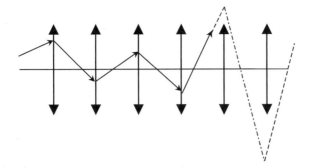

Figure 3.A.4. Unstable system. Starting with the fifth lens the light rays fall outside the lens, furthermore, they would surely not fulfill Gauss conditions.

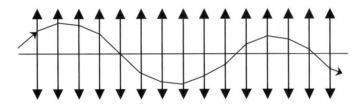

Figure 3.A.5. Stable system. The light rays oscillate on both sides of the axis.

should remain in the close vicinity of some average trajectory. An interesting application is also to be found when studying the stability of a laser resonator.

The arrangements of Figures 3.A.4 and 3.A.5 are made of a periodic succession of identical elementary patterns; usually a pattern will be made of two converging lenses of respective focal lengths, f_1 and f_2. A matrix T of the kind given by formula (3.A.2) characterizes each pattern. r_0 and u_0 are, respectively, the distance to the axis of the point where the first ray hits the first lens and the angle of the ray with the axis; r_n and u_n are the same parameters for the emerging ray, we have

$$\begin{pmatrix} r_n \\ u_n \end{pmatrix} = (T)^n \begin{pmatrix} r_0 \\ u_0 \end{pmatrix}, \tag{3.A.4}$$

where T^n is the matrix obtained by elevating the elementary matrix T to the nth power. It is shown in mathematics that r_n and u_n can be expressed using the eigenvalues, λ_1 and λ_2, of the matrix T, the following formulas are obtained:

$$r_n = \alpha \lambda_1^n + \beta \lambda_2^n, \quad u_n = \gamma \lambda_1^n + \delta \lambda_2^n, \quad \alpha, \beta, \gamma, \text{ and } \delta \text{ are from the initial values } r_0 \text{ and } u_0.$$

The matrix is written $T = \begin{pmatrix} A & B \\ C & D \end{pmatrix}$.

Its determinant $\begin{Vmatrix} A & B \\ C & D \end{Vmatrix} = (AD - BC) = 1$ is equal to one since the matrices are unitary.

The eigenvalues are obtained by cancelling the following determinant:

$$\begin{Vmatrix} (A - \lambda) & B \\ C & (D - \lambda) \end{Vmatrix} = 0 \quad \rightarrow \quad \lambda^2 + (A + D)\lambda + 1 = 0. \qquad (3.A.5)$$

Since its coefficients are real, the solutions of equation (3.A.5) are complex conjugate, their product is equal to one, and their sum is equal to $(A + D)$ and thus real. The following two situations may be encountered:

- The two solutions are real, the absolute value of one of the solutions is necessarily larger than one, λ_1, for example; λ_1^n goes to infinity if n increases indefinitely: the system is then *unstable*.
- The two solutions are complex conjugate, their modules are equal and equal to one; they may be written as $e^{\pm j\theta}$. After some algebraic manipulations it is obtained that $r_n = \alpha' \cos n\theta + \beta' \sin n\theta$ and $u_n = \gamma' \cos n\theta + \delta' \sin n\theta$.

u_n and r_n no longer go to infinity, the *system is stable*.

Annex 3.B

Optical Prisms

3.B.1. Definition and Description of Optical Prisms

Optical prisms are components that are commonly met in optical experimental arrangements and optical instruments; so far their main role has been to disperse the light in spectrometers, now they have been universally replaced by diffraction gratings. However, they remain very important in changing the direction of a beam, thanks either to refraction or, more often, to total internal reflection. Following the mathematical definition an optical prism is a triangular-based prism. The sides are carefully polished with a precision of the order of a small fraction of a micrometer (a good standard is $\lambda/10$ of the sodium wavelength). The precision of the angles is usually of one minute; once a prism has been polished the angles can be measured with extremely high precision (one second).

Optical prisms are very often cut from a piece of glass. For a given transparent material the knowledge of the index of refraction is very important. The more accurate method of measuring the index of refraction is to fashion a prism with the medium under consideration, and to measure the angle of deviation of a parallel monochromatic light beam. Index measurements using a refractometer are very accurate and easily give the value of refractive indices to five figures.

3.B.2. Light Propagation Inside a Prism

Demonstrations concerning the formulas that govern the propagation of light in a prism rely on rather simple geometrical considerations and will not be given here.

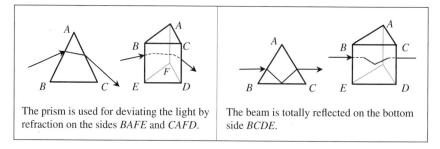

| The prism is used for deviating the light by refraction on the sides *BAFE* and *CAFD*. | The beam is totally reflected on the bottom side *BCDE*. |

Figure 3.B.1. Two typical propagation schemes inside a prism.

Conditions of Emergence for a Light Ray

The constructions in Figure 3.B.2 are valid only if the angle of incidence r' on the second side of the prism is smaller than the critical angle $\hat{C} = \sin^{-1}(1/n)$. The conditions for a light ray to emerge after having been refracted twice are indicated in Figure 3.B.3.

Minimum Deviation

We refer to Figure 3.B.4, it's important to notice that the bisecting plane of the angle $\hat{A}$ is a plane of symmetry for the prism. To a given angle of incidence i is associated an angle of emergence i' and an angle of deviation D; because of the principle of reversibility and because of the symmetry, if the incidence angle on the first side is made equal to i', then the angle of emergence is equal to i and the deviation has the same value in both cases. Finally, there are two values of the angle of incidence that give the same deviation and, consequently, the deviation should exhibit a maximum or a minimum for some value $i_{\min}$ of the angle of incidence. It's a minimum $D_{\min}$ that is obviously obtained

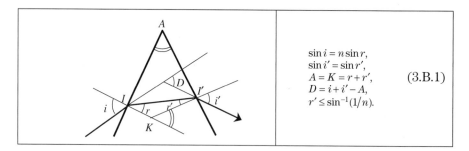

$$\sin i = n \sin r,$$
$$\sin i' = \sin r',$$
$$A = K = r + r', \qquad (3.B.1)$$
$$D = i + i' - A,$$
$$r' \le \sin^{-1}(1/n).$$

Figure 3.B.2. Ray tracing in a prism. The deviation is the angle D of the prolongation of the incident light ray with the emerging ray.

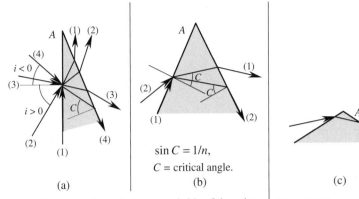

$\sin C = 1/n$,
C = critical angle.

(a)

(b)

(c)

A ray will emerge through the second side of the prism, under the condition that $r' \le C$, which is possible only if $A' \le 2C$. All the incident rays that give a ray inside the prism which is not totally reflected are inside the hatched triangles.

If $A > 2C$, because of total internal reflection, no ray is allowed to emerge through the second side of the prism.

Figure 3.B.3. Conditions of emergence through the second side of a prism.

when the light ray inside the glass is orthogonal to the bisecting plane. When the deviation is minimum, we have

$$i_{\min} = i'_{\min} \to r_{\min} = r'_{\min} = \frac{A}{2} \quad \to \quad D_{\min} = (i_{\min} + i'_{\min} - A),$$

$$i_{\min} = \frac{(A + D_{\min})}{2},$$

$$n = \frac{\sin\left(\dfrac{A + D_{\min}}{2}\right)}{\sin\left(\dfrac{A}{2}\right)}.$$

(3.B.2)

Equation (3.B.2) is quite useful for very accurate determinations of the index of refraction, since the determination of the minimum is very sensitive and since angular measurements can be extremely accurate.

Figure 3.B.4. Ray path in a prism at the minimum deviation; the figure is symmetric with respect to the bisector plane.

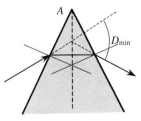

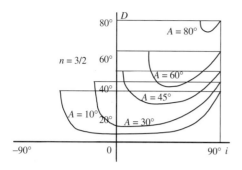

Figure 3.B.5. Variations of the deviation versus angle of incidence for different values of the prism angle. When the angle $\hat{A}$ is small the deviation is almost independent of the incidence.

In Figure 3.B.5 have been drawn the variations of the deviation versus the angle i of incidence on the first interface. For each value of the prism angle, i can vary from the value i_0 to which corresponds a grazing emergence ($i' = 90°$) up to a grazing incidence ($i = 0$). The tangent is vertical for $i = i_0$, and makes an angle of $45°$ with the axis when $i = 90°$. When the prism angle is small ($\hat{A} < 10°$) the Snell-Descartes law is replaced by the Kepler law, the deviation is then almost independent of the angle of incidence and is given by

$$D = (n-1)A. \tag{3.B.3}$$

Prism Spectrometer

In Figure 3.B.6 is shown the scheme of a spectrometer where the light is dispersed by a prism. A light source emits two rays of different colors (λ_1, λ_2) and is placed in front of a thin slit disposed in the object focal plane of a collimating lens. The prism receives a beam of parallel rays and gives two parallel beams of different colors and different directions. The second lens focuses these two beams at two different secondary focal points.

3.B.3. Reflecting Prisms

Reflection is very convenient for changing the direction of a beam. Rather than a plane mirror, it is often preferred to use reflecting prisms; first because,

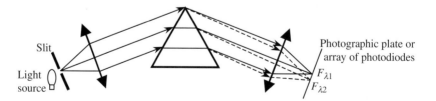

Figure 3.B.6. Principle of a spectrometer using a prism.

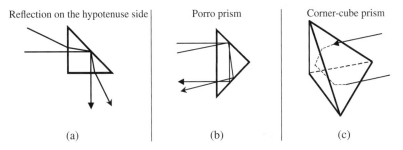

Figure 3.B.7. Optical prisms.

in the case of total internal reflection, the reflection coefficient is very close to unity, whatever the wavelength, which is not always the case with ordinary mirrors. And second, because maintenance is less severe in the case of a prism since the reflection occurs inside the glass, whilst the surface of a mirror is quite sensitive to dust and atmospheric pollution hazards. Opticians have been very imaginative in conceiving many different schemes using prisms; we will describe only a few of them.

Under the conditions of total internal reflection, the phase shift between the incident and reflected beams is not the same if the electric field is parallel, or orthogonal, to the plane of incidence. A polarization, which doesn't coincide with one of the previous directions, will be modified after total internal reflection: *a linearly polarized light beam will, in general, give an elliptically polarized reflected beam.*

The cross section of the prism of Figure 3.B.7(a) is a right angled isosceles triangle, total internal reflection occurs on the hypotenuse: incident and reflected rays are symmetric with regard to a plane orthogonal to the hypotenuse; in the case of normal incidence on one side of the right angle, the incident and reflected beams are orthogonal.

The prism of Figure 3.B.7(b) also has a right angle and isosceles triangle cross section; the emerging beam leaves the prism after two total internal reflections. The path followed by the ray is very familiar to billiard players: if rays propagate in a cross section, the reflected and incident beams are parallel. This property is not true outside a cross section.

The arrangement of Figure 3.B.7(c) is called a corner cube prism. It is a generalization of the Porro prism, three reflections, one on each side, are involved. The interesting property is that retrodirectivity is achieved, whatever the incident beam.

Annex 3.C

Gradient Index Devices—Light Optics and Electron Optics

When traveling inside a volume where the electric and magnetic fields are equal to zero, an electric charge follows a rectilinear path. The existence of a nonconstant electric voltage, which creates an electric field, bends the trajectory. In the same way, a light beam is rectilinear in a material where the index of refraction is constant and is bent in the presence of a gradient index. W.R. Hamilton first pointed out the analogy between optical rays and electric charge trajectories in 1831. For about one hundred years this "Hamiltonian analogy" was considered as a remarkable aesthetic curiosity and it was only in 1925 that H. Busch got the idea of using electric and magnetic fields to focus beams of charged particles. Since then, electron optics and, more generally, the optics of charged particles, was born; almost immediately the method of geometrical optics was transposed to electron optics, especially the matrix analysis of centered systems.

It was also around 1920 that Quantum Mechanics was invented. E. Schrödinger took advantage of the Hamiltonian analogy to obtain the approximations that are required to reduce Wave Optics to Geometrical Optics; then he suggested that the relationship between Quantum Mechanics and Classical Mechanics involved the same kind of approximation. De Broglie's wavelength, in Mechanics, plays the same role as the optical wavelength in Optics. As early as 1940, electron microscopes and accelerators of particles were available.

In spite of the theoretical identity between the optics of light rays and the optics of beam particles, the associated technologies are completely different. While light propagates just as easily in a vacuum or in air, beam particles require a high vacuum. In both cases microscopes can be fabricated, giving magnified images. In both cases the ultimate resolution is fixed by the wavelength. The value λ of De Broglie's wavelength is obtained from the energy of the particles and, in the case of electrons, is given by (3.C.1), where ϕ is the accelerating voltage

$$\lambda_{\text{angstrom}} = \frac{h}{\sqrt{2m\phi}} = \frac{12.5}{\sqrt{\phi_{\text{volt}}}}. \tag{3.C.1}$$

In microscopes, $\phi \approx 10^4$–10^5 V, the theoretical resolution is smaller than 1 Å. Of course the image is spoiled by geometric and chromatic aberrations. The chromatic aberrations come from the fact that all particles don't have exactly the same energy. The correction of the geometric aberrations, which have the same origin as the usual optics of light rays, is more difficult in the case of electron microscopy.

3.C.1. The Eikonal Equation

The eikonal equation governs the propagation of light rays in a gradient index material, which means that the index of refraction is not constant and varies with position. To be fair, the eikonal equation (from the Greek, $\varepsilon\iota\kappa\sigma\nu$ = image) is not very convenient; its main interest comes from the fact that it emphasizes the approximations that are made when Wave Optics is reduced to Geometrical Optics. We will not develop in great detail all the calculations, which can be found in the book, *Principles of Optics* by Born and Wolf.

We start with Maxwell's equations. The dielectric and magnetic constants ε and μ are not constant and vary in function of x, y, z; however, they are supposed to vary slowly at the scale of the wavelength. We are looking for a harmonic solution that is different from a planar wave solution. The electric and magnetic fields are written as

$$\begin{cases} \boldsymbol{E}_{(r,t)} = \boldsymbol{e}_{(r)}e^{-jk_0S(r)}e^{j\omega t}, \\ \boldsymbol{H}_{(r,t)} = \boldsymbol{h}_{(r)}e^{-jk_0S(r)}e^{j\omega t}, \end{cases} \quad \text{with} \quad k_0 = \omega/c,$$

where $\boldsymbol{e}_{(r)}$ and $\boldsymbol{h}_{(r)}$ are unknown vectors; $S(r)$ is an unknown scalar, which is called the optical path. After some calculation, the following expressions are obtained:

$$\boldsymbol{K}(e, S, n) - \frac{1}{jk_0}\boldsymbol{L}(e, S, n, \mu) + \frac{1}{(jk_0)^2}\boldsymbol{M}(e, \varepsilon, \mu) = 0, \qquad (3.C.2)$$

where

$$\boldsymbol{K}(e, S, n) = \left(n^2 - (\text{grad } S)^2\right)\boldsymbol{e}, \quad n \text{ is the index of refraction,} \qquad (3.C.3)$$

$$\boldsymbol{L}(e, S, n, \mu) = (\text{grad } S \,\text{grad}(\log \mu) - \nabla^2 S)\boldsymbol{e} \\ - 2(\boldsymbol{e}\,\text{grad}(\log n))\text{grad } S - 2(\text{grad } S \,\text{grad})\boldsymbol{e},$$

$$\boldsymbol{M}(e, \varepsilon, \mu) = \text{curl } \boldsymbol{e} \wedge \text{grad}(\log \mu) - \nabla^2 \boldsymbol{e} - \text{grad}(\boldsymbol{e}\,\text{grad}(\log \varepsilon)).$$

The only pleasant thing about the above formulas is that, in the case of a homogeneous medium, the gradient's terms vanish and they coincide with those of the homogeneous case.

If $1/k_0 = \lambda_0/2\pi$ is considered to be very small, formula (3.C.2) is a serial development. The geometrical approximation consists in omitting the small

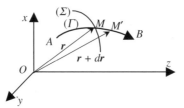

Figure 3.C.1. Bent light ray in an inhomogeneous medium.

terms of the series; in other words, *Geometrical Optics is the limit of Wave Optics when the wavelength goes to zero.* Equations (3.C.2) and (3.C.3) then become

$$\mathbf{K}(\mathbf{e}, S, n) = \left(n^2 - (\mathrm{grad}\, S)^2\right)\mathbf{e} = 0,$$

$$n^2 - (\mathrm{grad}\, S)^2 = 0, \tag{3.C.4}$$

$$n = |\mathrm{grad}\, S|, \quad \text{eikonal equation.}$$

Direct Demonstration of the Eikonal Equation

We refer to Figure 3.C.1 on which we have drawn:

- A light ray (Γ) going from point A to point B.
- Two points M and M' belonging to (Γ).
- The surface (Σ) of constant index containing point M.

By definition of the optical path $S(\mathbf{r})$ we have $S(B) - S(A) = \int_A^B n\, ds$, s is the curvilinear abscissa, measured along the light ray, which is bent when the propagation material is not homogeneous. The variation dS of the optical path, when going from M to M', is given by $dS = \mathrm{grad}\, S\, d\mathbf{r}\mathbf{u}$ ($\mathbf{u}$ is a unit vector perpendicular to (Σ) at point M). As $d\mathbf{r}$ may have any orientation, it is concluded that

$$n\mathbf{u} = \mathrm{grad}\, S \quad \rightarrow \quad n = |\mathrm{grad}\, S|. \tag{3.C.5}$$

3.C.2. Differential Equation of Light Rays

The main difference between the propagation in homogeneous and inhomogeneous materials is that in the last case the light rays are bent, instead of being rectilinear. This property obviously comes from the Fermat principle. We describe some experimental demonstrations of this phenomenon.

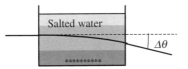

Figure 3.C.2. Propagation of light inside a gradient index material. The tank is filled with salted water, the concentration increases with depth, diluted near the surface and saturated at the bottom. The index of refraction varies from 1.33 to 1.38. An incident horizontal light ray is bent toward the region of higher index. $\Delta\theta$ is about 10°.

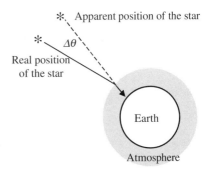

Figure 3.C.3. Stellar aberration due to the gradient index of the atmosphere. The index of refraction increases from one at the upper limit of the atmosphere up to 1.0003 at the surface of the Earth. $\Delta\theta$ is typically of 1 min, which is much larger than the accuracy of usual astronomic measurements.

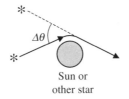

Figure 3.C.4. Deviation of a light ray by a heavy mass. According to the General Theory of Relativity the presence of a heavy mass modifies the property of the surrounding space and especially the index of refraction. A light ray coming from some star is deviated when passing in the neighborhood of the Sun. According to the fact that the light travels on one side of the Sun or on the other, the sign of the deviation is changed. $\Delta\theta$ is very small (seconds), but still measurable, which comforts the theory of General Relativity.

At Grazing Incidence Total Internal Reflection
May Occur on a Gradient Index

When the surface of the Earth is overheated by the Sun, the temperature of the soil is often much higher than the temperature of the air layers that are just above. As a consequence, an important gradient of temperature and, correlatively, a gradient of air density, and finally of an index of refraction are

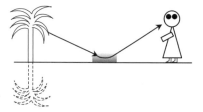

Figure 3.C.5. Bending of rays through the heated layers of air produces a mirage. The observer sees a palm tree and its image on a planar horizontal mirror. Planar mirrors that are most commonly met at the surface of the Earth are ponds or lakes; the step is narrow to dream of the presence.

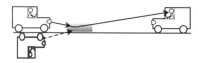

Figure 3.C.6. Tar on the road, locally heated by the Sun, produces a kind of mirage. The corresponding mirror is far from perfect.

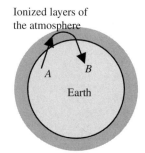

Ionized layers of the atmosphere

A   B

Earth

Figure 3.C.7. Reflection of radio waves by the ionized layers of the atmosphere. The upper layers of the atmosphere contain charged particles (ions and free electrons) that can be set in motion by the electromagnetic field of the wave. The plasma frequency is much lower than the frequency of a light beam, which propagates without interacting; the situation is different for radio waves, for which the index of refraction of this part of the atmosphere is quite high. Thanks to total reflection, links are possible between removed points for which a direct communication is not possible.

generated. Light rays may be deviated and possibly totally reflected, see Figures 3.C.5 and 3.C.6.

We refer to Figure 3.C.8; x, y, z are the respective unit vectors of the referential Ox, Oy, Oz. We consider a point M belonging to a light ray ($OM = r = xx + yy + zz$); s is the curvilinear abscissa of M. u is the unit vector of the tangent to the light ray at point M where the local index of refraction is equal to n. We start from equation (3.C.5) and calculate the derivative of nu, with respect to s.

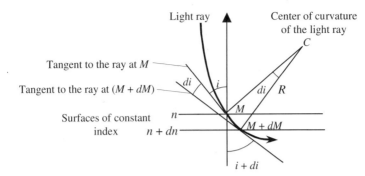

Figure 3.C.8. Propagation of a light ray in an inhomogeneous material. We have drawn the two planes that are tangent to the surfaces of constant index (respectively, n and $n + dn$). The light ray is bent and intersects the index surfaces at points M and M'. C and $R = CM$ are, respectively, the center and the radius of curvature of the ray at point M.

According to (3.C.5), we have $n\boldsymbol{u} = \text{grad } S$ (S is the optical path),

$$\frac{d(n\boldsymbol{u})}{ds} = \frac{d}{ds}(\text{grad } S) = \frac{d}{ds}\left(\frac{\partial S}{\partial x}\boldsymbol{x} + \frac{\partial S}{\partial y}\boldsymbol{y} + \frac{\partial S}{\partial z}\boldsymbol{z}\right),$$

$$d\left(\frac{\partial S}{\partial x}\right) = \text{grad}\left(\frac{\partial S}{\partial x}\right)\boldsymbol{dr} \quad \rightarrow \quad \frac{d}{ds}\left(\frac{\partial S}{\partial x}\right) = \text{grad}\left(\frac{\partial S}{\partial x}\right)\frac{\boldsymbol{dr}}{ds},$$

$$\frac{\boldsymbol{dr}}{ds} = \boldsymbol{u} \quad \rightarrow \quad \frac{d}{ds}\left(\frac{\partial S}{\partial x}\right) = \text{grad}\left(\frac{\partial S}{\partial x}\right)\boldsymbol{u} = \text{grad}\left(\frac{\partial S}{\partial x}\right)\left[\frac{1}{n}\text{grad } S\right].$$

After development of the scalar product we obtain

$$\frac{d}{ds}\left(\frac{\partial S}{\partial x}\right) = \frac{1}{2n}\frac{\partial}{\partial x}(n^2) = \frac{\partial n}{\partial x} \quad \text{and similar expressions for } y \text{ and } z.$$

Finally, the differential equation of a light ray is

$$\frac{d}{ds}(n\boldsymbol{u}) = \left(\frac{\partial n}{\partial x}\boldsymbol{x} + \frac{\partial n}{\partial y}\boldsymbol{y} + \frac{\partial n}{\partial z}\boldsymbol{z}\right) = \text{grad}(n). \tag{3.C.6}$$

In a homogeneous medium the index of refraction is constant and the gradient is equal to zero: $(d/ds)(n\boldsymbol{u}) = n(d\boldsymbol{u}/ds) = 0 \rightarrow \boldsymbol{u} = \text{constant} \rightarrow$ the light ray is a straight line.

The Snell-Descartes Law of Refraction in an Inhomogeneous Medium

In an inhomogeneous medium the surfaces of the constant index of refraction play the same role as the interfaces in homogeneous media. We refer to Figure 3.C.8; let us call $\boldsymbol{u}_1$ and $\boldsymbol{u}_2$ the unit vectors of the tangents to the ray, respectively, at points M and M', and call $\mathbf{N}$ a unit vector orthogonal to the surface of constant index. After integration of equation (3.C.6) from M to M', we obtain

$$(n+dn)\mathbf{u}_2 - n\mathbf{u}_1 = \mathbf{N}\,\mathrm{grad}(n)\,ds,$$

and after projection on the surface of the constant index,

$$(n+dn)\sin(i+di) = n\sin i. \tag{3.C.7}$$

Radius of Curvature of a Light Ray

We consider a curve, a light ray, for example, and some point M of this curve. Let $\mathbf{u}$ be the unit vector of the tangent to the curve at point M. The derivative of the vector $\mathbf{u}$ with regard to the curvilinear abscissa s is a vector that is orthogonal to $\mathbf{u}$ and which defines the direction of what is called the *principal normal* to the curve. The center of curvature C is located on this principal normal at a distance of M equal to the radius of curvature $R = CM$. For any curve, we have the following formula:

$$\frac{d\mathbf{u}}{ds} = \frac{\nu}{R} \quad (\nu \text{ is the unit vector of the principal normal}),$$

from (3.C.6), we obtain

$$\frac{dn}{ds}\mathbf{u} + \frac{n}{R}\nu = \mathrm{grad}(n). \tag{3.C.8}$$

After scalar multiplication by ν, we have

$$\frac{dn}{ds}\frac{n}{R}\nu = \nu\,\mathrm{grad}(n). \tag{3.C.9}$$

As a radius of curvature is always positive, the angle of the vector ν with the gradient is smaller than $90°$, which implies that the concavity of a light ray is on the same side as the gradient. If i is the angle of incidence on the surface of constant index, the radius of curvature is given by

$$\frac{1}{R} = -\frac{1}{n}\frac{dn}{ds}\tan i. \tag{3.C.10}$$

3.C.3. Centered Optical System with a Nonconstant Index of Refraction

We refer to Figure 3.C.9 and consider a medium that is not homogeneous but has a radial symmetry around the Oz axis. The index of refraction is a continuous function of the coordinates. In such material the notion of rays is, of course, still valid; however, the rays are no longer rectilinear. A point M

Figure 3.C.9. Refraction of a light ray on a surface of constant index, which is assimilated to the osculating sphere at point S (center Γ, radius $\rho = \Gamma S = \Gamma M$). M is the point of incidence; C and CM are, respectively, the center and the radius of curvature of the light ray. φM is tangent to the light ray. For the sake of simplicity the ray is contained in a meridian plane.

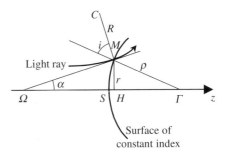

Surface of constant index

belonging to a ray is referenced either by its three coordinates x, y, z, or by its curvilinear abscissa s measured along the ray. As we did in the case of more usual centered systems made of successive lenses, we will restrict ourselves to paraxial light rays that propagate in the vicinity of the Oz axis. For the sake of simplification, we will only consider rays that are in a meridian plane, with the following consequences:

- The tangent at point M intersects the Oz axis, point of intersection Ω.
- The determination of the position of point M needs only two coordinates: the abscissa z, and the distance r to the Oz axis.

The surfaces on which the index keeps a constant value ($n(x, y, z) = $ constant, surfaces of constant index) play a special role; in a sense, they can be considered as analogous to interfaces between two transparent materials, of respective indices n and $n + dn$. As we remain very near to the Oz axis, the surfaces of constant index will be replaced by osculating spheres (center Γ, radius ρ). Because of the symmetry of the problem, the index is only a function of the two variables r and z; the law of variation along the Oz axis is written as $n(z)$, instead of $n(z, r = 0)$. Let us call i the angle of incidence (see Figure 3.C.9), because of symmetry it can be shown that the differential equation of a ray is

$$\frac{d}{dz}\left(n\frac{dr}{dz}\right) + \frac{dn}{dz}\frac{r}{\rho} = 0. \tag{3.C.11.a}$$

Changing slightly the notation by setting $a(z) = n(z)$, $b(z) = dn/dz$, $c(z) = (1/\rho)\, dn/dz$, we obtain

$$a(z)\frac{d^2r}{dz^2} + b(z)\frac{dr}{dz} + c(z) = 0, \tag{3.C.11.b}$$

where $a(z)$, $b(z)$, and $c(z)$ are functions of z, therefore the trajectories of the light rays obey a second-order linear differential equation. The optical properties of the centered optical system are just a consequence of the mathematical properties of the solutions of this kind of equation. We consider two specific functions $u(z)$ and $v(z)$ that are solutions of (3.C.12); any other solution can be written as

$$r(z) = \lambda u(z) + \mu v(z),\tag{3.C.12}$$

where λ and μ are two constants that are characteristic of the special solution (which means the special light ray) under consideration.

Properties of the Light Rays in a Centered Optical System

The light rays that propagate inside a centered optical system make a family with two degrees of freedom, as shown by formula (3.C.12). The number of degrees of freedom is lowered to one if all the rays of a given family should intersect at a given point A: the two constants, λ and μ, should obey an equation characteristic of the point A.

We consider first the case where A is on the Oz axis (abscissa $z = z_1$), we have

$$\lambda u(z_1) + \mu v(z_1) = 0 \quad \rightarrow \quad \frac{\lambda}{\mu} = -\frac{v(z_1)}{u(z_1)} = k_A(z_1) = \text{constant} \quad \forall \lambda \text{ and } \mu,$$

where $k_A(z_1)$ is a constant characteristic of point A. All the rays passing point A have the same kind of equation $r(z) = \mu[k_A(z_1)u(z) + v(z)]$.

Let us now look for the abscissa of the different points at which the ray again intersects the Oz axis, they are solutions of the equation

$$[k_A(z_1)u(z) + v(z)] = 0.\tag{3.C.13.a}$$

This equation is the same for all rays that pass point A. The different solutions of this equation correspond to the different intermediate images, A_1, A_2, ..., and to the final image, A'; which means that the system is stigmatic for points A and A'.

We now examine the case of point C that is out of the Oz axis (coordinates z_1, r_1). Two rays emitted by this point are, respectively, defined by the two couples of constants (λ_a, μ_a) and (λ_b, μ_b), and their equations are

$$\begin{aligned} r_a(z) &= \lambda_a u(z) + \mu_a v(z), \\ r_b(z) &= \lambda_b u(z) + \mu_b v(z), \end{aligned} \quad \text{with} \quad r_a(z_1) = r_b(z_1) = r_1.$$

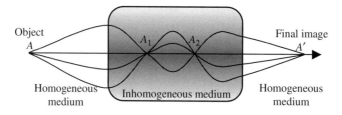

Object
A A_1 A_2 Final image
A'

Homogeneous medium   Inhomogeneous medium   Homogeneous medium

Figure 3.C.10. Trajectories of different light rays emitted by the same object A. A_1, A_2, and A' are images of A (intermediate and final images).

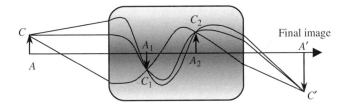

Figure 3.C.11. Trajectories of light rays emitted by point C that is not on the Oz axis.

Let us introduce the auxiliary function

$$\rho_{ab}(z) = r_a(z) - r_b(z) = \lambda_{ab} u(z) + \mu_{ab} v(z),$$

with $\lambda_{ab} = (\lambda_a - \lambda_b)$ and $\mu_{ab} = (\mu_a - \mu_b)$; $\rho_{ab}(z)$ is also a solution of the differential equation (3.C.12). As we have $\rho_{ab}(z_1) = 0$, the two constants λ_{ab} and μ_{ab} are related by

$$\frac{\lambda_{ab}}{\mu_{ab}} = -\frac{v(z_1)}{u(z_1)} = k_A(z_1).$$

The abscissas of the different points where the two rays intersect are again solutions of the equation

$$\rho_{ab}(z) = \mu_{ab}[u(z)k_A(z) + v(z)] = 0. \tag{3.C.13.b}$$

Equations (3.C.13.a) and (3.C.13.b) are identical; which means that the images C_1, C_2, and C' have, respectively, the same abscissa as points A_1, A_2, and A': the optical system is thus stigmatic and also aplanetic. We are now going to show that the linear magnification $\gamma = A'C'/AC$ keeps a constant value, whatever the position of C in a plane orthogonal to the Oz axis. The linear magnification is given by

$$\gamma = \frac{A'C'}{AC} = \frac{u(z_2) + k_A v(z_2)}{u(z_1) + k_A v(z_1)}.$$

As the correspondence between z_2 and z_2 is a one-to-one correspondence, γ is characteristic of the couple of conjugate planes.

As a conclusion, the correspondence *object* ↔ *image* has exactly the same properties for a gradient index centered optical system as for the more usual centered system, made of a succession of spherical interfaces. As a consequence, it will be possible to introduce the notion of cardinal elements (focus, focal planes, principal planes, . . .) and to obtain the Descartes and Newton formulas.

3.C.4. Optics of Charged Particle Beams

The purpose of this section is to show that the trajectories of charged particles, submitted to the action of an electrostatic field, obey an equation that is
identical to equation (3.C.11.b), and that the notion of cardinal elements
should be transposed to the optics of charged particle beams.

3.C.4.1. *Differential Equations of the Trajectories*

The electrostatic field under consideration is created by electrodes with radial
symmetry around the Oz axis. We admit that the particles remain in close
proximity to the Oz axis and that, as a consequence, the kinetic energy in a
direction perpendicular to Oz is negligible in comparison with the kinetic
energy parallel to Oz. For the sake of simplicity, we will only consider trajectories that are in a meridian plane.

As long as the distance r to the Oz axis remains small enough, $E_z(r, z)$
remains constant and equal to $E_z(0, z)$, which will be simplified as $E_z(z)$. The
radial component $E_r(r, z)$ cancels for $r = 0$; a simple expression of $E_r(r, z)$ is
obtained, thanks to the Gauss theorem for electrostatics,

$$E_r(r, z) = -\frac{r}{2}\frac{\partial E_z}{\partial z} = +\frac{r}{2}\frac{\partial^2 V}{\partial z}, \quad V \text{ is the electrostatic potential.} \quad (3.C.14)$$

The possibility of focusing a beam of particles is a direct consequence of
the fact that the radial component of the field is proportional to the distance
r. The situation is analogous to the action of a thin lens on an optical light
ray, the angle of deviation being proportional to the distance to the axis (see
Figure 3.A.1). The equations of the motion of a particle (charge q, mass m)
are

$$\begin{cases} m\dfrac{d^2z}{dt^2} = qE_z = -q\dfrac{dV}{dz}, \\ m\dfrac{d^2r}{dt^2} = qE_r = +q\dfrac{r}{2}\dfrac{d^2V}{dz^2}. \end{cases}$$

The differential equation of the trajectory is obtained after the elimination
of time. We take advantage of the fact that the radial component of the velocity is negligible as compared to the longitudinal component

$$\frac{1}{2}m\left[\left(\frac{dz}{dt}\right)^2 + \left(\frac{dr}{dt}\right)^2\right] \approx \frac{1}{2}m\left(\frac{dz}{dt}\right)^2 = qV \quad \rightarrow \quad \frac{dz}{dt} \approx \sqrt{\frac{2qV}{m}}.$$

Using the identity

$$\frac{d^2r}{dt^2} = \frac{d^2r}{dz^2}\left(\frac{dz}{dt}\right)^2 + \frac{dr}{dz}\frac{d^2z}{dt^2},$$

we have

$$2V\frac{d^2r}{dz^2} - \frac{dV}{dz}\frac{dr}{dz} - \frac{r}{2}\frac{d^2V}{dz^2} = 0. \qquad (3.C.15)$$

Equation (3.C.15) is formally identical to equation (3.C.11.b) (second-order linear equation). It is possible to show that if we replace the index of refraction n by the square root of the potential in equation (3.C.11.b), we obtain equation (3.C.15). We will give a direct demonstration of this proposition.

3.C.4.2. Light Optics and Electron Optics Are Equivalent if $n \leftrightarrow \sqrt{V}$

A grounded source (see Figure 3.C.12) emits charged particles (q, m) in the direction of two pairs of parallel metallic plates. The two electrodes of the first pair are at the same potential V_1, the two other plates being at V_2. Emitted with no initial velocity, the particles are first accelerated from the potential zero up to the potential V_1 and follow a straight line until they leave the first pair of electrodes with a vector velocity v_1. Inside the gap between the two pairs of electrodes, the particles are accelerated and follow a parabolic trajectory; between the second pair of electrodes they again have a linear trajectory with a vector velocity v_2,

$$v_1 = v_{1x}\boldsymbol{x} + v_{1y}\boldsymbol{y}, \quad v_2 = v_{2x}\boldsymbol{x} + v_{2y}\boldsymbol{y}.$$

Inside the gap, the electric field is parallel to $\boldsymbol{x}$, the force is parallel to $\boldsymbol{x}$, and thus the y-component of the velocity remains constant:

$$v_{1y} = v_{2y} \quad \rightarrow \quad v_1 \sin i_1 = v_2 \sin i_2,$$

$$\tfrac{1}{2}mv_1^2 = qV_1 \quad \text{and} \quad \tfrac{1}{2}mv_2^2 = qV_2,$$

$$\sqrt{V_1}\sin i_1 = \sqrt{V_2}\sin i_2, \quad \text{the Snell-Descartes law with } n = \sqrt{V}.$$

It's worthwhile noticing that in light optics, the Snell-Descartes law expresses the conservation of the projection of the wave vector on the interface; while in electron optics it expresses the conservation of the projection of the momentum on the plane of discontinuity of the potential.

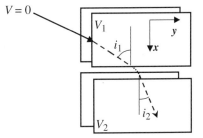

Figure 3.C.12. Refraction of the electron trajectory on an abrupt variation of potential.

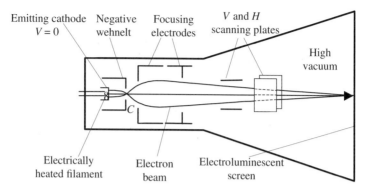

Figure 3.C.13. General arrangement of a cathode ray tube. A heated cathode emits electrons perpendicular to its surface. A first negatively polarized electrode called wehnelt, concentrates the electrons at a point that acts as an object C. The voltage of the wehnelt controls the number of emitted electrons; the more negative, the less electrons. The focusing electrodes image the point C on the screen, which is coated with an electroluminescent material.

3.C.4.3. *The Fermat and Maupertuis Principles*

Starting from the equations of Mechanics it can be shown that once it is launched in a field of forces a particle of mass m, going from point A to point B, follows the path along which the following integral, also called the *action integral*, is minimum:

$$\int_A^B mv(s)\,ds; \quad s \text{ is the curvilinear abscissa, } V(s) \text{ is the velocity.}$$

This principle, which was first introduced by Maupertuis as early as 1764, is also called the *Principle of Least Action*. The striking similarity with the Fermat Principle has played a very important role in the development of Quantum Mechanics by Schroedinger.

We come back to the motion of a charged particle in a field of force that comes from a potential; by a proper choice of the origin of the potential the kinetic energy can be written as

$$\tfrac{1}{2}mv^2 = qV; \quad \text{the action integral is thus proportional to } \int_A^B \sqrt{V}\,ds.$$

Because of the equivalence $n \leftrightarrow \sqrt{V}$, the Fermat Principle implies the Maupertuis Principle, and vice versa.

4

Polarized Light—Laws of Reflection

4.1. Light Vibration Is a Vector

4.1.1. *Elliptical, Rectilinear, and Circular Vibrations*

Light vibration is a vibrating vector. The oscillating aspect of light can be expected after some observation, its subtler vector character remains to be discovered. If the study of Optics is started from Maxwell's equations, the optical vibration is immediately introduced as a vector. The direction of the vector may be along any direction of the wave plane and is determined by the light source. x and y being two orthogonal unit vectors of the wave plane, and Oz being the direction of propagation, the light vibration and its two components can be written as

$$\boldsymbol{E} = \mathrm{Re}\{(\boldsymbol{x}a + \boldsymbol{y}be^{-j\varphi})e^{j(\omega t - kz - \Phi)}\}, \tag{4.1}$$

$$\begin{aligned} E_x &= a\cos(\omega t - kz - \Phi), \\ E_y &= b\cos(\omega t - kz - \Phi). \end{aligned} \tag{4.2}$$

Using a suitable change of the time origin, (4.2) can be written

$$\begin{aligned} E_x &= a\cos(\omega t - \varphi), \\ E_y &= b\cos(\omega t - \varphi). \end{aligned} \tag{4.2a}$$

Figure 4.1 indicates that, as times progresses, the endpoint of vector $\boldsymbol{E}$ describes an ellipse, this is the reason why it is said that the more general polarization is elliptical. According to the phase difference between E_x and E_y the ellipse will take different aspects and will be described clockwise or anticlockwise.

Chapter 4 has been reviewed by Dr. François Méot, Senior Physicist at the CEA (Commissariat à l'Energie Atomique).

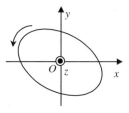   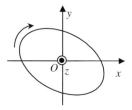

Left-handed elliptical polarization   Right-handed elliptical polarization

Figure 4.1. General right- and left-handed elliptical polarizations. A vibration is said to be right-handed (left-handed) when it rotates toward the right (left) side of an observer who receives the light.

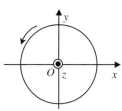   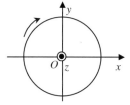

Left-handed circular polarization,   Right-handed circular polarization,
$E_x = a \cos \omega t, \quad E_y = b \cos \omega t.$ $E_x = a \cos \omega t, \quad E_y = -a \cos \omega t.$

Figure 4.2. Left- and right-handed circular polarizations. The two components have equal amplitudes. The phase difference is $\pi/2$. If the two amplitudes are different, the polarization would be elliptical, Ox and Oy being the axes of the ellipse.

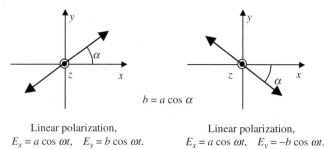

$b = a \cos \alpha$

Linear polarization,   Linear polarization,
$E_x = a \cos \omega t, \quad E_y = b \cos \omega t.$ $E_x = a \cos \omega t, \quad E_y = -b \cos \omega t.$

Figure 4.3. Rectilinear polarizations. The two components have the same phases, or opposite phases, $\varphi = (2p + 1)\pi$. The angle α with the Ox axis is given by $\tan \alpha = b/a$.

Any Polarization Is the Superposition of a
Right- and Left-Handed Polarization

As shown by formula (4.2), any polarization can be considered as the superposition of two orthogonal rectilinear polarizations. In the same way, any polarization can be considered as the addition of two opposite circular polarizations. We will only establish that the important result, according to which a rectilinear polarization may be obtained, is by superposing two opposite circular polarizations of equal amplitudes; the orientation is determined by the phase difference between the two circular polarizations.

Let us consider two opposite circular polarizations and their superposition:

$$\text{left} \begin{cases} E_x^l = a \cos \omega t, \\ E_y^l = a \sin \omega t, \end{cases} \quad \text{right} \begin{cases} E_x^r = a \cos(\omega t - \varphi), \\ E_y^l = -a \cos(\omega t - \varphi), \end{cases}$$

$$\begin{cases} E_x^r + E_x^l = 2a \cos(\omega t - \varphi/2) \cos \varphi/2, \\ E_y^r + E_y^l = 2a \cos(\omega t - \varphi/2) \sin \varphi/2. \end{cases} \tag{4.3}$$

Equations (4.3) represent a rectilinear polarization making the angle $\varphi/2$ with the Ox axis.

Intensity of an Elliptically Polarized Wave

According to formula (4.2) an elliptically polarized wave is the superposition of two orthogonal and linearly polarized waves. We now want to evaluate its intensity, that's to say, the amount of electromagnetic energy flowing, each second, across a unit surface disposed perpendicularly to the wave vector. The same kind of calculation has already been made for a linearly polarized wave; it was found that the intensity of a wave having an amplitude E_0 was easily given using the *wave impedance* Z:

$$I = \frac{E_0^2}{2Z}. \tag{4.4}$$

Using Pythagoras' theorem, we obtain

$$E_x^2 + E_y^2 = a^2 \cos^2 \omega t + b^2 \cos^2(\omega t - \varphi) = \tfrac{1}{2}[a^2 + b^2 + a^2 \cos 2\omega t + b^2 \cos 2(\omega t - \varphi)].$$

Since the response time of any photodetector is very long, as compared to the light frequency, we have to consider the time-averaged value of the electric field. Under such conditions, the terms $\cos 2\omega t$ and $\cos 2(\omega t - \varphi)$ disappear and we finally obtain $(a^2 + b^2)/2Z$. Apart from its immediate interest, formula (4.4) contains an important physical result: two waves with orthogonal polarizations are unable to interfere, since their phase difference disappears during the time-averaging process.

4.1.2. Unpolarized Light

Usual light sources, such as the Sun, a candle, or an electrically heated wire, emit light waves that are said to be unpolarized. We have already considered such sources and shown that their light was not coherent, see Section 1.6.3. The notion of incoherency will now be extended to polarization. Let us go back to formula (4.2.a):

$$E_x = a \cos \omega t,$$
$$E_y = b \cos(\omega t - \varphi).$$

In the case of unpolarized light, the end of the electric vector moves errat-ically and the light shows any preferential direction inside a plane perpen-dicular to the direction of propagation. A simple description may be obtained by considering that the quantities a, b, and φ are random functions of time exhibiting the following properties:

(i) They vary very *slowly*, at the time scale of the light period. So it can be considered that they keep a constant value during periods lasting for many hours.
(ii) They vary very *quickly*, if the time scale is now the response of a light detector.

An exhaustive description of the statistical properties of unpolarized light waves is a difficult subject; we will limit ourselves to a very simple model and only consider light beams having the following properties:

- Monochromatic and coherent (ω = constant).
- Rectilinear polarization ($\varphi = 0$).
- Constant amplitude ($a^2 + b^2$ = constant).
- Random variation of the polarization direction (a and b are random time functions following the above criteria (i) and (ii)).

4.2. Analyzers—Polarizers

(These two words have almost the same meaning.)

4.2.1. Linear Dichroism, Malus' Law

Linear dichroic media are used to make analyzers, they have the following properties:

- They are almost perfectly transparent for a linearly polarized beam having a polarization parallel to a specific direction Δ, called the *direction of the analyzer*.
- They have an important absorption coefficient for light beams that are not linearly polarized along Δ. The experiment shows that the absorption is maximum when the polarization is orthogonal to Δ; if the beam is com-pletely absorbed, the analyzer is then said to be *perfect*.

Receiving an unpolarized light beam, a dichroic sample will transmit a linearly polarized beam, the transmitted polarization being parallel to its direction Δ; the corresponding device is then called *a polarizer*. If the incident beam is already polarized, the transmitted beam will be the projection of the initial polarization on the direction Δ, the device is then said to be an *analyzer*.

Since 1928 when Edwin H. Land, a young Harvard College student, invented the first dichroic sheet polarizer, most polarizers were of the *polaroid* kind. They were made of plastic plates with two parallel faces; the direction Δ was parallel to the faces. For a thickness between 0.5 mm and 1 mm, it can be considered that a beam polarized orthogonal to Δ is practically not transmitted. The plastic material of a polaroid is obtained by polymerization of polyvinyl alcohol impregnated with iodine; dichroism is a consequence of an excellent alignment of long organic molecules, this alignment is obtained by applying strength to the material during polymerization, Δ is parallel to the direction of the strength.

We now consider a planar wave, linearly polarized and propagating along a direction Oz orthogonal to the faces of some polarizer. The angle of the polarization with the direction Δ of the polarizer is equal to θ. We are going to determine the characteristics of the transmitted beam. We call $\boldsymbol{\Delta}$ the unit vector of the direction of the polarizer, $\boldsymbol{x}$ and $\boldsymbol{y}$ are orthogonal unit vectors used to represent the polarizations of the incident and transmitted beams, see Figure 4.4.

Direction of the polarizer: $\Delta = \boldsymbol{x}\cos\beta + \boldsymbol{y}\sin\beta$.
Incident vibration: $\boldsymbol{E}_{\text{incident}} = A_{\text{incident}}(\boldsymbol{x}\cos\alpha + \boldsymbol{y}\sin\alpha)\cos\omega t$.
Vibration after the polarizer:

$$\boldsymbol{E}_{\text{transmit}} = (\boldsymbol{E}_{\text{incident}}\boldsymbol{\Delta})\boldsymbol{\Delta} = A_{\text{incident}}[(\boldsymbol{x}\cos\alpha + \boldsymbol{y}\sin\alpha)(\boldsymbol{x}\cos\beta + \boldsymbol{y}\sin\beta)]\Delta\cos\omega t.$$

(4.5)

$$\boldsymbol{E}_{\text{transmit}} = A_{\text{incident}}\Delta\cos(\alpha - \beta)\cos\omega t = A_{\text{incident}}\Delta\cos\theta\cos\omega t.$$

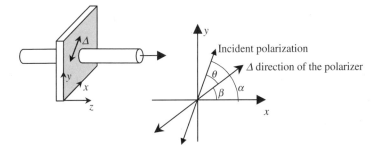

Figure 4.4. An analyzer transmits a polarization that is the projection of the incident polarization on its direction Δ. $A_{\text{transmit}} = A_{\text{incident}}\cos\theta \rightarrow I_{\text{transmit}} = I_{\text{incident}}\cos^2\theta$.

Malus' Law

Incident and transmitted amplitudes are related by

$$A_{\text{transmit}} = A_{\text{incident}} \cos\theta. \tag{4.6}$$

Incident and transmitted intensities are related by

$$I_{\text{transmit}} = I_{\text{incident}} \cos^2\theta \quad \text{ideal polarizer.} \tag{4.7}$$

For real polarizers the extinction of polarizations orthogonal to Δ is very efficient; on the other hand, polarizers are never perfectly transparent for polarizations parallel to Δ, a transmission coefficient τ should be introduced, τ^2 is usually between 50% and 90%:

Amplitude: $A_{\text{transmit}} = \tau A_{\text{incident}} \cos\theta,$

Intensity: $I_{\text{transmit}} = \tau^2 I_{\text{incident}} \cos^2\theta \quad \text{real polarizer.}$ $\tag{4.8}$

Formulas (4.7) and (4.8) are known as *Malus' law*.

4.2.2. Transmission of a Beam with any Polarization

4.2.2.1. Unpolarized Light Beams

An unpolarized light beam is considered to have constant amplitude and a polarization varying at random from time to time. After a polarizer, the transmitted polarization is of course rectilinear and parallel to the direction of the polarizer. To obtain the transmitted intensity we use formula (4.5) and take the time-averaged value of the following expression $(A_{\text{incident}} \cos\theta \cos\omega t)^2$.

Given the difference of the rhythms of variation of θ on one hand and of ωt on the other, we have the following expressions, where angle brackets indicate time-averaging:

$$I_{\text{transmit}} = A_{\text{incident}}^2 \langle\cos^2\theta\rangle\langle\cos^2\omega t\rangle = \tfrac{1}{2}\cdot\tfrac{1}{2}\cdot A_{\text{incident}}^2 = \tfrac{1}{2} I_{\text{incident}}. \tag{4.9}$$

The transmitted light intensity is only one-half of the incident intensity, the polarizer absorbs half of the incident power; in the case of a dichroic polarizer, half of the power is transformed into heat.

4.2.2.2. Crossed Polarizers, Parallel Polarizers

Two polarizers are said to be crossed when their respective directions, Δ and Δ', are mutually orthogonal. They are said to be parallel when Δ and Δ' are parallel.

Using real polarizers, the intensity between two crossed polarizers is not strictly equal to zero. Let us consider the set-up of Figure 4.5, and let J and J' be the respective intensities before and after the second polarizer. The quality

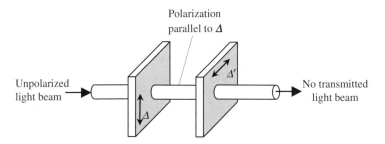

Figure 4.5. Crossed polarizers. The incident is polarized at random and has an intensity I_0. After the first polarizer the light is linearly polarized and has an intensity $J = I_0/2$. The second polarizer doesn't transmit any light.

of the polarizer is defined by J'/J, expressed in decibels, $10\,\mathrm{Log}_{10}(J'/J)$, it varies from $-20\,\mathrm{db}$ for usual polarizers and up to $-40\,\mathrm{db}$ for excellent polarizers.

4.2.2.3. *Evaluation of the Transmission of an Elliptic Polarization*

Let us consider an elliptic polarization described by

$$\boldsymbol{E}_{\text{incident}} = \boldsymbol{x}E_{x,\text{incident}} + \boldsymbol{y}E_{y,\text{incident}} = \boldsymbol{x}a\cos\omega t + \boldsymbol{y}b\cos(\omega t - \varphi).$$

We want to determine the polarization transmitted by a polarizer, the direction Δ of which makes an angle β with Ox,

$$\boldsymbol{E}_{\text{transmit}} = (\boldsymbol{E}_{\text{incident}}\boldsymbol{\Delta})\boldsymbol{\Delta} = [a\cos\beta\cos\omega t + b\sin\beta\cos(\omega t - \varphi)]\boldsymbol{\Delta},$$

$$\boldsymbol{E}_{\text{transmit}} = A_{\text{transmit}}\cos(\omega t - \psi)\boldsymbol{\Delta}.$$

We find a rectilinear vibration with amplitude, A_{transmit}, and a phase that are obtained after a rather tedious trigonometric calculation. It's probably more comfortable to use imaginary notations and to write

$$[a\cos\beta\cos\omega t + b\sin\beta\cos(\omega t - \varphi)] = \mathrm{Re}\{e^{j\omega t}[a\cos\beta + b\sin\beta\,e^{-j\varphi}]\},$$

$$\boldsymbol{E}_{\text{transmit}} = \mathrm{Re}\{A_{\text{transmit}}e^{j\omega t}e^{-j\psi}\}\boldsymbol{\Delta}.$$

We finally obtain $|A_{\text{transmit}}|^2 = (a\cos\beta + b\sin\beta\cos\varphi)^2 + b^2\sin^2\beta\sin^2\varphi$.

4.3. Reflection—Refraction

4.3.1. *General Considerations on Reflection and Refraction*

Refraction and reflection occur when an electromagnetic wave impinges on a surface separating two different transparent media, with two different indices of refraction.

The variation law of the refractive index versus position shows a discontinuity when crossing the separation surface. In fact, a perfect discontinuity is not needed, it's enough that the index variation is produced over a distance that is small, as compared to the wavelength.

The separation between two media having two different indices of refraction will be called an interface. Two different kinds of reflection should be distinguished according to the physical type of the materials located on both sides of the discontinuity:

- *Vitreous reflection*: The two media are made of dielectric materials. A good example is the interface air/glass or vacuum/glass.
- *Metallic reflection*: One medium is a dielectric, the other is a metal.

The two kinds of reflections can be studied using the same formalism, using a complex index of refraction $n = n' - jn''$:

- *Ideal dielectric*: $n'' = 0$, the index is a real number.
- *Practical dielectric*: The index is *almost* real, its imaginary part being small in comparison with the real part.
- *Real metal*: $n' = 0$, the index is purely imaginary.
- *Practical metal*: The index is *almost* imaginary, its real part being small in comparison with the imaginary part.

Reflection and refraction laws have been known for a long time; they were formulated as early as the seventeenth century, mainly from experimental observations. They give exact indications about the *directions* of the reflected and refracted beams, but they don't say anything about their relative intensities, nor do they say the way the incident power is shared between the two beams.

The First Snell-Descartes Law

The first Snell-Descartes law indicates that, in the case of isotropic media, the incident beam, the reflected beam, and the refracted beam are all in the same plane, called a *plane of incidence*.

Symmetry considerations are of preeminent importance in such problems. Let us consider the family of planes that is orthogonal to the interface. When the interface separates two isotropic materials, all the planes are strictly equivalent. However, the presence of an incident light ray causes the plane containing the incident ray (plane of incidence) to play a special role, for the sake of symmetry the reflected and refracted rays should belong to this plane.

The Second Snell-Descartes Law

The second Snell-Descartes law gives qualitative information about the respective directions of propagation of the different beams. The angles of the

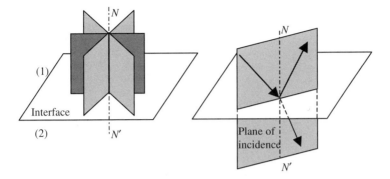

Figure 4.6(a). The first Snell-Descartes law. The normal to the interface, the incident, reflected, and refracted rays belong to the same plane, called the plane of incidence.

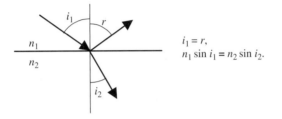

Figure 4.6(b). The second Snell-Descartes law. The figure has been drawn in the plane of incidence.

incident, reflected, and refracted (also called transmitted) beams, with the normal to the interface are, respectively, the angles of incidence (i_1), of reflection (r), and of refraction (i_2). The second law will be established later on, it is divided into two parts:

- The law of reflection, which says that the angles of incidence and refraction are equal: $i_1 = r$.
- The law of refraction, also designed as the *sine law*, which is a relation between the sine of the angles of incidence and of refraction:

$$n_1 \sin i_1 = n_2 \sin i_2.$$

Normal incidence: If the angle of incidence is equal to zero, the reflection and refraction angles are also equal to zero, the three beams are collinear, and the incidence is then said to be *normal*.

Physical Interpretation of the Generation of the Reflected and Refracted Beams

The electrons belonging to the atoms located along the interface are responsible for the generation of the reflected and transmitted waves. Set into

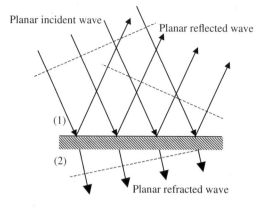

Figure 4.7. The electrons located near the interface are set in vibration by the electric field of the incident wave. Thanks to the electric dipolar radiation, wavelets are emitted on both sides, their interference corresponds to the reflected and refracted waves.

motion by the electric field of the incident wave, these electrons play the role of the children sitting along the swimming pool of Figure 1.8. Electromagnetic wavelets are generated; their superposition corresponds to the reflected and refracted beams.

When arriving at the interface the incident wave sets in vibration the electrons located in the hatched area of Figure 4.7. As the incident wave planes are not parallel to the interface, the vibrations are not in phase but have phase repartition that linearly varies with the position along the interface. The phase delay increases as the point is moving from left to right. The phase velocities of the wavelets are, respectively, equal to $v_1 = c/n_1$ and $v_2 = c/n_2$ in the upper and lower media (respective index of refraction, n_1 and n_2); under such conditions it can be shown that the wavelets interference gives planar waves propagating along directions obeying the second Snell-Descartes law.

4.3.2. Polarized Light Reflection

4.3.2.1. TE and TM Waves

To establish the first Snell-Descartes law we have taken advantage of the symmetry of the problem. In spite of the fact that it is part of the symmetry, we did not have to introduce the notion of light polarization, we shall do it now and show that the orientation of the electromagnetic vibration determines the repartition of the incident energy between the reflected and refracted beams.

Formulas (2.13) show that as soon as one of the two vectors E or H is known, then the other is also known. E or H can also represent what we call "the light vibration." In Optics, where the interaction between light and

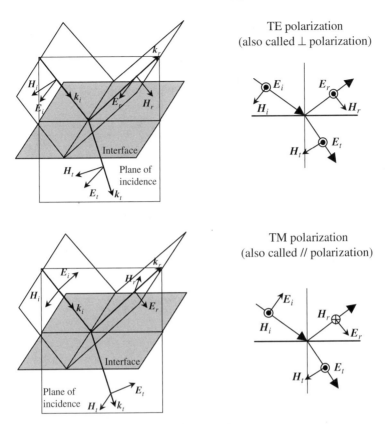

Figure 4.8. Relative positions of the different vectors for TE and TM waves (i = incident, r = reflected, t = transmitted).

material is almost exclusively due to electric dipolar mechanisms, the tradition is to assimilate the light vibration to the electric field.

The three wave vectors (incident, reflected, and refracted) are in the plane of incidence. In the case of planar waves E and H should be orthogonal to their respective wave vectors, which implies that they belong to a plane Π orthogonal to the associated wave vector (Π is a wave plane). Inside Π two directions play a special role:

- The direction orthogonal to the plane of incidence.
- The direction contained in the plane of incidence, which is also the intersection of Π with the plane of incidence.

The notations vary considerably from author to author; the most commonly used are shown overleaf:

// Parallel	TM transverse magnetic	P	E is parallel to the plane of incidence. H is perpendicular to the plane of incidence.
⊥ Perpendicular	TE transverse electric	S	E is perpendicular to the plane of incidence. H is parallel to the plane of incidence.

4.3.2.2. *Conservation of TE and TM Polarizations by Reflection or by Refraction*

It will be shown in Annex 4.A how TE and TM modes can be theoretically introduced, starting from Maxwell's equations. The most important consequence is the following: when a TE polarized wave meets some interface, the reflected and refracted waves are also TE polarized; in the same way, a TM wave only generates TM waves.

4.3.3. *Reflection and Refraction Coefficients*

4.3.3.1. *Brewster Phenomenon*

The amplitudes of the reflected and refracted waves vary with the angle of incidence, special experimental conditions may be found for which the amplitude of the reflected beam goes down to zero. David Brewster, in 1815, when observing the reflection on a windowpane, of the blue light of the sky which is partially polarized, first noted this extinction of a reflected beam. Measuring the variations of the reflected intensity versus the angle of incidence, the following results are obtained:

- *For TE polarization*: The reflected intensity permanently increases when the angle of incidence increases from $0°$ to $90°$ and is minimum at normal incidence.
- *For TM polarization*: The variation is not monotonic, a special angle of incidence exists (the *Brewster angle*), for which the intensity goes to a minimum equal to zero. The refracted beam is orthogonal to the direction that would take the reflected beam if its intensity were not equal to zero.

The fact that experimental conditions exist where the reflected intensity is equal to zero is proof of the fact that the electromagnetic vibrations are vectors orthogonal to the direction of propagation. The reflected intensity could not cancel if the vibrations were scalar numbers, or if they were vectors having a component along the direction of propagation.

The formula giving the Brewster angle will be established using Fresnel's expression for the reflection coefficient for TM waves, here it will be obtained from more physical considerations, considering how the reflected beam is generated by the motion of the electrons located along the interface. We will

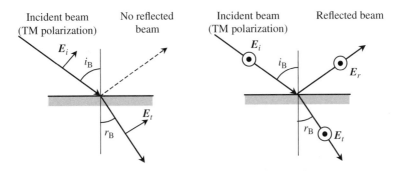

Figure 4.9. Brewster angle: The electric field of the transmitted beam sets in vibration the electrons of the interface; if the directions of the refracted and reflected beams are orthogonal and if the polarization is TM, then the motion of the electrons cannot generate any reflected wave, since its electric field is parallel to the direction of propagation.

limit ourselves to the case of a vacuum/dielectric interface ($n_1 = 1$), but the result is quite general. The oscillating electrons belong to the dielectric and are set in vibration by the electric field of the refracted wave; the direction of propagation of this latter wave is orthogonal to its electric field. We now consider the generation of the reflected wave; it's only the projection of the electron vibrations on the direction of the reflected wave planes that contributes to this generation. If the polarization is TM and if the reflected and refracted beams are *orthogonal*, then the projection of the dipole vibrations is equal to zero: no wave can be generated in the direction of the reflected beam. The only solution is that the reflected beam doesn't exist, all the incident energy being taken away by the transmitted beam.

The Brewster angle i_B is the angle of incidence for which the reflected and refracted beams are orthogonal, if r_B is the corresponding angle of refraction we have

$$i_B + r_B = \frac{\pi}{2} \quad \rightarrow \quad n_1 \sin i_B = n_2 \cos n r_B \quad \rightarrow \quad \tan i_B - \frac{n_2}{n_1}. \quad (4.10)$$

The preceding considerations do not hold for the TE case. A TE polarized incident beam always generates a reflected beam, whatever the angle of incidence.

4.3.3.2. *Fresnel Formulas*

Fresnel formulas give the reflected and refracted amplitudes, versus the incident amplitude. Their expressions are different for TE or TM polarizations.

The reference axes are $Oxyz$, the plane of incidence is (Ox, Oz) and (Oy, Oz) is the plane of the interface. We use the following notations, where the indices i, r, and t, respectively, stand for incident, reflected, and transmitted (refracted) waves:

Incident wave: $\boldsymbol{E}_i = \boldsymbol{E}_{i,0}e^{-jk_ir}$, $\boldsymbol{H}_i = \boldsymbol{H}_{i,0}e^{-jk_ir}$, $\boldsymbol{k}_i = -k_{ix}\boldsymbol{x} + k_{iy}\boldsymbol{y} + k_{iz}\boldsymbol{z}$,

Reflected wave: $\boldsymbol{E}_r = \boldsymbol{E}_{r,0}e^{-jk_rr}$, $\boldsymbol{H}_r = \boldsymbol{H}_{r,0}e^{-jk_rr}$, $\boldsymbol{k}_r = -k_{rx}\boldsymbol{x} + k_{ry}\boldsymbol{y} + k_{rz}\boldsymbol{z}$,

Refracted wave: $\boldsymbol{E}_t = \boldsymbol{E}_{t,0}e^{-jk_tr}$, $\boldsymbol{H}_t = \boldsymbol{H}_{t,0}e^{-jk_tr}$, $\boldsymbol{k}_t = -k_{tx}\boldsymbol{x} + k_{ty}\boldsymbol{y} + k_{tz}\boldsymbol{z}$.

We have to ask the following question: Is it possible to choose arbitrarily the nine vectors above? Of course the answer is no, they must satisfy equations (2.13) which are repeated below:

$$E = \frac{-1}{\omega\varepsilon}\boldsymbol{k} \wedge \boldsymbol{H} \quad \text{and} \quad H = \frac{1}{\omega\mu}\boldsymbol{k} \wedge \boldsymbol{E}, \tag{4.11}$$

where $(\boldsymbol{E}_i, \boldsymbol{H}_i, \boldsymbol{k}_i)$, $(\boldsymbol{E}_r, \boldsymbol{H}_r, \boldsymbol{k}_r)$, and $(\boldsymbol{E}_t, \boldsymbol{H}_t, \boldsymbol{k}_t)$ are orthogonal and positive, the field moduli are related to the impedances of their respective medium $(Z_0 = \omega/c)$:

$$\|\boldsymbol{E}_{i,0}\| = \frac{Z_0\|\boldsymbol{H}_{i,0}\|}{n_1}, \quad \|\boldsymbol{E}_{r,0}\| = \frac{Z_0\|\boldsymbol{H}_{r,0}\|}{n_1}, \quad \|\boldsymbol{E}_{t,0}\| = \frac{Z_0\|\boldsymbol{H}_{t,0}\|}{n_2}. \tag{4.12}$$

Boundary Conditions Along an Interface

We recall the boundary conditions along an interface separating two dielectric materials which don't contain any electric charges or any electric current:

- Tangential components (parallel to the interface) of both electric and magnetic fields should be continuous across the dielectric interface, at all points along the boundary.
- There are discontinuities for the normal components.

  In the incident medium: (1) the fields result from the superposition of the incident and reflected fields, in the transmission medium; and (2) we only have the transmitted field. The Snell-Descartes law as well as the Fresnel formulas are simply obtained by writing the equality of the addition of the incident and reflected fields on one hand and of the transmitted field on the other.

Demonstration of the Snell-Descartes Laws

Along the interface x is equal to zero, the continuity of tangential components gives

$$(E_{i0})_{yOz}e^{-j(k_{iy}y+k_{iz}z)} + (E_{r0})_{yOz}e^{-j(k_{ry}y+k_{rz}z)} = (E_{t0})_{yOz}e^{-j(k_{ty}y+k_{tz}z)}. \tag{4.13}$$

As equation (4.13) must be fulfilled for all values of y and z, the arguments of the exponential functions should be equal:

$$k_{iy} + k_{iz} = k_{ry} + k_{rz} = k_{ty} + k_{tz} \quad \text{or} \quad (\boldsymbol{k}_i)_{yOz} = (\boldsymbol{k}_r)_{yOz} = (\boldsymbol{k}_t)_{yOz}. \tag{4.14}$$

The equality of the projection of the three wave vectors on the plane of the interface directly implies the Snell-Descartes laws:

- *First law*: The three vectors lie in the same plane (plane of incidence).
- The *second law* is also called the *phase matching condition*: The tangential components of the wave vectors are conserved by reflection and refraction, this simply means that the three waves have equal phase velocities along the interface.

To obtain the famous sine law, we just have to introduce the modulus of the wave vectors ($k_i = k_r = n_1\omega/c$, $k_t = n_2\omega/c$) and the angles of incidence, reflection, and refraction $\rightarrow i_1 = r$ and $n_1 \sin i_1 = n_2 \sin i_2$.

Fresnel Formulas for Reflection and Refraction

To establish the Snell-Descartes laws we have only taken advantage of the invariance of planar waves when a translation is operated, this is the reason why the same result was obtained for TE or TM polarizations. We are now going to express qualitatively the conditions of continuity for the electric and magnetic fields and obtain relations between the incident, reflected, and refracted amplitudes. We will admit the existence of a transmitted beam, the case of total internal reflection will be considered later.

The calculations will be fully developed only for the TE case. A reflection coefficient ρ_{TE} and a transmission (or refraction) coefficient τ_{TE} are introduced using the following expressions in which $k_{iz} = k_{ix} = k_z$:

$$E_i = A\boldsymbol{y}e^{-j(-k_{ix}x+k_zz)}, \qquad H_i = \frac{n_1}{Z_0}A(-\boldsymbol{x}\sin i_1 - \boldsymbol{z}\cos i_1)e^{-j(-k_{ix}x+k_zz)},$$

$$E_r = \rho_{\text{TE}}A\boldsymbol{y}e^{-j(k_{rx}x+k_zz)}, \qquad H_r = \rho_{\text{TE}}\frac{n_1}{Z_0}A(-\boldsymbol{x}\sin i_1 + \boldsymbol{z}\cos i_1)e^{-j(k_{rx}x+k_zz)},$$

$$E_t = \tau_{\text{TE}}A\boldsymbol{y}e^{-j(-k_{tx}x+k_zz)}, \qquad H_t = \tau_{\text{TE}}\frac{n_1}{Z_0}A(-\boldsymbol{x}\sin i_2 - \boldsymbol{z}\cos i_2)e^{-j(-k_{tx}x+k_zz)}.$$

TE polarization   TM polarization

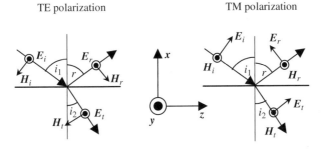

Figure 4.10. Definition of the notations used to establish Fresnel formulas.

For tangential components, the boundary conditions at $x = 0$ may be written as

$$E_{iy} + E_{ry} = E_{ty} \quad \text{and} \quad H_{iz} + H_{rz} = H_{tz},$$

and

$$(1 + \rho_{\text{TE}}) = \tau_{\text{TE}} \quad \text{and} \quad n_1(\rho_{\text{TE}} - 1)\cos i_1 = -n_2(\rho_{\text{TE}} + 1)\cos i_2.$$

After some calculations we finally obtain:

The TE-Fresnel Formula

$$\rho_{\text{TE}} = \frac{n_1 \cos i_1 - n_2 \cos i_2}{n_1 \cos i_1 + n_2 \cos i_2} = -\frac{\sin(i_1 - i_2)}{\sin(i_1 + i_2)},$$

$$\tau_{\text{TE}} = \frac{2n_1 \cos i_1}{n_1 \cos i_1 + n_2 \cos i_2} = \frac{2\sin i_2 \cos i_1}{\sin(i_1 + i_2)} = (1 + \rho_{\text{TE}}). \tag{4.15}$$

The TM-Fresnel Formula

$$\rho_{\text{TM}} = \frac{n_1 \cos i_2 - n_2 \cos i_1}{n_1 \cos i_2 + n_2 \cos i_1} = -\frac{\tan(i_1 - i_2)}{\tan(i_1 + i_2)},$$

$$\tau_{\text{TM}} = \frac{2n_2 \cos i_1}{n_1 \cos i_2 + n_2 \cos i_1} = \frac{2\sin i_2 \cos i_1}{\sin(i_1 + i_2)\cos(i_1 - i_2)} = \frac{n_1}{n_2}(1 + \rho_{\text{TM}}). \tag{4.16}$$

Variations of the Reflection and Transmission Coefficients

Phase shifts at reflection: As long as we are not in the condition of total internal reflection (see Section 4.3.4), the reflection and refraction coefficients are real numbers, either positive or negative; their signs are related to the relative phases of the reflected and refracted waves, with regard to the phase of the incident wave. A negative coefficient corresponds to a phase shift of π. If the coefficient is positive the corresponding wave is in phase with the incident one.

Normal incidence: When the angles i_1 and i_2 are small enough, formulas (4.15) and (4.16) show that, for both polarizations, the reflection coefficients are equivalent to $-(i_1 - i_2)/(i_1 + i_2)$ and takes the same value at normal incidence:

$$\rho_{\text{TE}(i_1=0)} = \rho_{\text{TM}(i_1=0)} = -\frac{n_2 - n_1}{n_2 + n_1}. \tag{4.17}$$

The case of normal incidence is degenerated; the difference between TE and TM loses its significance.

A material is said to be all the more refringent as its refractive index has a higher value.

Reflection at normal incidence on a more refringent medium,

$n_2 > n_1 \rightarrow \rho < 0,$ incident and reflected waves have opposite phases.

Reflection at normal incidence on a less refringent medium,

$n_2 < n_1 \rightarrow \rho > 0,$ incident and reflected waves are in phase.

Example: $n_1 = 1.5$, $n_1 = 1 \rightarrow \rho_{TE(i=0)} = \rho_{TM(i=0)} = -0.2$. The ρ coefficients correspond to the amplitudes of the oscillations; to obtain the reflection coefficients for intensities they have to be squared $\rightarrow R_{(i=0)} = 0.04 = 4\%$. In the case of a glass plate with two parallel interfaces, the global transmission losses due to reflection are equal to 8%.

When the incident medium is more refringent, ρ_{TE} is always negative whatever the angle of incidence: TE reflected and incident waves have opposite phases. For TM polarization the phase shift changes from π to 0, when the angle of incidence becomes greater than the Brewster angle, this sudden change of phase is not at all dramatic, since it occurs when the reflected amplitude is equal to zero.

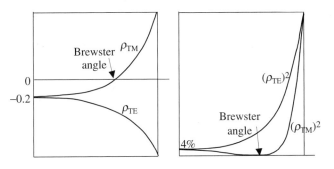

$$n_1 = 1, \quad n_2 = 1.5, \quad i_B = \tan^{-1}(n_2/n_1) = 56°$$

Figure 4.11. Variation of the various reflection coefficients versus angle. The reflection occurs on an air/glass interface, the second medium is more refringent. At normal incidence ρ_{TE} and ρ_{TM} are negative corresponding to a phase shift of π. The TM coefficient cancels and changes sign at the Brewster incidence.

Variation of the Polarization by Reflection or Refraction

Let us consider first a linearly polarized incident beam, its electric field can be projected in TE and TM directions: the two components have equal (or opposite) phases. The reflection and transmission coefficients being real, the TE and TM components of the two reflected and refracted beams have equal (or opposite) phases, which means that they are also linearly polarized. An elliptic incident polarization will of course generate elliptically polarized reflected or refracted beams; the axis ratios of the ellipses will however be different since $\rho_{TE} \neq \rho_{TM}$ as well as $\tau_{TE} \neq \tau_{TM}$.

Mathematical interpretation of Brewster's angle: If we consider formula (4.16), it is seen that if $i_1 + i_2 = \pi/2$, then $\tan(i_1 + i_2)$ is infinite and ρ_{TM} is equal to zero.

We consider the experiment of Figure 4.12, after the two interfaces of the first plate the ratio between the TE/TM amplitudes is equal to the squared ratio of the reflection coefficients; the ratio of the intensities is the fourth power. If N plates are used the intensity ratio is given by the $4N$th power. It's easy to establish the following formulas:

$$\left(\frac{\tau_{TE}}{\tau_{TM}}\right)_{1 \text{ interface}} = \cos(i_1 - i_2) = \sin 2i_2 = \frac{2\tan i_1}{1 + \tan^2 i_1} = \frac{2n}{1 + n^2} \quad \text{with } n = n_2/n_1,$$

intensity ratio for one plate: $\left(\dfrac{\tau_{TE}}{\tau_{TM}}\right)_{2N \text{ interfaces}} = \left(\dfrac{2n}{1 + n^2}\right)^{2N}$,   (4.18)

intensity ratio for N plates: $\left(\dfrac{I_{TE}}{I_{TM}}\right)_{2N \text{ interfaces}} = \left(\dfrac{2n}{1 + n^2}\right)^{4N}$.

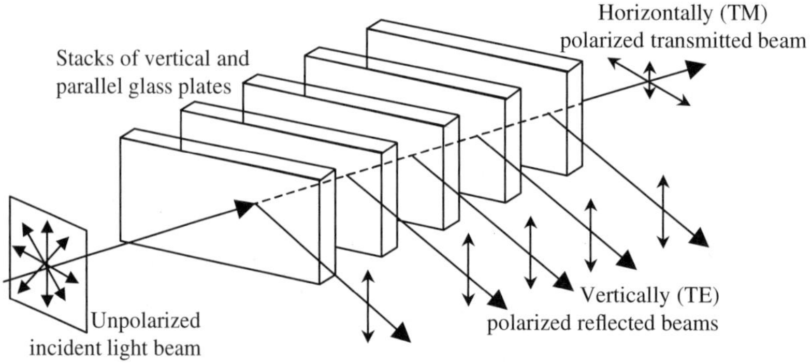

Figure 4.12. A parallel beam of unpolarized light is incident at the Brewster angle on a stack of parallel glass plates. TM polarization is integrally transmitted, while TE polarization is only partly transmitted. After many plates the transmitted beam is almost entirely TM polarized.

For $n = 3/2$ and $N = 8$, formula (4.18) gives a value equal to 0.077 for the intensity ratio, transmitted light should almost perfectly be linearly polarized. The reality is not so good, first the incidence cannot be exactly equal to the Brewster angle and, second, because the plates are never perfectly polished: small irregularities remain and produce some depolarization; the practical result is about 10^{-2}.

Energy Conservation

In Figure 4.13 have been illustrated the variations, versus angle of incidence, of the squared values of the reflection and refraction coefficients and of their sums. The sums are never equal to unity, which is not at all paradoxical and is not in contradiction with the energy conservation principle as explained in Figure 4.14. An incident pencil of cross section S_i creates a reflected pencil of some cross section $S_i' = S_i$ and a transmitted pencil having a cross section S_t. To obtain the energy conservation we must take care that the cross section of the transmitted pencil is not equal to the incident and reflected cross sections, and also that the wave impedances are not equal on both sides of the interface.

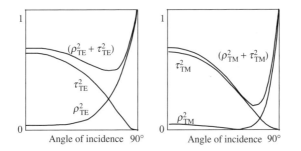

Figure 4.13. The sum of the squared values of the reflection and refraction coefficients is not equal to unity.

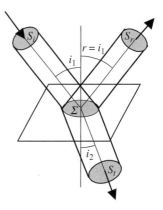

Figure 4.14. Energy conservation by reflection and refraction.

Referring to Figure 4.14 we may write $\Sigma = S_i \cos i_1 = S_r \cos i_1 = S_t \cos i_2$. The expressions of the light intensities should take into account the wave impedances of the various media: $Z_1 = Z_0/n_1$ and $Z_2 = Z_0/n_2$. Let us call A, $B = \rho A$ and $C = \tau A$, respectively, the amplitudes of the incident, reflected, and transmitted waves (ρ and τ can as well correspond to TE or TM polarizations). The different light intensities are given by

$$P_{\text{incident}} = n_1^2 \frac{A^2}{Z_0^2}, \quad P_{\text{reflected}} = n_1^2 \rho^2 \frac{A^2}{Z_0^2}, \quad P_{\text{transmitted}} = n_2^2 \tau^2 \frac{A^2}{Z_0^2}.$$

Using the relation between the cross sections, the expressions of the Snell-Descartes law and the Fresnel formulas, it can be seen that

$$P_{\text{incident}} = P_{\text{reflected}} + P_{\text{transmitted}}.$$

4.3.4. *Total Internal Reflection*

The Snell-Descartes law is a relation between the sine of the incident and refraction angles. Conditions may be encountered where the application of the formula gives a value greater than one for the sine of the angle of refraction. The question to be asked is: What about the significance of an angle having a sine greater than one? If we stay at an elementary level the answer is simple: the refraction angle doesn't exist, there is no more refracted beam, and there only remains the reflected beam. For the sake of energy conservation the reflected beam takes all the incident energy; this is completely in accordance with experimental observations.

At a less elementary level in mathematics, the notion of angle can be generalized and angles with sine greater than one can be imagined; if we keep the basic relation ($\sin^2 + \cos^2 = 1$), these angles should have cosine that are purely imaginary. We can now go back to the Fresnel formulas and calculate the reflection and transmission coefficients: complex values are found for which a physical interpretation should be given:

- The reflection coefficient is a complex number, its modulus is equal to unity, which is indeed in good agreement with the notion of *total reflection*.
- The transmission coefficient is purely imaginary. To understand what happens, a new kind of wave, called an *evanescent wave*, must be invented.

To treat, at the same time, the case of electric and magnetic fields, we will consider a vector V representing one or the other. We analytically describe three planar waves (incident [i], reflected [r], and transmitted [t]), using the following notations:

Incident wave (directed toward negative x and positive z):

$$V_i(x, y, z) = V_i e^{-jk_i r} \quad \text{with} \quad \boldsymbol{k}_i = -k_{ix}\boldsymbol{x} + k_{iy}\boldsymbol{y} + k_{iz}\boldsymbol{z}.$$

Reflected wave (directed toward positive x and positive z):

$$V_r(x, y, z) = V_r e^{-jk_r r} \quad \text{with} \quad k_r = +k_{rx}x + k_{ry}y + k_{rz}z.$$

Transmitted wave (directed toward negative x and positive z):

$$V_t(x, y, z) = V_t e^{-jk_t r} \quad \text{with} \quad k_t = -k_{tx}x + k_{ty}y + k_{tz}z.$$

The continuity of the tangential components at $x = 0$ is written as

$$(V_i)_{yOz}^{-j(k_{iy}y+k_{iz}z)} + (V_r)_{yOz}^{-j(k_{ry}y+k_{rz}z)} = (V_t)_{yOz}^{-j(k_{ty}y+k_{tz}z)}. \tag{4.19}$$

In an isotropic medium and for a given frequency, the wave vectors have the same modulus k whatever their direction ($k = nk_0 = n\omega/c$).

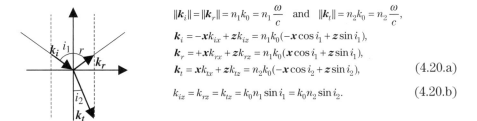

$$\|k_i\| = \|k_r\| = n_1 k_0 = n_1 \frac{\omega}{c} \quad \text{and} \quad \|k_t\| = n_2 k_0 = n_2 \frac{\omega}{c},$$

$$k_i = -xk_{ix} + zk_{iz} = n_1 k_0(-x\cos i_1 + z\sin i_1),$$

$$k_r = +xk_{rx} + zk_{rz} = n_1 k_0(x\cos i_1 + z\sin i_1),$$

$$k_t = xk_{tx} + zk_{tz} = n_2 k_0(-x\cos i_2 + z\sin i_2), \tag{4.20.a}$$

$$k_{iz} = k_{rz} = k_{tz} = k_0 n_1 \sin i_1 = k_0 n_2 \sin i_2. \tag{4.20.b}$$

Figure 4.15. Description of the incident, reflected, and transmitted wave vectors.

4.3.4.1. *Graphical Illustration of the Snell-Descartes Law*

The second Snell-Descartes law expresses the conservation of the tangential components of the wave vectors; it can be given a geometric illustration, usually called *Descartes' construction* of the reflected and refracted beams. Two concentric circles are drawn, centered at the point of incidence O and having radius, respectively, equal to the refractive indices. The incident ray is prolonged until it intersects at point I the circle having the incident index, n_1, as radius; a line is then drawn orthogonal to the interface, R and T are the points of intersections, respectively, with the circles of radii n_1 and n_2. The reflected ray is OR and the refracted ray is OT.

If the incidence medium is less refringent than the second medium ($n_1 < n_2$), whatever the angle of incidence, the sine law gives a value that is smaller than one for the sine of the refracted angle. The largest value for the angle of incidence is $90°$, the incidence is then called the *grazing incidence*; the associated angle of refraction is called the *critical angle*, often labeled as λ (not to be confused with the wavelength) and is given by

$$\lambda = \sin^{-1}(n_1/n_2). \tag{4.21}$$

If the incident medium is more refringent ($n_1 > n_2$), the construction of the refracted ray is only possible if the angle of incidence is less than the critical

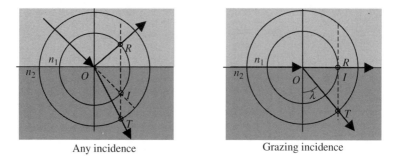

Any incidence   Grazing incidence

Figure 4.16. Descartes' construction when $n_1 < n_2$. This is just a graphical interpretation of the formula $n_1 \sin i_1 = n_2 \sin i_2$. If the upper medium is less refringent, the construction is always possible.

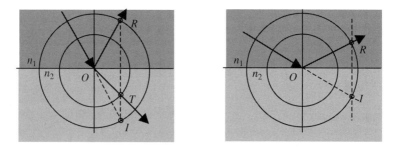

Figure 4.17. Descartes' construction when $n_1 > n_2$. It is always possible to draw the reflected beam. The refracted beam can be obtained only if the angle of incidence is smaller than the critical angle.

angle. For an angle of incidence equal to λ, the transmitted ray is parallel to the interface, *grazing emergence*.

4.3.4.2. *Total Internal Reflection, Evanescent Waves*

We consider the situation where Descartes' construction doesn't give any transmitted ray. The second medium is less refringent and the angle of incidence i_1 is larger than the critical angle λ. The mathematical formulas can still be formally written for a given value i_1 of the angle of incidence:

$$n_1 \sin i_1 = n_2 \sin i_2 \quad \rightarrow \quad \sin i_2 = \frac{n_1}{n_2} \sin i_1 = \gamma,$$

where γ is a real number larger than unity; the cosine of i_2 is obtained from

$$\cos^2 i_2 = 1 - \sin^2 i_2 = 1 - \gamma^2,$$

$$\cos i_2 = \pm j\sqrt{\gamma^2 - 1} = \pm j\delta \quad \text{with} \quad \delta^2 = \gamma^2 - 1 > 0.$$

From a mathematical point of view the problem is solved and a complex value is obtained for i_2. Correlatively, the wave vector of the transmitted wave is complex:

$$k = n_2 k_0 (-x \cos i_2 + z \sin i_2) = n_2 k_0 (\mp x j \delta + \gamma z).$$

Going back to formula (4.19) the transmitted plane wave is written as

$$V_t(x, y, z) = V_t e^{-j \gamma n_2 k_0 z} e^{\mp \delta n_2 k_0 x}, \quad \text{such a wave is an evanescent wave.}$$

As we are permanently juggling with real and complex numbers, it is probably safer to get back to the definitions at the moment of trying to give a physical interpretation of the transmitted wave in the case of total internal reflection. To do so we will reintroduce the factor $e^{j \omega t}$ in the formulas and use real expressions of the harmonic waves:

$$V_t(x, y, z) = \text{Re}\left[V_t e^{-j \gamma n_2 k_0 z} e^{\mp \delta n_2 k_0 x} e^{j \omega t}\right],$$
$$V_t(x, y, z) = \text{Re}[V_t] e^{\mp \delta n_2 k_0 x} \cos(\omega t - \gamma n_2 k_0 z). \tag{4.22}$$

The vector $[\text{Re}(V_t)]$ doesn't raise any special problem and may be obtained from V_i and the expression of the boundary conditions. Let us now examine the other terms of formula (4.22):

- $\cos(\omega t - \gamma n_2 k_0 z)$: This term simply describes a propagation along Oz at a phase velocity of c/n_2.
- $e^{\mp \delta n_2 k_0 x}$: This term is constant for a given value of x. The phase doesn't vary with x: no propagation along Ox. The amplitude of the field decays exponentially with the depth of penetration inside the second medium.

The ambiguity associated with the $\pm$ sign is not dramatic since it can easily be solved by energy considerations. The space domain in which formula (4.22) is valid extends to the negative side of Ox, in this region x may go to $-\infty$, only the positive sign should be kept.

In the lower medium of Figure 4.18, propagation is parallel to the interface; most of the energy is concentrated inside a layer having a thickness of the order of $\Delta = 1/\delta n_2 k_0$. Δ is usually quite small (a tiny fraction of the wavelength), this is the reason why *evanescent waves* are sometimes called *surface waves*.

Formally considered as interesting curiosities, evanescent waves play an important role in guided optics and optical fibers, inside which the light is guided thanks to total internal reflection.

4.3.4.3. *Interpretation of the Presence of Electromagnetic Energy in the Second Medium*

The existence of an evanescent wave of course corresponds to the presence of energy inside the second medium. How has this energy penetrated in this second medium if the incident beam is totally reflected? The answer to this

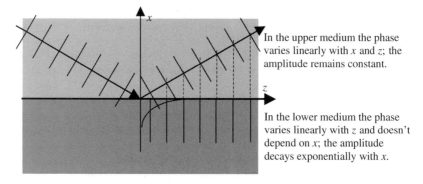

In the upper medium the phase varies linearly with x and z; the amplitude remains constant.

In the lower medium the phase varies linearly with z and doesn't depend on x; the amplitude decays exponentially with x.

Figure 4.18. Phase and amplitude repartitions for total internal reflection. The small lines perpendicular to the wave vectors are supposed to suggest wave planes along which the fields have the same phase (modulo 2π).

question is rather subtle. The representation of a harmonic signal by a complex exponential function of time corresponds to the replacement of the time derivative operator d/dt by the multiplication by $j\omega$, doing this we implicitly consider that a permanent state has been reached. We admit that the incident, reflected, and transmitted waves have always existed and will always exist. In fact, it should be considered that the incident wave has arrived at some initial time; a transient state then starts during which energy is accumulated inside the second medium. It's only at the end of the transient state that the light is totally reflected. The transient state should in principle last for an infinite time, but it is almost reached after a very short delay of the order of some (or even many) periods of the light signal (10^{-15} s).

4.3.4.4. *Fresnel Formulas for Total Internal Reflection*

$$\rho_{\text{TE}} = -\frac{\sin(i_1 - i_2)}{\sin(i_1 + i_2)} = \frac{\sin i_2 \cos i_1 - \sin i_1 \cos i_2}{\sin i_2 \cos i_1 + \sin i_1 \cos i_2} = \frac{\gamma \cos i_1 - j\delta \sin i_1}{\gamma \cos i_1 + j\delta \sin i_1},$$

$$\rho_{\text{TE}} = \frac{1 - j\dfrac{\delta}{\gamma}\tan i_1}{1 + j\dfrac{\delta}{\gamma}\tan i_1} = e^{-j\phi_{\text{TE}}} \quad \text{with} \quad \tan\frac{\phi_{\text{TE}}}{2} = \frac{\delta}{\gamma}\tan i_1.$$

The phase shift is neither zero nor π and is given by the above formula.

In the TM case the reflection coefficient also has a modulus equal to unity, the expression for the TM phase shift is a bit more complicated:

$$\rho_{\text{TM}} = -\frac{\tan(i_1 - i_2)}{\tan(i_1 + i_2)} = -\frac{1 - j\dfrac{\delta}{\gamma \tan i_1}}{1 + j\dfrac{\delta}{\gamma \tan i_1}} = \frac{1 - j\dfrac{\delta \tan i_1}{\gamma}}{1 + j\dfrac{\delta \tan i_1}{\gamma}} = e^{-j\phi_{\text{TM}}}.$$

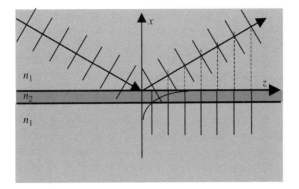

Figure 4.19. Optical tunnel effect or frustrated total internal reflection.

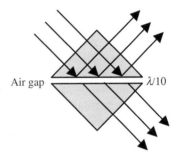

Figure 4.20. Beam splitter using a Lummer cube, this set-up is an experimental demonstration of the optical tunnel effect.

4.3.4.5. *Optical Tunnel Effect or Frustrated Total Internal Reflection*

We consider the arrangement of Figure 4.19 where a layer of a low index material is sandwiched between two more refringent media. If the sandwich is very thick, total internal reflection occurs on the first interface in the usual way. On the contrary, if the sandwich is thin enough, the tail of the evanescent wave will penetrate inside the lowest medium inside of which light will be transmitted. This is very similar to the tunnel effect in Quantum Mechanics, it's the reason why this phenomenon is often referred to as the *optical tunnel effect*; it's also called *frustrated total internal reflection*.

The optical tunnel effect is used in an optical component called the *Lummer cube*, which is a beam splitter. The two hypotenuse faces of two isosceles rectangular glass prisms are put in close proximity. If the air gap is small ($\cong \lambda/10$) light is partially transmitted, and the ratio between the transmitted and reflected beams can be adjusted by playing with the thickness. In commercially available splitters dielectric layers have replaced the air gap.

Annex 4.A

TE Modes—TM Modes

4.A.1. Scalar Nature of Two-Dimensional Electromagnetic Problems

We will say that a problem is a two-dimensional problem if one of the geometrical coordinates, z, for example, is not involved. We also say that the problem is invariant in a translation parallel to Oz. We intend to show, in the case of an electromagnetic problem exhibiting the Oz invariance, that the set of the electromagnetic vectors is the union of two independent subsets that are called the TE and TM modes.

Introducing the six components $(E_x, E_y, E_z, H_x, H_y, H_z)$ of an electromagnetic field and the unit vectors $(\boldsymbol{x}, \boldsymbol{y}, \boldsymbol{z})$ of the coordinate axis, TE and TM are defined by

$$(\boldsymbol{EM})_{\text{TE}} = \begin{pmatrix} (E_x \boldsymbol{x} + E_y \boldsymbol{y}) \\ H_z \boldsymbol{z} \end{pmatrix} \quad \text{and} \quad (\boldsymbol{EM})_{\text{TM}} = \begin{pmatrix} E_z \boldsymbol{z} \\ (H_x \boldsymbol{x} + H_y \boldsymbol{y}) \end{pmatrix}.$$

Any field can be considered as the addition of a TE and TM field.

The independence versus z is simply introduced by cancelling all z derivatives, $(d/dz = 0)$. We write the two first Maxwell equations for harmonic waves in a vacuum:

$$\text{curl } \boldsymbol{H} = j\omega\varepsilon_0 \boldsymbol{E} = jkc\varepsilon_0 \boldsymbol{E}, \quad k = \omega/c, \tag{4.A.1}$$

$$\text{curl } \boldsymbol{E} = -j\omega\mu_0 \boldsymbol{H} = -jkc\mu_0 \boldsymbol{H}. \tag{4.A.2}$$

Equations (4.A.1) and (4.A.2) are relations between vectors, and represent in fact six equations between the components of the electromagnetic field,

$$\left\{ \begin{aligned} \frac{\partial E_z}{\partial y} &= -jkc\mu_0 H_x, \\ \frac{\partial E_z}{\partial x} &= jkc\mu_0 H_y, \\ \frac{\partial E_y}{\partial x} - \frac{\partial E_x}{\partial y} &= -jkc\mu_0 H_z, \end{aligned} \right. \qquad \left\{ \begin{aligned} \frac{\partial H_z}{\partial y} &= jkc\varepsilon_0 E_x, \\ \frac{\partial H_z}{\partial x} &= -jkc\varepsilon_0 E_y, \\ \frac{\partial H_y}{\partial x} - \frac{\partial H_x}{\partial y} &= jkc\varepsilon_0 E_z. \end{aligned} \right. \tag{4.A.3}$$

The six equations (4.A.3) may be assembled in another way:

$$
\text{TM} \quad
\begin{cases}
\dfrac{\partial E_z}{\partial y} = -jkc\mu_0 H_x, \\[2mm]
\dfrac{\partial E_z}{\partial x} = jkc\mu_0 H_y, \\[2mm]
\dfrac{\partial H_y}{\partial x} - \dfrac{\partial H_x}{\partial y} = jkc\varepsilon_0 E_z,
\end{cases}
\tag{4.A.4}
$$

$$
\text{TE} \quad
\begin{cases}
\dfrac{\partial H_z}{\partial y} = jkc\varepsilon_0 E_x, \\[2mm]
\dfrac{\partial H_z}{\partial x} = jkc\varepsilon_0 E_y, \\[2mm]
\dfrac{\partial E_y}{\partial x} - \dfrac{\partial E_x}{\partial y} = -jkc\mu_0 H_z.
\end{cases}
\tag{4.A.5}
$$

For a TE mode ($E_z = 0$, $H_x = 0$, $H_y = 0$), the three (4.A.4) equations are automatically fulfilled, as well as the (4.A.5) equations for a TM mode ($H_z = 0$, $E_x = 0$, $E_y = 0$). As a consequence, on the occasion of the various transformations of a field during its propagation (reflection, refraction, diffraction) the TE (or TM) nature is kept.

Deriving equations (4.A.4) and (4.A.5) and doing suitable linear combinations, we can obtain the Helmholtz equations for TE and TM modes:

$$
\text{TE: } \frac{\partial^2 H_z}{\partial x^2} + \frac{\partial^2 H_z}{\partial y^2} + k^2 H_z = 0 \quad \text{and} \quad \text{TM: } \frac{\partial^2 E_z}{\partial x^2} + \frac{\partial^2 E_z}{\partial y^2} + k^2 E_z = 0.
$$

Each solution may be described using only one function of z, E_z for TM and H_z for TE. As a matter of fact, for TE and TM modes, we can forget the vector nature of electromagnetic waves and consider that they are scalar quantities following a Helmholtz equation.

Annex 4.B

Determination of an Unknown Polarization

The diagram on the following page indicates a procedure to be followed in order to determine the characteristics of the unknown polarization state of a light beam. A simple analysis, using a linear polarizer, will not usually be sufficient, except in the case where minima with zero intensity are observed. If the minima are not equal to zero, the light can as well be elliptically polarized or be the superposition of a fully polarized component (linear or elliptical) and of a nonpolarized component.

To remove the uncertainty a quarter wave should be placed upstream of the linear analyzer. The measurements of the ratio between maxima and minima that are observed when the analyzer is rotated, give information about the polarized and unpolarized components.

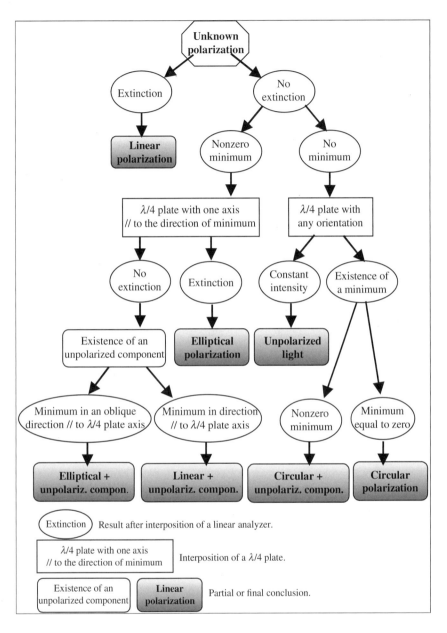

5

Birefringence

5.1. Double Refraction

When some incident beam hits the interface separating two transparent media, it may happen that *two transmitted beams are generated*; when this is the case, at least one of the two media is anisotropic. This phenomenon was first observed in 1669 by the Dane Erasmus Bartholimus who called it *double refraction*. Media in which double refraction occurs are said to be birefringent. As early as 1690, using his famous construction of refracted beams, Huygens could give an interpretation of the principal aspects of birefringence.

When crossing an anisotropic material limited by two planar interfaces, a parallel beam of natural (unpolarized) light generates two transmitted beams, they are linearly polarized along two mutually orthogonal directions which are labeled (1) and (2) in Figures 5.1 and 5.2, and which are determined by the orientation of the material.

When the two sides of the anisotropic sample are planar and parallel, the two emerging beams are parallel to the incident beam. In the case of Figure 5.1(a), the intensities of the two transmitted beams are equal. In the case of Figure 5.1(b), a polarizer has been introduced before the plate; in general two transmitted beams are observed, their relative intensities varying with the orientation of the polarizer. Figure 5.2 show the existence of two special orientations of the polarizer for which only one transmitted beam is observed, while the other beam is extinguished. We will call these very special orientations *privileged directions of vibrations*, labeled (1) and (2) in Figures 5.1 and 5.2. The orientations of (1) and (2) depend of course on the anisotropic material under consideration, but they also depend on the direction of propagation, this is the reason why we will speak of *privileged directions of vibrations associated to a given direction of propagation*.

Chapter 5 has been reviewed by Dr. François Méot, Senior Physicist at the CEA (Commissariat à l'Energie Atomique).

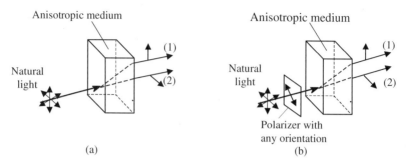

Figure 5.1. Transmission of a parallel beam of natural (unpolarized) light by an anisotropic plate. Two transmitted beams are observed, they have two mutual linear and orthogonal polarizations, along directions that are determined by the orientation of the plate. Since the two sides of the plate are parallel, the two transmitted beams and the incident beam are also, respectively, parallel.

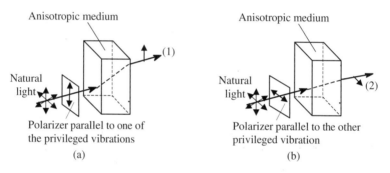

Figure 5.2. Transmission of a parallel beam with a linear polarization parallel to one of the privileged directions of vibration. Only one transmitted beam is observed.

5.2. Permittivity Tensor

Tensors are mathematical tools generalizing the notion of *vectors*, in the same way as vectors are the generalization of scalar numbers. Tensor relationships are a generalization of proportionality relations between vectors. It's often on the occasion of birefringence that young students will meet tensors for the first time.

5.2.1. *The Relationship Between Electric and Displacement Fields Is a Tensor*

In an isotropic material the electric displacement vector D is proportional to the electric field E, which means that the three components (D_x, D_y, D_z) of D are proportional to the corresponding component (E_x, E_y, E_z) of the electric field. The situation is different in an anisotropic material where (D_x, D_y, D_z) are *linear combinations* of (E_x, E_y, E_z).

Relationship between $\boldsymbol{D}$ and $\boldsymbol{E}$		
Isotropic material	Anisotropic material	
Any coordinates	Any coordinates	Principal axes of tensor $[\varepsilon]$
$D_x = \varepsilon E_x$	$D_x = \varepsilon_{xx}E_x + \varepsilon_{xy}E_y + \varepsilon_{xz}E_z$	$D_X = \varepsilon_X E_X$
$D_y = \varepsilon E_y$	$D_y = \varepsilon_{yx}E_x + \varepsilon_{yy}E_y + \varepsilon_{yz}E_z$	$D_Y = \varepsilon_Y E_Y$
$D_z = \varepsilon E_z$	$D_z = \varepsilon_{zx}E_x + \varepsilon_{zy}E_y + \varepsilon_{zz}E_z$	$D_Z = \varepsilon_Z E_Z$

A relation between a tensor and a vector is symbolically written as

$$\boldsymbol{D} = [\varepsilon]\boldsymbol{E}, \quad [\varepsilon] \text{ is called the dielectric permittivity tensor.} \tag{5.1}$$

In the same way that, in three-dimensional space, a vector is represented by three numbers (the components on the coordinate axes), a vector is represented by nine numbers arranged in a 3×3 matrix. The values of the matrix elements are associated to a given set of coordinate axes.

When a change of coordinate axes is operated, the matrix elements take new values that are obtained from the initial values thanks to well-established formulas. If the coordinate axes have no special orientation, the matrix elements will be different from zero; however, except for very special situations requiring a rather fanciful mathematical imagination, it is always possible to find a special system of reference for which the matrix is diagonal (all the elements are equal to zero, except the diagonal elements). The corresponding coordinate axes are called the *principal axes* of the sample, their orientation is of course correlated with the symmetry elements of the material,

$$[\varepsilon] = \begin{bmatrix} \varepsilon_X & & \\ & \varepsilon_Y & \\ & & \varepsilon_Z \end{bmatrix}, \quad \text{setting} \quad \varepsilon_{XX} = \varepsilon_X, \quad \varepsilon_{YY} = \varepsilon_Y, \quad \varepsilon_{ZZ} = \varepsilon_Z. \tag{5.2}$$

The mathematical operation leading to the principal axes is called diagonalization of the matrix, the rules of this game have been well established by our mathematical colleagues. The values of the matrix elements, once the matrix has been diagonalized, are called the *principal dielectric constants of the material*.

The relationship $\boldsymbol{D} = \varepsilon\boldsymbol{E}$ is just the expression of the interaction between the electric field of the light wave with electric charges of the material. Figure 5.3 shows a simplified model of this interaction: the bound electrons of an

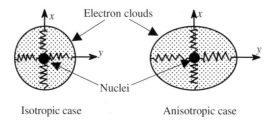

Figure 5.3. Oversimplified case of isotropic and anisotropic atoms.

atom make an electron cloud that is assimilated to a spherical shell to which the nucleus is linked by six springs (only four have been represented).

In the case of an isotropic material, the six springs have the same stiffnesses. In the absence of any electric field, the respective centers, G_+ and G_-, of the positive and negative charges, coincide with the center of the sphere. When an electric field is applied, G_+ and G_- become separated in such a way that the vector $\boldsymbol{G_+G_-}$ is collinear with the field, this is the reason why the electric field and the electric displacement vectors remain collinear $\rightarrow \boldsymbol{D} = \varepsilon\boldsymbol{E}$, where ε is a scalar number.

The model for an anisotropic material is roughly the same, except that the stiffnesses of the different springs are not equal. The vector $\boldsymbol{G_+G_-}$ is no longer collinear with the applied electric field, as well as the electric displacement vector $\rightarrow \boldsymbol{D} = [\varepsilon]\boldsymbol{E}$.

Natural and Artificial Birefringence

Tensor relationships are met whenever the space, in which the physical phenomenon occurs, loses its symmetry.

Natural birefringence: In the case of transparent crystalline materials, the arrangement of the atoms inside the elementary cell is responsible for the anisotropy of the space inside of which the light is propagating. The properties of the tensor $[\varepsilon]$ are directly connected to the symmetry of the lattice.

Artificial (or induced) birefringence: A naturally isotropic material can be made anisotropic if the symmetry of the environment is modified, for example, if an electric or a magnetic field or mechanical stresses are applied. One of the coordinate axes that diagonalizes the tensor then coincides with the direction of the field.

5.2.2. Principal Dielectric Constants, Principal Indices of Refraction

Using the formulas that are given in the following table, *principal velocities and principal indices of refraction* are associated to the principal dielectric constants ε_X, ε_Y, and ε_Z:

Principal dielectric constants	Associated phase velocities	Principal indices of refraction
ε_X	$\varepsilon_X \mu_0 V_X^2 = 1, \quad V_X = \sqrt{\dfrac{1}{\varepsilon_X \mu_0}}$	$n_X = n_1 = \dfrac{c}{V_X} = \sqrt{\dfrac{\varepsilon_X}{\varepsilon_0}}$
ε_Y	$\varepsilon_Y \mu_0 V_Y^2 = 1, \quad V_Y = \sqrt{\dfrac{1}{\varepsilon_Y \mu_0}}$	$n_Y = n_2 = \dfrac{c}{V_Y} = \sqrt{\dfrac{\varepsilon_Y}{\varepsilon_0}}$
ε_Z	$\varepsilon_Z \mu_0 V_Z^2 = 1, \quad V_Z = \sqrt{\dfrac{1}{\varepsilon_Z \mu_0}}$	$n_Z = n_3 = \dfrac{c}{V_Z} = \sqrt{\dfrac{\varepsilon_Z}{\varepsilon_0}}$

5.3. Planar Waves Obeying Maxwell's Equations in an Anisotropic Material

5.3.1. *Maxwell's Equations for Planar Harmonic Waves*

Maxwell's equations are formally the same in isotropic or anisotropic materials; however, the existence of a tensor relation between E and D considerably modifies the properties of the solutions.

As in the isotropic case, we start from a planar harmonic wave for which the four fields are described by the following expressions:

$$E = E_0 e^{j\omega t} e^{-jkr}, \quad D = D_0 e^{j\omega t} e^{-jkr},$$

$$H = H_0 e^{j\omega t} e^{-jkr}, \quad B = B_0 e^{j\omega t} e^{-jkr}.$$

The results have already been established for the isotropic case (see Section 2.4). E_0, H_0, D_0, and B_0 cannot be chosen at random if we want E, H, D, and B to be allowed solutions of Maxwell's equations. There are two kinds of conditions:

- The first are concerned with the relative orientations of the different vectors.
- The other is a relation (see formula (2.14)), between the values of the frequency, of the dielectric constants, and the modules of the wave vectors.

Maxwell's Equations for Harmonic Planar Waves
in an Anisotropic Medium

In an anisotropic medium Maxwell's equations for harmonic planar waves are exactly the same as for the isotropic case:

$$\begin{cases} \boldsymbol{k} \wedge \boldsymbol{E} = \omega\mu_0 \boldsymbol{H}, \\ \boldsymbol{k} \wedge \boldsymbol{H} = -\omega \boldsymbol{D}, \\ \quad kB = \mu_0 kH = 0, \\ \quad kD = 0, \end{cases} \rightarrow \begin{cases} \boldsymbol{E} \perp \boldsymbol{H} \text{ and } \boldsymbol{k} \perp \boldsymbol{H}, \\ \boldsymbol{D} \perp \boldsymbol{H} \text{ and } \boldsymbol{k} \perp \boldsymbol{D}. \end{cases} \tag{5.3}$$

We will consider that the medium is isotropic from a magnetic point of view and is anisotropic from an electric point of view. In other words, the relation between $\boldsymbol{B}$ and $\boldsymbol{H}$ is just a proportionality, while the relation between $\boldsymbol{D}$ and $\boldsymbol{E}$ involves a tensor:

$$\boldsymbol{B} = \mu_0 \boldsymbol{H}, \quad \boldsymbol{D} = [\varepsilon]\boldsymbol{E}, \quad \begin{cases} D_x = \varepsilon_x E_x, \\ D_y = \varepsilon_y E_y, \\ D_z = \varepsilon_z E_z. \end{cases} \tag{5.4}$$

Relations (5.3) show that the vectors $\boldsymbol{D}$ and $\boldsymbol{k}$ are orthogonal and thus we can write

$$k_x D_x + k_y D_y + k_z D_z = \varepsilon_x k_x E_x + \varepsilon_y k_y E_y + \varepsilon_z k_z E_x = 0,$$

it is then easily deduced that $(k_x E_x + k_y E_y + k_z E_z)$ is not equal to zero, which shows that the electric field $\boldsymbol{E}$ is not orthogonal to the wave vector $\boldsymbol{k}$.

In an anisotropic material the Poynting vector has the same definition as in an isotropic material, it is defined as the vector product of the electric field

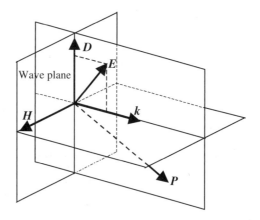

Figure 5.4. Positions of the different vectors in the case of an anisotropic material. $\boldsymbol{D}$ and $\boldsymbol{H}$ are orthogonal to the wave vector $\boldsymbol{k}$ and, consequently, in the wave plane. $\boldsymbol{E}, \boldsymbol{D}, \boldsymbol{P}$, and $\boldsymbol{k}$ are in a same plane which is orthogonal to $\boldsymbol{H}$.

by the magnetic field: $P = E \wedge H$. An important difference is that P *is no longer parallel to the wave vector*, this implies that the *light rays no longer coincide with the normal to the wave surfaces.*

Which vector should be chosen to represent the light vibration? Any of the three vectors D, H, or E could, a priori, be chosen, since if one vector is known the other two are easily deduced. However, D or E should be preferred because the light propagation in a material is mostly concerned with electric dipolar interactions. In the isotropic case where D and E are parallel there is no reason to take one or the other; in the anisotropic case we will choose D, because it is parallel to the wave planes and orthogonal to the wave vector.

The above considerations about the positions of the different vectors are nothing but necessary conditions, and not at all necessary and sufficient conditions. More detailed calculations are developed in Annex 5.B. In the following section we will give an illustration and type of *recipe* of the main results of those calculations.

5.3.2. *Comparison of Propagation Laws for Isotropic and Anisotropic Conditions*

$$\left(\begin{cases} k \wedge E = \omega \mu_0 H, \\ k \wedge H = -\omega[\varepsilon]E, \end{cases} \right) \quad \rightarrow \quad k \wedge (k \wedge E) + \omega^2 \mu_0 [\varepsilon]E = 0. \tag{5.5.a}$$

In the isotropic case, starting from the above equations, we arrive at the fact that the only condition that is required for the direction of the electric vector is that it be orthogonal to the wave vector. For formula (5.5.a) to be satisfied, we must have

$$k^2 - \varepsilon \mu_0 \omega^2 = 0.$$

This last relation is nothing other than the law of dispersion of the material. It should be emphasized that the dispersion law doesn't depend on the direction of propagation, i.e., on the direction of the wave vector.

In the anisotropic case, the situation is made more complicated by the presence of the tensor $[\varepsilon]$. We obtain

$$-(kE)k + (k^2 - \omega \mu_0 [\varepsilon])E = 0. \tag{5.5.b}$$

Formula (5.5.b) implies three linear and homogeneous equations between the (x, y, z) components of the electric field and of the wave vector, these three equations can be concisely written as

$$\begin{pmatrix} (\omega^2 \varepsilon_X \mu_0 - k_y^2 - k_z^2) & k_x k_y & k_x k_z \\ k_y k_x & (\omega^2 \varepsilon_Y \mu_0 - k_x^2 - k_z^2) & k_y k_z \\ k_z k_x & k_z k_y & (\omega^2 \varepsilon_Z \mu_0 - k_y^2 - k_x^2) \end{pmatrix} \begin{pmatrix} E_x \\ E_y \\ E_z \end{pmatrix} = 0. \tag{5.6}$$

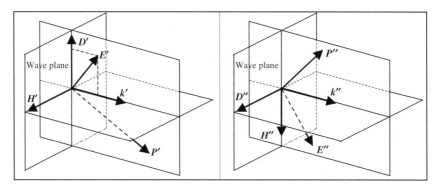

Figure 5.5. The two privileged directions of vibration. For a given direction of propagation there are two possible speeds of propagation, V' and V'', the wave vector thus has two possible moduli: $k' = \omega/V'$ and $k'' = \omega/V''$. To each value is associated a specific disposition of the trihedral (D, H, k) and of the electric field.

The three components of the electric field are thus obtained from the set of three equations (5.6). If we don't want to limit ourselves to the trivial solution ($E_x = E_y = E_z = 0$) the following conditions should be fulfilled:

- A proper choice for the direction of the electric field.
- The determinant of the set of equations should be equal to zero (this will give the law of dispersion of the medium).

5.3.3. *Main Characteristics of a Planar Wave in an Anisotropic Medium*

We now consider a harmonic planar wave (angular frequency ω, wave vector k, propagation speed V) propagating along a direction defined by some unit vector s (s_x, s_y, s_z); we have the following relations:

$$k = ks = \frac{\omega}{v}s \quad \text{with} \quad s_x^2 + s_y^2 + s_z^2 = 1. \tag{5.7}$$

Calculations developed in Annex 5.B and formula (5.6) show that, if V_X, V_Y, and V_Z are the *principal phase velocities* introduced in Section 5.2.3, the speed of propagation V along the direction s (s_x, s_y, s_z) is given by the following equation:

$$\frac{s_x^2}{V^2 - V_X^2} + \frac{s_y^2}{V^2 - V_Y^2} + \frac{s_z^2}{V^2 - V_Z^2} = 0, \tag{5.8}$$

where V^2 is the solution of a second degree equation which has two solutions, V'^2 and V''^2, finally we find four possible values for V: $\pm V'$ and $\pm V''$. The $\pm$ signs correspond to waves propagating either in the direction of the unit vector s or in the opposite direction.

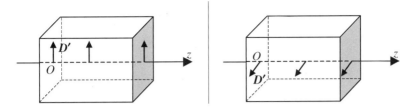

Figure 5.6(a). A linear vibration parallel to one of the privileged directions keeps its polarization while propagating. The two privileged vibrations, D' and D'', do not propagate at the same speed.

Figure 5.6(b). Propagation of an elliptical vibration. The vibration is projected along the two privileged directions, the two components don't propagate synchronously giving a new elliptical vibration, the axis of which progressively rotates during propagation.

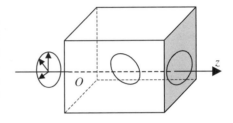

To the waves propagating, respectively, at the speed V' or at the speed V'', are associated to two specific sets of vectors (E', D', H') and (E'', D'', H''). D' and D'' are the privileged directions of vibrations that have been experimentally introduced at the beginning of this chapter. The way to obtain D' and D'' is not at all straightforward, it will be explained later on, and only in the simplest case of a uniaxial material.

A linear vibration, parallel to one of the privileged directions, D' or D'', keeps its polarization during propagation. If the initial polarization is linear, but not parallel to one of the privileged directions of vibration associated to the considered direction of propagation, we will consider its projections on D' and D''. Initially in phase the two components will accumulate an outphasing as propagation occurs since they don't propagate with the same phase velocity: the vibration becomes elliptical. The shape of the ellipse changes during propagation and periodically coincides with a straight segment, when the two components are again *in phase*.

5.3.4. *Light Rays and Normal to Wave Surfaces*

Refraction of the Normal

We consider a planar wave which is refracted from an isotropic medium onto an anisotropic one, see Figure 5.8. Two refracted waves are generated in the second medium, they propagate at two different speeds with two different wave vectors, k' and k'', which determines the directions of the wave planes that are, respectively, normal to k' and k''.

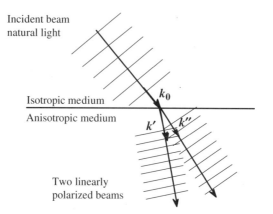

Incident beam
natural light

Isotropic medium

Anisotropic medium

Two linearly
polarized beams

Figure 5.7. When arriving at an interface separating an isotropic medium from an anisotropic one, a planar wave generates two planar waves that propagate at two different speeds. If the incidence is oblique, of course the two refracted waves propagate along two different directions.

The representations of Figures 5.7 and 5.8 are far too schematic, since the tree wave vectors of the incident and refracted waves are not always lying in the same plane. The two refracted waves are linearly polarized along two directions D' and D'' that have not been represented; the rules for obtaining D' and D'' will be given later.

5.4. Constructions of the Refracted Beams

The mathematical formulas are rather complicated, this is the reason why graphical constructions play an important role in anisotropic geometrical optics, they largely take advantage of the Huygens and Descartes constructions. Both constructions are just a graphic solution of the Descartes-Snell equation. However, we would like to recall that those constructions are associated to some physical interpretation:

- Conservation of the tangential component of the wave vector in the case of the Descartes construction.
- Envelope of the wavelets emitted by the different points of the interface in the case of the Huygens construction.

5.4.1. *The Descartes and Huygens Constructions for Isotropic Media*

The constructions are far simpler when the two media are both isotropic, since the surfaces that are used have only one sheet which is spherical, with the consequence that rays and wave vectors coincide.

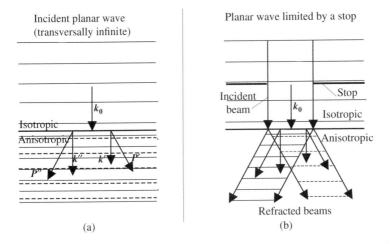

Figure 5.8. Illustration of the difference between the directions of propagation of the phase on one hand and of the energy on the other hand. The incidence is normal: the incident wave planes, as well as the refracted wave planes, are parallel to the interface. In the second medium two waves are propagating at two different speeds: their wave planes remain parallel to the interface, while the Poynting vectors lie along two different directions. In (b) a stop limits the cross section of the incident beam: two different refracted light beams are observed, each of them being parallel to the corresponding Poynting vector.

For Descartes' construction, we start with two concentric circles having radii that are, respectively, equal to the indices of refraction of the two media. From the common center O of the two circles a line is drawn parallel to the direction of the incident beam which propagates in the medium of index n_1, let I_1 be the intersection point with the circle of radius n_1. A line I_1H is then drawn orthogonal to the interface, let I_2 be the intersection point with the circle of radius n_2, the direction of the refracted beam is given by OI_2.

For Huygens' construction we also start with two concentric circles, their radii are now, respectively, equal to the inverse values, $1/n_1$ and $1/n_2$, of the indices of refraction. From the common center O, a line is drawn parallel to the incident beam direction; it intersects the circle of radius $1/n_1$ at point T_1, the tangent line to the circle at this point intersects the interface at point T. From T we draw the tangent TT_2 to the circle of radius $1/n_2$, OT_2 gives the direction of the refracted beam.

5.4.2. *Shape of the Wavelets in an Anisotropic Medium*

In the Huygens method, or preferably the Huygens-Fresnel method, the reflected and refracted waves that are generated by an incident wave are considered to be produced by the interference of wavelets emitted by the different points of the interface. The refracted and reflected wave surfaces are the envelopes of the different wavelets that have been emitted at the same time.

Isotropic material index n_1

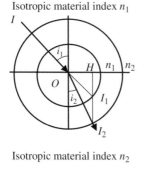

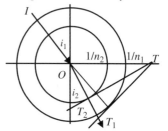

Isotropic material index n_1

Isotropic material index n_2

Isotropic material index n_2

$$OH = n_1 \sin i_1 = n_2 \sin i_2$$

$$OT = \frac{1/n_1}{\sin i_1} = \frac{1/n_2}{\sin i_2}$$

Descartes' construction

Huygens' construction

Figure 5.9. Descartes' and Huygens' constructions.

In the case of a parallel beam (planar wave) the amplitudes of the wavelets are all equal and proportional to the amplitude of the incident beam, the initial phase of a wavelet is governed by the phase of the incident wave at the point under consideration. The amplitude proportionality coefficient, as well as the phase shift, are functions of the indices and of the angle of incidence (see the Fresnel formulas for refracted and reflected beams).

When the two media are both isotropic, the propagation speed doesn't depend on the direction of propagation: the wavelets are spherical. The situation must be reconsidered in the anisotropic case. To do so, let us imagine an oscillating electric dipole which would be located at point O and which would have started oscillating at time $t = 0$. By definition a wavelet is the set of points reached by the oscillation after a given time, which is usually taken equal to one second. The trouble in an anisotropic material is that the propagation speed varies with the direction of propagation and also with the orientation of the vibrations.

We thus place at point O of an anisotropic medium, an electric dipole that vibrates along a given direction. We then consider, see Figure 5.10, waves propagating parallel to some vector $\boldsymbol{k}$, their polarizations should be parallel to one or the other directions of the two privileged polarizations, $\boldsymbol{D'}$ and $\boldsymbol{D''}$, associated to the direction of $\boldsymbol{k}$. As those two polarizations do not propagate at the same speed, after one second, two different points P' and P'' will have been reached: *in an anisotropic medium the wave surface has two sheets.*

5.4.3. *The Huygens and Descartes Constructions in Anisotropic Media*

In the case of isotropic media, two kinds of spherical surfaces have been used, their radii were equal either to the indices or to the inverse of the indices of

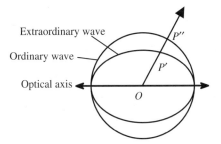

Figure 5.10. Aspect of wavelets in an anisotropic medium. The figure corresponds to a positive uniaxial material. A wavelet is the set of points reached after one second by a wave having left the origin at time $t = 0$. Two polarizations and two speeds of propagation are associated to a given direction such as $OP'P''$. After one second one polarization will have reached P', and the other reached P'': the wavelet has two sheets.

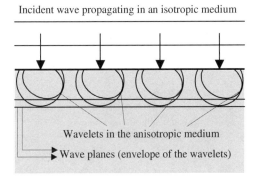

Figure 5.11. Wave planes and Huygens' wavelets in the case of normal incidence. Wavelets simultaneously emitted at four different points of the interface have been illustrated. Each wavelet has two sheets. The incidence being normal, the three wave vectors are normal to the interface. The envelopes of the wavelets are two planes parallel to the interface.

refraction. These notions can be generalized to the case of anisotropic media in which are introduced: (i) a *surface of the indices*; and (ii) a *surface of the inverse of the indices*. Each of the previous surfaces has two sheets that are not necessarily spherical.

The surface of the indices is used to determine the direction of the wave vectors of the refracted waves, thanks to Descartes' construction.

The surface of the inverse of the indices is used to determine the direction of the light rays, thanks to Huygens' construction.

Descartes' Construction of the Normal

We refer to Figure 5.13. Starting from the origin O a line is drawn parallel to the direction of the wave vector of the incident wave plane and is prolonged

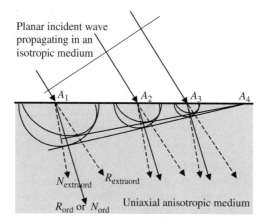

Figure 5.12. Wavelets and wave planes for a nonnormal incidence. The incident wave first reaches point A_1, and then A_2, A_3, and A_4, ... There are two different directions for the refracted wave planes. One sheet of the wavelets being spherical, the corresponding wave vector (AN_{ord}) and light ray (AR_{ord}) coincide. The situation is different for the other sheet which is elliptical, AN_{extraord} and AR_{extraord} are distinct.

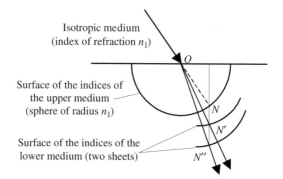

Figure 5.13. Descartes' construction of the wave vectors. The lower medium is anisotropic and its indices of refraction are higher than the index of the upper isotropic medium: the two sheets of the surface of the indices are outside the sphere of the indices of the upper medium. ON' and ON'' indicate the directions of the refracted wave vectors.

until its intersection, at point N, with the surface of the indices of the incident medium (a sphere of radius n_1). The line, drawn from point N and orthogonal to the interface, intersects the two sheets of the surface of the indices of the anisotropic medium at points N' and N''. ON' and ON'' give the directions of the two refracted waves.

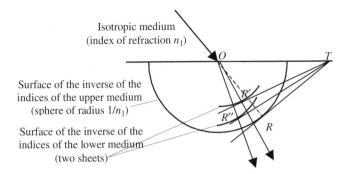

Figure 5.13(i). Huygens' construction of the light rays. The lower medium is anisotropic and its indices of refraction are higher than the index of the upper isotropic medium. The two sheets of the surface of the inverse of the indices are inside the sphere of the indices of the upper medium. OR' and OR'' indicate the directions of the refracted light rays.

Huygens' Construction of the Refracted Light Rays

We refer to Figure 5.13(i). Starting from the origin O a line is drawn parallel to the direction of the incident light rays and is prolonged till its intersection, at point R, with the surface of the inverse of the indices of the incident medium (a sphere of radius $1/n_1$). A line drawn tangent to the previous sphere intersects the interface at point T. Starting from T two lines are drawn, respectively, tangent at points R' and R'', to the two sheets of the surface of the inverse of the indices: OR' and OR'' indicate the directions of the refracted light rays.

5.5. Aspect of the Surfaces of the Indices and the Inverse of the Indices

5.5.1. *Uniaxial and Biaxial Materials*

The shape of the characteristic surfaces (indices or inverse of the indices) is directly linked to the symmetry of the material; it is all the more simple that this symmetry is higher. The classification of the different birefringent materials is obtained from the properties of their dielectric permittivity tensor. Let us recall that the *principal axes of a material* are the coordinate axes in which the tensor is diagonalized. The equations of the characteristic surfaces are the simplest when the principal axes are used. In the most general case, when the symmetry is the lowest possible, each principal axis meets the characteristic surfaces at two different points, the abscissa of which are called the two principal indices (or inverse of the principal indices).

The orientations of the principal axis are, of course, in close connection with the orientation of the crystalline lattice. The classification of anisotropic materials follows the shape of the ellipsoid of the indices that will be described in Section 5.B.5 of the annex to this chapter, where an accurate distinction between uniaxial and biaxial crystals will be given.

Biaxial materials: The principal axes have fixed positions inside the lattice; however, for lower symmetry, the orientation of two axes (monoclinic) or, possibly, of three axes (triclinic), will vary with the color of the light. When the symmetry increases (trigonal, tetragonal, and hexagonal) one axis is fixed and has a well-defined orientation, the two others axes being freely rotatable and then indeterminate.

Optically isotropic materials: The simplest case, after isotropic materials, is the case of cubic crystals: in both cases the principal indices are equal. The two sheets of the characteristic surfaces are spherical and coincide. *From an optical point of view, a cubic crystal behaves as an isotropic material.*

Uniaxial materials: In the case of hexagonal, tetragonal, and trigonal crystals, two of the three principal indices are equal; the corresponding axis can freely rotate about the third axis which is called the "optical axis" of the material. One sheet of the characteristic surface is a sphere, the second sheet is an ellipsoid of revolution about the optical axis.

Crystal system	Dielectric axes		Ellipsoid of the indices	Optical classification
Triclinic		3 axes with color dispersion	general ellipsoid	biaxial
Monoclinic		1 fixed axis 2 axes with color dispersion	general ellipsoid	biaxial
Orthorhombic Trigonal		3 fixed axes	general ellipsoid	biaxial
Tetragonal Hexagonal		1 fixed axis 2 indeterminate axes	spheroid	uniaxial
Cubic		3 indeterminate axes	sphere	isotropic

Material	n_x	n_y	n_z	Crystal system
Sodium D light ($\lambda = 589.29$ nm)				
Mica, Na and K aluminosilicate	1.560	1.594	1.598	monoclinic
Aragonite, $CaCO_3$	1.531	1.682	1.686	orthorhombic
Lithargite, PbO	2.512	2.610	2.710	—
Stibnite, Sb_2S_3 ($\lambda = 762$ nm)	3.194	4.046	4.303	—
Anhydrite, $CaSO_4$	1.569	1.575	1.613	orthorhombic
Gypse, $CaSO_4$, $2H_2O$	1.529	1.523	1.530	monoclinic
Sulfur	1.950	2.043	2.240	orthorhombic
Topaze, $(2AlO)FSiO_2$	1.619	1.620	1.627	orthorhombic
Turquoise copper aluminophosphate	1.520	1.523	1.530	—
Tartaric acid, $(COOH)_2$	1.496	1.535	1.604	monoclinique

5.5.2. *Uniaxial Material*

Uniaxial materials are the most important materials and also the most commonly met. The two sheets of a characteristic surface are, respectively, spherical and elliptical. The constructions involving the *spherical sheet* are very similar to the constructions in an isotropic material: rays have the same directions as the wave vectors and are orthogonal to the wave surfaces. This is the reason why the corresponding elements are said to be *ordinary*: ordinary sheet of the surface, ordinary rays and wave vectors, ordinary polarization. In opposition, the other sheet and the associated elements are said to be *extraordinary*.

In a uniaxial material two of the three principal indices are equal, their common value is called the *ordinary index* n_o, the third principal index is the *extraordinary index* n_e. According to the sign of the difference $(n_e - n_o)$ a uniaxial material will be said to be positive or negative:

- Positive uniaxial material: $(n_e - n_o) > 0$.
- Negative uniaxial material: $(n_e - n_o) > 0$.

The two most important materials for elaborating anisotropic optical components are calcite and crystalline quartz. Calcite is one of the most birefringent crystals, unfortunately its relatively high price and poor mechanical properties are a limitation.

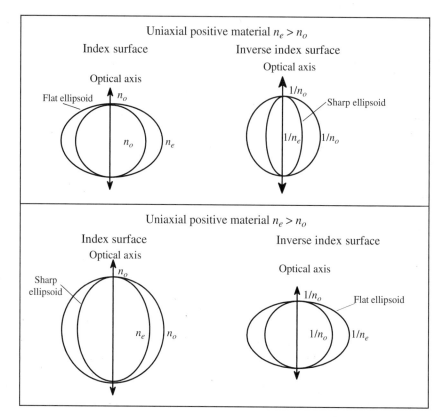

Figure 5.14. In a uniaxial material the two sheets of a characteristic surface (a sphere and a revolution ellipsoid) are tangent at the points of intersection with the optical axis. A flat ellipsoid looks like a frying pan, a sharp ellipsoid looks like a rugby ball.

Principal indices of some uniaxial crystals (sodium D ray, $\lambda = 589\,\text{nm}$)				
Positive uniaxial			n_o	n_e
Quartz	SiO_2 crystallized	hexagonal	1.5442	1.5533
Calomel	$HgCl$	quadratic	1.9732	2.6559
Zircone	$SiO, 2ZrO_2$	quadratic	1.92	1.97
Rutile	TiO_2	biaxial	2.6131	2.9089
Negative uniaxial				
Calcite	$CaCO_3$	rhombohedral	1.6584	1.4865
Sodium nitrate	$NaNO_3$	rhombohedral	1.5874	1.3361
Tourmaline		rhombohedral	1.639	1.620
KDP	KH_2PO_4	$\bar{4}2m$	1.51	1.47
Lithium niobate	$LiNbO_3$	$3m$	2.29	2.20

5.5.3. *Light Propagation in a Uniaxial Material*

5.5.3.1. *Polarization of the Refracted Beams*

We have given methods to graphically obtain on one hand the directions of the wave vectors (and, consequently, of the wave planes) and, on the other hand, the directions of the light rays. We have now to determine the directions of the associated vibrations, that is to say, the directions of the electric displacement vectors of the ordinary and extraordinary waves. The method relies on rather elaborated graphical constructions involving the index surface that will be introduced in Annex 5.B. From a practical point of view, the following important results should be kept in mind:

- Ordinary and extraordinary polarizations (electric displacement vectors) are parallel to the wave planes.
- The ordinary polarization is orthogonal to the optical axis.
- The extraordinary polarization is parallel to the projection of the optical axis on the wave plane.

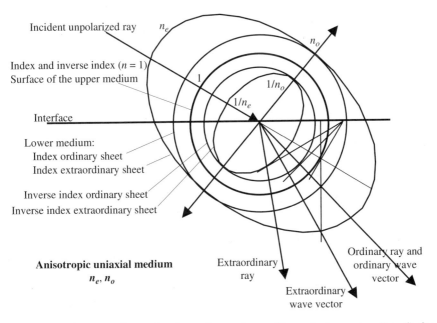

Figure 5.15. In this illustration have been shown Descartes' construction (index surfaces) and Huygens' construction (inverse index surfaces). Ordinary rays and wave vectors coincide. The ordinary polarization is orthogonal to the plane of the figure, the extraordinary polarization lies in the plane of the figure.

In the case of hexagonal, tetragonal, and trigonal crystals, two of the three
principal indices are equal characteristic surfaces.

5.6. Circular Birefringence

5.6.1. *Introduction to Circular Birefringence*

The phenomenon of circular birefringence was discovered at the beginning
of the nineteenth century by Arago on one hand and by Biot on the other. For
historical reasons, it is also called *optical activity*. Materials in which this
phenomenon can be observed are said to be optically active.

Circular birefringence experiments are no more complicated than the
experiments involving linear birefringence. However, the microscopic interpretation at the atomic level is more complicated in the case of optical
activity.

Figure 5.16 describes an experimental arrangement demonstrating the
existence of optical activity. The cell having been removed, the analyzer and
polarizer are first adjusted to have crossed positions, so that the transmitted
light is extinguished; in the presence of the cell, light is transmitted again,
extinction can be restored by a suitable rotation of the analyzer. Thus it can
be said that the emerging beam is still linearly polarized, but the polarization
has rotated during propagation in the optically active material. This kind of
polarization is not modified by optical activity: an elliptical polarization will

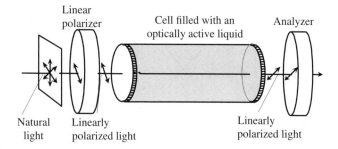

Figure 5.16. Basic experiment on optical activity. A cell, filled with an optically active
liquid, is lighted by a parallel beam of linearly polarized light. The emerging beam
is still linearly polarized, but the output polarization is not parallel to the input
polarization and keeps a constant direction as the cell is rotated about its axis. If the
incident linear polarization is rotated, the emerging polarization rotates as well, the
angle between the two polarizations remaining constant and proportional to the cell
length.

lead to an emerging elliptical polarization, the ratio of the two axes of the ellipse remains the same, their orientation is only rotated. Of course, an incident beam of natural light gives an emerging beam of the same nature.

An optically active material causes the polarization of an incident beam to rotate: if, for an observer receiving the light, the polarization appears to have revolved clockwise, the material is said to be dextrorotatory; if the rotation is anticlockwise the material is levorotatory.

Circular birefringence, as linear birefringence, is due to the symmetry of the environment in which are found the electrons of the atoms of the material under consideration. In the case of gases, liquids, or amorphous solids, circular birefringence is met any time that the elementary molecules may have one of two different configurations that are mirror images of one another. In the case of a crystal, the elementary crystal cell will also have two different mirror image configurations. In both cases the material will take two different forms that are said to be enantiomorphs of each other. In the case of crystals the two enantiomorphs will correspond to two different crystallographic structures; a well-known example is crystalline quartz in which the molecules are arranged along a helix which can rotate either clockwise or anticlockwise. Very often during a chemical synthesis the two enantiomorphs are obtained in equal proportions, the mixture which, by compensation, has no optical activity, is called a racemic. The two enantiomorphs have very similar physical and chemical properties and are, of course, very difficult to separate. Studying the optical activity of a material is quite important from a chemical point of view.

Following Fresnel we will use a purely phenomenological description of circular birefringence. We consider that optically active materials are special anisotropic media where the electromagnetic vibrations that remain *unchanged* during propagation are *left- and right-handed circular polarizations*. We will admit that right or left circular waves have different wave velocities that will be quoted as V_{left} and V_{right} and to which will be associated two indices of refraction and two wave vectors:

$$n_{\text{left}} = c/V_{\text{left}} \quad \text{and} \quad n_{\text{right}}t = c/V_{\text{right}},$$

$$k_l = \frac{\omega}{V_{\text{left}}} = n_{\text{left}}\frac{\omega}{c} = n_{\text{left}}k_0 \quad \text{and} \quad k_r = \frac{\omega}{V_{\text{right}}} = n_{\text{right}}\frac{\omega}{c} = n_{\text{right}}k_0,$$

(5.9)

where ω is the angular frequency and k_0 is the vacuum wave vector.

It is important to notice that circular birefringence is less important, by at least two orders of magnitude, than linear birefringence. The circular birefringence which is defined as the index difference, $\Delta n_{\text{circul}} = (n_{\text{left}} - n_{\text{right}})$, is always smaller than 10^{-2}, while the linear birefringence, $\Delta n_{\text{linear}} = (n_e - n_o)$, is about 10^{-2} for quartz and can reach 0.18 in the case of calcite.

The first observation of circular birefringence was achieved by Arago in a quartz crystal. As already mentioned, quartz exhibits an important linear birefringence by which circular birefringence is usually hidden. Arrangements

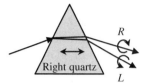
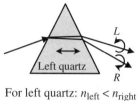

For right quartz: $n_{\text{right}} < n_{\text{left}}$ For left quartz: $n_{\text{left}} < n_{\text{right}}$

Figure 5.17. Separation of the right- and left-hand circular polarizations using circular birefringence. Two prisms have been cut, respectively, in a right and in a left quartz crystal, the optical axis being orthogonal to the bisector plane. A nonpolarized light beam is sent into the prisms in the conditions for minimum deviation: for the right quartz prism the right-hand polarized beam is less deviated, it's the opposite for the left quartz prism.

where the propagation is parallel to the optical axis should be selected to observe the effects of optical activity, light beams can then be considered as ordinary beams, whatever the polarization; see, for example, Figure 5.17 which clearly illustrates the expression "circular birefringence." It is well known in crystallography that quartz crystals are found in two different forms, mirror images of one another, referring to the optical activity these two forms are called, respectively, right-quartz and left-quartz.

Asymmetrical Atoms of Carbon in Organic Chemistry

An important example of optical activity is given by organic molecules where a carbon atom is bound to four atoms of a different species, such an atom of carbon is said to be asymmetrical. The molecule to which this carbon atom belongs has two possible different forms, one a mirror image of the other.

Corresponding compounds may exist in two different enantiomorph forms. A given enantiomorph is optically active. If dissolved in some solvent an enantiomorph gives a solution which is all the more optically active as the solution is more concentrated. The most famous example is probably given by sugar syrups which are a mixture of d-glucose and l-glucose, the proportions of each component are usually measured using a polarimeter.

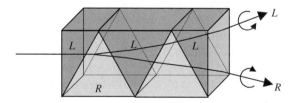

Figure 5.18(a). Arrangement imagined by Fresnel to increase the effect of circular birefringence, it is made of a succession of alternate right and left quartz prisms.

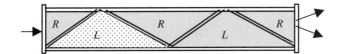

Figure 5.18(b). In spite of its weakness, the optical activity of liquid can be demonstrated by using liquid prisms alternately filled with dextrorotatory or levorotatory enantiomorphs.

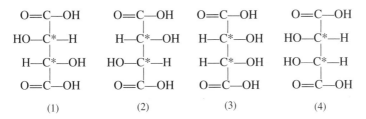

Figure 5.19. The four isomers of tartaric acid. Tartaric acid has four carbon atoms, two alcohol functions, and two acid functions. The two most central carbon atoms are asymmetrical since they are bound to four different atoms or radicals. Tartaric acid may have four different forms, among which two are optically active (dextrorotatory and levorotatory) and two are not.

Birefringence of Tartaric Acid and Tartarate Salts

The story of the optical activity of tartaric acid is a very nice story, as it was the first scientific contribution by Louis Pasteur on the occasion of his PhD thesis. The asymmetrical character of tartaric acid is even enhanced in a salt called double tartarate of sodium and ammonium, where one acid function has been neutralized by ammonia and the other by soda: one carboxyl radical is bound to an NH_4^+ ion and the other to an Na^+ ion. Tartarate of sodium and ammonium may give nice small crystals, dextrorotatory or levorotatory crystals have slightly different aspects; using a microscope and a sharp needle, Louis Pasteur was able to separate the left microcrystal from the right one. After dissolution in water, he finally obtained two solutions, a dextrorotatory one and a levorotatory one. As he was also a biologist, he remarked that a culture of penicillium glaucum on a racemic of sodium-ammonium tartarate, selectively destroys, by moisture, the dextrorotatory compound, leaving only the levorotatory enantiomorph.

5.6.2. *Description of the Propagation in an Optically Active Medium*

Any vibration can be considered as the superposition of a left-hand circular polarization, V_L, and of a right-hand circular polarization, V_R. A linear polar-

Left-hand circular vibration

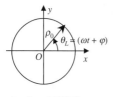

Right-hand circular vibration

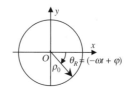

$\rho = \rho_0 = $ constant

$\theta_L = +wt + \varphi,$

$V_L = \rho_0[\boldsymbol{x}\cos\theta_L + \boldsymbol{y}\sin\theta_L],$

$V_L = \rho_0[\boldsymbol{x}\cos(wt+\varphi)] + [\boldsymbol{y}\sin(wt+\varphi)],$

$V_L = \mathrm{Re}\{\rho_0 e^{j\varphi}(\boldsymbol{x} - j\boldsymbol{y})e^{jwt}\},$

$\rho = \rho_0 = $ constant

$\theta_R = -wt + \varphi,$

$V_R = \rho_0[\boldsymbol{x}\cos\theta_R + \boldsymbol{y}\sin\theta_R],$

$V_R = \rho_0[\boldsymbol{x}\cos(-wt+\varphi)] + [\boldsymbol{y}\sin(-wt+\varphi)],$

$V_L = \mathrm{Re}\{\rho_0 e^{j\varphi}(\boldsymbol{x} - j\boldsymbol{y})e^{-jwt}\}.$

Figure 5.20. Analytical expressions of circular vibrations.

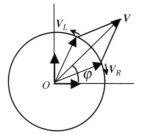

Figure 5.21. A linear vibration V is the superposition of two opposite circular vibrations V_R and V_L of the same amplitude. The direction of V is along the bisector of the angle (V_R, V_L).

ization V is the superposition of two inverse circular vibrations of the same amplitudes:

$$V = V_R + V_L = \mathrm{Re}\{\rho_0 e^{j\varphi}(\boldsymbol{x} - j\boldsymbol{y})(e^{j\omega t} + e^{-j\omega t})\},$$

$$V = \mathrm{Re}\{2\rho_0 e^{j\varphi}(\boldsymbol{x} - j\boldsymbol{y})\cos\omega t\} = 2\rho_0(\boldsymbol{x}\cos\varphi + \boldsymbol{y}\sin\varphi)\cos\omega t. \tag{5.10}$$

The orientation of the linear vibration described by (5.10) is determined by the phase difference φ between its two circular components: V makes with the Ox axis an angle which is equal to φ. Referring to Figure 5.21, it is seen that V is directed along the bisector of the angle between the two vectors V_L and V_R,

$$\theta_L = (\omega t - k_l z + \varphi) \quad \text{and} \quad \theta_R = (-\omega t + k_r z + \varphi), \tag{5.11.a}$$

$$\theta = \frac{(\theta_L + \theta_R)}{2} = \frac{(k_R - k_L)}{2}z = \frac{k_0}{2}(n_{\text{right}} - n_{\text{left}})z. \tag{5.11.b}$$

During propagation in an optically active transparent material, left and right circular waves keep constant amplitudes, while the angles θ_L and θ_R vary according to (5.9), in which k_l and k_r have been defined in formula (5.9). The polar angle θ between the two rotating vectors is given by (5.11.b), the polarization remains *linear* but its *direction rotates* at the same time as propagation occurs.

5.6.3. *Rotatory Power*

The angle of rotation of the polarization is thus proportional to the propagation length inside the optically active material; the proportionality coefficient $[\alpha]$ is called the rotatory power of the material and is given by (5.11.c):

$$[\alpha] = k_0 \frac{(n_{\text{right}} - n_{\text{left}})}{2} = \frac{\pi}{\lambda}(n_{\text{right}} - n_{\text{left}}). \tag{5.11.c}$$

In the definition of rotatory power, anticlockwise rotations are considered as positive; other authors use another possibility, which changes the sign of $[\alpha]$. The rotatory power of quartz has a specially high value of the order of $20°/\text{mm}$, the corresponding value of the circular birefringence is equal to $\Delta n = |(n_{\text{right}} - n_{\text{left}}| \approx 10^{-5}$. Usually the rotatory power of liquids is smaller and of the order of a few degrees per centimeter.

The rotatory power of a solution is proportional to the number of asymmetric molecules per volume unit, and thus to the concentration. In the case of a mixture of several optically active compounds, the rotatory power is given by formula (5.12), where $[\alpha_i]$ and c_i are, respectively, the rotatory power and the concentrations of the different components,

$$[\alpha] = \sum [\alpha_i]c_i. \tag{5.12}$$

5.7. Induced Birefringence

5.7.1. *Definition of Induced Birefringence*

A material can see its index of refraction modified if one exerts some external influence on its environment, such as mechanical stresses, or a magnetic or electric field. If these external influences are not isotropic, the optical birefringence will be modified:

- An initially isotropic medium will then acquire induced birefringent properties.
- If the medium is already birefringent, the shape of its characteristic surfaces is modified.

According to the electronic dipolar model for the interaction of a light wave with a material, the value of the index of refraction is determined by the local electric field E_{atomic} existing in the electronic cloud surrounding the nucleus of the atoms. The forces that are responsible for the coherence of a piece of material are quite strong, which implies that E_{atomic} is high; the induced index variations are quite small and can only be demonstrated thanks to interference experiments.

Interference experiments using polarized light beams are very much simplified by the fact that problems due to spatial coherency are completely avoided since the ordinary and the extraordinary beams issue from a single initial beam.

The polarization of a light beam, after propagation in a birefringent material, is just the interference of two vibrations oriented along the directions of the privileged directions. As propagation occurs over distances that can be important (as compared to the wavelength), even in the case of a small-induced birefringence, rather important phase differences can be accumulated. This is the reason why induced birefringence can produce spectacular effects, and so provide useful methods for measuring the physical process responsible for the index variations.

5.7.2. *Stress Birefringence*

Stress birefringence effects, also known as photoelasticity, was discovered in 1813 by Seebeck and was studied by Brewster around 1816. When a material is submitted to compressing forces, its density d and its index of refraction n increase simultaneously, the ratio $(n - 1)/d$ follows Gladstone's law and remains roughly constant.

Let us consider some isotropic material, a piece of glass for example, its dielectric permittivity tensor is diagonal, all nonzero elements being equal. When external mechanical forces are applied to a sample that has been cut in such a material, stresses appear inside the material. The field of stresses is described by a tensor, the mechanical stress tensor. Consequently, the material becomes birefringent, the permittivity tensor has the same symmetry as the stress tensor and its elements take new values: the variation of a given element is proportional to the value of the corresponding element of the mechanical stress tensor.

5.7.2.1. *Case of a Uniaxial Stress*

We refer to Figure 5.22, an isotropic sample is submitted to the action of two opposite forces and becomes anisotropic, uniaxial with the optical axis parallel to the direction of the forces.

The thickness and length of the sample are, respectively, equal to e and l, the surface of its face is $S = le$. The direction of the force makes an angle of $45°$ with the direction of the analyzer. To again obtain the extinction, the compensator should introduce an optical length difference δ between the ordinary and extraordinary beams. Let us call, respectively, n_e and n_o the extraordinary and ordinary indices of the stressed material, the birefringence Δn is related

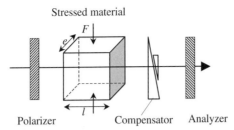

Stressed material

Polarizer   Compensator   Analyzer

Figure 5.22. Experimental set-up for studying stress birefringence. A parallelepiped has been cut into an isotropic transparent material. Two opposite forces applied on parallel sides generate a uniaxial stress repartition inside the material. In the absence of any applied forces no light is transmitted by the crossed polarizer-analyzer. Some light is transmitted again when a force is applied, the direction of the force should not be parallel to the direction of the polarizer. Using a compensator, the extinction can be obtained again and quantitative measurements of the birefringence can be made.

to the pressure $P = F/S$ by formula (5.13), in which k is a constant which can be positive or negative; the experiment indicates that k is roughly independent of the wavelength λ,

$$\Delta n = (n_e - n_o) = k\lambda P, \tag{5.13}$$

$$\delta = (n_e - n_o)l = k\frac{\lambda F}{e}. \tag{5.14}$$

The optical length difference δ is simply given by (5.14), its value is independent of the thickness and only determined by the amount of force per unit length F/l, in the visible and for the usual glass k is negative and of the order of 10^{-2} if the lengths are expressed in millimeters and the forces in kilograms.

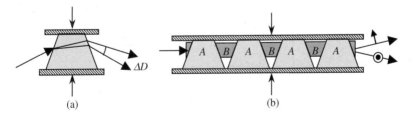

(a)   (b)

Figure 5.23. Direct demonstration of stress birefringence. In (a) the angle ΔD is very small, a few seconds for a pressure of $1\,\text{kg/mm}^2$. In (b) the effect is amplified by cascading stressed A-prisms and unstressed B-prisms cemented by glue having the same index.

5.7.2.2. *Photoelasticity*

As polarized light beams make it very easy to obtain interferences and to demonstrate small optical delay, photoelasticity is used as a technique for studying the stresses inside transparent materials. If one wants to know the repartition of the forces inside some fixed or mobile mechanical structure (shaft of a crane, driving shaft of a motor, . . .), a transparent scale model is made out of a suitable transparent material having a high proportionality coefficient between the stress and permittivity tensor elements. PMMA is a suitable material. Starting from the orientation of the privileged vibrations and from local optical birefringence, it is possible to obtain qualitative information about stresses.

It must however be said that, since the existence of very powerful computing tools, the experimental determination of the stress repartition inside a solid has lost part of its interest: mechanical equations can now be integrated even with complicated boundary conditions.

5.7.2.3. *Some Remarks About the Presence of Residual*
Stresses in a Piece of Glass

Glass is of course a very important material in Optics; it is used to make a large variety of components that are usually molded from molten glass. Glass being a rather poor heat conductor, it is difficult to maintain a homogeneous temperature during cooling; external layers getting colder faster, the most inner parts are submitted to stresses, which sometimes remain *frozen*. For example, it is possible that some light could be transmitted through a thick glass plate disposed between crossed polarizers. Lenses to be used for interference experiments using polarized light beams should have been annealed (reheated and then slowly and carefully cooled). The proportionality coefficients between stress and birefringence being positive for some glasses and negative for others, composite glasses can be made with a low residual birefringence.

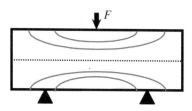

Figure 5.24. Basic experiment of photoelasticity. A sample made of PMMA put onto two prismatic blocks and submitted to a force F is placed between two crossed polarizers. A light beam propagating perpendicular to the figure produces interferences: dark and clear lines appear in the sample, they are a representation of the lines along which the stress keeps a constant value. If a white light beam is used, the stress variations will be suggested by nice color variations.

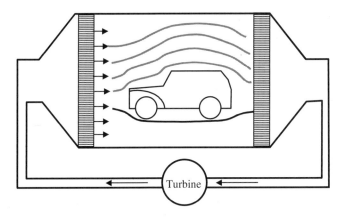

Figure 5.25. To obtain the aerodynamic profile of some object a scale model is disposed inside a wind tunnel, between two very large crossed Polaroid sheets. The interference allows a very good visualization of the air streams. This method, which has long been used to improve airplanes and rockets, is now superseded by numerical simulation techniques.

5.7.3. *Flow Birefringence*

At the scale of one wavelength of visible light, a fluid, liquid, or gaseous material, is a priori homogeneous. A cubic element of the order of 0.4 to $0.8\,\mu m^3$, always contains an enormous number of molecules submitted to Brownian agitation; the distribution of their speed vectors is thus statistically isotropic. When this fluid is flowing, a drift motion is superimposed to the anarchic agitation; the medium becomes birefringent with the same symmetry as the field of speed vectors. Studying this so-called *flow birefringence* brings important information about either the speed repartition of the molecules or the symmetry of the flowing molecules.

5.7.4. *Electrical Birefringence*

Electrical birefringence corresponds to a situation where the anisotropy of a material is created or modified by the application of an electric field.

5.7.4.1. *The Kerr Effect*

5.7.4.1.1. *Orientation of Molecular Electric Dipoles by an Electric Field*

Observed for the first time by Kerr in 1875, this effect is mostly observed in liquids and possibly in gases. Let us consider a liquid made of molecules in which the centers of positive and negative charges do not coincide, each molecule can be considered as an electric dipole. Good examples of such a liquid are given by mono-nitrobenzene ($C_6H_5NO_2$) and carbon disulfide CS_2 and $C_6H_5NO_2$, unfortunately both are toxic and explosive materials.

Materials		Kerr constant (SI units)
		Sodium D light
Benzene	C_6H_6	0.67×10^{-14}
Carbon disulfide	CS_2	3.56×10^{-14}
Chloroform	$CHCl_3$	-3.88×10^{-14}
Water	H_2O	5.1×10^{-14}
Nitrotoluene	$C_6H_5CH_3NO_2$	1.37×10^{-12}
Nitrobenzene	$C_6H_5NO_2$	2.44×10^{-12}

In the absence of any applied electric field, every tiny cell with a volume of the order of the cube of the wavelength has a global electric momentum equal to zero since the orientation of the molecules is at random, because of thermal agitation. An external electric field will try to orient all the molecules, an anisotropy will be created, and finally the liquid becomes birefringent, uniaxial, with the optical axis parallel to the direction of the electric field.

The experiment shows that Kerr-induced birefringence is proportional to the square value of the electric field, this is not at all surprising since the phenomenon should remain the same if the direction of the field is reversed. For a given wavelength λ, the birefringence Δn is given by

$$\Delta n = (n_e - n_o) = B\lambda E^2, \qquad (5.15)$$

where B, a constant specific of the liquid under consideration, is called the *Kerr constant*; it is usually positive and of the order of 10^{-13} to 10^{-12} (MKS units). B, as well as the index of refraction, varies with the wavelength λ. If n, n_e, and n_o are, respectively, the refractive index in the absence of a field, and of the extraordinary and ordinary indices when a field is applied, we have the following relations that have been established by Havelock:

$$(n_e + 2n_o) = 3n, \qquad (5.16.a)$$

$$\frac{B\lambda n}{(n^2 - 1)^2} = \text{constant.} \qquad (5.16.b)$$

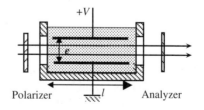

Polarizer   Analyzer

Figure 5.26. The Kerr cell. A Kerr cell is made of a cuvette with two parallel and transparent windows and filled with a Kerr active material in which are immersed the two plates of a planar capacitor. The electric field $E = V/e$ orients the molecules of the liquid which becomes birefringent and uniaxial.

5.7.4.1.2. Order of Magnitude of the Field Necessary for Producing the Kerr Effect

In its attempts to increase the degree of order in a liquid by giving a common orientation to all the molecular electric dipoles, an electric field is disturbed by the action of thermal agitation. Let us call M the electric dipole momentum of an individual molecule, its energy *in* an electric field E is equal to the scalar product $W_{elec} = ME$. The effectiveness of the orientation effect of the field is obtained by comparing W_{elec} to the thermal agitation energy W_{therm} of the molecule at temperature T K. To evaluate W_{therm}, we must examine the number of degrees of freedom of a molecule under consideration, as we just need an order of magnitude we will take $W_{therm} = kT = 25\,\text{meV}$ (room temperature). If the electric momentum is taken equal to 10^{-26} (MKS), the equality between the two energies is obtained for a field of 4×10^4 V/m, which corresponds to a voltage V of a few kilovolts if the distance between the two electrodes of Figure 5.27 is 1 cm.

5.7.4.1.3. Optical Kerr Switches

From a practical point of view, an important property of the Kerr effect is certainly the rapidity with which the birefringence is established after the application of an electric field, allowing optical switches with rise times as short as picoseconds. The first Q-switched lasers used nitrobenzene Kerr cells.

We now consider Figure 5.26, the Kerr cell is placed between a crossed polarizer and analyzer and the electric field is oriented at 45° in the direction of the polarizer. In the absence of any electric field applied to the Kerr cell, the liquid remains isotropic: the switch is "open," which means that it is opaque and that no light is transmitted. The switch is "closed" and becomes fully transparent if the Kerr cell behaves as a half-wave plate, the emerging

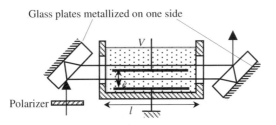

Figure 5.27. Measurement of Kerr-induced birefringence using a Jamin interferometer. A first parallel light beam generates two separate beams: one propagates between the two plates of the capacitor and the other one outside. A polarizer sets the polarization to be either ordinary or extraordinary inside the Kerr material. The two beams are then recombined and interfere on a photodetector.

vibration, which is symmetrical to the incident vibration, being parallel to the direction of the analyzer.

Numerical application: Calculate the voltage $V_{\lambda 2}$ necessary for a Kerr cell to be equivalent to a half-wave plate

$$\frac{2\pi}{\lambda}(n_e - n_o)l = \pi \quad \rightarrow \quad B\lambda\frac{1}{\lambda}\left(\frac{V_{\lambda/2}}{e}\right)^2 l = \frac{l}{2} \quad \rightarrow \quad V_{\lambda/2} = \frac{e}{\sqrt{2Bl}}. \quad (5.17)$$

As a first approximation, $V_{\lambda 2}$ doesn't depend on the color. For a cell with a reasonable size ($e = 1\,\mathrm{cm}$, $l = 10\,\mathrm{cm}$) filled with nitrobenzene: $V_{\lambda 2} = 14 \times 10^3$ V, corresponding to an electric field of 14×10^5 V/m.

5.7.4.1.4. Kerr Effect Rise Time

P being the electrical momentum per unit volume, it can be considered, as a first approximation, that its time evolution is a first-order phenomenon related to the electric field **E** by a first-order differential equation of the following kind:

$$\frac{d\boldsymbol{P}}{dt} + \frac{\boldsymbol{P}}{\tau} = \chi\boldsymbol{E} \quad \rightarrow \quad \boldsymbol{P} = \tau\chi\boldsymbol{E}(1 - e^{-t/\tau}). \quad (5.18)$$

According to (5.18), it is seen that if the electric field is initially equal to zero and is suddenly raised to some value **E**, then **P** is progressively oriented parallel to **E** and reaches, with a relaxation time τ, a constant limit value equal to $\tau\chi E$; χ is a proportionality coefficient characteristic of the Kerr activity of the liquid.

According to the molecules, two different mechanisms may be associated to the Kerr effect, the order of magnitude of the time constants are quite different. In the case of the first mechanism, the electrical molecular dipoles exist before the application of the electric field, the role of which is just to orient the dipoles. All the dipoles are coupled together by electrostatic forces, which are responsible for the viscosity of the liquid and, consequently, for the time constant τ. Mono-nitrobenzene, $C_6H_5NO_2$, is a typical example of this first category of Kerr active liquid, the time constant is of the order of a few nanoseconds (10^{-9} s).

Self-induced Kerr effect: For the second category of liquids, the electric dipole momentum is equal to zero in the absence of any applied electric field. The electric field then plays a double role: first, it creates electric dipoles by separating the positive and negative centers of charges in each molecule and, second, it orients all the dipoles. The appearance of electric dipole momentum is due to the deformation of the electronic clouds of the molecules; in this case, it is spoken of as the *self-induced Kerr effect*. In this second case Kerr constants, B, as well as the rise times, τ, are much smaller; carbon disulfide, CS_2, is a typical example, τ is of the order of a few picoseconds (10^{-12} s).

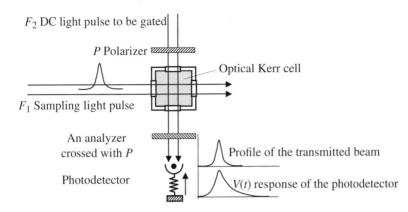

Figure 5.28. Principle of an optical gate. The two light beams F_1 and F_2 cross inside a cell filled with a Kerr active liquid. F_2 has a constant intensity and doesn't usually reach the photodetector, because of the crossed polarizer and analyzer. F_1 is a powerful and short light pulse which makes the cell birefringent, and so allows F_2 to reach the photodetector during the time of the pulse.

Static Kerr effect: The orientating electric field is a DC field or, as well, a *slowly varying* field. By *slowly* we mean that the value and the orientation of the field do not vary appreciably during a lapse of time of the order of the relaxation time; the electric field of radiowave or even of a microwave are considered to produce a static Kerr effect.

Optical Kerr effect: The orientating field is the electric field of a linearly polarized *light wave*. Characterization and utilization of the optical Kerr effect require rather important fields and had been made possible only after the appearance of lasers.

Optical sampling: Once we know how, rapidly and periodically, to open a gate, it becomes possible to perform optical sampling and to characterize optical signals having rise times shorter than the response time of a photodetector.

If the beam F_1 in Figure 5.28 is made of a succession of periodic short light pulses (period T_{signal}), and under the condition that the Kerr relaxation time (τ_{Kerr}) should be shorter than the duration of the light pulses, the beam transmitted from F_2 will be a succession of short pulses having the same periodicity. If the period of the light pulses is longer than the response time $\tau_{\text{photodetector}}$ of the photodetector, the electric signal V will be a succession of pulses of duration $\tau_{\text{photodetector}}$ and a periodicity of T_{signal}.

Autocorrelation Measurement of Picosecond Light Pulses

Mode synchronized lasers deliver periodic light pulses having a duration of the order of 1 ps or less. The response time of a photodetector and of the associated circuitry is always much longer. Hopefully the successive light pulses are very identical and produced with an accurate periodicity, which allows the determination of their time profile by way of optical sampling.

We refer to Figure 5.29; a light beam, made of periodic light pulses is first linearly polarized by a polarizer P and then split into two beams, F_1 and F_2. A half-wave plate makes the polarization of F_2 orthogonal to the polarization of F_1. The two beams cross inside a Kerr optical gate. An optical delay line, made of two translatable reflecting prisms, allows a careful tuning of the delay θ between F_1 and F_S light pulses, so that the pulses will coincide a little, completely, or not at all. . . . For a given adjustment of the delay line, the photodetector receives short periodic bursts of light that are shorter than its response time and thus integrates the signal and delivers a photocurrent I which is proportional to the hatched areas of Figure 5.29(c). The optical delay

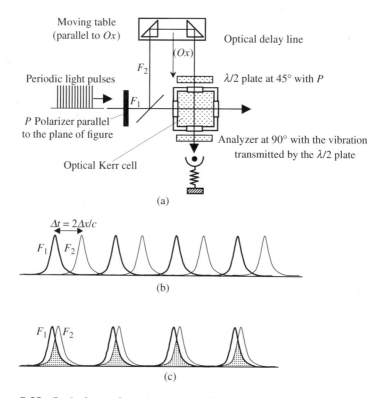

Figure 5.29. Optical sampler using an optical delay line. In the case of (a), the delay between the pulses of the two beams is so important that there is almost no superposition. In the case of (c), the pulses coincide over a long time; the photodetector is illuminated during the periods corresponding to the hatched areas and delivers a DC signal, which is proportional to these areas.

θ is easily varied by a translation of the prisms, a translation of Δx produces an augmentation (or a diminution) of the delay of $\Delta\theta = 2\Delta x/c$ ($\Delta\theta/\Delta x = 2/c$ = 6.7 ps/mm). The law of variation $I(\theta)$ is nothing other than the autocorrelation function of the time variation law $I(t)$ of an individual light pulse.

5.7.4.2. The Pockels Effect

5.7.4.2.1. Definition of the Pockels Effect

The German physicist Pockels discovered the Pockels effect, also called the electrooptic effect, in 1893. It is a consequence of the deformation of the electron clouds of the atoms under the action of an external electric field which produces a modification of the polarization of the medium and thus of the index of refraction. If the material was initially isotropic, the application of an electric field will make it uniaxial, with the optical axis parallel to the field. In an initially birefringent material the shape of the characteristic surfaces will change; a uniaxial material may become biaxial. The main properties of the electrooptic effect are the following:

- It's a linear effect: The deformation of the characteristic surfaces is proportional to the modulus of the field and not its squared value.
- Because of its "electronic" origin the electrooptic effect has a very short rise time, always subpicoseconds.
- There is no possibility of an electrooptic effect in centrosymmetric materials and especially in isotropic materials.

5.7.4.2.2. Mathematical Description of the Electrooptic Effect

In order to describe the electrooptic effect we will use the ellipsoid of the indices (see Annex 5.B), using a general axis of coordinates and in the absence of any external electric field, its equation is written as

$$\frac{x^2}{n_{xx}^2} + \frac{y^2}{n_{yy}^2} \frac{z^2}{n_{zz}^2} + 2\frac{yz}{n_{yz}} + 2\frac{zx}{n_{zx}} + 2\frac{xy}{n_{xy}} = 1. \qquad (5.19.a)$$

Using the following convention for contracting the indices (Voigt notations):

$$xx = 1, \quad yy = 2, \quad zz = 3, \quad yz = zy = 4, \quad zx = xz = 5, \quad xy = yx = 6,$$

the equation becomes

$$\frac{x^2}{n_1^2} + \frac{y^2}{n_2^2} \frac{z^2}{n_3^2} + 2\frac{yz}{n_4} + 2\frac{zx}{n_5} + 2\frac{xy}{n_6} = 1. \qquad (5.19.b)$$

5.7.4.2.3. *The Electrooptic Tensor*

When an external electric field E is applied, equations (5.19) remain quite similar except that the values of the coefficients will depend on the values of the projections of the field on the coordinate axis. As the variations are small we will admit that the variations $\Delta(1/n_i^2)$ of the various coefficients are related to the field components (E_x, E_y, E_z) through linear formulas. Using the following conventions for indexing the field components: $1 = x$, $2 = y$, $3 = z$; $\Delta(1/n_i^2)$ can be written using a matrix formulation and $[r_{ij}]$ is called the electrooptic tensor:

$$\Delta\left(\frac{1}{n_i^2}\right) = \sum_{j=1}^{j=3} r_{ij} E_j = \begin{bmatrix} \Delta(1/n_1^2) \\ \Delta(1/n_2^2) \\ \Delta(1/n_3^2) \\ \Delta(1/n_4^2) \\ \Delta(1/n_5^2) \\ \Delta(1/n_6^2) \end{bmatrix} = \begin{bmatrix} r_{11} & r_{12} & r_{13} \\ r_{21} & r_{22} & r_{23} \\ r_{31} & r_{32} & r_{33} \\ r_{41} & r_{42} & r_{43} \\ r_{51} & r_{52} & r_{53} \\ r_{61} & r_{62} & r_{63} \end{bmatrix} \begin{bmatrix} E_1 \\ E_2 \\ E_3 \end{bmatrix}. \tag{5.20}$$

As an example let us evaluate $\Delta(1/n_1^2)$ and $\Delta(1/n_6^2)$:

$$\Delta\left(\frac{1}{n_1^2}\right) = r_{11} E_1 + r_{12} E_2 + r_{13} E_3 = r_{xxx} E_x + r_{xxy} E_y + r_{xxz} E_z,$$

$$\Delta\left(\frac{1}{n_6^2}\right) = r_{61} E_1 + r_{62} E_2 + r_{63} E_3 = r_{xyx} E_x + r_{xyy} E_y + r_{xyz} E_z.$$

With eighteen elements the 6×3 electrooptic tensor is, a priori, rather complicated. Hopefully, and as is often the case in such situations, many elements are equal to zero, while the other elements have either equal or opposite values. The form of the r_{ij} tensor is determined uniquely by the point-group symmetry of the crystal. When the symmetry is high, only a few elements are not equal to zero. Triclinic crystals are the only crystals in which none of the eighteen elements is equal to zero.

In a centrosymmetric crystal, because of the existence of inversion symmetry, the properties should remain the same when the electric field is reversed. From equation (5.20) it is seen that all the elements of the tensor are equal to zero,

$$\Delta\left(\frac{1}{n_i^2}\right) = \sum_{j=1}^{j=3} r_{ij} E_j = \sum_{j=1}^{j=3} r_{ij}(-E_j) \quad \rightarrow \quad r_{ij} = 0.$$

The most important electrooptic materials are lithium niobate, $LiNbO_3$, potassium dehydrogenated phosphate, KH_2PO_4 (often designated as KDP), and potassium dideuterium phosphate, KD_2PO_4 (designated as KD*P).

Electrooptic Tensors of Some Crystals

(The values of the tensor elements are given in 10^{-12} m/V.)

$$\begin{bmatrix} 0 & 0 & 0 \\ 0 & 0 & 0 \\ 0 & 0 & 0 \\ a & 0 & 0 \\ 0 & a & 0 \\ 0 & 0 & b \end{bmatrix}$$

KDP ($\lambda = 0.633\,\mu$m).

Tetragonal $\overline{4}2m$ (m parallel to x),

$n_o = 1.5074, \quad n_e = 1.4669,$

$r_{41} = r_{52} = 8, \quad r_{63} = 11,$

KD*P ($\lambda = 0.633\,\mu$m).

Tetragonal $\overline{4}2m$ (m parallel to x),

$n_o = 1.502, \quad n_e = 1.462,$

$a = r_{41} = r_{52} = 8.8, \quad b = r_{63} = 24.1.$

$$\begin{bmatrix} 0 & a & b \\ 0 & -a & b \\ 0 & 0 & c \\ 0 & d & 0 \\ d & 0 & 0 \\ a & 0 & 0 \end{bmatrix}$$

LiNbO$_3$ ($\lambda = 0.633\,\mu$m).

Trigonal $3m$, m orthogonal to x,

$n_o = 2.286, \quad n_e = 2.200,$

$a = r_{12} = -r_{22} = -6.8, \quad b = r_{13} = r_{23} = 9.6,$

$c = r_{33} = 30.9, \qquad d = r_{51} = r_{42} = 32.6,$

$$\begin{bmatrix} 0 & 0 & 0 \\ 0 & 0 & 0 \\ 0 & 0 & 0 \\ a & 0 & 0 \\ 0 & a & 0 \\ 0 & 0 & a \end{bmatrix}$$

GaAs (measured at $\lambda = 0.9\,\mu$m).

Cubic $\overline{4}3m$, isotropic with no applied field,

$n_o = n_e = 3.6,$

$a = r_{41} = r_{52} = r_{63} = 1.1.$

Exercise: Write the equation of the index ellipsoid in a KDP crystal in some electric field: $\boldsymbol{E} = E_x \boldsymbol{x} + E_y \boldsymbol{y} + E_z \boldsymbol{z}$. The symmetry of the crystal is $\overline{4}2m$, Oz is a fourfold axis of symmetry and corresponds to the optical axis, and Ox and Oy are chosen to make an orthogonal trihedral. The initial ellipsoid index equations are

$$\frac{x^2}{n_o^2} + \frac{y^2}{n_o^2} + \frac{z^2}{n_e^2} = 1, \quad \text{with no applied field,}$$

$$\frac{x^2}{n_o^2} + \frac{y^2}{n_o^2} + \frac{z^2}{n_e^2} + 2r_{41}E_x yz + 2r_{41}E_y xz + 2r_{63}E_z xy = 1, \quad \text{with the field } \boldsymbol{E},$$

$$\frac{x^2}{n_o^2} + \frac{y^2}{n_o^2} + \frac{z^2}{n_e^2} + 2r_{63}E_z xy = 1, \quad \text{with a field parallel to } Oz.$$

We now have to determine the principal axes of the last ellipsoid, that is to say the referential for which the ellipsoid equation has rectangular terms. This needs well-known and rather tedious mathematical manipulations; we will only consider the special situation where the electric field is parallel to the optical axis Oz. The new referential is $(Ox'y'z)$, Ox' and Oy' are obtained after a rotation of $45°$ of Ox and Oy,

$$x = \frac{\sqrt{2}}{2}(x'+y'),$$
$$y = \frac{\sqrt{2}}{2}(-x'+y'),$$
$$\rightarrow \quad \left(\frac{1}{n_o^2} - r_{63}E_z\right)x'^2 + \left(\frac{1}{n_o^2} + r_{63}E_z\right)y'^2 + \frac{z^2}{n_e^2} = 1. \quad (5.21)$$

Equation (5.21) shows that the medium is now biaxial. As the value of $r_{63}E_z$ is small as compared to $1/n_o^2$ the three principal indices are given by

$$n_{x'} = n_o + \frac{n_o^3 r_{63}}{2} E_z, \quad n_{y'} = n_o - \frac{n_o^3 r_{63}}{2} E_z, \quad n_z = n_e. \quad (5.22)$$

Equation (5.22) contains an important general result according to which $n^3 r_{ij}$ is an important parameter to be used in such cases.

5.7.4.2.4. *Devices Using an Electrooptic Effect*

The Pockels effect, because of its linearity and because it requires lower voltages, is much more convenient than the Kerr effect for modulating a light beam. If the applied electric field is constant, the indices will vary and produce a phase and/or a polarization modulation of the light. If the crystal is placed between crossed polarizers, a polarization modulation is readily transformed into an amplitude modulation. Thanks to the very short time constant of the Pockels effect, very high frequencies of modulation can be reached.

The first thing to do in applying an electric field is to evaporate electrodes on two opposite sides of a crystal. The electric field can be perpendicular near to the direction of propagation of the light or collinear with it. The second arrangement, Figure 5.30(b) requires transparent electrodes.

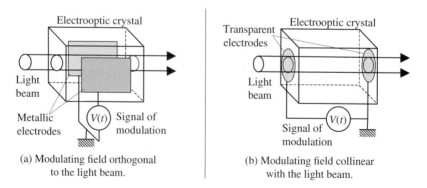

(a) Modulating field orthogonal
to the light beam.

(b) Modulating field collinear
with the light beam.

Figure 5.30. The two main arrangements for applying an electric field inside a crystal.

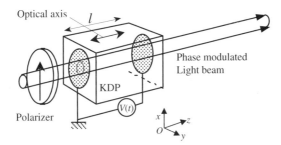

Figure 5.31. Phase modulator. A voltage $V(t)$ is applied to a KDP crystal; the index of refraction and the optical length are not constant, their variations follow the variation of the voltage $\rightarrow$ phase modulation of the transmitted beam.

Phase Modulator

We refer to Figure 5.31; an electric field is applied in the direction of the optical axis, which is also parallel to the propagation of an incident beam. Ox and Oy are the directions of the privileged polarizations; Oz is the direction of propagation. A polarizer fixes the incident polarization to be parallel to Ox, for this polarization the index of refraction is $n_x = n_o + (n_o^3 r_{63}/2)/(V(t)/l)$. If ω is the angular frequency of the light wave and $E_{\text{optic,in}}\,\boldsymbol{x}\sin \omega t$ is the *incident* optical vibration, the output vibration $\boldsymbol{E}_{\text{optic,out}}$ can be written as

$$\text{E}_{\text{optic,out}} = E_{\text{optic,in}}\,\boldsymbol{x}\sin\left[\omega\left(t - n_x\frac{l}{c}\right)\right],$$

$$E_{\text{optic,out}} = \boldsymbol{x}E_{\text{optic,in}}\,\sin\left[\omega t - \frac{\omega n_o^3 r_{63}}{2c}V(t) + \phi\right], \qquad (5.23)$$

$$E_{\text{optic,out}} = \boldsymbol{x}E_{\text{optic,in}}\,\sin[\omega t - \pi\lambda V(t) + \phi].$$

Equation (5.23) represents a light wave (vacuum wavelength λ), phase modulated by the voltage $V(t)$.

Amplitude Modulator

The arrangement of Figure 5.32 allows the transformation of a phase modulation into an amplitude modulation. A KDP crystal, with its optical axis parallel to the direction of propagation Oz, is placed between a polarizer and a crossed analyzer. The polarizer has been oriented so as to be parallel to one of the privileged directions (Ox) of vibration in the KDP crystal when no voltage is applied. According to equation (5.21), the privileged vibrations are along Ox' and Oy' in the presence of an electric field parallel to Oz. Before reaching the analyzer the beam crosses a quarter wavelength plate, the neutral axes of which are, respectively, parallel to Ox' and Oy'.

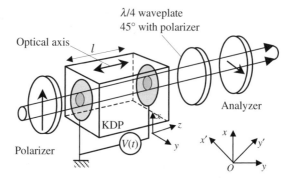

Figure 5.32. Amplitude modulator. The privileged vibrations in KDP are along Ox and Oy if no voltage is applied, and along Ox' and Oy' in the presence of an electric field parallel to Oz, $(Ox, Ox') = 45°$. The polarizer and the analyzer are, respectively, parallel to Ox and Oy. The neutral axes of the $\lambda/4$ waveplate are parallel to Ox' and Oy'.

The electric field of the optical wave arriving at the KDP crystal is written as

$$E_{\text{optic,in}} = E_0 \boldsymbol{x} = \frac{1}{\sqrt{2}} E_0 (\boldsymbol{x'} + \boldsymbol{y'}).$$

The $\boldsymbol{x'}$ and $\boldsymbol{y'}$ components of the incident optical vibration are in phase, after the crystal there is a phase difference Γ which, according to equation (5.22), is given by

$$\Gamma = \frac{2\pi l}{\lambda}(n_{x'} - n_{y'}) = \frac{2\pi}{\lambda} n_o^3 V_{(t)} = \pi \frac{V_{(t)}}{V_\pi} \quad \text{with} \quad V_\pi = \frac{\lambda}{2 n_o^3}. \tag{5.24}$$

The electric field and the light intensity after the crystal are given by

$$\boldsymbol{E}_{\text{optic,out}} = \frac{1}{\sqrt{2}} E_0 (\boldsymbol{x'} e^{-j\Gamma} + \boldsymbol{y'}) = \boldsymbol{y} \frac{E_0}{2}(1 - e^{-j\Gamma}),$$

$$I_{\text{out}} = \boldsymbol{E}_{\text{optic,out}} \boldsymbol{E}^*_{\text{optic,out}} = \frac{E_0^2}{4}(1 - e^{-j\Gamma})(1 - e^{+j\Gamma}), \tag{5.25}$$

$$I_{\text{out}} = E_0^2 \sin^2 \frac{\Gamma}{2} = E_0^2 \sin^2 \left(\frac{\pi}{2} \frac{V_{(t)}}{V_\pi} \right).$$

The graph of I_{out} versus V is plotted in Figure 5.33(b). Near the origin the curve can be assimilated to a parabola, which is not well adapted for a linear modulation. Let us evaluate the output signal in the case of a sine signal $V(t) = aV_\pi \sin \omega t$ of small amplitude (a is supposed to be much less than unity):

$$V(t) = aV_\pi \sin \omega t \quad \rightarrow \quad \frac{I_{\text{out}}}{E_0^2} = \sin^2 \left(\frac{\pi a}{2} \sin \omega t \right) \approx \left(\frac{\pi a}{2} \right)^2 \sin^2 \omega t. \tag{5.26}$$

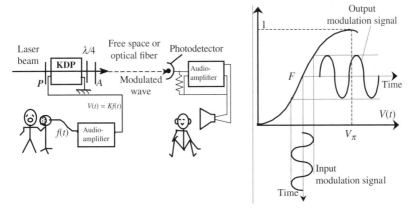

Figure 5.33. Carrying information by an amplitude modulated light beam.

According to equation (5.26), the variations of I_{out} do not reproduce the modulating signal $V(t)$; this drawback could be overcome by using an electrical bias equal to $V_\pi/2$ (the working point being in coincidence with the turning point F). We are going to show how a $\lambda/4$ wave plate provides an elegant optical solution, the phase difference Γ is increased by $\pm\pi/2$, the choice of the sign corresponds to the fact that the polarizer is parallel to one bisector, or the other, of the angle (x', y'); in formula (5.25), Γ should be replaced by $(\Gamma \pm \pi/2)$:

$$I_{out} = E_0^2 \sin^2\left(\frac{\Gamma}{2} \pm \frac{\pi}{4}\right) = E_0^2[1 \pm \sin\Gamma] \approx \frac{E_0^2}{2}[1 \pm \pi a \sin\omega t]. \qquad (5.27)$$

5.7.5. Magnetic Birefringence

5.7.5.1. The Faraday Effect

The application of a magnetic field makes a material birefringent. A magnetic field involves a cross product and its definition needs a referential: for that reason, vectors of that kind are called *axial vectors* (or also *pseudo vectors*), in opposition to *ordinary* vectors, such as an electric field, which are called *polar vectors*. The symmetries of the two kinds of field vectors are quite different, and so are the properties of electrical and magnetic birefringence.

Faraday studied the action of a magnetic field on the optical properties of a transparent material for the first time in 1845. The basic experiment performed by Faraday is described in Figure 5.34: a linearly polarized light beam propagates along the direction of a magnetic field to which is submitted a glass rod, *the polarization remains linear but its direction rotates* through an angle α which is proportional to the length l of the medium traversed and also to the magnetic field H. The direction of rotation remains the same if the

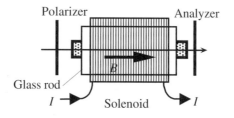

Figure 5.34. Basic Faraday effect experiment. A transparent material is put along the axis of a solenoid fed with an electric current I. A rectilinear polarized light beam propagates parallel to the magnetic field. During the propagation, the polarization remains rectilinear, but rotates through an angle, which is proportional to the length of the traversed medium and to the magnetic field. The sign of the rotation changes with the direction of the field. The optical vibration rotates in the same direction as the current in the coil, whatever the beam propagation direction along the axis.

propagation is reversed and is attached to the direction of the magnetizing current. α is given by equation (5.28):

$$\alpha = \rho l H. \tag{5.28}$$

The proportionality coefficient ρ, designated as the Verdet constant, is considered positive when the rotation occurs in the direction of the current. In most cases ρ is positive; materials in which it is negative can be found, the most famous being iron salts and potassium bichromate, which often correspond to materials with paramagnetic atoms. The Verdet constant is very sensitive to the color of light; an order of magnitude is one degree per tesla and per centimeter for liquids and solids (at $0.5\,\mu\text{m}$). In a field of 1 T (10^4 Gauss), which is rather an important value, a 1 cm long cell, filled with water produces a rotation of about 2°. This angle is multiplied by a factor of about seven if the water cell is replaced by a 1 cm long piece of heavy flint glass containing lead oxide.

5.7.5.2. *Comparison with Natural Optical Activity*

The Faraday effect bears a resemblance to natural optical activity, the main difference comes from the axial character of the magnetic field and has already been mentioned: it's a *nonreciprocal effect*, which simply means that the direction of rotation is fixed by the magnetic field and is independent of the direction of propagation. The polarization behaves differently in the case of natural optical activity where the direction of propagation is reversed with the direction of propagation.

The best illustration of the difference between the two situations is observed when light is reflected by a mirror and makes a double transit inside the sample. In Figure 5.35(a) a levorotatory active sample is crossed by a light beam, the polarization rotates at some angle; the light is reflected by a mirror

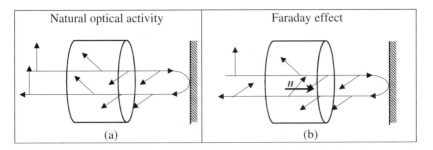

Figure 5.35. Comparison between natural optical activity and the Faraday effect. After a double transit, the input and return polarizations are parallel in the case of natural activity and orthogonal in the case of the Faraday effect.

and again crosses the sample, the polarization rotates the same angle, but in the other direction and is finally parallel to its initial orientation. In Figure 5.35(b) the sample is submitted to a magnetic field, the second transit produces a rotation α in the same direction as the first: the initial and final polarizations don't coincide and make an angle equal to 2α.

5.7.5.3. *Optical Faraday Isolators*

In order that a source emitting coherent electromagnetic waves works correctly, which means with a constant power and stable frequency, any feedback, even very small, should be carefully avoided. This problem is very general and is also met in microwave where the generators (klystron, carcinotron, or Gunn diode, . . .) are always efficiently decoupled from their load by a *uniline component*. In the case of lasers, as in the case of the previous oscillators, the working conditions are determined from a stationary wave pattern inside some resonator. A part of the signal coming back to the source contributes to the stationary waves process, the properties of which then depend on the phase of the return signal.

A Faraday cell is placed between a polarizer and an analyzer making an angle of 45°. The current is tuned for a 45° Faraday rotation for one transit, and 90° for a double pass. The emitted light is polarized at 45° with regard to the analyzer. The feedback light, which is polarized by the analyzer and which takes, after a 45° Faraday rotation, a polarization orthogonal to the polarizer is blocked and doesn't reach the laser.

5.7.5.4. *The Faraday Effect Ammeter*

The Faraday effect proves to be useful for measuring electric currents thanks to the magnetic field produced in a coil. The main interest of such a device comes from the high electric insulation between the wire in which the current is circulating and the meter. A basic arrangement for measuring a current is

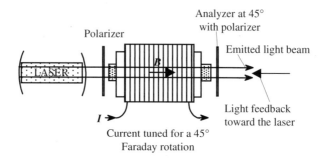

Figure 5.36. Optical Faraday isolator.

described in Figure 5.34, the electric wire is wound around a glass rod with parallel end faces; the measurement of the rotation of the polarization and equation (5.28) give the value of the intensity.

Currents in a lead raised to a very high voltage can be easily measured using a Faraday ammeter, the corresponding arrangement is shown in Figure 5.37. An optical fiber is wound around the lead, in this case the current is not the same at the different points of the Faraday material: an integral is involved. Hopefully, Ampère's theorem shows that the rotation of the polarization is proportional to the current. If N is the number of turns made by the optical fiber around the wire, we have

$$\alpha = \int_{\text{fiber}} \rho H \, dl = \rho \int_{\text{fiber}} H \, dl = \rho N I. \qquad (5.29)$$

An important effort in developing optical fiber ammeters has been produced between 1980 and 1990; unfortunately, they appear to be very sensitive

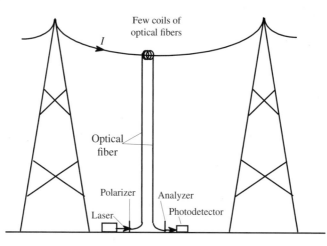

Figure 5.37. Optical fiber ammeter for measuring the current in a high voltage wire.

to temperature variations and also to mechanical vibrations (50 or 60 Hz) that are often met in electrotechnical equipment. Nevertheless, they remain quite attractive because of their large dynamics and their short rise time.

5.7.5.5. *The Cotton-Mouton Effect, the Voigt Effect*

The Voigt and Cotton-Mouton effects have been, respectively, discovered in 1902 and 1905 and have taken the names of their discoverers. The difference with the Faraday effect is that *the magnetic field is now applied perpendicular to the propagation of the light.*

The Voigt and Cotton-Mouton effects are very similar; the first is concerned with gases and the second with liquids. As for the optical Kerr effect, the birefringence is a consequence of the orientation of the molecules by the

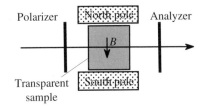

Figure 5.38. The Voigt and Cotton-Mouton effects. A transparent isotropic material is made birefringent (uniaxial) after the application of a magnetic field. The Voigt effect (gas) and the Cotton-Mouton effect (liquid) correspond to propagation along a direction orthogonal to the field.

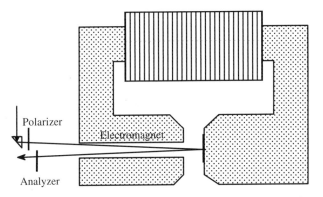

Figure 5.39. The Magnetooptic Kerr effect. During the reflection on a mirror made of a ferromagnetic material the polarization of the reflected beam is controlled by the presence of a magnetic field.

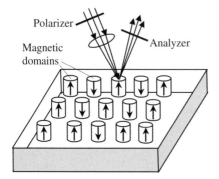

Figure 5.40. Optical reading of magnetic memories. A lens focuses the reading beam on a magnetic domain; the polarization of the reflected beam carries information, which allows the determination of the magnetization at the focus. Mechanical scanning allows reading the different domains.

magnetic field. In the absence of any external field the material is isotropic and becomes uniaxial after the application of a magnetic field. The difference between the values of the principal indices is proportional to the square of the field with a formula quite similar to the Kerr effect:

$$n_1 - n_2 = C\lambda H^2 = C\lambda \frac{B^2}{\mu_0}. \qquad (5.30)$$

For organic materials such as mono-nitrobenzene, the proportionality coefficient C is of the order of 10^{-14} MKS units. A field B of 4 T (40,000 Gauss) is necessary to produce a phase difference of $\pi/4$ after 25 cm. Weaker, by about three orders of magnitude, the Voigt effect has mainly a theoretical interest because it allows us to deduce a magnetooptical effect from the Zeeman effect. In the same way, electrooptical effects are connected to the Stark effect.

5.7.5.6. *The Magnetooptic Kerr Effect*

The magnetooptic Kerr effect, which should not be confused with the electrooptic Kerr effect, occurs when a linearly polarized light beam is reflected at about normal incidence on a ferromagnetic material having its magnetization normal to the plane of the mirror: the reflected beam is elliptically polarized. The ellipse is very flat (ratio of the two axes $\approx 10^{-3}$). The angle of the main axis with the direction of the linear incident polarization is proportional to the magnetization, for saturated iron this angle is small but easily measurable (≈ 20 min). Using appropriate materials, such as YIG (Yttrium Iron Garnet) more important angles can be obtained (a few degrees). The magnetooptic Kerr effect is used to optically read magnetic memories in which the information is stored in magnetic domains having their magnetization perpendicular to the plane of the sample; an upside magnetization being, for example, associated to a *zero*, and a downside to a *one*.

Annex 5.A

Ray Tracing in Uniaxial Media

5.A.1. Construction of the Refracted Beam on an Isotropic/Uniaxial Interface

We will limit ourselves to the most simple and frequent case where the optical axis is parallel to the plane of incidence. In all the following examples the medium is negative uniaxial ($n_e - n_o < 0$). In all cases the ordinary ray and the normal ray coincide and may be obtained either by Descartes' or Huygens' constructions, this not true for the extraordinary ray and normal ray. The polarizations are represented by the electric displacement vectors and D_o; both vectors are parallel to the wave plane, the extraordinary polarization D_e is the projection of the optical axis on the wave plane and the ordinary polarization is perpendicular to the optical axis.

In the case of normal incidence, the directions of the ordinary and extraordinary rays in Figure 5.A.1 coincide, however, they don't propagate at the same speed: a linear incident polarization will give an elliptical refracted polarization. For a general angle of incidence, there are two refracted rays with orthogonal polarizations (D_e and D_o in Figure 5.A.1), they can also be traced using the Descartes or Huygens constructions.

At normal incidence ordinary and extraordinary rays remain collinear, a linear incident polarization giving an elliptical refracted polarization. For an oblique incidence there are two different refracted rays, polarized at right angles, they are different from the normals, which have not been shown in Figure 5.A.2. Rays have been obtained by the Huygens construction; normals would be obtained by the Descartes construction.

When the optical axis is perpendicular to the interface and parallel to the plane of incidence (Figure 5.A.3), any ray arriving at normal incidence can be considered as ordinary, whatever its polarization: the incident and refracted beams have exactly the same polarization. An oblique incidence with any polarization gives two refracted beams having orthogonal polarizations.

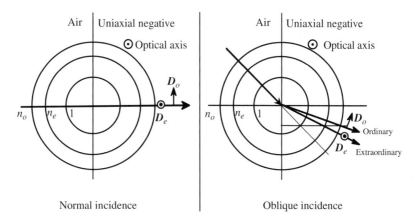

Figure 5.A.1. Ray tracing when the optical axis is orthogonal to the interface and parallel to the plane of incidence. Intersections of the characteristic surfaces are just circles. The second medium is equivalent to some isotropic medium with two indices n_o and n_e. Rays and normals coincide.

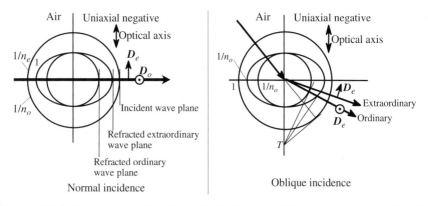

Figure 5.A.2. Ray tracing when the optical axis is parallel to the interface and to the plane of incidence.

In Figure 5.A.4 the optical axis is still parallel to the plane of incidence, but is now oblique with regard to the interface. Even at normal incidence there are two different refracted rays, the ordinary ray being in the prolongation of the incident ray.

5.A.2. Refraction of a Monochromatic Beam by a Birefringent Prism

All the prisms considered in this section have been cut in a uniaxial negative material ($n_e < n_o$). The incident rays are monochromatic, orthogonal to the first interface and are not polarized (natural light).

The case described in Figure 5.A.5 is very simple, the optical axis is perpendicular to the plane of incidence and the constructions of the refracted rays are made as if we had two isotropic materials of respective indices n_o and n_e. After the first interface, ordinary and extraordinary rays are collinear but they propagate at different speeds, the second refraction orientated the two families of rays in two different directions. Either Descartes' or Huygens' constructions can be used.

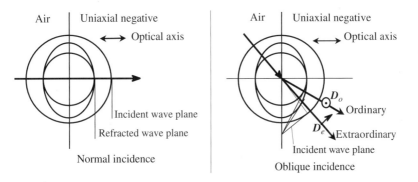

Figure 5.A.3. Ray tracing when the optical axis is parallel to the plane of incidence and perpendicular to the interface.

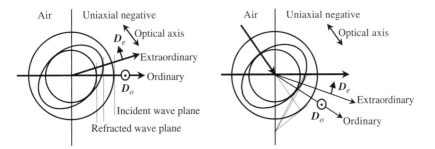

Figure 5.A.4. Ray tracing when the optical axis is parallel to the plane of incidence and has no special orientation with regard to the interface.

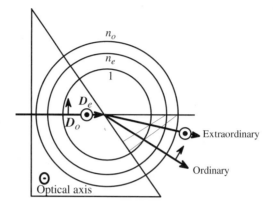

Figure 5.A.5. Normal incidence, optical axis parallel to the first interface and orthogonal to the plane of incidence. The behavior of the race is the same as if we had two isotropic materials of refractive indices n_e and n_o. Descartes' construction has been used.

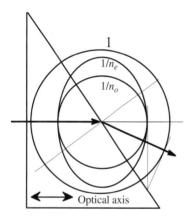

Figure 5.A.6. Normal incidence, optical axis orthogonal to the first interface and parallel to the plane of incidence. Whatever its direction, the incident polarization is orthogonal to the optical axis and should be considered as ordinary, the second refraction gives only one refracted beam with the same polarization.

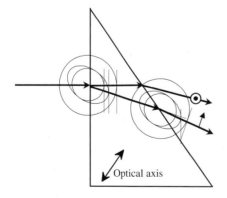

Figure 5.A.7. Normal incidence on the first interface, optical axis parallel to the plane of incidence in any direction. For the sake of simplicity the construction of the emerging ordinary ray has not been shown.

5.A.3. Separation of Extraordinary Rays from Ordinary Rays by a Birefringent Plate

Figure 5.A.8 shows an unpolarized ray arriving at a parallel birefringent plate made of calcite; the problem is to calculate the distance Δ between the extraordinary and ordinary emerging rays. The plate has been cut so that the optical axis is parallel to the plane of incidence and makes an angle of $45°$ with the surface of the plate. We introduce two axes, Ox and Oy, respectively, parallel and orthogonal to the optical axis. Let us, respectively, call θ and α the angles of the extraordinary ray OO' with the normal to the plate and with Oy, we have $\theta = (\alpha - 45°)$.

The equation of the extraordinary sheet of the inverse index surface is

$$n_o^2 x^2 + n_e^2 y^2 - 1 = 0, \quad \text{by differentiation} \rightarrow n_o^2 x\, dx + n_e^2 y\, dy = 0,$$

the slope of the tangent at point (x, y) is easily deduced:

$$\frac{dy}{dx} = -\frac{n_o^2}{n_e^2} \frac{x}{y}.$$

Let (X, Y) be the coordinates of the point of intersection of the extraordinary ray with the extraordinary sheet, N is defined by the fact that the tangent is parallel to the plate and has a slope of -1:

$$-\frac{n_o^2}{n_e^2} \frac{X}{Y} = -1 \quad \rightarrow \quad \tan \alpha = \frac{Y}{X} = \frac{n_e^2}{n_o^2} \quad \rightarrow \quad \tan \theta = \tan(\alpha - 45°) = \frac{n_o^2 - n_e^2}{n_o^2 + n_e^2},$$
$$\tag{5.A.1}$$

$$\Delta = e \tan \theta = e \frac{n_o^2 - n_e^2}{n_o^2 + n_e^2}.$$

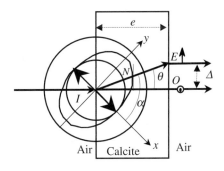

Figure 5.A.8. A parallel plate of calcite of thickness e, has been cut and polished so that the optical axis makes an angle of $45°$ with the end faces. An incident ray at normal incidence is split into an ordinary (IO) and an extraordinary (IE) ray and finally gives two parallel transmitted beams.

5.A.4. Realization of Polarizers

5.A.4.1. *Nicol Polarizer*

Because of the noticeable difference between the values of the ordinary and extraordinary indices in calcite, it is possible to find conditions where the ordinary beam is totally reflected, while the extraordinary beam is fully transmitted. The first arrangement of this kind had been proposed in 1828 by Nicol, and for many years a polarizer was simply designated as a "*Nicol*," an exhaustive description is not so easy and requires a good knowledge of calcite crystals and the way they can be cleaved; the development of a Nicol polarizer has been made possible by the large size of calcite samples than can be found (several tens of cm³).

Conditions for total internal reflection should be obtained from the direction of the normal and obtained using the Descartes construction. The

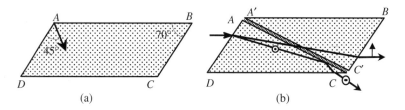

Figure 5.A.9. (a) shows a natural rhombus of calcite ($AD \approx 2\,\mathrm{cm}$, $AB \approx 5\,\mathrm{cm}$). In (b) it is shown how to make a polarizer: the sample is sawed along the diagonal plane AC, the two faces are carefully polished and reassembled using a suitable glue (initially Canada balsam) with a refractive index (1.53) intermediate between the two indices of calcite (1.49, 1.66). The ordinary ray is totally reflected by the glue, the extraordinary ray being transmitted.

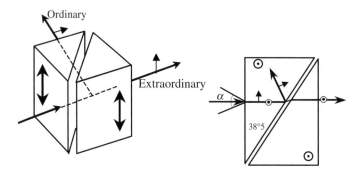

Figure 5.A.10. Glan prisms. The gap between the two half-prisms is filled with glue for the Glan-Thomson prism and with air for the Glan-Foucault prism.

thickness of the film of glue should be very thin but, however, thick enough to avoid an optical tunnel effect. The lower face *DC* is painted black to absorb the light. The incident and transmitted beams are parallel.

5.A.4.2. *Glan-Thomson, Glan-Foucault, and Wollaston Prisms*

The Nicol prism is only of historical interest, other arrangements such as the Glan prisms are now preferred, they are based on the same principle and also made of calcite, they need less material, and the orientation of the optical axis is easier. A Glan polarizer is made of two identical half-prisms separated by a thin film of glue (Glan-Thomson) or, more simply, of air (Glan-Foucault). To be totally reflected on the intermediate film a ray must lie inside a cone having its axis orthogonal to one face of the prism, at an angle of 30° for a Glan-Thomson prism, and at 10° for a Glan-Foucault prism. The latter prism having the advantage of accepting more powerful beams, which is important in the case of laser beams.

The Wollaston prism is a polarizing beam-splitter, which orientates one polarization in one direction and the orthogonal polarization in another direction. It is made of calcite or of quartz of two half-prisms assembled as indicated in Figure 5.A.11, they can be cemented or optically contacted. The trick is that an ordinary ray in the first section becomes an extraordinary ray in the second, and vice versa. According to the prism angle the separation between the emerging beams ranges from 10° to 45°.

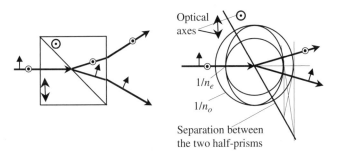

Figure 5.A.11. Wollaston beam splitter. The constructions have been drawn for the case of quartz, which is a positive uniaxial material.

5.A.5. Birefringence and Dispersion

Both ordinary and extraordinary indices vary with color. A spectroscope using a birefringent prism gives two spectra. The angular separation between ordinary and extraordinary spectra is much larger than the separation between the blue and red light rays. The dispersion of calcite is normal, angular deviation is more important for blue light than for red light.

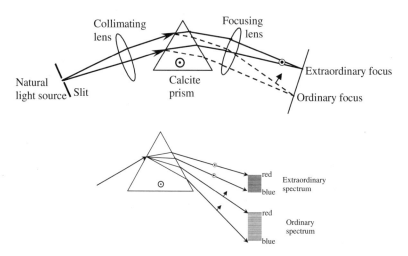

Figure 5.A.12. Birefringence and dispersion.

Annex 5.B

Characteristic Surfaces in Anisotropic Media

5.B.1. Fresnel Formulas in Anisotropic Media

The relative positions of the four basic vectors $\boldsymbol{E}, \boldsymbol{D}, \boldsymbol{H}, \boldsymbol{B}$ were easy to find from the elementary geometrical properties of a vector product, we now want to establish quantitative relationships between the frequency of the waves, the module, and the direction of the wave vector. We also want to find the rules that give the directions of the privileged vibrations $\boldsymbol{D}'$ and $\boldsymbol{D}''$.

We start from Maxwell's equations for planar harmonic waves and obtain the wave equation for the electric field

$$\boldsymbol{k} \wedge \boldsymbol{E} = \omega \mu_0 \boldsymbol{H} \quad \text{and} \quad \boldsymbol{k} \wedge \boldsymbol{H} = -\omega[\varepsilon]\boldsymbol{E},$$

$$\boldsymbol{k} \wedge (\boldsymbol{k} \wedge \boldsymbol{E}) + \omega^2 \mu_0 [\varepsilon]\boldsymbol{E} = 0. \tag{5.B.1}$$

To use this tensor equation we choose the principal axes of the tensor as the axes of coordinates,

$$\begin{pmatrix} \omega^2 \varepsilon_X \mu_0 - k_y^2 - k_z^2 & k_x k_y & k_x k_z \\ k_x k_y & \omega^2 \varepsilon_Y \mu_0 - k_x^2 - k_z^2 & k_y k_z \\ k_x k_z & k_y k_z & \omega^2 \varepsilon_Z \mu_0 - k_x^2 - k_y^2 \end{pmatrix} \begin{pmatrix} E_x \\ E_y \\ E_z \end{pmatrix} = 0. \tag{5.B.2}$$

Equation (5.B.2) represents a set of three homogeneous and linear equations that allow the determination of the three components of the electric field,

$$\begin{cases} (\omega^2 \varepsilon_X \mu_0 - k_y^2 - k_z^2)E_x + k_x k_y E_y + k_x k_z E_z = 0, \\ k_x k_y E_x + (\omega^2 \varepsilon_Y \mu_0 - k_x^2 - k_z^2)E_y + k_y k_z E_z = 0, \\ k_x k_z E_x + k_y k_z E_y + (\omega^2 \varepsilon_Z \mu_0 - k_x^2 - k_y^2)E_z = 0. \end{cases} \tag{5.B.3}$$

If we want the set of equations to have a solution other than the trivial solution ($E_x = E_y = E_z = 0$), the determinant should be equal to zero:

$$\begin{vmatrix} (\omega^2 \varepsilon_X \mu_0 - k_y^2 - k_z^2) & k_x k_y & k_x k_z \\ k_x k_y & (\omega^2 \varepsilon_Y \mu_0 - k_x^2 - k_z^2) & k_y k_z \\ k_x k_z & k_y k_z & (\omega^2 \varepsilon_Z \mu_0 - k_x^2 - k_y^2) \end{vmatrix} = 0. \quad (5.B.4)$$

In the three-dimensional space of vectors $\boldsymbol{k}$ (k_x, k_y, k_z) equation (5.B.4) represents a sixth-degree surface that is sometimes called the *surface of the normals*, since $\boldsymbol{k}$ is orthogonal to the wave surfaces. It's more convenient to introduce the *surface of the indices* which is just homothetic with a ratio equal to c/ω (c is the speed of light and ω is the angular frequency).

Let us now express the wave vector $\boldsymbol{k}$ versus its unit vector $\boldsymbol{s}$ (s_x, s_y, s_z), the angular frequency, the vacuum wave vector modulus $k_0 = \omega/c$, the speed of propagation V, and the refractive index $n = c/V$ associated with the planar wave under consideration,

$$\boldsymbol{k} = k\boldsymbol{s} = (\omega/V)\boldsymbol{s} = nk_0\boldsymbol{s}.$$

Taking into account the fact that $s_x^2 + s_y^2 + s_z^2 = 1$, the following two equivalent equations are obtained; they are called the *Fresnel equations for normals*:

$$s_x^2 \frac{n_X^2}{n^2 - n_X^2} + s_y^2 \frac{n_Y^2}{n^2 - n_Y^2} + s_z^2 \frac{n_Z^2}{n^2 - n_Z^2} = 0, \quad (5.B.5)$$

$$\frac{s_x^2}{n^2 - n_X^2} + \frac{s_y^2}{n^2 - n_Y^2} + \frac{s_z^2}{n^2 - n_Z^2} = \frac{1}{n^2}. \quad (5.B.6)$$

Equation (5.B.6) is a second-degree equation if we take n^2 as the unknown; we will admit that for a given direction $\boldsymbol{s}$ (s_x, s_y, s_z), two positive solutions are found for n^2 and, consequently, two real solutions, n' and n'', for the index of refraction.

5.B.2. Surface of the Normals (or Surface of the Indices)

The surfaces that we are now introducing are useful for geometrical constructions of the normals to the wave surfaces (Descartes' construction).

Given a planar wave having wave planes orthogonal to some unit vector $\boldsymbol{s}$ of (s_x, s_y, s_z) in some coordinate axis ($Oxyz$), starting from the origin O and along the direction of $\boldsymbol{s}$, we plot two points P' and P'' such as

$$OP' = n' \quad \text{and} \quad OP'' = n''. \quad (5.B.7)$$

When the direction of $\boldsymbol{s}$ is varied, the points P' and P'' move along a double sheet surface which is the *index surface*. The planes drawn orthogonal to OP' and OP'' are wave planes: there are two families of plane waves.

Figure 5.B.1. The two sheets of the index surface.

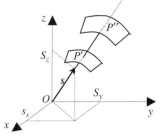

Equation of the index surface: The coordinates of points P' and P'' are given by

$$x = s_x n, \quad y = s_y n, \quad z = s_z n \quad \rightarrow \quad x^2 + y^2 + z^2 = n^2.$$

The equation of the index surface is

$$x^2 n_x^2 (n^2 - n_y^2)(n^2 - n_z^2) + y^2 n_y^2 (n^2 - n_x^2)(n^2 - n_z^2) + z^2 n_z^2 (n^2 - n_y^2)(n^2 - n_x^2) = 0.$$

The equation with the (xOy) plan is

$$(x^2 + y^2 - n_z^2)[x^4 n_x^2 + y^4 n_y^2 + x^2 y^2 (n_x^2 + n_y^2) - n_x^2 n_y^2 (x^2 + y^2)] = 0,$$

$$(x^2 + y^2)(x^2 + y^2 - n_z^2)(x^2 n_x^2 + y^2 n_y^2 + n_x^2 + n_y^2) = 0.$$

$$\text{"circle"} \qquad\qquad \text{"ellipse"}$$

Figure 5.B.2 shows the general aspect of the index surface which, except for a center of symmetry, is rather complicated. Because of the existence of two umbilical points, D' and D'', the most general birefringent material is said to be *biaxial*, OD' and OD'' are called the two optical axes of the material. For a propagation occurring along the optical axes the two refractive indices are equal.

5.B.3. Wave Surface (or Surface of the Inverse of the Indices)

In Section 5.4.2 we have described Huygens' wavelets inside an anisotropic material, these wavelets had been defined as the set of points reached, after some propagation time τ, by a vibration that had been coming from the origin of coordinates. For a given homogeneous material all the wavelets are homothetic, initially we had decided to make $\tau = 1\,\text{s}$, thus giving a simple interpretation of OP which was simply equal to the speed of light V along the light ray.

Instead of plotting a length equal to V, we can also plot a length equal to V/c, which is just the inverse of the index of refraction. We will call the *surface of the inverse of the indices* this new surface; the expression is not very elegant but it avoids any confusion. Other expressions can be found in text-

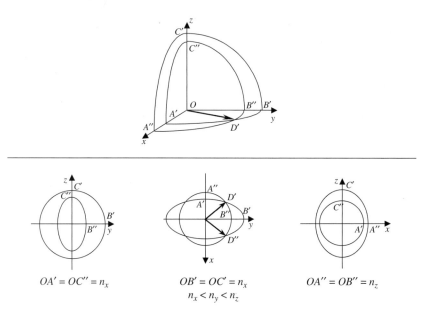

$$OA' = OC'' = n_x$$

$$OB' = OC' = n_x$$
$$n_x < n_y < n_z$$

$$OA'' = OB'' = n_z$$

Figure 5.B.2. Surface of the indices.

books, such as *ray surfaces*, because they are used for the geometrical construction of light rays (Huygens' construction).

Relation Between Index and Inverse Index Surfaces

The relative positions of the two surfaces correspond to a well-known, but rather complicated, geometrical transformation which is shown in Figure 5.B.3 and clearly illustrates the fact that the wave surface (*n*-surface) is the envelope of the wave plane.

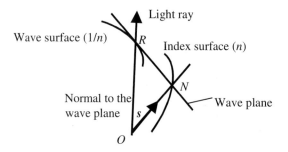

Figure 5.B.3. Transformation of the surface wave into the index surface. A light ray intersects the surface wave at point R, the wave plane intersects the wave surface at point R; the projection of the origin on the wave plane belongs to the wave surface.

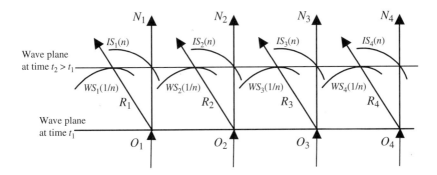

Figure 5.B.4. Propagation of a plane wave with the horizontal wave in an anisotropic material. WS_i are wave surfaces (Huygens' wavelets). IS_i are index surfaces. R_i and N_i are, respectively, the light rays and the normals. A wave plane is the envelope of the different Huygens wavelets.

5.B.4. Index Ellipsoid

We consider a biaxial material, and the axis in which the permittivity tensor is diagonal. We introduce the indices of refraction associated to the principal dielectric constants:

$$[\varepsilon] = \begin{bmatrix} \varepsilon_X & 0 & 0 \\ 0 & \varepsilon_X & 0 \\ 0 & 0 & \varepsilon_X \end{bmatrix}, \quad n_X = \sqrt{\frac{\varepsilon_X}{\varepsilon_0}}, \quad n_Y = \sqrt{\frac{\varepsilon_Y}{\varepsilon_0}}, \quad n_Z = \sqrt{\frac{\varepsilon_Z}{\varepsilon_0}}.$$

By definition the equation of the index ellipsoid is

$$\frac{x^2}{n_X^2} + \frac{y^2}{n_Y^2} + \frac{z^2}{n_Z^2} = 1. \tag{5.B.8}$$

If the material is uniaxial, it is a revolution ellipsoid around the optical axis.

The index ellipsoid is mainly used to find the directions of the privileged vibrations associated to a given wave vector or, on the contrary, to find the direction of propagation of a wave that carries a given light vibration (the following results come from rather tedious calculations that are not given here and can be found in the book *Principles of Optics* by Born and Wolf):

(a) In Figure 5.B.5(a) the direction of the wave vector $\boldsymbol{k}$ is known. To obtain the privileged light vibration we consider the ellipse of intersection of a

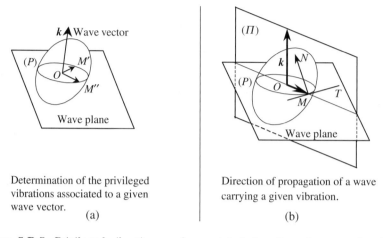

Determination of the privileged vibrations associated to a given wave vector.

(a)

Direction of propagation of a wave carrying a given vibration.

(b)

Figure 5.B.5. Privileged vibrations and associated direction of propagation in the most general case of a biaxial material.

wave plane (P) with an index ellipsoid having its center O on (P): the two axes, ON' and ON'', of this ellipse are, respectively, parallel to the privileged vibrations and the associated indices are equal to the length of the two segments.

(b) In Figure 5.B.5(b) the direction of vibration, OM, is given. We consider first the normal MN to the ellipsoid at point M and the plane (Π) that contains OM and MN; the direction of the wave plane that carries the vibration OM is parallel to the plane that contains OM and is orthogonal to (Π).

Example of a Uniaxial Material

In the case of a uniaxial material the constructions are, of course, easier.

Figure 5.B.6 corresponds to Figure 5.B.5(a) in the simpler case of uniaxial material. $N'N''$ which is orthogonal to the optical axis is the ordinary vibration, the refractive index, ON', is always equal to n_o, whatever the orientation of the wave vector. The extraordinary vibration $M'M''$ is the projection of the optical axis on the wave plane, the corresponding index is $n_{\text{extra}} = OM' = OM'$. Let us call θ the angle between the optical axis and the wave vector, and (x, y) the coordinate of M'; we can write

$$\frac{x^2}{n_o^2} + \frac{y^2}{n_e^2} = 1, \quad x = n_{\text{extra}} \cos\theta, \quad y = -n_{\text{extra}} \sin\theta,$$

$$n_{\text{extra}}^2 \left(\frac{\cos^2\theta}{n_o^2} + \frac{\cos^2\theta}{n_e^2} \right) = 1.$$

(5.B.9)

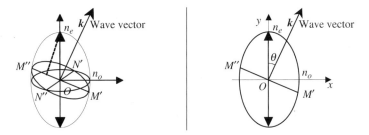

Figure 5.B.6. Privileged vibrations and corresponding indices in a uniaxial material.

It is left as an exercise to use the method of Figure 5.B.5(a) and to show that a vibration orthogonal to the optical axis can propagate in any direction and to examine what happens when the vibration is not orthogonal to the optical axis. (*Hint*: The normal to the ellipsoid index always intersects the optical axis.)

Annex 5.C

Interference Using Polarized Light Beams. Wave Plates

5.C.1. Orthogonally Polarized Beams Are Unable to Interfere

Let us first consider two scalar signals, (S_1, S_2), with sinusoidal time variations of the same frequency ω, let (A_1, A_2) and (I_1, I_2), respectively, be their amplitudes and intensities. The two signals are considered to *interfere* if the response of a detector, receiving them simultaneously, is not equal to the addition of the responses obtained for separate receptions, but depends on their phase difference. More accurately, we can write

$$S_1 = A_1 \cos \omega t = \text{Re}(A_1 e^{j\omega t}) \quad \text{and} \quad S_2 = A_2 \cos(\omega t - \varphi) = \text{Re}(A_2 e^{-j\varphi} e^{j\omega t}).$$

The superposition of the two signals is written as

$$S_1 + S_2 = \text{Re}[e^{j\omega t}(A_1 + A_2 e^{-j\varphi})].$$

The corresponding intensity is

$$
\begin{aligned}
I &= (S_1 + S_2)(S_1 + S_2)^* = (A_1 + A_2 e^{-j\varphi})(A_1 + A_2 e^{j\varphi}), \\
I &= A_1^2 + A_2^2 + 2A_1 A_2 \cos \varphi = I_1 + I_2 + 2\sqrt{I_1 I_2} \cos \varphi.
\end{aligned}
\tag{5.C.1}
$$

The last term of (5.C.1) describes the interference phenomenon.

Instead of scalar signals let us now consider vector signals V_1 and V_2,

$$V_1 = A_1 \cos \omega t \quad \text{and} \quad V_2 = A_2 \cos(\omega t - \varphi).$$

The intensity corresponding to the superposition of the two signals V_1 and V_2 is given by

$$
\begin{aligned}
I &= (S_1 + S_2)(S_1 + S_2)^* = (A_1 + A_2 e^{j\varphi})(A_1 + A_2 e^{j\varphi})^*, \\
I &= (A_1 + A_2 e^{-j\varphi})(A_1 + A_2 e^{j\varphi}) = A_1^2 + A_2^2 + 2A_1 A_2 \cos \varphi.
\end{aligned}
\tag{5.C.2}
$$

If the two vectors A_1 and A_2 are parallel, formulas (5.C.1) and (5.C.2) are identical, but if they are orthogonal, their scalar product is equal to zero and

Figure 5.C.1. A polarizer allows the interference of orthogonal vibrations.

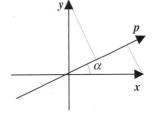

the interference term disappears: $I = I_1 + I_2$. *Two orthogonal vector signals cannot interfere.*

A Polarizer Allows the Interference of Two Orthogonal Vector Vibrations

Two polarized light beams, having any initial polarization, after having been transmitted by a polarizer, are parallel to the direction of the polarizer and are thus able to interfere. In Figure 5.C.1 have been represented two orthogonal vibrations V_1 and V_2, respectively, parallel to the unit vectors $\boldsymbol{x}$ and $\boldsymbol{y}$ and a direction parallel to the unit vector $\boldsymbol{p}$ which makes an angle a with the direction of $\boldsymbol{x}$; omitting the $e^{j\omega t}$ factor the vibrations are written $V_1 = A_1\boldsymbol{x}$ and $V_2 = A_2 e^{-j\varphi}\boldsymbol{x}$.

After the polarizer, the addition of the two vibrations V_1 and V_2 is described by

$$\boldsymbol{P} = [(V_1 + V_2)\boldsymbol{p}]\boldsymbol{p} = (A_1 \cos\alpha + A_2 e^{-j\varphi} \sin\alpha)\boldsymbol{p}.$$

The intensity is given by

$$I = (A_1 \cos\alpha + A_2 e^{-j\varphi} \sin\alpha)(A_1 \cos\alpha + A_2 e^{+j\varphi} \sin\alpha),$$
$$I = A_1^2 \cos^2\alpha + A_2^2 \sin^2\alpha + A_1 A_2 \sin 2\alpha \cos\varphi, \tag{5.C.3}$$

where $A_1 A_2 \sin 2\alpha \cos\varphi$ corresponds to interference.

5.C.2. Interference Using Polarized Monochromatic Light Beams

Wave plates, also called retarders, are optical components that are very often met in many optical arrangements. They are very thin plates (10–$100\,\mu m$) of a birefringent material with polished and parallel faces. Most of the time they are used at normal or quasi-normal incidence. As the thickness is very small the lateral separation between ordinary and extraordinary rays is negligible (see Section 5.A.3).

We call the *neutral lines* of such plates the two orthogonal privileged directions of vibration associated with a propagation perpendicular to the faces. To each neutral line is associated a refractive index (ordinary and extra-ordinary) and a speed of propagation. The neutral line with the highest speed

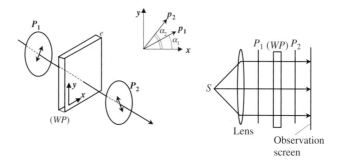

Figure 5.C.2. The light of a natural source is collimated by a lens and polarized by a first polarizer P_1 and then crosses a wave plate *WP* and then a polarizer P_2. The interference pattern is examined on a screen.

(and the smallest index) is often referred as the *fast axis* and the direction perpendicular to it is the *slow axis*.

We consider Figure 5.C.2; x, y, p_1, and p_2 are, respectively, the unit vectors of the neutral lines of the wave plate and of the two polarizers. n_x and n_y are the indices for vibrations along x or y. Before and after P_1 the vibrations can be written as

$$A\boldsymbol{p}_1 = A(\boldsymbol{x}\cos\alpha_1 + \boldsymbol{y}\sin\alpha_1),$$

$$A\left(x e^{-j\frac{2\pi n_x e}{\lambda}} \cos\alpha_1 + y e^{-j\frac{2\pi n_y e}{\lambda}} \sin\alpha_1 \right)$$

$$= A e^{-j\frac{2\pi n_x e}{\lambda}} \left(x\cos\alpha_1 + y e^{-j\frac{2\pi(n_y - n_x e)}{\lambda}} \sin\alpha_1 \right).$$

Before (*WP*) the x and y components of the vibration are in phase, after that they have a phase difference equal to $\phi = (2\pi e/\lambda)(n_x - n_z) = 2\pi e\Delta n/\lambda$ and the vibration is $A(\boldsymbol{x}\cos\alpha_1 + \boldsymbol{y}e^{-j\phi}\sin\alpha_1)$. The vibration and the final intensity I_{out} after the last polarizer are

$$A\boldsymbol{p}_2(\cos\alpha_1\cos\alpha_2 + e^{-j\phi}\sin\alpha_1\sin\alpha_2), \tag{5.C.4.a}$$

$$I_{\text{out}} = A^2(\cos\alpha_1\cos\alpha_2 + e^{-j\phi}\sin\alpha_1\sin\alpha_2)(\cos\alpha_1\cos\alpha_2 + e^{+j\phi}\sin\alpha_1\sin\alpha_2), \tag{5.C.4.b}$$

$$I_{\text{out}} = I_0(1 + \cos 2\alpha_1\cos 2\alpha_2 + \sin 2\alpha_1\sin 2\alpha_2\cos\phi), \tag{5.C.4.c}$$

$$I_{\text{out}} = I_0(1 + \cos 2(\alpha_1 - \alpha_2) - 2\sin 2\alpha_1\sin 2\alpha_2\sin^2\phi/2). \tag{5.C.4.d}$$

5.C.3. Wave Plates

A wave plate, or retarder, is said to be λ/n, when the phase difference $\Delta\phi$ between the x and y components is equal to $2\pi/n$. Since retarders modify the

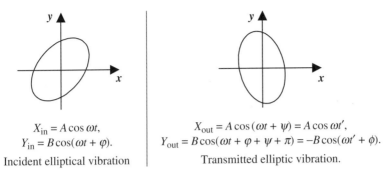

$$X_{in} = A\cos\omega t,$$
$$Y_{in} = B\cos(\omega t + \varphi).$$

Incident elliptical vibration

$$X_{out} = A\cos(\omega t + \psi) = A\cos\omega t',$$
$$Y_{out} = B\cos(\omega t + \varphi + \psi + \pi) = -B\cos(\omega t' + \phi).$$

Transmitted elliptic vibration.

Figure 5.C.3. Description of an elliptic vibration, before and after a $\lambda/2$ wave plate. The transmitted ellipse is symmetric to the incident ellipse, with regard to the neutral lines.

phase difference between the two orthogonal projections of a given vibration, they alter the state of polarization of a beam; this is their main role.

Note that a λ plate, also called a *full-wave plate*, doesn't at all change the polarization ($\Delta\phi = 2\pi$).

Half-Wave Plate ($\lambda/2$ plate) $\rightarrow \phi = \pi$

We refer to Figure 5.C.3 and to the formulas below the illustrations. The subscripts in and out just mean *before* and *after* the wave plate. The elliptical character of the incident vibration is due to the presence of the term φ. The phase shift ψ that exists in both components of the transmitted vibration is not important and can be canceled by a suitable change of the time origin; we can symbolically consider that, in the case of a half-wavelength plate, we have

$$\begin{cases} X_{out} = X_{in}, \\ Y_{out} = -Y_{in}. \end{cases}$$

In other words, the transmitted polarization is *symmetric* to the incident polarization, with respect to the neutral lines. As a consequence, a linear polarization remains linear after a $\lambda/2$ wave plate. An easy way to adjust the orientation of the polarization of a beam is to rotate a $\lambda/2$ wave plate around an axis orthogonal to the plate.

Quarter-Wave Plate ($\lambda/4$ Plate $\rightarrow \phi = \pi/2$). Circular Polarizers

$\lambda/4$ are mostly used, starting from linear vibrations, to obtain elliptical vibrations with predeterminate properties, and especially circular polarizations.

The transformation of a linear polarization in a circular one is the most important application of $\lambda/4$ plates; the angle of the linear polarization with

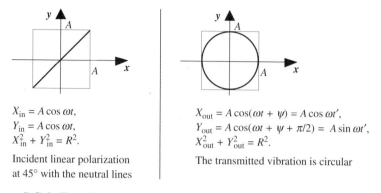

$X_{in} = A \cos \omega t,$
$Y_{in} = A \cos \omega t,$
$X_{in}^2 + Y_{in}^2 = R^2.$

Incident linear polarization
at 45° with the neutral lines

$X_{out} = A \cos(\omega t + \psi) = A \cos \omega t',$
$Y_{out} = A \cos(\omega t + \psi + \pi/2) = A \sin \omega t',$
$X_{out}^2 + Y_{out}^2 = R^2.$

The transmitted vibration is circular

Figure 5.C.4. Transformation of a linear polarization into a circular polarization.

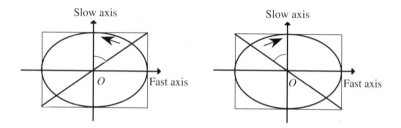

Figure 5.C.5. The elliptic polarization rotates in the same direction as the rotation that puts in coincidence the initial linear polarization with the slow axis with an angle less than 90°.

the neutral lines should be equal to 45°. If the angle is different, an elliptical polarization is obtained, the axes of the ellipse always coincide with the neutral lines, and the ratio of the two axes is fixed by the angle of the linear polarization with the neutral lines.

A recipe for obtaining the sign of the rotation is given in Figure 5.C.5.

Order of Magnitude of the Wave Plate Thickness

The most common materials for making wave plates are quartz ($\Delta n \approx 0.009$) and calcite ($\Delta n \approx 0.18$). For a half-wave plate at a wavelength of $0.5\,\mu m$ the thicknesses are the following: $e_{\lambda/2,quartz} = 14\,\mu m$ and $e_{\lambda/2,calcite} = 1.4\,\mu m$; in the case of calcite the plates would be too thin, in this case it would be better to make plates with a thickness of $\lambda/2 + p\lambda$. Some transparent organic materials, such as cellophane, or scotch tape if submitted to some stress during polymerization, are birefringent; the birefringence is small, which means that the wave plates will be thicker (hundreds of μm) and easier to manipulate and much cheaper.

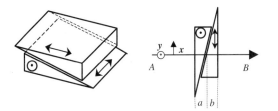

Figure 5.C.6. Babinet compensators. Two geometrically identical prisms have been cut in a piece of quartz with their optical axes oriented as shown in the figure. The two hypotenuses are put in contact and can smoothly and accurately slide one against the other.

Compensators

Babinet compensator: Compensators are a type of retarder in which the phase difference is adjustable and finely tunable. We refer to Figure 5.C.6: for a light beam propagating along AB, a vibration having its polarization parallel to y is ordinary in the first prism and extraordinary in the second. It's exactly the opposite for a vibration parallel to x. The respective thicknesses, a and b, covered by the light in each prism will vary when the prisms are shifted. The phase difference between the two polarizations is given by

$$\phi = \frac{2\pi}{\lambda}(b-a)\Delta n. \tag{5.C.5}$$

A high sensitivity for tuning the phase difference is readily obtained if a small value is given to the acute angle of the prisms. To work properly a Babinet compensator must receive beams with narrow enough cross sections, if not, the phase difference and, hence, the polarization will not be constant from one point to another, inside a given cross section. This drawback is avoided with the Soleil compensator.

Soleil compensator: This second type of compensator is described in Figure 5.C.7. The phase difference introduced by the Soleil compensator is the same for all rays and is given by

$$\phi = \frac{2\pi}{\lambda}[e-(b+a)]\Delta n. \tag{5.C.6}$$

Lyot Optical Filter

A Lyot optical filter, Figure 5.C.8, is made of a succession of parallel polarizers between which birefringent plates ($L_0, L_1, L_2, \ldots$) have been inserted. The neutral lines (x and y) are, respectively, parallel to one another and make an angle of $45°$ with the unit vector p of the polarizers.

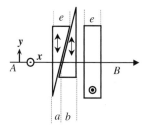

Figure 5.C.7. Soleil compensator. Again two sliding prisms are used, but their optical axes are now parallel to one another and parallel to the plane of incidence. A parallel plate has been cut in the same material as the two prisms, its thickness is equal to the small side of each prism, and its optical axis is orthogonal to that of the prisms.

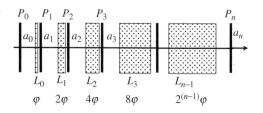

Figure 5.C.8. Scheme of a Lyot optical filter.

n_x and n_y being, respectively, the refractive indices of the plates for vibrations polarized along $\boldsymbol{x}$ or $\boldsymbol{y}$, and $\varphi = 2\pi/\lambda(n_x - n_y)$ being the phase difference introduced by the first plate, the phase differences introduced by the plates are equal to

$$\varphi_0 = \varphi, \quad \varphi_1 = 2\varphi, \quad \varphi_2 = 2^2\varphi, \quad \varphi_3 = 2^3\varphi, \ldots, \varphi_n = 2^n\varphi.$$

The vibrations after the different polarizers P_1, P_2, P_3, $\ldots$, P_n are, respectively, written as $a_1\boldsymbol{p}$, $a_2\boldsymbol{p}$, $a_3\boldsymbol{p}$, $\ldots$, $a_n\boldsymbol{p}$.

Vibration after the polarizer P_i: $\dfrac{a_i}{\sqrt{2}}(\boldsymbol{x}+\boldsymbol{y})$.

Vibration after the plate L_i that follows P_i: $\dfrac{a_i}{\sqrt{2}}(\boldsymbol{x}+\boldsymbol{y}e^{j\varphi_i})$, $\varphi_i = 2^i\varphi$.

Vibration after the polarizer P_{i+1}: $a_{i+1}\boldsymbol{p} = \dfrac{a_i}{2}(1+e^{j\varphi})\boldsymbol{p}$.

Vibration transmitted by the Lyot filter:

$$a_n\boldsymbol{p} = \frac{a}{2^n}(1+e^{j\varphi})(1+e^{j2\varphi})(1+e^{j2^2\varphi})(1+e^{j2^3\varphi})\ldots(1+e^{j2^{n-1}\varphi})\boldsymbol{p}.$$

It can be shown that

$$(1+e^{j\varphi})(1+e^{j2\varphi})(1+e^{j2^2\varphi})(1+e^{j2^3\varphi})\ldots(1+e^{j2^{n-1}\varphi}) = \sum_{m=0}^{m=2^n-1} e^{jm\varphi} = \frac{1-e^{j2^n\varphi}}{1-e^{j\varphi}}.$$

If I and I_0 are the incident and transmitted intensities, we have

$$\frac{I}{I_0} = \frac{1}{2^{2n}} \frac{1-e^{j2^n\varphi}}{1-e^{j\varphi}} \frac{1-e^{-j2^n\varphi}}{1-e^{-j\varphi}} = \frac{1}{2^{2n}} \frac{(1-\cos 2^n \varphi)}{(1-\cos\varphi)} = \frac{1}{2^{2n}} \frac{\sin^2(2^n\varphi/2)}{\sin^2(\varphi/2)}. \tag{5.C.7}$$

According to (5.C.7) the ratio I/I_0 differs from zero only if φ is very near to an integer multiple of 2π, and the graph of the variation of I/I_0 versus wavelength is a succession of very narrow maximums equal to one.

Numerical Application

Thickness of the first plate $e = 250\,\mu m$, $(n_x - n_y) = 10^{-2}$; four plates $(n = 3)$. The well-transmitted wavelengths λ are given by

$$\varphi = \frac{2\pi(n_x - n_y)e}{\lambda} = p2\pi \quad \rightarrow \quad \lambda = \frac{(n_x - n_y)e}{p} = \frac{2.5}{p}.$$

If λ belongs to the visible: $0.4 < 2.5/p < 0.8\,\mu m \rightarrow p = \{4, 5, 6\} \rightarrow \lambda = \{0.625, 0.500, 0.417\}$.

The frequencies of a Lyot filter quickly vary with thickness and birefringence, as these two parameters are strongly temperature dependent, it's necessary to stabilize accurately the temperature, at the same time a fine-tuning of the filter can be obtained by adjusting the temperature.

5.C.4. Interference with Polarized White Beams

We now come back to the arrangement of Figure 5.C.2 and to formulas (5.C.4.a,b,c,d) which give the intensity of the light on the screen, this intensity is uniform over all of the screen and depends on two different parameters:

- The phase difference ϕ introduced by the wave plate.
- The angles α_1 and α_2 between the neutral lines and the directions of the polarizer and of the analyzer.

For the sake of simplification we will only consider the cases where the analyzer and the polarizer are parallel ($\alpha_1 = \alpha_2$) or orthogonal ($\alpha_1 = \alpha_2 + 90°$),

$$\frac{I_{out}^=}{I_0} = \left(1 - \sin^2 2\alpha_1 \sin^2 \frac{\phi}{2}\right), \quad P_1 \,/\!/\, P_2, \tag{5.C.8.a}$$

$$\frac{I_{out}^\perp}{I_0} = \sin^2 2\alpha_1 \sin^2 \frac{\phi}{2}, \quad P_1 \perp P_2. \tag{5.C.8.b}$$

The phase difference ϕ depends on the frequency and thus on the color of the light. If the initial light is white, the aspect of the screen will result in the superposition of the various interference patterns corresponding to all the

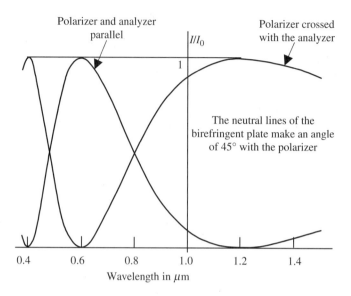

Figure 5.C.9. Interference using polarized white light beams. The curves go up to one and down to zero, because the angle between the polarizer and the neutral lines is 45°.

colors of the white spectrum. The radiations for which $\phi = 2p\pi$ give a maximum, while the radiations for which $\phi = (2p + 1)\pi$ are completely faded.

To make things simpler, we will forget the effect of dispersion on birefringence and consider that $\Delta n = (n_x - n_y)$ is constant whatever the wavelength. Using the expression of the phase difference $\phi = (2\pi/\lambda)(n_x - n_y)e = 2\pi\pi\Delta n/\lambda$, we obtain

$$\frac{I^=_{\text{out}}}{I_0} = \left(1 - \sin^2 2\alpha_1 \sin^2 \frac{\pi\Delta ne}{\lambda}\right), \quad P_1 /\!/ P_2, \qquad (5.C.9.a)$$

$$\frac{I^+_{\text{out}}}{I_0} = \sin^2 2\alpha_1 \sin^2 \frac{\pi\Delta ne}{\lambda}, \quad P_1 \perp P_2. \qquad (5.C.9.b)$$

The screen of Figure 5.C.2 is not illuminated by a monochromatic light, but by a white light source that has passed through a filter having a profile described by equations (5.C.4.a,b,c,d). The curves of Figure 5.C.10 correspond to a plate for which $\Delta ne = 0.6\,\mu m$. For a wavelength of $0.6\,\mu m$, and parallel analyzer and polarizer, the reinforcement of the light by interference is maximum for a wavelength of $0.6\,\mu m$; the screen appears to be orange-red. When the analyzer and polarizer are crossed, on the contrary, the light is destroyed by interference and the intensity is minimum, this minimum is equal to zero if the neutral lines make an angle of 45° with the polarizer and analyzer.

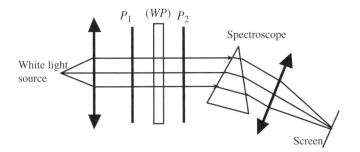

Figure 5.C.10. Channeled spectrum. A prism spectroscope disperses a light beam that has previously crossed a polarizer, a birefringent wave plate, and an analyzer. The wave plate is thick, the spectrum is crossed by light and dark stripes which correspond to reinforcement or extinction of the light by interference.

As we have $I_{out}^{=} + I_{out}^{+} = I_0$, whatever the wavelength, the two colors taken by the screen in Figure 5.C.2, for a parallel or crossed polarizer and analyzer, are complimentary.

In principle the colors taken by the screen are the same as interference colors obtained from a double-beam interference experiment using a thin film of air. In fact, because of the dispersion of the birefringence the colors are very similar although not exactly identical.

The number and position of maximums and minimums within the visible domain characterize a spectrum like that of Figure 5.C.9. If the birefringent plate is thin enough, as is the case for the curves of the figure, there is only one maximum or one minimum; the color of the screen is then brilliant and pure. If the plate is thick, many wavelengths will give maximums and minimums, the global color will be white. If the light enters the slit of a spectrometer the spectrum is crossed by light and dark bands and is called a channeled *spectrum*.

Numerical Application

We refer to Figure 5.C.10, P_1 and P_2 are parallel, the thickness of (WP) is $e = 50\,\mu m$ and its birefringence is $\Delta n = 0.18$; we are looking for the positions of the light and dark bands. The light bands are obtained when $e\Delta n = p\lambda_p$; let us calculate the values of $e\Delta n/\lambda$ for the two extremities of the visible spectrum:

$$\lambda = 0.8\,\mu m \quad \rightarrow \quad e\Delta n/\lambda = 11.25; \quad \lambda = 0.4\,\mu m \quad \rightarrow \quad e\Delta n/\lambda = 22.5.$$

Therefore eleven light bands, corresponding to the eleven integers between 12 and 22, will be observed: $\lambda_{11} = 9/12 = 0.75\,\mu m$ is a red stripe and $\lambda_{22} = 9/22 = 0.41\,\mu m$ is a blue stripe.

For the dark bands we have $e\Delta n = (2p + 1)\lambda_p/2 \rightarrow 11 < p < 21 \rightarrow 11$ dark bands, the wavelengths of the external stripes are $0.78\,\mu m$ and $0.42\,\mu m$.

Annex 5.D

Liquid Crystals

5.D.1. Introduction

Liquid crystals are certainly one of the nicest examples of crossed fertilization between basic and applied research. The first theoretical introduction of the notion of an organized state of matter in the case of liquids and the first experimental observations were made by Reinitzer in 1888, the expression *liquid crystal* was introduced for the first time by O. Lehmann in 1890. At that time scientists were merely concerned in studying a phase transition between two states of matter and surely didn't think of display or of signal processing devices.

Once the high technological potentialities of liquid crystals were identified, it became necessary to identify and understand the physical mechanisms responsible for their behavior. This necessitated a large research effort, in both applied and basic research. The French Physics Nobel Prizewinner, P.G. de Gennes, was among those who, as early as 1960, had anticipated the importance of liquid crystals in technology and who has largely contributed in developing the physics of the related phenomena.

One purpose of liquid crystal devices is to compete with cathode ray tubes, their main advantage is their extremely low-energy consumption which has allowed them to take an overwhelming position on the market of small-size display devices (watch, pocket computer and, more recently, personal computer screens).

5.D.2. Physical-Chemistry of Liquid Crystals

From a thermodynamic point of view a liquid crystal is intermediate between a crystal where a quasi-perfect order is found and a liquid where no long-range order remains. A crystal is a regular arrangement of an enormous number (Avogadro number) of identical groups of several atoms or ions. These groups may have anisotropic properties, because of the existence of electric or magnetic dipoles or, in a simpler way, because of their geometric shape which can be elongated or flattened.

The long-range order in crystals first concerns the positions of the groups. The sites occupied by the groups can be deduced one from the other by translations, the vectors of which define the elementary crystal cell: we then speak of *translation order*. The long-range order in a crystal may also be concerned by the orientation of the groups, it is then spoken of as *orientation order*. When the transition *solid/liquid* (fusion) occurs the order collapses; generally speaking, the two types of order disappear simultaneously.

In the case of liquid crystals, on the one hand, the *translation order* disappears before the *orientation order* and, on the other hand, the anisotropy takes its origin in the shape of the elementary groups. After fusion, the material is in a new state called *mesophase* which is intermediate between a crystal and a liquid: the groups have become mobile but they keep the same orientation. In the mesophase the fluidity is that of a liquid (the material can be poured), but the anisotropy is that of a solid.

Liquid crystals are always organic materials, the elementary groups are made of organic molecules which have an elongated shape. There are several possibilities for mesophases, they have been classified by a French chemist, G. Friedel, in smectic, nematic, and cholesteric. Oversimplifying the phenomena, it can be considered that, as the temperature is raised, the following phase transitions are met, corresponding to a diminution of the order in the system: solid crystal → smectic liquid crystal → nematic liquid crystal → ordinary liquid. During the two first transitions the orientation order is preserved, the molecules remaining parallel.

In the smectic phase, the groups are parallel and lie along planar layers inside which they have an erratic distribution. The layers can easily slip with

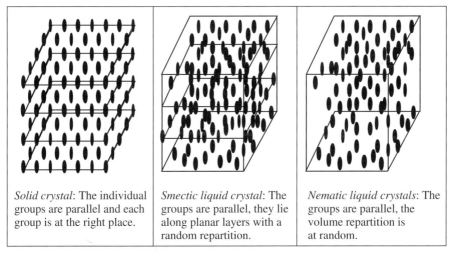

| *Solid crystal*: The individual groups are parallel and each group is at the right place. | *Smectic liquid crystal*: The groups are parallel, they lie along planar layers with a random repartition. | *Nematic liquid crystals*: The groups are parallel, the volume repartition is at random. |

Figure 5.D.1. The different kinds of liquid crystals.

regard to one another, the name smectic comes from this property, *smectos* meaning *soap* in Greek. In the nematic phase (*nematos* = filament) the repartition along planar layers diappears, the groups remain parallel but with a random volume distribution. Both smectic and nematic crystals behave as a uniaxial material which is highly birefringent, typically $\Delta n = n_e - n_o$ is of the order of 0.1 to 0.3.

The cholesteric phase looks very much like the nematic phase, except that the molecules are of the same kind as the molecules that we have met on the occasion of circular birefringence (two possible forms that are not mirror images of one another). As was the case for SiO_2 molecules in quartz, in a cholesteric phase the molecules are arranged along helical filaments. Cholesteric liquid crystals have a high rotatory power.

For display devices, nematic crystals, added to a small part of cholesteric, are used. An electric field is used to change the orientation of the molecules and hence the birefringence Δn, and also the orientation of the neutral lines.

5.D.3. Orientation of Molecules in a Nematic Phase

In a liquid crystal all the molecules are supposed to be parallel, the next question is: Parallel to which direction? A liquid crystal is a fluid and has no shape of its own, it must be contained inside a vessel where it has usually a milky appearance which is due to the existence of tiny liquid microcrystals having any orientation and diffusing the light.

If liquid crystals are to be used in some devices, the equivalent of single crystals should be made. An electric field can be used to impose a common orientation, more often and more simply it is obtained by the creation of suitable mechanical boundary conditions on the inner faces of the vessel. As boundary conditions play an important role, the surfaces should be extremely clean and free of any remaining adsorbed molecules.

Orientation of Molecules Perpendicular to a Surface

We consider molecules that can be assimilated to tiny elongated ellipsoids of revolution (they look like rugby balls). The experiment shows that, when they are put in contact with a well-cleaned surface, such molecules try to minimize the area of contact and orient their axes perpendicular to the surface. Special detergent compounds, called *surfactants*, can be used to enhance this effect.

Orientation Along Grooves Drawn on a Surface

This method is the result of a very simple experimental observation: if, after careful cleaning, a surface is gently wiped along a given direction with a soft

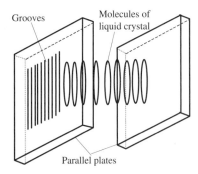

Grooves

Molecules of liquid crystal

Grooves

Parallel plates

Twisted nematic crystal

(a) Near the brushed face, the molecules align parallel to the grooves; the orientation is then progressively extended to the entire volume. The material is uniaxial with the optical axis parallel to the molecules axes.

(b) Grooves have been brushed along two orthogonal directions on two parallel plates separated by 5 to 10 μm. The orientation of the molecules progressively varies from one orientation to the other.

Figure 5.D.2. Orientation of the molecules by the boundary conditions.

tissue (usually a type of velvet which acts like a brush), fine grooves are drawn on the surface along which the molecules will then align. After this preliminary observation a careful technological development has defined the rules for obtaining grooves, in a reproducible way. The method is still very empirical and the physical and chemical surface interactions, at the interface between the substrate and the liquid crystal, are not completely understood; a lot of progress is still possible.

Figure 5.D.2(b) represents a useful arrangement for display devices. Over a distance of 5 to 10 μm the orientation of the molecules is rotated by 90°. From an optical point of view the arrangement is equivalent to a uniaxial material inside of which the direction of the optical axis would continuously rotate. If a light beam has its polarization parallel to the grooves of the first plate, its polarization will remain linear and follow the local optical axis. An artificial optically active material has been obtained, the rotatory power is very high: a rotation of the polarization can be 90° over a distance of a few micrometers. The addition of a small amount of some cholesteric considerably enhances the phenomenon.

Application of an Electric Field

Because of their elongated shape, the molecules of a liquid crystal behave as electric dipoles that are easily oriented parallel to some applied electric field. Using the brushing technique the liquid crystal is first put into a state where all the molecules are parallel to the surface; if an electric field is then applied perpendicular to the surface, the molecules will follow the field.

In the presence, as well as in the absence, of an electric field the liquid crystal is birefringent. The application of the electric field changes the orientation of the optical axis and controls the birefringence of the liquid crystal cell. The next problem is that of the electrodes, since they must be, at the same time, transparent to the light and the conductor of electricity. In most cases they are made of an oxide of indium and tin (ITO—Indium Tin Oxide).

The required voltages are small (volts), however the electric field is quite high (0.1 MV/m) since the electrodes separation is measured in micrometers. Liquid crystals, being organic material, the impedances of liquid crystal cells are very high and need almost no power to maintain the molecular orientation. This is very important for applications.

5.D.4. Liquid Crystal Display

Liquid Crystal Display Using a Mirror

This kind of display is very popular and, for example, is used in watches and pocket calculators, as there in no internal light source the energy required is small. They are illuminated with a white unpolarized light source, reflection occurs only at places where the electrodes apply an electric field, the shape of the electrodes then appear as if they where painted on the screen.

In the absence of any electric field, the molecules are parallel to the grooves that have been brushed parallel to the electrodes. For a light beam propagating perpendicular to the plates, the cell is a wave plate having its neutral lines parallel and perpendicular to the grooves. The cell is given such a thickness that it is a $\lambda/4$ wave plate; because of the mirror, the light makes a double passage in the cell which becomes equivalent to a $\lambda/2$ wave plate. The polarizer is oriented at 45° with regard to the grooves, and so polarizes

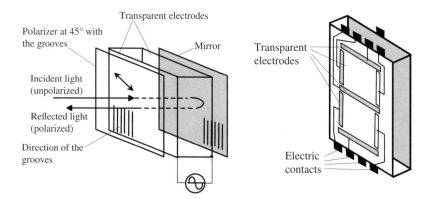

Figure 5.D.3. In the absence of any electric polarization, molecules are parallel to the grooves → the optical axis is vertical. The electric field orients the molecules and makes the optical axis perpendicular to the plates.

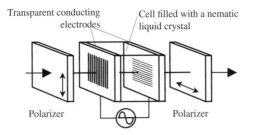

Transparent conducting electrodes

Cell filled with a nematic liquid crystal

Polarizer

Polarizer

Figure 5.D.4. With zero voltage on the cell, the molecular orientation follows the indications of the grooves: the cell is optically active, the polarization is flipped 90° and the light is fully transmitted. The electric field orientates the molecules parallel to the direction of propagation making the rotatory power disappear, the cell becomes opaque.

the light arriving at the cell. The polarization of the light coming back is flipped 90° and is thus blocked by the polarizer. The electric field makes the optical axis parallel to the direction of propagation, the light rays are ordinary, whatever the polarization, and the returned light is fully transmitted by the polarizer.

Transmission Nematic Liquid Crystal Modulator

We refer to Figure 5.D.4, the final orientation of the molecules is the result of two contradictory actions: the grooves and the electric field. If the electric field is strong enough, its orientation action is predominant; after several molecular layers of transition all the molecules become parallel to the field and the cell completely loses its circular birefringence. The transparency of the cell goes from unity for zero voltage on the electrodes, to zero for a sufficient value. If the voltage varies versus time, the transmitted light intensity is modulated. The time constant is of a few milliseconds, and the frequency of modulation ranges in the kilohertz domain.

Matrix Display

Liquid crystal cells may have a very small size, allowing a local, and almost punctual, control of the birefringence of a transparent plate.

Figure 5.D.5 illustrates the principle of a matrix of elementary cells with an electrically controlled birefringence. Each cell is a liquid crystal sandwiched between two transparent electrodes, the molecule's orientation is addressed by a voltage applied between the front and back sides of the cell. All the front electrodes of the cells belonging to the same line (designated by m) are connected to the same wire, in the same way that all the back electrodes of the same column (n) are connected together. The (mn) cell is addressed by modifying the voltage between conductors m and n.

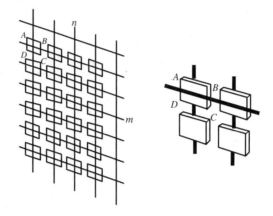

Figure 5.D.5. Matrix of addressable liquid crystal cells.

The elementary cells are called *pixels* (picture elements). The size of a pixel and the matrix step are typically of a few hundred nanometers. Two major technological difficulties have to be overcome. First, the realization of a matrix with a million elements, knowing that the human eye doesn't accept a failure rate larger than 1% and, second, the electrical addressing of the pixels. We will forget the computing side of the problem that is easily solved and will focus on the addressing voltage.

If we don't want to cope with high voltage, the different pixels must be made electrically independent of one another. The arrangement of Figure 5.D.6 shows how this obtained. All the back electrodes are connected together and earthed; the front electrodes are connected to the *column* electrodes by a field-effect transistor the grid of which is connected to the *line* electrodes. A transistor should be made for each pixel; this is easy work in microelectronics, the size of each transistor is of the order of 1 μm.

Color display can be made using a three-color representation. A matrix of holes is inserted between the light source and the liquid crystal cell matrix; the holes are filled with a colored gelatin. One pixel is now made of four sub-pixels, three would be enough but, for the sake of symmetry, it's easier to have four, then it has been decided to double the number of green subpixels and, accordingly, to increase the opacity of the green filters.

Local Control of the Optical Thickness of a Transparent Plate

We consider a matrix analog to that of Figure 5.D.5; each elementary cell is filled with a nematic liquid crystal and the grooves on the two plates are parallel. In the absence of any applied voltage the molecules lie parallel to the grooves, the crystal is birefringent and uniaxial. If an electric field is applied, the molecules will be submitted to an orientating torque; the number of molecules that become parallel to the field increases with its strength. The crystal

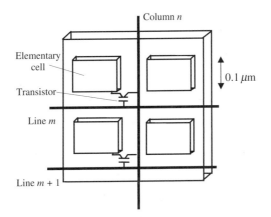

Figure 5.D.6. Transistors may be used to lower the addressing voltage.

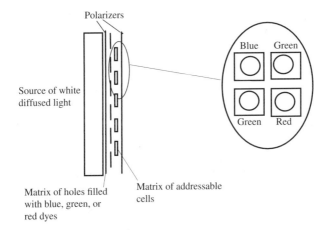

Figure 5.D.7. Principle of a color display. Each pixel is made of four cells, each giving one color for a trichromatic representation.

is now biaxial; one privileged polarization is parallel to the grooves, the associated index decreases with increasing fields and its value is a function of the field module: in other words, for a light beam polarized parallel to the grooves, the optical thickness of the cell varies with the voltage and can be addressed thanks to the matrix arrangement.

5.D.5. Light Valve

The notion of *valve* has been initially introduced in the technology of tires and is a device that allows air to inflate tires or balloons. The word is also

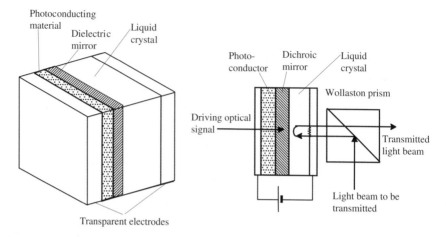

Figure 5.D.8. Optically driven optical valve.

used in electronics for components that conduct the current in only one direction, in this case the word diode is also used. In electronics the notion of valve has been enriched by the invention of *driven valves*. A driven valve still allows the current to go only in one direction, but only if some suitable driving signal is applied; the situation becomes very interesting if the power of the driving signal is much smaller than the main current in the diode.

Optical valves have been proposed, they are usually opaque and become transparent when they receive a small optical signal. Optical valves are elegant and promising devices, it seems too early to predict any important development.

Figure 5.D.8 describes the principle of an *optically driven optical valve*. The trick is to apply the polarizing voltage to the liquid crystal across a photoconducting transparent material. This material is BSO ($Bi_{12}SiO_{20}$); an electrical insulator in darkness, this compound becomes a conductor when suitably illuminated. At places where BSO receives the driving light signal, a voltage is applied to the liquid crystal, and the optical axis becomes parallel to the direction of propagation of the light to be transmitted.

The exact way the device works uses a special state of liquid crystals that has not yet been introduced and for which the liquid crystals have a milky aspect and diffuse the light. In the absence of any electric field, the so-called *light to be transmitted* is diffused and only a small part reaches the Wollaston prism. With an electric field the crystal becomes transparent and birefringent, and the field is adjusted so that the liquid crystal cell is a $\lambda/4$ wave plate for a single transit (and thus $\lambda/2$ for a double passage). Having a polarization orthogonal to the polarization of the incident light, the light reflected by the mirror is transmitted by the Wollaston prism.

6

Interference

Because it is a basic property of every process involving vibrations, interference has been introduced as early as the first chapter of this book. Furthermore, the notion of interference is usually generalized to other domains, quite often far from Physics, such as Economy or Psychology. . . . Two phenomena are said to interfere if their simultaneous actions have new consequences, and if compared with the superposition of the consequences of their independent actions.

General conditions for two light beams to interfere have already been given: they should have exactly, and with high accuracy, the same frequency; their polarizations should not be strictly orthogonal and, last but not least, they must be *coherent with one another*. The fact that two light beams, coming from two independent sources, cannot interfere is rather a good thing, if this were not the case everyday life would occur in a vast field of interference where everything would be striped with interference fringes.

All the experimental set-ups that we are going to describe are supposed to work with incoherent sources. To give the photodetectors the impression that they receive coherent beams, we will always use arrangements in which *all the different interfering beams originate from a unique point source.*

6.1. Wave Front Division Interferometers

The interest of wave front division arrangements is mostly didactic. The prototype of such arrangements was proposed by Young.

6.1.1. *The Young Experiment*

A lamp illuminates, with a light which, for the moment, will be considered as monochromatic (wavelength λ), a tiny hole S with a diameter δ of the order

Chapter 6 has been reviewed by Dr. Dominique Persegol from Schneider Electric.

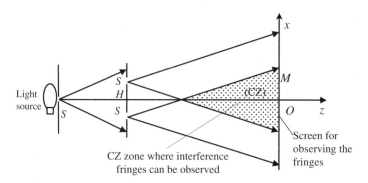

Figure 6.1. Young's experiment. A tiny hole S is illuminated by an extended light source; some light is diffracted inside the first cone inside of which two other holes are placed in close vicinity. S_1 and S_2 can be considered as two coherent sources, interference can be produced in the shadowed area CZ, which is the common zone (CZ) between the two cones of light diffracted from S_1 and S_2. The interference fringes are visible on a screen.

of fifty to hundred times the wavelength. The light wave is diffracted through the hole inside a sharp cone (angle of the order of λ/δ). The light is then diffracted again through the two holes S_1 and S_2 (about the same size as S) which can be considered as two sources emitting the same frequency. If S is at the same distance from S_1 and S_2 the two sources are synchronous, if not, the phase difference is equal to $2\pi(SS_1 - SS_2)/\lambda$. At each point M of the *common zone* (CZ) arrive two beams which have traveled along two different paths SS_1M and SS_2M and have a phase difference equal to $2\pi(SS_1M - SS_2M)/\lambda$. If the difference between the times taken to cover the two paths is smaller than the coherence time of the source S, S_1 and S_2 can be assimilated to coherent sources: interference will then be observable at point M, with maxima if the waves arrive in phase (constructive interference) and minima if they have opposite phases (destructive interference). If the two holes are identical the two sources have the same amplitude and the minima are equal to zero.

To simplify the formulas we consider that $S_1M = S_2M$ and the positions of the maxima and minima are given by

- Maxima: $\psi = \dfrac{2\pi}{\lambda}(MS_2 - MS_1) = (2p+1)\pi \quad \rightarrow \quad (MS_2 - MS_1) = p\lambda.$

- Minima: $\psi = \dfrac{2\pi}{\lambda}(MS_2 - MS_1) = (2p+1)\pi \ \rightarrow \ (MS_2 - MS_1) = (2p+1)\lambda/2.$

The above formulas define a family of revolution hyperboloids; each hyperboloid is labeled by the integer p which is called the *order of interference* along the considered hyperboloid. The intersection of the common zone

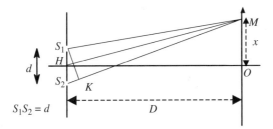

Figure 6.2. Definition of the notations used to describe the Young experiment.

(CZ) and the observation screen seems to be painted with alternately brilliant and dark fringes that are the intersections of a family of hyperboloids with the plane of the screen.

In order to make the fringes brighter, the holes are replaced by narrow parallel slits perpendicular to the plane SS_1S_2. The revolution hyperboloids are then replaced by hyperbolic cylinders, the fringes are parallel to the slits.

To evaluate the state of interference at a given point $M(x)$ we refer to Figure 6.2 in which S_1 and S_2 are two parallel slits. The difference between the phases of the waves arriving at point M corresponds to the propagation of the light from S_2 to K (point of intersection of S_2M with the circle centered at M and of radius MS_1). In all practical situations the distance $d = S_1S_2$ is small in comparison with the distance D to the screen, K can be considered as the projection of S_1 on S_2M:

$$S_2K = x\frac{d}{D}, \quad S_1M = HM - x\frac{d}{2D}, \quad S_2M = HM + x\frac{d}{2D}. \tag{6.1}$$

If S is equidistant from S_1 and S_2, the interferences are *constructive* or *destructive* at points of respective abscissas x_{Max} and $x_{\min}$, given by

$$x_{\text{Max}} = p\frac{\lambda D}{2d} \quad \text{and} \quad x_{\min} = (2p+1)\frac{\lambda D}{2d}, \tag{6.2}$$

$$i = x_{\text{Max}(p+1)} - x_{\text{Max}(p)} = x_{\min(p+1)} - x_{\min(p)} = p\frac{\lambda D}{d}. \tag{6.3}$$

The interference pattern is made of bright and dark bands called interference fringes, they are equidistant and orthogonal to the plane SS_1S_2. The spacing, i, between the centers of two adjacent brilliant, or dark, fringes (see formula (6.3)) is made of the order of a few millimeters by the magnifying ratio D/d; d is a small fraction of one millimeter while D is one meter or more.

We now want the expression of the variation of the light intensity versus abscissa x. We need the expressions of the complex amplitudes of the vibrations at points S_1 and S_2. Of course, we omit the $1/r$ attenuation of the waves and use the expressions $E_1 = \alpha e^{-jkS_1M}$ and $E_2 = \alpha e^{-jkS_2M}$, their superposition gives

$$E_1 + E_1 = \alpha(e^{-jkS_1M} + e^{-jkS_2M}) = \alpha e^{-jkHM}\left(e^{jk\frac{xd}{2D}} + e^{-jk\frac{xd}{2D}}\right),$$

$$E_1 + E_2 = 2\alpha e^{-jkHM}\cos k\frac{xd}{2D}.$$

The intensity is then given by

$$I = (E_1 + E_2)(E_1 + E_2)^* = 4\alpha^2\cos^2\left(k\frac{xd}{2D}\right), \tag{6.4}$$

$$I = 2\alpha^2\left(1 + \cos k\frac{xd}{D}\right) = 2\alpha^2\left(1 + \cos 2\pi\frac{x}{i}\right). \tag{6.5}$$

6.1.2. *Other Arrangements Using Wave Front Division*

In front of some extended light source is disposed a small hole which, of course, transmits only a small amount of light but which can be considered as a source S of spatially coherent light. The spherical wave diverges and then its wave front is divided into two parts that will follow two different optical paths. In the experimental arrangements that we are going to examine, two optical systems, mirrors and lenses, form two different images of S, these two images play the same role as the Young holes.

The Lloyd mirror arrangements, or some derivative arrangements, are often used for making holographic diffraction gratings. A photoresist is deposited on the mirror. A photoresist is an organic resin made of molecules

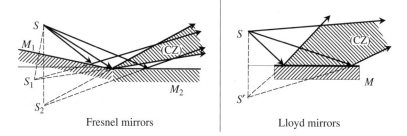

Fresnel mirrors   Lloyd mirrors

Figure 6.3. Fresnel and Lloyd mirrors. Mirrors are used to duplicate the initial point source S. The point source is in fact replaced by a slit perpendicular to the plane of the figure. Interferences are localized inside the common zone (CZ). In the case of Lloyd mirrors, an extra difference of phase should be added because of the reflection of a dielectric mirror, the zero-order fringe (achromatic) is black.

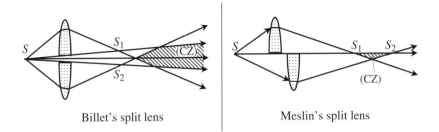

Billet's split lens   Meslin's split lens

Figure 6.4. Billet's and Meslin's split lenses. A convex lens is sawn along an equatorial plane and the two halves are separated by a small distance, at a right angle to the optic axis for Billet's lenses and along the axis for Meslin's lenses. Two real images are produced from a single point source, S_1S_2 being smaller than the coherence length of S, the two secondary sources can be considered as coherent. In the case of the Meslin arrangement, as one wave passes through a focus, and not the other, the zero-order fringe is dark.

(monomers) that easily polymerize (photopolymerize) if illuminated by a suitable light, giving a solid material. Polymerization is of course more active at the places of brilliant fringes than at the places of dark fringes, after development of the resin the interference pattern has been permanently transferred onto the substrate. The advantage of this arrangement comes from the fact that it is very insensitive to mechanical vibrations, since the light source S and the mirror can be firmly attached together.

6.1.3. *Some Physical Remarks About the Preceding Interference Arrangements*

Two Waves Interference

Along the interference field the law of variation of the light intensity versus position is a sine. Such a law is characteristic of a two-beam interference phenomenon. Let us go back to formulas (6.4) and (6.5), and suppose that one source has been switched off, the light intensity would be constant all over the screen and equal to $I = \alpha^2$, and the energy received per unit of surface proportional to $I = \alpha^2$. When the two beams are sent simultaneously the energy is not constant but spatially modulated as indicated by (6.5), the average value of the energy per unit surface is now $2\alpha^2$, which is consistent with energy conservation: more energy is concentrated in the clear fringes and less in the dark fringes.

Nonlocalized Fringes

Because of the existence of a vast zone (CZ) inside of which interference phenomena may be observed, it is said that the fringes are nonlocalized. Later on we will meet other arrangements where fringes are observable only within restricted areas.

Effect of the Geometric Width of the Slit Sources

To avoid, or at least to limit, the inconvenience of spatial incoherency of the usual light sources, the slits should be as narrow as possible. The slits cannot obviously be completely closed, the question is then: What happens when the slits are progressively enlarged? An exhaustive answer is difficult and needs a description of the electromagnetic boundary conditions at the level of the slit edges. Hopefully the result is not that difficult. On the observation screen the fringes keep their positions, the illumination of dark fringes progressively increases and, at the same time, the clear fringes become less and less brilliant. When the slits become too broad, the screen becomes homogeneously illuminated: the fringes have faded and disappear.

Fringes with Quasi-Monochromatic and White Light, Achromatic Fringe

The wavelength is explicitly written in formulas (6.2) and (6.3) that give the positions of the fringes, which implies that we are using monochromatic beams. At a given point M the two waves coming from points S_1 and S_2 have accumulated a phase difference of geometric origin, the value of which is equal to $2\pi(S_1M - S_2M)/\lambda$. If the arrangement is fully symmetric, it's this phase difference that should be considered for the determination of the state of the interference; if not, some extra phase difference may have to be introduced, this is the case for Lloyd's mirrors (reflection or dielectric mirror) and for Meslin's lenses (passing through a focus), where the phase difference is to be increased of π.

Let us now come back to the symmetrical case, the clear fringes are defined from $(S_1M - S_2M) = p\lambda$, where p is an integer that allows us to label the fringes. The fringe labeled zero, the *zero-order fringe* or the *central fringe*, has the same position whatever the color, this is the reason why it's also designated as the *achromatic fringe*. It is always interesting to see where the achromatic fringe of a given arrangement is; from an experimental point of view the achromatic fringe is easily observed with a white light source, since it is the only noncolored fringe. It's easy to see that for the Lloyd and Meslin arrangements the achromatic fringe is black. The zero-order fringe is the only one to be achromatic, since the spacing between adjacent fringes is proportional to the wavelength.

Far enough away from the achromatic fringe, there is a superposition of the fringes associated with the different colors of the white light: the screen has a white appearance. In fact, some colors are lacking. If the entrance slit of a spectrometer, see Figure 6.6, is at a point where different wavelengths are extinguished, the corresponding colors will not appear and will be

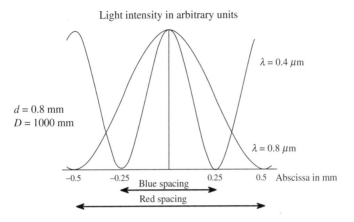

Figure 6.5. Aspect of the achromatic fringe. The red spacing ($\lambda = 0.8\,\mu$m) is twice as broad as the blue spacing ($\lambda = 0.4\,\mu$m). The central part of the fringe which has all the colors is white, the edges seem to be colored in red.

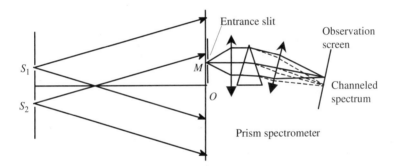

Figure 6.6. Spectral analysis of a channeled spectrum. A spectrometer reveals the presence of brilliant and dark bands parallel to the slits.

replaced by dark bands; the colors with a maximum of intensity will give (brilliant bands. Such a spectrum is called a *channeled spectrum.*

6.2. Amplitude Splitting Interferometers

6.2.1. *Fringes Localization*

Amplitude splitting interferometers are experimental arrangements that allow the observation of interference fringes using extended sources. Of course, a way should be found to alleviate the lack of spatial coherency.

In Figure 6.7 is described the general organization of experimental arrangements of most amplitude splitting interferometers. An incident light

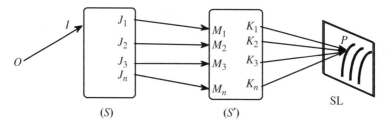

Figure 6.7. Fringe localization. The different rays arriving at *P* are coherent and may interfere. If a screen has been placed on (SL) it seems that fringes are painted on it.

ray arrives at a first optical system S which generates several different transmitted beams J_1M_1, J_2M_2, J_3M_3, ..., J_nM_n. A second optical system S' then recombines all these beams to converge toward some point P: so, *to each light ray OI, is associated a point P*. If we now consider a set of incident rays such as *OI*, we obtain a set of points P, all belonging to a surface SL. On condition that the difference between the times taken by the light in going from O to P are smaller that the coherence time of the source, it can be considered that all the rays arriving at P are coherent and may interfere.

The amplitude of the light vibration at point P is determined by the amplitudes and phases of the vibrations associated to the different rays K_iP. SL is called the surface of localization of the fringes, *it is the only surface along which interferences are observable*. If this surface is real, by opposition to the virtual, it can be covered by an observation screen on which fringes seem to have been painted.

Let Φ_1, Φ_2, Φ_3, ..., Φ_n be the respective phases of the different rays K_iP; very often the arrangement will have been chosen in such a way that the phase difference $\psi = (\Phi_i - \Phi_{i+1})$ is independent of i and is only determined by the initial light ray *OI*; then at each point P of SL is associated some value of ψ. If $\psi = 2p\pi$ (p is an integer), all the rays arriving at P will interfere constructively: all the points corresponding to the same value of p will be distributed along a curve drawn on SL and called the pth brilliant fringe of interference. In the same way the set of points for which $\psi = (2p + 1)\pi$, ψ will define the pth dark fringe.

Let ξ and η be two coordinates of a point P on the surface SL and let λ be the wavelength, ψ can always be written as a function of ξ and η,

$$\psi = \frac{2\pi}{\lambda} f_{(\xi,\eta)}, \qquad (6.6)$$

where $f_{(\xi,\eta)}$ is homogeneous to a length and represents the difference between the optical of two consecutive rays arriving at point P.

It is convenient to classify the interferometers according to the total number n of light rays that are generated by the optical system S, the most interesting cases are:

- $n = 2$, dual beam interferometer.
- $n > 2$, and usually $n \gg 2$, multiple beam interferometer.

6.2.2. *Dielectric Films, Double Beam Interference*

A parallel dielectric plate of refractive index n and thickness e is surrounded by a medium of index equal to one, it is illuminated by an extended light source S, wavelength λ. An incident ray SA, because of the numerous reflections on each side of the film, generates two families $(R, R', R'', \ldots)$ and $(T, T', T'', \ldots)$, interferences are possible, and the localization surfaces are rejected at infinity. The fringes may then be observed in the focal plane of a lens, see Figure 6.9, these are often called *fringes of equal inclination.*

A parallel plate is, a priori, a multiple beam device, however, a simple evaluation of the amplitudes of the successive different reflected beams will show that usually it should be considered only as a dual beam arrangement. The reflection coefficient, at normal incidence, on a dielectric interface is of the order of 4%, under such conditions it's easy to see that the ratios to the initial intensity of ray SA of the respective intensities of the rays $R, R', R'', \ldots$ are equal to 4%, 3.8%, and 0.15%: only the first two reflected beams will thus be considered.

Starting from points B and K of Figure 6.8, the two beams R and R' cover equal distances, their phase difference just comes from the difference necessary to travel along ADB which is air and AK which is glass, the optical length difference δ and the phase difference are given by the following expressions, in which i and r are the angles of incidence and of refraction:

$$\delta = n(AD + DB) - AK, \quad \sin i = n \sin r,$$

$$AD = DB = \frac{e}{\cos r}, \quad AB = 2e \tan r, \quad \cos r = \frac{e}{AD},$$

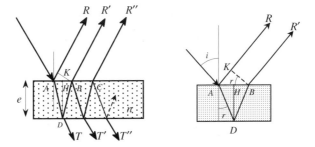

Figure 6.8. The two families of parallel rays generated by reflection on the two sides of a thin film. The localization surfaces are rejected at infinity.

$$AK = AB \sin i = nAB \sin r = 2ne \tan r \sin r = 2ne \frac{\sin^2 r}{\cos r},$$

$$\delta = 2ne \frac{1 - \sin^2 r}{\cos r} = 2ne \cos r,$$

$$\delta = 2ne \frac{1 - \sin^2 r}{\cos r} = 2ne \cos r + \frac{\lambda}{2} \quad \rightarrow \quad \phi = \frac{2\pi\delta}{\lambda} + \pi = \frac{4\pi ne \cos r}{\lambda} + \pi. \tag{6.7}$$

In equation (6.7) ϕ has been increased by a factor of π, and δ by $\lambda/2$, because of the phase differences occurring at reflection on the two sides of the film.

If δ is equal to an integer number of wavelengths the two reflected beams R and R' interfere constructively, on the contrary they interfere destructively if δ is an odd number of half-wavelengths. Generally speaking, δ is neither odd nor integer, two important parameters are then the *order of interference p* and the *fractional order of interference* ε which are defined as

$$\delta = (p + \varepsilon)\lambda. \tag{6.8}$$

For given values of δ and λ, p is the largest possible integer and ε is smaller than unity. The state of interference is fixed by the fractional order ε.

Nothing very general can be said about the state of interference at normal incidence, the first thing to do is to determine the fractional order of $2ne + \lambda/2$; the order of interference, p, is maximum at normal incidence.

Figure 6.9 shows an experimental set-up that allows the observation of the interference pattern between two waves, respectively, reflected on the two sides of a parallel thin plate at normal, or almost normal, incidence. The light coming from an extended source is sent to the plate using a semitransparent mirror which, on one hand, sends the rays toward the plate and, on the other, from transmits the reflected rays to a lens and then to a screen on which the interference pattern is observed. The fringes are localized in the focal plane of the lens; at each point of this plane two rays arrive, one has been reflected the upper face of the plate and the other the lower face. According to formulas (6.7) and (6.8) their phase difference is the same for all pairs of rays making the same angle with the normal to the plate. The axis of the lens which is orthogonal to the plate is an axis of symmetry: the phase difference and interference state are the same for all rays making the same angle with the axis.

We refer to Figure 6.10, the rays making angle i with the axis are focused at points located at the same distance, fi, from the focal point.

Let $r_1, r_2, \ldots, r_p$ and $i_1, i_2, \ldots, i_p$ be, respectively, the angles of refraction and of incidence for which the phase difference is a multiple of the wavelength: corresponding rays interfere constructively at the point of recombination by the lens where a family of circular clear fringes will be found. Between two adjacent clear rings, there is a dark ring for which the phase difference is an odd multiple of a half-wavelength. P and ε are, respectively, the

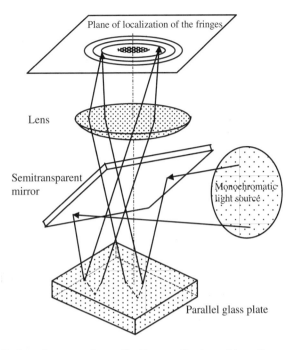

Figure 6.9. Interference after reflection on the two sides of a parallel plate.

order of interference and the fractional excess at normal incidence, the case of clear fringes corresponds to

$$i_1 = nr_1, \quad i_2 = nr_2, \ldots, i_p = nr_p,$$

$$2ne + \frac{\lambda}{2} = (P + \varepsilon)\lambda,$$

$$2ne \cos r_1 + \frac{\lambda}{2} \approx 2ne\left(1 - \frac{r_1^2}{2}\right) + \frac{\lambda}{2} = P\lambda,$$

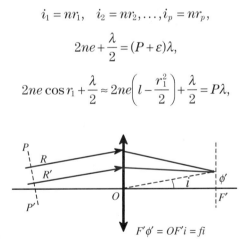

$$F'\phi' = OF'i = fi$$

Figure 6.10. Recombination of two parallel rays R and R'. PP'' is the trace of a plane orthogonal to R and R'. After the lens, the transmitted beams intersect at the focal point ϕ'. Because of stigmatism, the two paths $P\phi'$ and $P'\phi$ take the same time: when they arrive at ϕ' the two rays have the same phase difference as in P and P'.

$$2ne\cos r_2 + \frac{\lambda}{2} \approx 2ne\left(1 - \frac{r_2^2}{2}\right) + \frac{\lambda}{2} = (P-1)\lambda,$$

$$2ne\cos r_p + \frac{\lambda}{2} \approx 2ne\left(1 - \frac{r_p^2}{2}\right) + \frac{\lambda}{2} = (P-p+1)\lambda,$$

$$r_p^2 = \frac{(P-1+\varepsilon)\lambda}{ne} \quad \rightarrow \quad i_p^2 = n^2 r_p^2 = n\frac{(P-1+\varepsilon)\lambda}{e}.$$

Let us suppose that the fractional excess is just equal to zero at normal incidence, the center of the field pattern is clear. The radii of the different fringes, $\rho_1, \rho_2, \ldots, \rho_p$, are obtained by multiplying the angle of incidence by the focal length and they are proportional to the square roots of the successive integers:

$$\rho_1 = 0, \quad \rho_2 = f\sqrt{\frac{ne}{\lambda}}, \ldots, \rho_P = f\sqrt{\frac{ne}{\lambda}}\sqrt{(P-1)}. \tag{6.9}$$

6.2.3. *Fringes of Slides with a Variable Thickness*

6.2.3.1. *Prismatic Plate*

Interferences given by the prismatic plate of Figure 6.11 can been easily observed by the naked eye of some observer accommodating on the plate, and making the two rays, IR and KR', converge and interfere on the retina. The retina image will be clear at points where the two rays are in phase and dark at points where the phase difference is an odd multiple of π. Finally, the observer gets the impression that the fringes *are painted on the plate*.

The points I, J, and K of Figure 6.11 are very close, $IJ \approx JK$ can be considered as the thickness e of the prism at this point. The difference in the optical length δ of the two rays at point M is the result of two contributions,

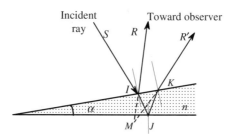

Figure. 6.11. Fringe localization for a prismatic plate. An incident ray SI generates two main reflected rays IR and KR' that are mutually coherent. The determination of the surface of localization is simple only in the case when IR and KR', at some point M that in the case of quasi-normal incidence, remain in close vicinity with the two sides of the prism: the fringes seem to have been painted on the plate.

one has a geometrical origin and the other comes from the difference of the type of reflection of the two beams

$$\delta = 2ne\cos r + \lambda/2 \approx 2ne + \lambda/2.$$

For a clear fringe, the thickness is such that

$$2ne + \frac{\lambda}{2} = p\lambda \quad \rightarrow \quad e = \left(p - \frac{1}{2}\right)\frac{\lambda}{2n}.$$

The fringes are lines parallel to the edges of the prism, the difference in thickness $(e_2 - e_1)$ corresponding to two adjacent fringes is obtained by incrementing p of one unit. The fringe separation, i, is constant and is given by

$$(e_2 - e_1) = \frac{\lambda}{2n} \quad \rightarrow \quad i = \frac{(e_2 - e_1)}{\tan \alpha} = \frac{\lambda}{2n \tan \alpha} \approx \frac{\lambda}{2n\alpha}. \tag{6.10}$$

6.2.3.2. *Fringes of an Air-Filled Prism*

In the experimental arrangements of Figures 6.12 or 6.8, an incident ray generates several reflected and transmitted beams, we have only drawn those who correspond to air/glass or glass/air reflections. However, four reflected beams still remain; we consider that the thicknesses of the plates are greater than the length of coherence of the light source, which eliminates the interference between the following pairs of rays: RR_1, $R'R'_1$, and $R_1R'_1$. Finally, we only keep the rays R and R': the arrangement becomes equivalent to an air prism which would be limited by two semitransparent mirrors with reflection coefficients of about 4% (intensity). If the wavelength is $0.6\,\mu m$ and the angle $0.1' = 3 \times 10^{-5}$ rad, formula (6.10) gives a fringe separation of $5\,mm$.

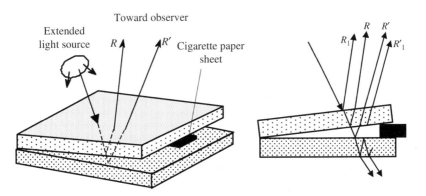

Figure 6.12. Fringes of an air prism. A thin air prism is readily obtained by inserting a thin sheet of paper between the two glass plates.

6.2.3.3. *Fringes of Equal Thickness, Thin Plates Colors*

A prismatic plate, illuminated near its edge, is a special case of a thin transparent medium of nonconstant thickness. It's usual in Physics to use thin dielectric plates, having a good transparency and a thickness of a few micrometers: oil films on a water surface, soap bubbles, faintly oxidized metallic surfaces. . . . Current observation shows that illumination by a white light source produces very pretty interference colors. These colors resemble the wings of some butterflies, the colors of which are also due to interference.

If the film of Figure 6.13 is very thin, the order of interference is not very high and interference fringes may be observed with a white light source. A white source is the superposition of an infinity of monochromatic vibrations, each of which has an infinitesimal amplitude. Let us call $\alpha_\lambda\, d\lambda$ the amplitude of the component of wavelength λ; the different colors cannot interfere and the global intensity, which is the sum of the intensities of the different colors, is proportional to the integral

$$\int_{red}^{blue} \alpha_\lambda^2\, d\lambda = \int_{red}^{blue} I_\lambda\, d\lambda, \tag{6.11}$$

where I_λ is a function of the wavelength, the law of variation versus λ determines the special shade of the white light under consideration.

The two rays IR and $I'R'$ of Figure 6.13 that interfere, after having been reflected at a point where the thickness is e, have a difference of optical paths equal to $\delta = 2ne + \lambda/2$; the complex amplitude of the vibration $da_{(\lambda)}$ and the intensity $dI_{(\lambda)}$ are given by

$$da_{(\lambda)} = a_\lambda\left(1 + e^{-jk(2ne+\lambda/2)}\right)d\lambda, \tag{6.12.a}$$

$$dI_{(\lambda)} = a_\lambda a_\lambda^* = 4a_\lambda^2 \sin^2\left(\frac{2\pi ne}{\lambda}\right)d\lambda. \tag{6.12.b}$$

If the two reflections are of the same kind it is not necessary to add a phase difference of $\pi/2$ and we then have

$$dI_{(\lambda)} = 4a_\lambda^2 \cos^2\left(\frac{2\pi ne}{\lambda}\right)d\lambda = 4I_\lambda \cos^2\left(\frac{2\pi ne}{\lambda}\right)d\lambda. \tag{6.12.c}$$

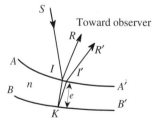

Figure 6.13. The material between the surfaces AA' and BB' is transparent (index of refraction n), its thickness e varies from one point to another. The two rays IR and $I'R'$ combine on the retina of the observer who gets the impression that lines of equal thickness are painted on the film. Dark fringes are obtained when $2ne + \lambda/2 = (2p + 1)/\lambda/2$.

If the plate is thin enough, there will be only a few wavelengths, or eventually only one wavelength, for which the argument of the integrals of (6.11) are equal to zero: illuminated by a white source, the plate takes a color which is characteristic of its thickness. If the thickness is not constant, the fringes of equal thickness are colored. In addition to the aesthetic aspect, this method gives an elegant way to evaluate the thickness of a thin plate; the colors change from straw-yellow to purple as the thickness varies from 280 to 560 nm. The colors obtained by interference of two white light waves is also met on the occasion of the interference between polarized white light beams, and the corresponding calculations are exhaustively developed in Section 5.C.2 of Annex 5.C.

Interference Colors

The colors corresponding to formulas (6.11) and (6.12) are conventionally represented using the following notations:

$$\int I_\lambda \sin^2 \frac{2\pi n e}{\lambda} d\lambda \quad \text{and} \quad \int I_\lambda \cos^2 \frac{2\pi n e}{\lambda} d\lambda,$$

$$\int I_\lambda \sin^2 \frac{2\pi n e}{\lambda} d\lambda + \int I_\lambda \cos^2 \frac{2\pi n e}{\lambda} d\lambda = \int I_\lambda \, d\lambda. \qquad (6.13)$$

Formula (6.13) shows that the two different colors, respectively, associated to the same delay δ, are complementary, in the same sense of the word as used by painters. Interference colors are sometimes useful for evaluating, at a glance, the thickness of a thin film. For small delays colorations are bright (as an example, the coloration of butterfly wings are, more or less, the result of an interference process). As the delay is increased the colors get duller to become what is called a *higher-order white*, when the geometric path difference is larger than a few visible wavelengths. The colors that are indicated below are obtained within the two following conditions:

- The spectral composition I_λ of the light source is very similar to the light of the Sun.
- The refractive index dispersion of the film material is negligible; this second condition is fully satisfied in the case of a film of air.

There are many situations in life where interference colors may be observed. Illuminated by a white source, the soap shells of blown bubbles, which have a thickness of a fraction of a micrometer, provide a magnificent illustration of interference colors. In the same way, under white light illumination, the thin films that are obtained when some oil spreads on a wet surface are an illustration of interference colors; because of its nonmiscibility with water, grease tends to have an area as extended as possible and produces extremely thin films, possibly monomolecular. In the case of very small thicknesses, no coloration will appear.

$\delta = ne\ nm$	$\int I_\lambda \sin^2 \dfrac{2\pi\delta}{\lambda} d\lambda$		$\int I_\lambda \cos^2 \dfrac{2\pi\delta}{\lambda} d\lambda$	
0	First	black	First	white
40	order	iron gray	order	white
97		lavender gray		yellowish white
158		gray blue		brownish white
218		lighter gray		yellowish brown
234		greenish white		brown
259		white		light red
267		yellowish white		carmine
275		pale straw yellow		dark reddish brown
281		straw yellow	Second	dark violet
306		light yellow	order	indigo
332		bright yellow		blue
430		brown yellow		gray blue
505		reddish orange		bluish green
536		deep red		pale green
551		darker red		yellowish green
565	Second	purple		lighter green
575	order	violet		greenish yellow
589		indigo blue		golden yellow
664		sky blue		orange
728		greenish blue		brownish orange
747		green		light carmine red
826		lighter green		purple
843		yellowish green	Third	purplish
866		greenish yellow	order	violet
910		pure yellow		indigo blue
948		orange		dark blue
998		bright reddish orange		greenish blue
1101		dark purple red		green

6.3. Dual-Beam Interference

6.3.1. *The Michelson Interferometer*

The Michelson interferometer is a very nice optical device, its construction requires the greatest care; the accuracy of the angular and translation position measurements are, respectively, 1 s and 50 nm. One of its most spectac-

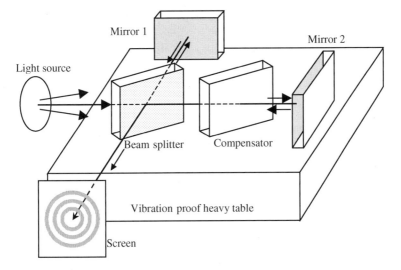

Figure 6.14. The Michelson interferometer.

ular utilizations was made by the physicists Michelson and Morley who gave experimental proof of the fact that the value of the speed of light was independent of the coordinate axis and so brought a demonstration of the validity of Einstein's theory of relativity.

A Michelson interferometer, see Figures 6.14 and 6.15, is made of two mirrors, M_1 and M_2, with high reflection coefficients disposed along two ver-

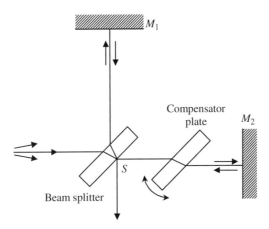

Figure 6.15. Top view of a Michelson interferometer. One interest of this device is to allow the observation of zero- and low-order fringes. M_1 and M_2 can be accurately shifted, the compensator plate is finely rotatable to equalize the length of the two arms SM_1 and SM_2.

tical and orthogonal planes. An important component is the *beam splitter*, another vertical semitransparent mirror making an angle of 45° with M_1 and M_2; one side is antireflection coated, the reflection and transmission coefficients of the other side are equal to 50%. The interferometer is illuminated by an extended monochromatic source. The beam splitter plays two roles: first, it separates the incident beam into two identical beams that are sent to M_1 and M_2 and, second, it recombines them so that they interfere on a screen or a photodetector. The purpose of this interferometer is the observation fringes with a *low order* of interference; the beam that is reflected from mirror M_1 makes an extra double transit inside the beam splitter, this is the reason why a fourth vertical plate, called a compensator plate, is inserted in front of mirror M_2. The compensator plate is almost identical to the beam splitter except that it is antireflection coated on both sides. Rotating finely the compensator plate around a vertical axis allows a fine tuning of the length of SM_1 and makes it equal to SM_2.

A Michelson interferometer has many degrees of freedom, in Figure 6.16 are shown two interesting ways of using it. Let us consider the image M'_2 of the mirror M_2 in the beam splitter, in Figure 6.16(a) M'_2 and M_1 are parallel and separated by a distance e that can be adjusted by a translation of one of the mirrors. The interference pattern is the same as for a parallel plate of the same thickness and filled with air. In Figure 6.16(b), M'_2 and M_1 make a small and adjustable angle, the fringes are those of a wedge plate. If a transparent and weakly inhomogeneous plate P is placed in one arm of the interferome-

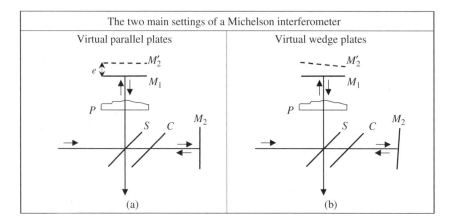

The two main settings of a Michelson interferometer	
Virtual parallel plates	Virtual wedge plates
(a)	(b)

Figure 6.16. M'_2 is the image of M_2 through the beam splitter. According to the adjustment of S, M'_2 and M_1 can be parallel or make a small angle. P is a transparent inhomogeneous plate. In the absence of P, the fringes are concentric rings (virtual parallel plates) or parallel lines (virtual wedge plates). The order of interference being almost zero, the fringes are observable with a white illumination giving typical colored fringes.

ter, the fringe pattern (rings or parallel lines) is modified, and the optical characteristics of the plate are easily obtained from the modifications of the fringes. The method is so sensitive that the variation of the index of refraction of air should be considered.

6.3.2. *The Twyman-Green Interferometer*

The Twyman-Green interferometer is a direct consequence of the Michelson interferometer; its main objective is the characterization of high-quality optical components, such as optical prisms, lenses, or objectives. Disposed in one arm of a Michelson interferometer, the element to be tested is crossed twice by the light rays. In the case of perfect optical quality of the element under test, the return wave and the incident wave are both planar waves, and, according to the setting, usual fringes (parallel plate or wedge plate) are observed. If the optical element has some defects, the return wave is no longer planar, the shape of the fringes is modified, and their modifications give interesting information about the optical imperfections.

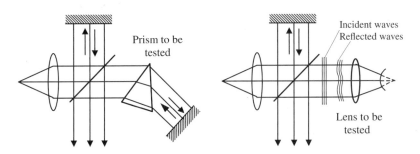

Figure 6.17. Test of optical components using a Twyman-Green interferometer.

6.3.3. *The Mach-Zehnder Interferometer*

The Mach-Zehnder interferometer is described in Figure 6.18; it's a two-wave amplitude splitting interferometer using two beam splitters and two Porro prisms, which can be replaced by two 100% reflecting mirrors. The two arms *ABC* and *ADC* can be given the same optical length and the interference order is almost equal to zero. The two interfering beams travel along two well-separated paths, one is used as a reference (*ADC*), the other (*ABC*) is used for testing some equipment. A Mach-Zehnder interferometer is very convenient for testing refractive index variations occurring inside volumes that can be quite large.

The two emerging beams, *CR* and *CR′*, can be parallel or make a small angle; the fringe pattern is that of a thin parallel plate in the first case, and of a wedge plate in the second case. Before the appearance of laser sources it

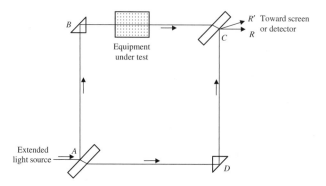

Figure 6.18. Mach-Zehnder interferometer.

was hard work to tune a Mach-Zehnder interferometer; using sources having a long length of coherence considerably simplifies the problem.

In the experimental arrangement of Figure 6.19 the Mach-Zehnder interferometer works as a wedge plate; it is used for characterizing a transparent plate with a step variation of thickness. A lens is used to image the plate on a screen on which the interference fringes of a wedge are superimposed on the image of the plate. Counting the number of fringes allows an accurate measurement of the thickness of the step.

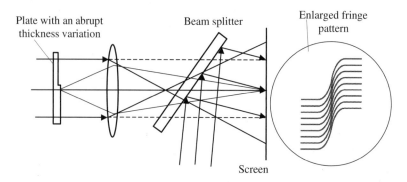

Figure 6.19. Characterization of a thickness variation using a Mach-Zehnder interferometer. Nine fringes may be counted, which corresponds to an optical thickness variation of nine wavelengths.

6.4. The Fabry-Perot Interferometer

It was in an attempt to increase the visibility of the interference fringes that were obtained with thin plates (see Section 6.2.2) that the French physicist Raymond Boullouch, at the end of the nineteenth century, proposed to partially silver the two faces of a glass plate, the purpose being to increase the reflection coefficient and to give it a value higher than 4% which is typical of an air/glass interface. He opened the way to multiple wave interference. Charles Fabry immediately guessed all the spectroscopic possibilities offered by the arrangement of two parallel mirrors and, with the help of Alfred Perot, a very skilled mechanical engineer, gave an experimental demonstration of this fascinating device, which is now universally known as the *Fabry-Perot resonator*, often designated as an FP resonator. The FP resonator is an essential component of most lasers.

6.4.1. *Description of a Fabry-Perot Resonator*

An FP resonator is mainly made of two parallel mirrors, their reflection coefficients are usually high (80–90%) and the parallelism is excellent (10 s $\approx$ 10^{-6} rad).

In Figure 6.20 are shown the two main arrangements of an FP resonator. In Figure 6.20(a) the FP resonator is made of two separate mirrors, each of which is made of a wedge glass plate, one side is coated to have a high reflection coefficient; the two faces are not parallel and make an angle of a few degrees, so that the beams that are reflected on the noncoated faces don't participate in the interference process. A high precision screw allows an accurate translation of the mirrors, which remain parallel while their separation is smoothly varied.

In the arrangement of Figure 6.20(b), reflecting layers are deposited each side of a transparent plate, the faces of which are parallel with a high accuracy. This arrangement, which is known as a *Fabry-Perot etalon*, is very convenient although the separation of the two mirrors is, of course, not tunable.

The performances of an FP resonator come from the quality and the parallelism of the mirrors. The roughness of the faces should be very low, the mean quadratic error with regard to an ideal plane is always better than $\lambda/10$ (λ is some optical wavelength, usually the sodium wavelength) and usually of

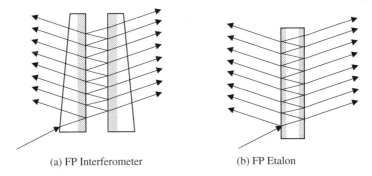

(a) FP Interferometer   (b) FP Etalon

Figure 6.20. Examples of an FP resonator.

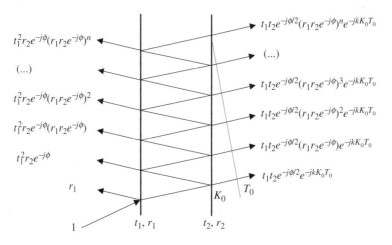

Figure 6.21. Simplified representation of an FP resonator.

$\lambda/100$. Initially, the mirrors were made by metal evaporation, dielectric coatings are now exclusively used, since their absorption coefficients are very low ($0.1\% = 10^{-3}$).

In Figure 6.21 is shown a simplified model that will be used for a theoretical analysis, the glass plates supporting the mirrors are omitted: the resonator is made of two reflective planes separated by a medium having a refractive index equal to one. To each mirror are associated a reflection, a transmission, and an absorption coefficient. Two kinds of coefficients are used corresponding, respectively, to the complex amplitudes or to the intensities of the waves. For complex amplitudes we use lowercase italic letters (r_1, r_2, t_1, t_2, a_1, a_2) and for intensities capital italic letters (R_1, R_2, T_1, T_2, A_1, A_2). The amplitude coefficients are complex numbers, the argument of which is determined by the phase shift occurring at reflection. The intensity coefficients are real and equal to the squared modulus of the amplitude coefficients,

$$R_p = r_p r_p^*, \quad T_p = t_p t_p^*, \quad A_p = a_p a_p^*, \quad R_p + T_p + A_p = 1. \qquad (6.14)$$

Referring to Figures 6.21 and 6.22, it is seen that an incident ray generates two families of parallel rays that, respectively, propagate on the left and right sides of the resonator. Since all the rays belonging to a given family are parallel they interfere at infinity and the interference pattern should be observed in the focal plane of a lens L. An FP resonator is usually illuminated by an extended source emitting rays in all directions. To each direction Δ is associated a focus $P(\Delta)$. If the lens is stigmatic, the phase repartition between the different rays is kept on their arrival at $P(\Delta)$ and determines the state of interference. The phase difference between the following two rays is fixed by the angle of incidence i, and remains the same for all the rays making the same angle with the normal to the mirrors. This normal to the mirrors is an axis of

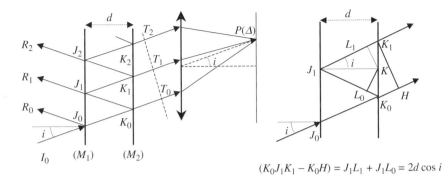

$$(K_0 J_1 K_1 - K_0 H) = J_1 L_1 + J_1 L_0 = 2d \cos i$$

Figure 6.22. Ray trajectories inside an FP resonator. All rays coming from the initial ray $I_0 J_0$ are focused at the same point $P(\Delta)$ where they interfere with the same phases that they had, respectively, at points T_0, T_1, T_2. During the time taken by the ray, which leaves the FP resonator at point K_1 to travel along the optical path $K_0 J_1 K_1$, the ray leaving at point K_0 will only have covered $K_0 H$.

symmetry, which means that *the interference fringes are concentric circular rings* centered on the focus of the lens corresponding to the direction of the normal.

We must be aware of the fact that the previous treatment is, a priori, not valid since its uses, explicitly and extensively, the notion of light rays to analyze an interference phenomenon, which should be described by waves. The result is however quite correct and there is no paradox at all; the reason is that the resonator is illuminated by planar waves, and a ray like $I_0 J_0$ of Figure 6.22 is just the wave vector of the incident wave, in the same way $K_i J_{i+1}$ and $K_i T_i$ are the wave vectors of reflected and transmitted waves that are generated on the mirrors.

A planar monochromatic wave (angular frequency $\omega = 2\pi\nu$, vacuum wavelength $\lambda = c/\nu$, vacuum vector $k = 2\pi/\lambda$) illuminates the devices of Figures 6.21 and 6.22, the wave vector is oriented along $I_0 J_0$. The interaction with the first mirror generates a reflected wave $J_0 R_0$ and a transmitted wave $J_0 K_0$, this last wave generates in turn a planar transmitted wave $K_0 T_0$ and a reflected wave $K_0 J_1$, and so on. . . . Almost all analog treatments can be given for the two families of waves, the expressions are slightly simpler for the transmitted waves, because they all follow the same law of recurrence, while the first reflected wave $J_0 K_0$ has a different expression from the others. Although they have equal wave vectors and, consequently, parallel wave planes, all the waves of a given family don't constitute a unique planar wave since there is a phase discrepancy between two successive waves. The phase difference ϕ is evaluated in Figure 6.22:

$$\phi = k2d \cos i = \frac{4\pi d \cos i}{\lambda} = \frac{4\pi d \nu \cos i}{c}. \tag{6.15}$$

6.4.2. *Expressions of the Transmitted Waves, Airy Function*

From the complex amplitudes of the transmitted and reflected waves that are given in Figure 6.21, it is easy to calculate the amplitude $Y(\phi)$ of the wave that is transmitted in the direction which makes an angle i with the normal to the mirrors,

$$Y(\phi) = t_1 t_2 e^{-jkK_0T_0} e^{-j\phi/2} \sum_p \left(r_1 r_2 e^{-j\phi} \right)^p. \tag{6.16}$$

Formula (6.16) is quite general, however, if the reflection coefficients are small the factor $(r_1 r_2)^p$ rapidly becomes negligible as p increases. If $r_1 = r_2 = (4\%)^{1/2} \rightarrow (r_1 r_2)^3 = 6 \times 10^{-5}$, only the two first waves will have some importance in the interference process. Now if $r_1 r_2 = 0.8 \rightarrow (r_1 r_2)^{10} = 0.10$, after ten reflections the signal still has a meaningful value: the FP resonator is really a multiple wave interferometer.

The summation indicated in (6.16) is readily done and the result is

$$Y(\phi) = t_1 t_2 e^{-jkK_0T_0} e^{-j\phi/2} \frac{1}{(1 - r_1 r_2 e^{-j\phi})}. \tag{6.17}$$

If the incident wave is given an intensity equal to unity, (6.18) gives the transmitted intensity:

$$I_{t(\phi)} = Y_{(\phi)} Y_{(\phi)}^* = \frac{(t_1 t_2)^2}{(1 - r_1 r_2 e^{-j\phi})(1 - r_1 r_2 e^{+j\phi})}, \tag{6.18}$$

$$I_{t(\phi)} = \frac{(t_1 t_2)^2}{(1 + r_1 r_2)^2 - 2 r_1 r_2 (1 + \cos\phi)} = \frac{(t_1 t_2)^2}{(1 + r_1 r_2)^2 - 4 r_1 r_2 \cos^2 \phi/2}. \tag{6.18.a}$$

We use the reflection $(\rho = r_1 r_2)$ and transmission $(\tau = t_1 t_2)$ coefficients for intensities and we define the *coefficient of finesse F*, which is an important parameter of an FP resonator:

$$F = \frac{4\rho}{(1 - \rho)^2}, \quad \text{Finesse factor.} \tag{6.19}$$

Using a new function, called the *Airy function*, which is defined by formula (6.20), the intensity is given by formula (6.21):

$$A_{(\phi)} = \frac{1}{1 + F \sin^2 \dfrac{\phi}{2}}, \tag{6.20}$$

$$I_{t(\phi)} = \frac{\tau^2}{(1 - \rho)^2} A_{(\phi)}. \tag{6.21}$$

According to (6.15), ϕ is proportional to the frequency; the graph $I_{t(\phi)}$ can thus be considered as a representation of the spectral response of the FP resonator.

The Airy function is periodic (2π) and so is the spectral response with a period equal to $2d/c = \tau_{RT}$ (RT for *return time*), τ_{RT} is the time taken by the light to go back and forth between the mirrors. The *finesse factor* such as defined by formula (6.19) is a dimensionless parameter that becomes infinite when the two mirrors are perfectly reflecting. $A(\phi)$ is maximum and equal to one when ϕ is an integer multiple of 2π, and $A(\phi)$ is minimum and equal to $1/(1 + F)$ for odd multiples of π. When F is of the order of unity, or even smaller, the Airy function looks very much like a sine oscillating between 1 and $1/(1 + F)$: it can then be considered that only two beams are interfering. Multiple interferences correspond to high values of the finesse coefficient; as F goes to infinity, $A(\phi)$ becomes a succession of equidistant peaks, with height equal to one and width equal to zero. This is a general property: the higher the number of waves that interfere, the more accurately the condition of positive interference has to be fulfilled; for an infinite number of interfering waves, light is observed only in the directions for which the condition $\phi = p2\pi$ is strictly satisfied.

Finesse and Coefficient of Finesse

Instead of the coefficient of finesse F another parameter is often introduced, it is simply called the *finesse* F_i of the FP resonator and is defined by the following formula:

$$F_i = \frac{\pi}{2}\sqrt{F} = \pi\sqrt{\frac{\rho}{(1-\rho)^2}} = \frac{\pi\sqrt{r_1 r_2}}{(1-r_1 r_2)}, \quad \text{Finesse.} \tag{6.22}$$

Let $\Delta\phi_{1/2}$ be the half-width height of the teeth of the Airy function comb and let ν_p be the frequency of the pth tooth $(\phi = 2p\pi)$, for large enough values of the finesse we have

$$\Delta\phi_{1/2} = \frac{4}{\sqrt{F}} \quad \rightarrow \quad F_i = \frac{2\pi}{\Delta\phi_{1/2}}. \tag{6.23}$$

If the graph of Figure 6.23 is considered as the representation of the spectral response of the FP resonator as a function of the frequency ν, the finesse is easily written as a function of the half-width spectral height $\Delta\nu_{1/2}$ on the one hand, and of the frequency difference $(\nu_p - \nu_{p-1})$ between two adjacent teeth on the other hand; this frequency difference $(\nu_p - \nu_{p-1})$ has been given a name: the *free spectral range* (FSR) of the interferometer,

$$F_i = \frac{(\nu_p - \nu_{p-1})}{\Delta\nu_{1/2}}. \tag{6.24}$$

Following a rule of thumb, the finesse is considered to give an order of magnitude for the number of rays that really interfere, since their amplitude is not too small.

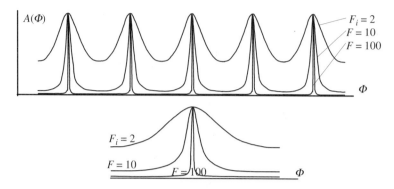

Figure 6.23. Combs of modes for different values of the finesse.

Quality Factor Q

The quality factor Q is a number that is mostly used in electronics to characterize an RLC circuit, or the selectivity of a frequency filter; it is defined as the ratio of the central frequency to the bandwidth, it is usually a large number. A Q factor can be defined for the teeth of the comb of mode

$$Q = \frac{\nu_p}{\Delta\nu_{1/2}} = F_i \frac{\nu_p}{(\text{FSR})} = F_i \frac{\nu_p}{(\nu_p - \nu_{p-1})}. \tag{6.25}$$

Numerical Application

We consider an FP resonator with the following optogeometric properties:

$$r_1 = r_2 = 0.96 \quad \rightarrow \quad \rho = 0.92, \quad t_1 = t_2 = 0.282 \quad \rightarrow \quad \tau = 0.08,$$
$$d = 6\,\text{mm}, \quad \tau_{\text{RT}} = 2d/c = 4.10^{-11}, \quad s = 40\,\text{ps}.$$

Ratio of vibrations, respectively, labeled $p = 0$ and $p = 40$ in formula (6.16):

$$\left(\frac{E_p}{E_0}\right)^p = \rho^p = (0.92)^{40} = 0.035 = 3.5\%.$$

Free Spectral Range: (FSR) = $(\nu_p - \nu_{p-1})$ = $1/\tau_{\text{RT}}$ = 2.5×10^{10} Hz = 25 GHz.

Finesse factor: $F = \dfrac{4\rho}{(1-\rho)^2} = 575$;   Finesse: $F_i = \dfrac{\pi}{2}\sqrt{F} = 37.7 \approx 40.$

Spectral width of a tooth belonging to the comb of modes,

$$\Delta\nu_{1/2} = \frac{(\text{FSR})}{F_i} = 625\,\text{MHz}.$$

Quality factor: $Q = \dfrac{\nu_p}{\Delta\nu_{1/2}} = F_i \dfrac{\nu_p}{(\text{FSR})} = 0.8 \times 10^6.$

The Fabry-Perot Resonator Using Spherical Mirrors

Up to now the waves propagating between the two mirrors were always supposed to be planar and to have an infinite transverse extension, with the nonformulated assumption that the mirrors also have infinite transverse dimensions. . . . If the mirrors don't have an infinite extension, oblique rays, after a great number of reflections, will eventually fall outside the mirrors; this is the reason why the finesse hardly exceeds 50. In the middle of the twentieth century the French physicist Pierre Connes proposed replacing the planar mirrors by spherical mirrors; under such conditions the light rays remain in the vicinity of the line joining the centers of the two mirrors, finesses as high as 1000 can then be reached. Lasers almost exclusively use spherical mirrored resonators.

6.4.3. *Size of the Rings*

We consider the rays that are transmitted through an FP resonator; observed on a screen disposed in the focal plane of a lens, their interference pattern is made of concentric circular fringes, which appear as bright rings on a dark background. Let $FP = x$ be the distance of some point P to the center of the rings and let i be the angle associated to the direction for which P is the focal point. If f is the focal length of the lens, we can write

$$i \approx \tan i = \frac{x}{f} \quad \text{and} \quad \phi = \frac{4\pi d}{\lambda} \cos i.$$

The thickness d of the FP resonator contains a large number of wavelengths, we can always write

$$2d = \lambda(p_0 + \varepsilon), \quad 0 \leq \varepsilon \leq 1, \quad p_0 \text{ is an integer.} \tag{6.26}$$

ε is a positive number smaller than unity and is called the *fractional excess*.

p_0 is the largest number of half-wavelengths $\lambda/2$, contained in the thickness d, and is called the *order of interference at the center* of the interference pattern.

i being the angle of incidence corresponding to some bright interference ring, we have

$$\sin^2\left(\frac{2\pi d}{\lambda} \cos i\right) = 0 \quad \rightarrow \quad \frac{2\pi d}{\lambda} \cos i = p\pi \quad (p \text{ integer}),$$

$$\cos i = \frac{p}{p_0 + \varepsilon}. \tag{6.27}$$

For a given thickness, according to formula (6.27), the order of interference is at most equal to p_0 and decreases with the distance to the center of the interference pattern. For the first clear ring the order of interference is

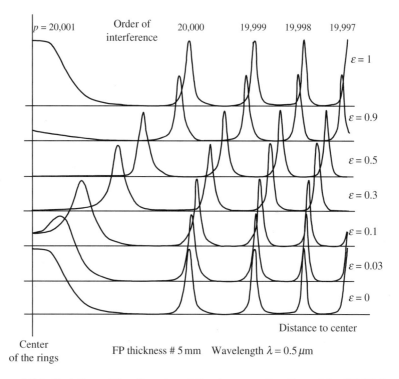

Figure 6.24. Variation of the diameter of the rings as a function of the FP thickness. When the fractional excess $\varepsilon = 0$, we observe a broad and bright spot at the center, surrounded by clear thin rings. As ε increases from 0 to 0.5, a new ring appears, starting from the center.

$p = p_0$, the following rings are obtained by decrementing p, one unit after the other. The angles of incidence of the successive rings are given by the following formulas, in which the angles are considered to be very small:

$$\text{First ring:}\quad \cos i_0 = \frac{p_0}{p_0 + \varepsilon} \approx 1 - \frac{\varepsilon}{p_0} \quad \to \quad i_0 \approx \frac{\sqrt{2\varepsilon}}{\sqrt{p_0}}.$$

$$\text{Second ring:}\quad \cos i_1 = \frac{p_0 - 1}{p_0 + \varepsilon} = \frac{p_0}{p_0 + \varepsilon} - \frac{1}{p_0 + \varepsilon} \approx 1 - \frac{\varepsilon + 1}{p_0} \quad \to \quad i_1 \approx \frac{\sqrt{2(\varepsilon + 1)}}{\sqrt{p_0}}.$$

$$q\text{th ring:}\quad \cos i_q = \frac{p_0 - q}{p_0 + \varepsilon} = \frac{p_0}{p_0 + \varepsilon} - \frac{q}{p_0 + \varepsilon} \approx 1 - \frac{\varepsilon + q}{p_0} \quad \to \quad i_q \approx \frac{\sqrt{2(\varepsilon + q)}}{\sqrt{p_0}}.$$

6.4.4. Chromatic Resolving Power

The main purpose of a spectroscope is the examination of the detailed structure of spectral lines; its performances can be described using either the frequency or the wavelength. The chromatic resolution is the smallest wavelength difference $\Delta\lambda_{min}$, or the smallest frequency difference $\Delta\nu_{min}$ that can be measured. The chromatic resolving power is the ratio of the wavelength to the chromatic resolution, or of the frequency to the chromatic resolution,

$$R = \frac{\lambda}{\Delta\lambda_{min}} = \frac{\nu}{\Delta\nu_{min}}, \tag{6.28}$$

where $\Delta\lambda_{min}$ is obtained from point C of Figure 6.25(b) and from the expression of the *Airy function* $A(\phi)$ for the two wavelengths λ_1 and λ_2, see formula (6.23),

$$\phi_1 = \frac{4\pi d}{\lambda_1}\cos i = p\pi + \frac{\Delta\phi_{1/2}}{2}, \quad \phi_2 = \frac{4\pi d}{\lambda_2}\cos i = p\pi - \frac{\Delta\phi_{1/2}}{2},$$

$$\frac{4\pi d}{\lambda_1} - \frac{4\pi d}{\lambda_2} \approx \Delta\phi_{1/2} = \frac{2\pi}{F_i} \quad \rightarrow \quad 2d\frac{\lambda_2 - \lambda_1}{\lambda_2\lambda_1} \approx \frac{2d}{\lambda}\frac{\Delta\lambda}{\lambda} = p\frac{\Delta\lambda}{\lambda} = \frac{1}{F_i},$$

$$R = \frac{\lambda}{\Delta\lambda} = pF_i. \tag{6.29}$$

The chromatic resolving power is proportional to the coefficient of finesse and to the order of interference.

It is sometimes convenient to introduce the *Free Spectral Range* (FSR) in the expression of $\Delta\lambda_{min}$:

$$(\text{FSR})_{\text{frequency}} = (\nu_{p+1} - \nu_p) = c\left(\frac{1}{\lambda_{p+1}} - \frac{1}{\lambda_p}\right) \approx c\frac{\lambda_p - \lambda_{p+1}}{\lambda_p^2},$$

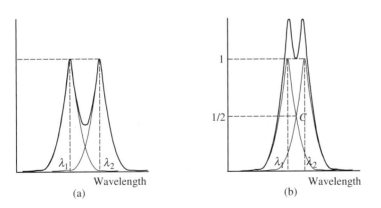

Figure 6.25. Separation of two adjacent rays. Here are shown the intensity profiles of two rings having the same order of interference but two different wavelengths. In (a) the wavelengths are quite different and easily resolved. In (b) the two curves intersect at the point C where the intensity is equal to one-half of the maximum; it is considered that this situation represents the limit of the resolution.

$$(\text{FSR})_{\text{wavelength}} = (\lambda_p - \lambda_{p+1}) \approx \frac{\lambda_p^2}{c} (\text{FSR})_{\text{frequency}},$$

$$\Delta\lambda_{\text{min}} = \frac{(\text{FSR})_{\text{wavelength}}}{F_i}, \quad \Delta\nu_{\text{min}} = \frac{(\text{FSR})_{\text{frequency}}}{F_i}, \tag{6.30}$$

where $\Delta\lambda_{\text{min}}$ is all the smaller as the FP resonator is thicker. Very important thicknesses are not however of practical use, since the different orders of interference would then overlap, making impossible the interpretation of the spectrograms.

Scanning the Fabry-Perot Interferometer

Taking photographs and then performing densitometric measurements has been, for some time, the only way to analyze the interferograms. Direct measurements, using photodetectors, are of course more comfortable; in Figure 6.26 is shown a very convenient arrangement, known as the *Scanning Fabry-Perot* arrangement. As illustrated in Figure 6.24, the diameter of the fringes increases with the distance d between the two mirrors; when d is increased by a half-wavelength, the ring labeled $(q + 1)$ takes the place of the ring labeled q. A translation of a fraction of one micrometer is readily obtained with a piezoelectric cell.

The interference pattern is analyzed by a photodetector through a tiny hole, the dimension of which is small as compared to the broadness of an interference fringe. Let us suppose that the light is monochromatic (wavelength λ) and that the hole is located on a clear ring; if the separation between the mirrors is increased, the diameter of the ring also increases: the intensity of the light transmitted through the hole diminishes according to a law which

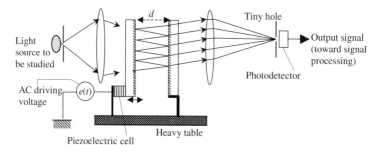

Figure 6.26. Scanning Fabry-Perot arrangement. One of the mirrors is fixed on a piezoelectric cell. The mirror separation, d, is modulated, thanks to a low-frequency (1 kHz) driving voltage $e(t)$. A tiny hole is placed in the focal plane of the second lens in front of a photodetector. The output signal is proportional to the light intensity at the location of the hole.

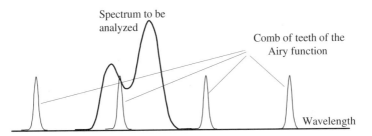

Figure 6.27. Sampling of the spectral profile of a radiation. When the mirror separation d is varied, the frequency distance between two adjacent teeth can be considered to keep a constant value, and the comb is shifted as a whole. The entire Free Spectral Range is swept as d is increased to $\lambda/2$.

is determined by the *Airy function*. We now consider the case of a light source that is no longer monochromatic, that's to say that its intensity is a function, $I(\lambda)$, of the wavelength; however, the variation is supposed to be smooth enough, so that $I(\lambda)$ remains constant inside the frequency band corresponding to the broadness of the ring of the FP resonator: the arrangement of Figure 6.27 allows the sampling of the function $I(\lambda)$ and a suitable signal processing, which takes account of the shape of the *Airy function*, finally giving the profile of I.

6.5. Interference Using Stacks of Thin Transparent Layers

6.5.1. *Considerations About the Technology of Thin Films*

As an answer to the demand in Optics and also in microelectronics, efficient, reproducible, and cheap methods have been developed for depositing thin films on the surface of various substrates. The films can be made of dielectric or metallic materials. Their thicknesses, which range from a few nanometers to a few micrometers, and their chemical composition are controlled with high accuracy. The layers that we will consider here are made of a dielectric material of constant composition and constant refractive index, with a thickness of the order of an optical wavelength (1000 to 10,000 nm).

In most cases, the aim is to elaborate stacks of alternately low-index layers (typically $n = 1.3$) and higher-index layers ($n = 2$), see Figure 6.28. The interesting property of such an arrangement is that its reflection and transmission coefficients vary with the wavelength according to a law that can be easily controlled, by playing on the thickness and on the index of refraction of the layers. A great variety of response curves can be obtained, going from very selective filters to a flat response function inside a given spectral band.

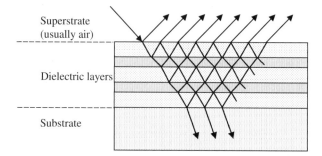

Figure 6.28. Stack of dielectric layers deposited on top of a substrate. An incident wave generates many reflected and transmitted waves. The reflected (or transmitted) wave is the result of the interference of the various partially reflected or transmitted waves.

6.5.2. Antireflection Coatings

Antireflection coatings were probably the first industrial applications of thin optical layers. Only one layer is used, its refractive index n_1 is intermediate between the index n_0 of the superstrate and the refractive index n_s of the substrate, see Figure 6.29. The purpose is to obtain that the two first reflected rays, R_1 and R_2, cancel by interference: all the incident energy is then transferred to the transmitted ray. The problem is to determine the index n_1 and the thickness e, so that R_1 and R_2 have the same amplitudes and opposite phases. The incidence is supposed to be quasi-normal, the moduli of the normal reflection coefficient are given by the following expressions:

$$\rho_{\text{superstrate/layer}} = \frac{n_1 - n_0}{n_1 + n_0} \quad \text{and} \quad \rho_{\text{layer/superstrate}} = \frac{n_s - n_1}{n_s + n_1},$$

$$\rho_{\text{layer/substrate}} = \rho_{\text{superstrate/layer}} \quad \rightarrow \quad n_1 = \sqrt{n_0 n_s}. \tag{6.31.a}$$

The optical path length difference δ between R_1 and R_2 is equal to

$$\delta = 2n_1 e \cos r \approx 2n_1 e, \quad \text{since the incidence is normal or almost normal.}$$

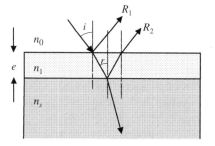

Figure 6.29. Antireflection coating. The incidence is supposed to be quasi-normal. If the index of the layer is equal to $(n_0 n_s)^{1/2}$, and if the thickness is such that a double transit inside the layer produces a phase difference equal to π, then the two reflected rays are mutually destroyed by interference.

The interference will be destructive at the wavelength λ, if the layer thickness is such that $2n_1e = (2p + 1)\lambda/2$, the smallest value of the thickness fulfilling this condition is

$$e = \lambda/4n_1. \tag{6.31.b}$$

Numerical application: Let us look for a material which would make an antireflection coating, if deposited on a piece of glass suspended in air.

$n_0 = 1$, $n_s = 1 \rightarrow (n_0n_s)^{1/2} = 1.35$; no convenient material exists with such a low index, the nearest one is cryolithe, AlF_6Na_3, the index of which is 1.35. The deposition of a layer of cryolithe with a suitable thickness on top of a glass slide reduces the normal reflection coefficient from 4% down to 0.3%. All the interfaces of optical instruments, including spectacles, are covered with antireflection coatings. Let us consider the case of a microscope objective made of five different lenses, the light rays will cross ten air/glass interfaces; in the absence of any treatment the ratio of the transmitted intensity to the incident lens is $0.96^{10} = 66\%$, while it becomes $0.997^{10} = 97\%$ with antireflection coatings. The gain of luminosity is not the only advantage; the unwanted reflected light rays constitute a parasitic light signal returning to the eyes of the observer.

If a treated surface is illuminated by a white source, it appears to have a faint blue-violet characteristic color, this comes from the fact that the reflection coefficient is practically equal to zero for a yellow light and has a small, but still appreciable, value for red and blue light rays.

6.5.3. *Generalization to the Case of n Layers*

6.5.3.1. *Matrix Representation of Reflection and Transmission by an Interface*

To determine the waves that are transmitted, or reflected, by an FP resonator, the method used in Section 6.4 consists in the superposition (interference) of the various waves generated on the two mirrors. Although it should be theoretically possible, this method can hardly be generalized to the arrangement shown in Figure 6.28. We are going to follow another way, a matrix will characterize each layer and the action of several cascaded layers will be described using a product of matrices. An excellent description of the matrix treatment of thin layers can be found in the book *Optical Waves in Layered Media* by Pochi Yeh (Wiley).

In Annex 4.A of chapter four it is shown how the invariance in a translation parallel to an interface leads to the conservation of TE and TM polarizations when a beam is reflected or refracted. We refer to Figure 6.30(a and b): the first interface receives a planar wave (wave vector, $\boldsymbol{k}_{1,R}$); a planar reflected wave (wave vector, $\boldsymbol{k}_{1,L}$) and a planar transmitted wave (wave vector, $\boldsymbol{k}_{2,R}$) are created on the first interface. Similarly, a planar reflected wave

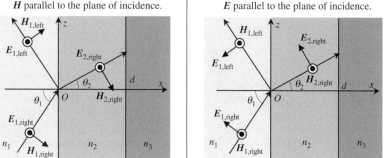

Figure 6.30. TE and TM waves keep their polarization by reflection or refraction.

($k_{2,L}$) and a planar transmitted wave ($k_{3,R}$) are created on the second interface. The labeling conventions are the following:

- 1, 2, 3 refers to the medium inside of which the propagation is made.
- R or L, respectively, mean toward the right- or the left-hand side.

$$k_{1,R} = k_{1,x}x + k_{1,z}z, \quad k_{1,L} = k_{1,x}x + k_{1,z}z,$$
$$k_{2,R} = k_{2,x}x + k_{2,z}z, \quad k_{2,L} = k_{2,x}x + k_{2,z}z. \tag{6.32}$$

If there is only one planar incident wave on the first interface, in the layer labeled α we will find two planar waves with the respective wave vectors $k_{\alpha R}$ and $k_{\alpha,L}$. These two wave vectors are obtained from the Snell-Descartes laws for refraction and reflection. Because of the orientation chosen for $k_{1,R}$ in Figure 6.30, all the vectors $k_{\alpha R}$ are parallel to the positive directions of the Ox and Oz axes, while all the $k_{\alpha,L}$ are directed parallel to the positive direction of Ox and the negative direction of Oz. To justify the above allegations, we can either be confident in our physical intuition, other waves being hardly conceivable, or we can say that the solution that we are going to exhibit does satisfy Maxwell's equations, since it's the superposition of planar waves and since it has been specially adjusted to fulfill the boundary conditions along the various interfaces.

According to the Descartes-Snell formula, the projections of all the wave vectors on the axis Oz have the same value β,

$$\beta = k_{1z,R} = k_{1z,L} = k_{2z,R} = k_{2z,L} = k_{3z,R} = k_{3z,L} = \cdots,$$
$$\beta = k_1 \sin \theta_1 = k_2 \sin \theta_2 = k_3 \sin \theta_3 = k_\alpha \sin \theta_\alpha = \cdots, \tag{6.33}$$
$$\beta = k_0 n_1 \sin \theta_1 = k_0 n_2 \sin \theta_2 = k_0 n_3 \sin \theta_3 = k_0 n_\alpha \sin \theta_\alpha = \cdots,$$

where $k_0 = \omega/c$ and k_α are, respectively, the moduli of the wave vectors in a vacuum and in a material of refractive n. The projections of the wave vectors on the Ox axis are given by the following formulas:

$$k_{\alpha x,R} = k_0 n_\alpha \cos \theta_\alpha \quad \text{and} \quad k_{\alpha x,L} = -k_0 n_\alpha \cos \theta_\alpha. \tag{6.34}$$

The electric and magnetic fields on both sides of the first interface of Figure 6.30 are given by

$$\left(E_{1,R}e^{-jk_1x\cos\theta_1} + E_{1,L}e^{+jk_1x\cos\theta_1}\right)e^{j(\omega t - \beta z)} \quad \text{for } x > 0, \tag{6.35.a}$$

$$\left(E_{2,R}e^{-jk_2x\cos\theta_2} + E_{2,L}e^{+jk_2x\cos\theta_2}\right)e^{j(\omega t - \beta z)} \quad \text{for } x > 0, \tag{6.35.b}$$

$$H = \frac{j}{\omega\mu_0}\nabla \wedge E. \tag{6.36}$$

For homogeneous and isotropic materials the boundary conditions are very simple: the tangential components of the electric and magnetic fields vary continuously when crossing an interface.

TE Wave

The electric field is along Oy; the magnetic field has two components along Ox and Oz. We set $E_{1,R} = E_{1,R}^{\mathrm{TE}}y$ and $E_{1,L} = E_{1,L}^{\mathrm{TE}}y$ and we obtain

$$E_{1,R}^{\mathrm{TE}} + E_{1,L}^{\mathrm{TE}} = E_{2,R}^{\mathrm{TE}} + E_{2,L}^{\mathrm{TE}},$$
$$n_1(E_{1,R}^{\mathrm{TE}} + E_{1,L}^{\mathrm{TE}})\cos\theta_1 = n_2(E_{2,R}^{\mathrm{TE}} + E_{2,L}^{\mathrm{TE}})\cos\theta_2. \tag{6.37}$$

If we introduce the matrix

$$\Delta_\alpha^{\mathrm{TE}} = \begin{pmatrix} 1 & 1 \\ n_\alpha\cos\theta_\alpha & -n_\alpha\cos\theta_\alpha \end{pmatrix}, \quad \alpha \in \{1; 2\},$$

equation (6.37) can be written

$$\Delta_1^{\mathrm{TE}}\begin{pmatrix} E_{1,R}^{\mathrm{TE}} \\ E_{1,L}^{\mathrm{TE}} \end{pmatrix} = \Delta_2^{\mathrm{TE}}\begin{pmatrix} E_{2,R}^{\mathrm{TE}} \\ E_{2,L}^{\mathrm{TE}} \end{pmatrix}. \tag{6.38}$$

TM Wave

The electric field has two components along Ox and Oz, the magnetic field is along Oyo; we write the continuity of E_z and of H_y:

$$(E_{1,R}^{\mathrm{TM}} + E_{1,L}^{\mathrm{TM}})\cos\theta_1 = (E_{2,R}^{\mathrm{TM}} + E_{2,L}^{\mathrm{TM}})\cos\theta_2,$$
$$n_1(E_{1,R}^{\mathrm{TM}} - E_{1,L}^{\mathrm{TM}}) = n_2(E_{2,R}^{\mathrm{TM}} - E_{2,L}^{\mathrm{TM}}), \tag{6.39}$$

$$\Delta_\alpha^{\mathrm{TM}} = \begin{pmatrix} \cos\theta_\alpha & \cos\theta_\alpha \\ n_\alpha & -n_\alpha \end{pmatrix} \quad \rightarrow \quad \Delta_1^{\mathrm{TM}}\begin{pmatrix} E_{1,R}^{\mathrm{TM}} \\ E_{1,L}^{\mathrm{TM}} \end{pmatrix} = \Delta_2^{\mathrm{TM}}\begin{pmatrix} E_{2,R}^{\mathrm{TM}} \\ E_{2,L}^{\mathrm{TM}} \end{pmatrix}, \quad \alpha \in \{1; 2\}. \tag{6.40}$$

Fresnel's Formula for Reflection and Refraction

If the second medium is infinitely extended on the positive side of the Ox axis $\rightarrow E_{2,L}^{\mathrm{TE}} = E_{2,L}^{\mathrm{TM}} = 0$; we can then define the reflection and refraction coefficients: $\rho^{\mathrm{TE}} = (E_{1,L}^{\mathrm{TE}}/E_{1,R}^{\mathrm{TE}})$ and $\tau^{\mathrm{TE}} = (E_{2,R}^{\mathrm{TE}}/E_{1,R}^{\mathrm{TE}})$, and equivalent formulas for the TM waves,

$$\rho^{\mathrm{TE}} = \frac{n_1 \cos\theta_1 - n_2 \cos\theta_2}{n_1 \cos\theta_1 + n_2 \cos\theta_2}, \quad \rho^{\mathrm{TM}} = \frac{n_1 \cos\theta_2 - n_2 \cos\theta_1}{n_1 \cos\theta_2 + n_2 \cos\theta_1},$$

$$\tau^{\mathrm{TE}} = \frac{2n_1 \cos\theta_1}{n_1 \cos\theta_1 + n_2 \cos\theta_2}, \quad \tau^{\mathrm{TM}} = \frac{2n_2 \cos\theta_1}{n_1 \cos\theta_2 + n_2 \cos\theta_1}. \tag{6.41}$$

Formulas (6.41) are the same as the Fresnel formulas that have already been established in Section 4.3.3.2, see formulas (4.15) and (4.16).

6.5.3.2. *Representation of a Thin Layer by a 2 × 2 Matrix*

We now come back to the arrangement of Figure 6.30. For a TE polarization there is no discontinuity for the electric vector $\boldsymbol{E}$, since it is purely tangential; the situation is different for the magnetic vector $\boldsymbol{H}$, the normal component of which is discontinuous. For a TM polarization, the magnetic vector is purely tangential and thus continuous, while the normal component of the electric field varies discontinuously.

The full calculation of the matrix method will only be made in the case of a TE polarization. As the electric fields are always parallel, we can forget that they are vectors and treat them as if they were scalar,

$$E_{\alpha(x,z,t)} = \left(E_{\alpha,R}e^{-jk_\alpha x \cos\theta_\alpha} + E_{\alpha,L}e^{+jk_\alpha x \cos\theta_\alpha}\right)e^{j(\omega t - \beta z)},$$

where $E_{\alpha,R}$ and $E_{\alpha,L}$ are constant values inside the layer labeled α; the amplitudes of the waves propagating toward the right- or left-hand side are, respectively, written as:

$$R_{\alpha(x)} = E_{\alpha,R}e^{-jk_\alpha x \cos\theta_\alpha} = E_{\alpha,R}e^{-j(k_\alpha)_x x}, \quad (k_\alpha)_x = k_\alpha \cos\theta_\alpha, \tag{6.42.a}$$

$$L_{\alpha(x)} = E_{\alpha,L}e^{+jk_\alpha x \cos\theta_\alpha} = E_{\alpha,L}e^{+j(k_\alpha)_x x}, \tag{6.42.b}$$

$$x < 0 \quad \rightarrow \quad (k_\alpha)_x = (k_1)_x = n_1 \frac{\omega}{c}\cos\theta_1 = n_1 k_0 \cos\theta_1,$$

$$0 < x < d \quad \rightarrow \quad (k_\alpha)_x = (k_2)_x = n_2 \frac{\omega}{c}\cos\theta_2 = n_2 k_0 \cos\theta_2,$$

$$x > d \quad \rightarrow \quad (k_\alpha)_x = (k_3)_x = n_3 \frac{\omega}{c}\cos\theta_3 = n_3 k_0 \cos\theta_3.$$

Dropping the label TE, formula (6.38) becomes

$$\Delta_\alpha = \begin{pmatrix} 1 & 1 \\ n_\alpha \cos\theta_\alpha & n_\alpha \cos\theta_\alpha \end{pmatrix},$$

$$\Delta_1 \begin{pmatrix} R_1(0^-) \\ L_1(0^-) \end{pmatrix} = \Delta_2 \begin{pmatrix} R_2(0^+) \\ L_2(0^+) \end{pmatrix},$$

$$\begin{pmatrix} R_1(0^-) \\ L_1(0^-) \end{pmatrix} = (\Delta_1)^{-1} \Delta_2 \begin{pmatrix} R_2(0^+) \\ L_2(0^+) \end{pmatrix}, \tag{6.43.a}$$

where 0^- and 0^+ are the abscissas of two points, respectively, located immediately before or immediately after the interface.

The phase difference $\phi_2 = (k_2)_x d$ due to the propagation from the plane
$x = 0^+$ to the plane $x = d^-$, and the phase difference $-\phi_2$ for a propagation in
the opposite direction, correspond to a propagation matrix P_2,

$$P_2 = \begin{pmatrix} e^{j\phi_2} & 0 \\ 0 & e^{-j\phi_2} \end{pmatrix} \rightarrow \begin{pmatrix} R_2(0^+) \\ L_2(0^+) \end{pmatrix} = P_2 \begin{pmatrix} R_2(d^-) \\ L_2(d^-) \end{pmatrix},$$

$$\begin{pmatrix} R_1(0^+) \\ L_1(0^+) \end{pmatrix} = \Delta_1^{-1} \Delta_2 P_2 \begin{pmatrix} R_2(d^-) \\ L_2(d^-) \end{pmatrix} = \Delta_1^{-1} \Delta_2 P_2 \Delta_2^{-1} \Delta_3 \begin{pmatrix} R_3(d^+) \\ L_3(d^+) \end{pmatrix}. \tag{6.43.b}$$

The above formula remains valid for a TM polarization, if the matrix Δ is
taken from formula (6.40).

6.5.3.3. *Matrix of an Arrangement of n Layers*

Formula (6.44), which is a generalization of (6.43), is not very convenient for
handmade calculations, but proves to be useful if a computer is used,

$$\begin{pmatrix} R_{in} \\ L_{in} \end{pmatrix} = \Delta_0^{-1} \left[\prod_1^N \Delta_l P_l \Delta_l^{-1} \right] \Delta_{N+1} \begin{pmatrix} R_{out} \\ L_{out} \end{pmatrix},$$

$$\begin{pmatrix} R_{in} \\ L_{in} \end{pmatrix} = \begin{pmatrix} M_{11} & M_{12} \\ M_{21} & M_{22} \end{pmatrix} \begin{pmatrix} R_{out} \\ L_{out} \end{pmatrix}. \tag{6.44}$$

The matrix elements are obtained from formula (6.34), instead of using
the sine or cosine of the angle θ_α, it may be more convenient to use
the modulus of the projection on the Ox axis, $k_{\alpha,x} = k_0 n_\alpha \cos \theta_\alpha$, of the wave

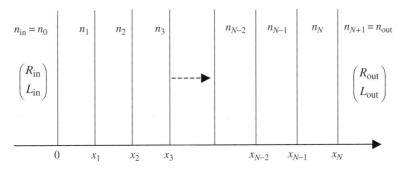

Figure 6.31. Stack of N multidielectric layers sandwiched between a substrate n_{out}
and a superstrate n_{in}.

vector

$$\Delta_\alpha^{\mathrm{TE}}=\begin{pmatrix} 1 & 1 \\ n_\alpha\cos\theta_\alpha & -n_\alpha\cos\theta_\alpha \end{pmatrix}=\begin{pmatrix} 1 & 1 \\ k_{\alpha,x}/k_0 & -k_{\alpha,x}/k \end{pmatrix},$$

$$\Delta_\alpha^{\mathrm{TM}}=\begin{pmatrix} \cos\theta_\alpha & \cos\theta_\alpha \\ n_\alpha & -n_\alpha \end{pmatrix}=\begin{pmatrix} k_{\alpha,x}/n_\alpha k_0 & k_{\alpha,x}/n_\alpha k_0 \\ n_\alpha & -n_\alpha \end{pmatrix}.$$

The products of matrices, $D_{l,l+1}=\Delta_l^{-1}\Delta_{l+1}$, that are found in (6.43) and (6.44) have the following expressions:

$$D_{l,l+1}^{\mathrm{TE}} = \begin{pmatrix} \dfrac{1}{2}\left(1+\dfrac{k_{l+1,x}}{k_{l,x}}\right) & \dfrac{1}{2}\left(1-\dfrac{k_{l+1,x}}{k_{l,x}}\right) \\ \dfrac{1}{2}\left(1-\dfrac{k_{l+1,x}}{k_{l,x}}\right) & \dfrac{1}{2}\left(1+\dfrac{k_{l+1,x}}{k_{l,x}}\right) \end{pmatrix}, \tag{6.45.a}$$

$$D_{l,l+1}^{\mathrm{TM}} = \begin{pmatrix} \dfrac{1}{2}\left(1+\dfrac{n_{l+1}^2 k_{l,x}}{n_l^2 k_{l+1,x}}\right) & \dfrac{1}{2}\left(1-\dfrac{n_{l+1}^2 k_{l,x}}{n_l^2 k_{l+1,x}}\right) \\ \dfrac{1}{2}\left(1-\dfrac{n_{l+1}^2 k_{l,x}}{n_l^2 k_{l+1,x}}\right) & \dfrac{1}{2}\left(1+\dfrac{n_{l+1}^2 k_{l,x}}{n_l^2 k_{l+1,x}}\right) \end{pmatrix}. \tag{6.45.b}$$

From a practical point of view, the projection β of the wave vector on the Oz axis will first be evaluated, and then the projection on the Ox axis will be obtained using the following formula (6.46):

$$\beta^2+k_{\alpha,x}^2=n_\alpha^2 k_0^2 \quad \rightarrow \quad k_{\alpha,x}=\sqrt{n_\alpha^2 k_0^2-\beta^2}, \tag{6.46}$$

if $k_{\alpha,x}$ is either real or complex, and if $k_{\alpha,x}$ is real the wave inside the layer of index n_α is a progressive wave. There are two situations for which $k_{\alpha,x}$ is not real: in the first one, the material is absorbing and the index of refraction n_α is a complex number. The second situation is more interesting, the material is fully transparent, n_α is real, but the conditions are those of total internal reflection; $k_{\alpha,x}$ is purely imaginary.

Stack of λ/4 Layers

We now consider the case where the stack of Figure 6.31 is made of the succession of N pairs of layers each having an optical thickness equal to $\lambda/4$ $(n_1 d_1 = n_2 d_2 = n_3 d_3 = \cdots = n_\alpha d_\alpha = \lambda/4)$.

According to formulas (6.38) and (6.39), the matrices have the same expression for TE and TM polarizations,

$$\begin{pmatrix} M_{11} & M_{12} \\ M_{21} & M_{22} \end{pmatrix}=\Delta_0^{-1}\left[\Delta_1 P_1 \Delta_1^{-1}\Delta_2 P_2 \Delta_2^{-1}\right]^N\Delta_{2N},$$

$$P_1=P_2=\begin{pmatrix} j & 0 \\ 0 & -j \end{pmatrix},$$

$$\begin{pmatrix} M_{11} & M_{12} \\ M_{21} & M_{22} \end{pmatrix} = \Delta_0^{-1} [\Delta_1 P_1 \Delta_1^{-1} \Delta_2 P_2 \Delta_2^{-1}]^N \Delta_{2N},$$

$$\Delta_1 P_1 \Delta_1^{-1} \Delta_2 P_2 \Delta_2^{-1} = \begin{bmatrix} -n_2/n_1 & 0 \\ 0 & -n_2/n_1 \end{bmatrix}.$$

Finally, we suppose that the last medium is infinitely extended on the positive side of the Ox axis, which means that there is no wave coming back from the right-hand side,

$$L_{2N}(x_{2N}^+) = 0 \quad \rightarrow \quad \begin{pmatrix} R_0(x_0^+) \\ L_0(x_0^+) \end{pmatrix} = \begin{pmatrix} M_{11} & M_{12} \\ M_{21} & M_{22} \end{pmatrix} \begin{pmatrix} R_{2N}(x_{2N}^+) \\ 0 \end{pmatrix},$$

$$R_0(x_0^+) = M_{11} R_{2N}(x_{2N}^+) \quad \text{and} \quad L_0(x_0^+) = M_{21} R_{2N}(x_{2N}^+).$$

We can now evaluate the reflection coefficients of the stack of dielectric layers for either the amplitude or the intensity of the waves,

$$r_{2N} = \frac{L_0(x_0^+)}{R_0(x_0^+)} = \frac{M_{21}}{M_{11}} \quad \text{and} \quad r_{2N} r_{2N}^* = \frac{\left| L_0(x_0^+) \right|^2}{\left| R_0(x_0^+) \right|^2} = \frac{\left| M_{21} \right|^2}{\left| M_{11} \right|^2},$$

$$r_{2N} r_{2N}^* = \left(\frac{1 - (n_{\text{out}}/n_{\text{in}})(n_1/n_2)^{2N}}{1 + (n_{\text{out}}/n_{\text{in}})(n_1/n_2)^{2N}} \right)^2.$$

$n_0 = n_{\text{in}}$ and $n_{2N+1} = n_{\text{out}}$ are the respective indices of the first and last mediums.

N	0	1	2	3	4	5	6	7
R_{2N}	0.04	0.207	0.425	0.621	0.765	0.860	0.919	0.953
N	8	9	10	11	12	13	14	15
R_{2N}	0.974	0.985	0.992	0.995	0.997	0.998	0.9991	0.9995

The above table gives the values of the reflection coefficient for different values of the number of pairs of layers. The first medium is air ($n_{\text{in}} = 1$), the indices of the two layers of each pair are, respectively, $n_1 = 2$ and $n_2 = 1.4$, the index of the substrate is $n_{\text{out}} = 1.5$.

An important application of dielectric multilayers is the fabrication of mirrors, totally or partly reflecting. The main advantage of using dielectric materials, rather than a metal, is the low value of their absorption coefficients, which is as small as 0.1% (compared with a few per cent for silvered mirrors).

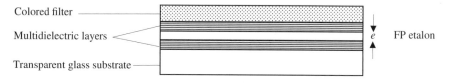

Figure 6.32. Interference filter. The FP etalon only transmits the frequencies of the teeth of the comb of modes. The colored piece of glass is a broader filter centered on the frequency to be selected; it keeps only one tooth and rejects all the others. The substrate is a high-quality piece of glass and ensures mechanical rigidity.

6.5.3.4. *Interference Filter*

An interference filter is used to select a very narrow spectral band of light emitted by a less monochromatic source, they are very common optical components. As shown in Figure 6.32, they are made of an FP etalon, made by depositing dielectric multilayers on both sides of a transparent plate, and of a colored glass plate. The FP etalon transmits all the wavelengths of the different teeth of the comb of modes, its thickness e is adjusted so that one of the teeth just coincides with the frequeqency which is to be selected; if e is small, the frequencies of the adjacent teeth will be quite different and easily removed by a broader band filter.

7

Diffraction

7.1. The Huygens-Fresnel Postulate

Diffraction has played an important role in the development of Optics and, more generally, of Physics. It should be remembered that the wave nature of light had been recognized only after Arago and Fresnel had made their experiments on diffraction by circular holes and disks.

The propagation of light had been experimentally analyzed well before any theoretical formulation (Maxwell's equations) or before the introduction of the necessary mathematical tools (vector analysis, Fourier development, . . .). The physicists working during the eighteenth century proved to be very imaginative and pragmatic when they described the phenomena that they were observing.

7.1.1. *Intuitive Approach of the Huygens-Fresnel Principle*

The Huygens-Fresnel principle is a typical example; the description of the propagation of light waves using wavelets is completely intuitive and is in full agreement with experimental results. Its validity is extended to many other domains and is not restricted to electromagnetic waves.

Diffraction is characteristic of any physical phenomenon leading to the propagation of waves, such as electromagnetic waves, acoustic waves, mechanical surface waves, De Broglie waves. . . . Diffraction occurs any time that some *obstacle* is interposed on the wave trajectory; its effects are especially noticeable when the cross section of a beam is limited by a diaphragm having dimensions that cannot be considered as infinite, that's to say large, as compared to the wavelength.

Chapter 7 has been reviewed by Dr. Ludovic Brassé from Teemphotonics.

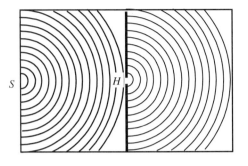

Figure 7.1. Diffraction of mechanical surface waves by a small aperture. The hole can be assimilated to a point source which, apart from a $\pi/2$ phase difference, has the same amplitude as the vibration produced, at point H, by the wave emitted by the source S.

To introduce the Huygens-Fresnel principle, we will start from the observation of the diffraction of mechanical surface waves propagating along the free interface of a liquid; the experimental arrangement is shown in Figure 7.1: a tank is divided into two parts by a wall inside of which a small hole H has been drilled. An electrically activated vibrator is disposed at point S, generating circular waves. All the walls are covered with a special foam to eliminate reflected waves. The two parts of the tank only communicate through the small hole H; the molecules of the liquid are set in motion by the wave coming from S. As far as the second half of the tank is concerned, everything occurs as if a vibrator is disposed at point H and generates circular waves.

According to the Huygens-Fresnel postulate, the vibrations of the molecules located at the hole are the same as the vibrations that would be created by the source S in the absence of the wall. In the arrangement of Figure 7.2, the hole is quite large and the effects of diffraction are only noticeable in the vicinity of the edges.

Diffraction and Interference

In most practical situations, the light that arrives at a given point has several origins: light coming directly from the source and light that has been diffracted by different obstacles. What is observed is the result of the interfer-

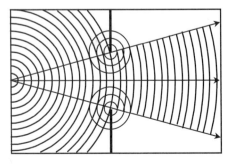

Figure 7.2. Diffraction of a circular wave by a large size obstacle. The effect of diffraction is only noticeable in the close vicinity of the edges of the diaphragm.

ence between all these radiations; if the initial source is coherent enough, fringes will be seen. The arrangement of Young's holes is an interesting example involving, simultaneously, diffraction and interference.

Qualitative Considerations on Diffraction Patterns

The basic properties of a planar wave are the following:

- Only one wave vector is involved.
- The wave surfaces have an infinite extension, since they are planar, with the mathematical meaning of this word.
- The complex amplitude of the vibrations remains constant and doesn't go to zero at infinity.

When the transverse extension of a planar wave is limited by some hole drilled in an opaque screen, the wave is said to be *stopped down*. The transmitted wave is no longer planar: the surface waves are not planar and the amplitude goes to zero on the border of the diffracted pattern. Such a wave (see Section 2.3.4) should be considered as the superposition of an infinity of elementary planar waves, having different wave vectors and different amplitudes. The amplitude of these planar waves is all the larger as the orientation of their wave vectors is closer to the wave vector of the incident planar wave. In Figure 7.3 has been drawn a curve, called an indicatrix, which is obtained by plotting, from the center of the diaphragm and in the direction of the wave vector, a segment having a length proportional to the intensity of the corresponding planar wave. All the diffracted wave vectors are practically inside the cones shown in the figure, the shape and size of the diaphragm determine their angle α. Practically $\alpha = K\lambda/a$, where a is of the order of the transverse

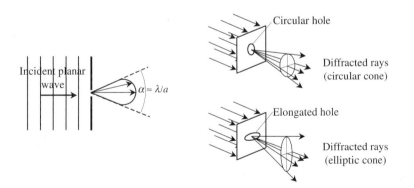

Figure 7.3. Diffraction of a planar wave by a hole. The smaller the hole, the wider the cone inside of which the diffracted light is spread. For an elongated hole, the diffraction pattern is more spread in a direction orthogonal to the larger size of the hole.

size of the diaphragm, and K is a proportionality coefficient of the order of 1. For a circular diaphragm having a diameter a, the coefficient K is 1.22. A small hole transmits only a little light, but the light is all the more spread the smaller it is. The indicatrix of a very small hole is a half-sphere.

Very often the diaphragm has the shape of a thin rectangular slot, the length of which can be considered as infinite compared to its width a. There is no diffraction in a direction parallel to the slot; it will be shown that in the other direction the angle of diffraction is $\alpha = \lambda/a$.

Huygens-Fresnel Wavelets

To solve a problem of electromagnetic propagation, we have to determine the electromagnetic vector at a given time and at all the points of the geometric space, knowing its repartition, over a given surface and at some previous time. The given surface can be the surface of the emitting source. The fields are supposed to be harmonic and can be expressed by the following formula:

$$A_{(x,y,z)}e^{-j\varphi(x,y,z)}e^{j\omega t}.$$

The source S of Figure 7.4 continuously produces harmonic waves with a period equal to T. In the same figure are shown, at time $t = 5T$, the six surface waves that have been emitted at the respective instants $t = 0, T, 2T, 3T, 4T, 5T$.

A first way of formulating the Huygens-Fresnel principle consists in the replacement of the source S by a set of point sources (called *auxiliary Huygens-Fresnel sources*), disposed over a wave surface (Σ) and emitting *in-phase* spherical waves (called *Huygens-Fresnel wavelets*) having the same amplitude that was produced at the same place as the source S. It is considered that the surface waves are the envelope of the wavelets emitted by all the auxiliary sources.

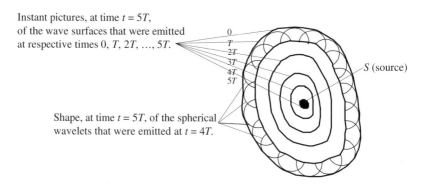

Instant pictures, at time $t = 5T$, of the wave surfaces that were emitted at respective times $0, T, 2T, ..., 5T$.

0
T
2T
3T
4T
5T

S (source)

Shape, at time $t = 5T$, of the spherical wavelets that were emitted at $t = 4T$.

Figure 7.4. Illustration of the Huygens-Fresnel principle in the case where the auxiliary sources are distributed over a surface.

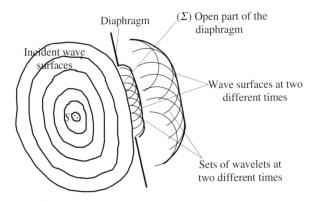

Figure 7.5. Interpretation of diffraction by diaphragm. To reconstitute the same field as produced by the source, it is necessary to let all the auxiliary sources interfere. If the diaphragm obliterates some of them, the reconstituted field is different as it was in the absence of a diaphragm.

As an example, we have drawn the wavelets that have been emitted at time $t = 4T$, their envelope, at time $t = 5T$, coincides with the wave surface that had been emitted at time $t = 0$.

Once the Huygens-Fresnel principle is admitted, the interpretation of diffraction is immediate: to reconstitute exactly a field identical to the one that is radiated by the source, we need to consider all the auxiliary sources; if a diaphragm obliterates some of them, the field is modified. The utilization of the wavelets for determining the diffraction pattern in the situation of Figure 7.5 is surely not rigorous, although commonly used.

7.1.2. *Huygens-Fresnel Mathematical Formulation*

Starting from the wave equation (2.3), Helmholtz and Kirchhoff gave a demonstration of the Huygens-Fresnel principle, which is then no longer a principle.

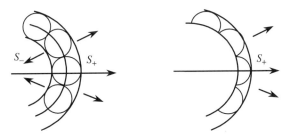

Figure 7.6. According to the Huygens-Fresnel principle, two waves S_+ and S_-, should be considered. Mathematicians could find good arguments to eliminate the wave, which has no experimental existence.

Their demonstration is valid in the case when the auxiliary sources are spread over a closed surface. Quite interesting from a theoretical point of view, this theory doesn't bring any additional physical features with regard to the intuitive approach; in addition, there are very few situations where rigorous calculations can be carried over. However, thanks to computers, numerical integration techniques have largely contributed to new developments of this aspect of electromagnetism.

Amplitude of Vibration of the Auxiliary Sources

A second formulation of the Huygens-Fresnel principle is to place over the open part (Σ) of the diffracting diaphragm, auxiliary sources that have complex amplitudes proportional to the amplitudes $F(M)$ of the vibration at the same place, but in the absence of the diaphragm. These elementary amplitudes become meaningful only after an integration over the surface (Σ), they are proportional to the area $d\sigma$ of an elementary surface drawn around point M and will be written as $KF(M)\,d\sigma$, K is a proportionality coefficient that must be evaluated. The elementary field dE created at a point P of a spherical wavelet coming from point M is given by

$$dE = KF(M)\frac{e^{jkPM}}{MP}d\sigma.$$

A full development of the theory gives, for the proportionality coefficient, a value, $K = j/\lambda$, which can be interpreted in the following way: dE and $F(M)$ have the same dimension, thus K is homogeneous to the inverse of a length; the only length already met in the problem is the wavelength, which can be considered as a good reason for making $K = 1/\lambda$. The existence of the imaginary factor $j = e^{j\pi/2}$ is easily understood if we think of a spherical wave that diverges from the center of the wavelet and if we remember that there is a π phase shift for a spherical wave crossing its focus.

To obtain a mathematical expression for the Huygens-Fresnel principle, we will use the two formulas (7.1) and (7.2), given in Figure 7.7.

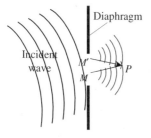

$F(M)$ = vibration created by the incident wave in M.
Elementary vibration sent from M to P:

$$dE = KF(M)\frac{e^{-jkPM}}{MP}d\sigma. \tag{7.1}$$

Global vibration in P:

$$E = \frac{j}{\lambda}\iint\limits_{\text{diaphragm}} F(M)\frac{e^{-jkPM}}{MP}d\sigma. \tag{7.2}$$

Figure 7.7. Interference of the wavelets emitted by the points located in the open part of the diaphragm.

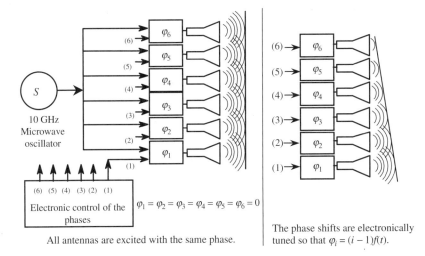

Figure 7.8. Concrete illustration of the Huygens-Fresnel principle: electronic angular scanning of the wave emitted by a microwave antenna network.

Figure 7.8 shows an arrangement often used in microwave to obtain an angular scanning of a radar electromagnetic wave. Antennas are regularly disposed along a plane; they are connected to a unique source through phase shifters, each of them shifting the phase of a specific amount that can be electrically tuned by an external signal.

If all the antennas are excited in phase, the superposition of all the waves that they emit makes a planar wave emitted parallel to the plane of the antennas. By a suitable choice of the phase shifts, wave surfaces of any shape can be synthesized. If the phase shifts vary according to an arithmetic progression, a planar wave is generated obliquely; if the step of the progression is varied versus time, an angular scanning is achieved.

7.1.3. *Resolution of a Problem of Diffraction*

7.1.3.1. *Notations*

Figure 7.9 summarizes any diffraction experimental arrangement: an incident wave illuminates an aperture opened in a diaphragm, which may have any shape but most often is a plane (xOy). The diffraction pattern is observed on a plane $(XO'Y)$ that is parallel to (xOy). Usual experimental conditions are such that, on one hand, the distance d between the two planes is far larger than the wavelength and that, on the other hand, the line MP joining any point of the diffracting diaphragm to any point of the plane observation makes a small angle with Oz; to be more precise, the two angles, u and v, of Oz with the projection of OM, respectively, on the plane xOz (or XOz) or on the plane yOz (or $YO'z$) are small enough to be assimilated to their sine or tangent.

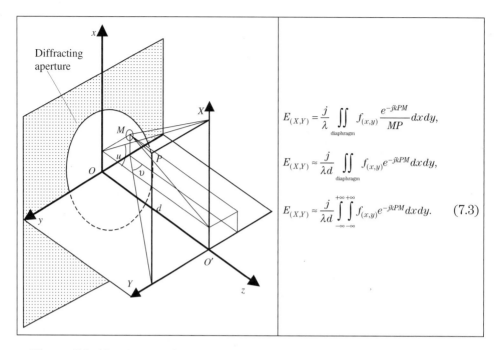

Figure 7.9. Notations used to study the diffraction by an aperture centered on the point O. $M(x, y)$ is a point inside the diffracting aperture. $P(X, Y)$ is the point where the diffracted field $F(X, Y)$ is observed. MP is assimilated to the distance $d = OO'$.

7.1.3.2. Near Field Diffraction (Fresnel), Far Field Diffraction (Fraunhofer)

Near field diffraction, preferably called *Fresnel diffraction*, is the most general and the most difficult problem of diffraction. The distance between the diffracting aperture and the point of observation (distance d between the two planes xOy and $XO'Y$ of Figure 7.9), is of course far larger than the wavelength, but can be comparable to the size of the aperture.

Far field diffraction is preferably called *Fraunhofer diffraction*. Let us suppose that the plane $XO'Y$ is covered by a screen on which the diffraction pattern is observed and let us see what happens when the distance d is increased: the diffraction pattern becomes larger and larger and less contrasted, but it tends to keep, whatever the distance, the same aspect that is called the *far field diffraction pattern*.

In fact, it's not necessary to go as far as infinity to observe a far field pattern, since infinity can be brought to the focal plane of a lens. Joseph von Fraunhofer was the first to propose this method of studying diffraction, it's probably the reason why the following epitaph has been written on his tombstone: "Approximavit sidera," "He made the stars closer."

7.1.3.3. *Approximate Expressions for the Diffraction Integrals*

Since the angles u and v are small, formula (7.3) can be simplified thanks to a series of developments in the arguments of the complex exponential, the method is very similar to what we did in Section 2.6,

$$MP = \sqrt{d^2 + (X-x)^2 + (Y-y)^2} \approx d\left[1 + \frac{(X-x)^2}{2d^2} + \frac{(Y-y)^2}{2d^2}\right].$$

The presence of the square root of polynomials inside the argument of an exponential is never pleasant, the development considerably simplifies the calculation

$$F(X,Y) = \frac{j}{\lambda d} \int\limits_{-\infty}^{+\infty} \int\limits_{-\infty}^{+\infty} f(x,y) e^{-j\frac{k}{2d}\left[(X-x)^2 + (Y-y)^2\right]} \, dx \, dy. \tag{7.4}$$

Let e_x, e_y, and e_z, respectively, be the three projections of a unit e vector parallel to MP, on the three axes Ox, Oy, and Oz we can write

$$e_x = \sin u \approx u \ll 1, \quad e_y = \sin v \approx v \ll 1,$$

$$e_z^2 = 1 - e_x^2 - e_y^2 \cong 1 - u^2 - v^2 \cong 1 \quad \forall (u,v).$$

Because of the assumption of small angles, the orientation of MP involves only the two parameters u and v. We now look at what happens when $d \to \infty$, the following asymptotic formulas can be written:

$$\frac{X}{d} \approx u, \quad \frac{Y}{d} \approx v, \quad \frac{x^2}{d} \to 0, \quad \frac{y^2}{d} \to 0,$$

$$\frac{(X-x)^2 + (Y-y)^2}{d} \approx \frac{X^2 + Y^2}{d} - 2(ux + vy). \tag{7.5}$$

Formula (7.4) then becomes

$$F(X,Y) = \frac{j}{\lambda d} e^{-jkd} e^{-jk\frac{X^2+Y^2}{d}} \int\limits_{-\infty}^{+\infty} \int\limits_{-\infty}^{+\infty} f(x,y) e^{-jk(ux+vy)} \, dx \, dy, \tag{7.6}$$

$$F(u,v) = K \int\limits_{-\infty}^{+\infty} \int\limits_{-\infty}^{+\infty} f(x,y) e^{-jk(ux+vy)} \, dx \, dy, \tag{7.7}$$

$$F(k_x, k_y) = K \int\limits_{-\infty}^{+\infty} \int\limits_{-\infty}^{+\infty} f(x,y) e^{-j(k_x x + k_y y)} \, dx \, dy, \tag{7.8}$$

$$K = \frac{j}{\lambda d} e^{-jkd} e^{-jk(X^2+Y^2)/2d},$$

where k_x and k_y are the two projections on the axes OX and OY of a wave vector parallel to the unit vector $\boldsymbol{e}$.

The modulus of the complex number K is easily obtained and is equal to $1/\lambda d$; on the other hand, the determination of this is more difficult and would need more information about the asymptotic behavior of $(X^2 + Y^2)/2d$. In fact we don't really care about the phase, since expressions like (7.4) and (7.6) are mostly designated to evaluate the intensity of the diffracted light: in the end, $F(X, Y)$ will be multiplied by its complex conjugate and the phase will disappear.

Relationship Between Diffraction and Fourier Transform

Formula (7.8) is very interesting for two reasons:

- It is the decomposition of the diffracted field in planar harmonic waves.
- $F(k_x, k_y)$ is the two-dimensional Fourier transform of $f(x, y)$.

A Fourier transform makes a correspondence between a function $f(x, y)$ defined in the Ox, Oy geometric space and a function $F(k_x, k_y)$ defined in the space of the wave vectors. A planar wave of vector (k_x, k_y) is focused by a lens at a focal point ϕ_k; let ξ and η be the coordinates of ϕ_k (see Figure 7.10), the repartition $F(\xi, \eta)$ of the complex amplitudes of the light vibration over the plane $(F\xi, F\eta)$ is the Fourier transform of $f(x, y)$.

7.2. Fraunhofer Diffraction

7.2.1. *Definition and Conditions of Observation*

We refer to Figure 7.10; because of the Malus theorem, the stigmatism of the lens implies that it takes the same time to cover the optical paths $MI\phi_k$, $H'I'\phi_k$, $H''I''\phi_{kk}$, and $H_0I_0\phi_k$: when arriving at the point ϕ_k the rays emitted by the

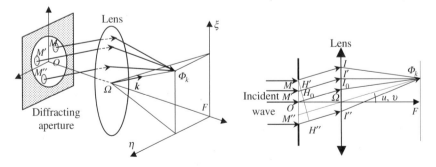

Figure 7.10. All the rays diffracted by the aperture in the direction (u, v) are focused at the same focal point ϕ_{uv} (coordinates $\xi = fu$, $\eta = fv$).

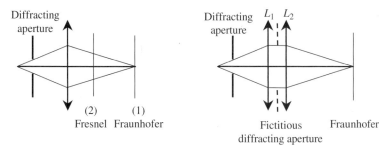

Figure 7.11. If the observation screen is in (2), we have Fresnel diffraction, while (1) corresponds to Fraunhofer diffraction. The lens of the left-hand figure (focal length f) is replaced by two separate lenses (focal lengths f_1 and f_2, such that $1/f_1 + 1/f_2 = 1/f$). The right-hand figure can then be considered as a far field diffraction arrangement through the fictitious diffracting aperture.

auxiliary sources M, M', O, and M'' have exactly the same phase differences that they had at points M, H', H_0, and M''.

We suppose that the incident wave is a planar wave propagating parallel to Oz; the phase difference φ_M between the two planar waves, respectively, is diffracted by the points O and M in the direction defined by the angles u and v,

$$\varphi_M = kOM = k(ux + vy).$$

Superposing, at the focal point ϕ_{uv}, all the planar waves diffracted in the (u, v) direction by the different points of the diffracting aperture gives the same result as expressed by formula (7.7).

Diffraction in the Vicinity of a Geometric Image

The case of Fraunhofer diffraction is more general than far field diffraction. We will say that we are in the situation of Fraunhofer diffraction every time that the diffraction pattern is observed in the vicinity of the image of an object through some optical imaging device. It is shown in Figure 7.10 how a Fraunhofer diffraction problem can be studied as a far field diffraction problem.

7.2.2. *Diffraction by One or Several Slits*

7.2.2.1. *One Slit*

Direct Determination of the Diffraction Pattern

The problem is to calculate the diffraction pattern of a rectangular aperture having a length L that is considerably larger than its width a. The notations and experimental conditions are described in Figure 7.12. Since the incident wave propagates orthogonal to the plane of the slit, all the auxiliary sources vibrate with the same phase, let b_0 be the amplitude of their vibrations.

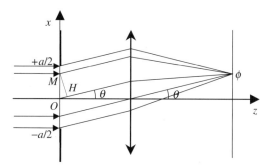

Figure 7.12. Fraunhofer diffraction by a thin slit. The diffraction pattern is observed in the focal plane of the lens.

We first consider all the rays that arrive at the focal point ϕ associated to the direction of propagation corresponding to the angle θ, the diffracted ray coming from point M has covered a shorter path than the ray coming from point O, the difference is equal to OH and the phase difference is equal to $\varphi = kOH = kOM\sin\theta$. The global vibration diffracted by the slit in the direction θ and at point ϕ is given by the integral

$$B(\theta) = \frac{jb_0 L}{\lambda} \int_{-a/2}^{+a/2} e^{-jk\sin\theta}\,dx,$$

$$B(\theta) = \frac{jb_0 L}{\lambda} \frac{1}{jk\sin\theta} \int_{(-jka\sin\theta)/2}^{(+jka\sin\theta)/2} e^{-u}\,du = -j\frac{b_0 L}{\pi\sin\theta}\sin\left(\frac{ka\sin\theta}{2}\right), \qquad (7.9)$$

$$B(\theta) = -jB_0 \frac{\sin\beta}{\beta} = \mathrm{sinc}\,\beta, \quad \text{with} \quad \beta = \frac{ka}{2}\sin\theta \quad \text{and} \quad B_0 = \frac{b_0 aL}{\lambda},$$

$$I(\theta) = B_0^2\,\mathrm{sinc}^2\,\beta.$$

Determination of the Diffraction Pattern Using the Fourier Transform

$$\mathrm{Rect}\left[\frac{x}{a}\right] = 1 \quad \text{if } |x| \le \frac{a}{2},$$
$$\mathrm{Rect}\left[\frac{x}{a}\right] = 0 \quad \text{if } |x| > \frac{a}{2}. \qquad (7.10)$$

It is established in mathematics that the Fourier transform $F(u)$ of a rectangle function is a *sinc* and that, omitting the proportionality coefficient, we have

$$F(u) = \mathrm{FT}\left\{\mathrm{Rect}\left[\frac{x}{a}\right]\right\} = \frac{\sin\left(\dfrac{kau}{2}\right)}{\dfrac{kau}{2}} = \mathrm{sinc}\left(\frac{kau}{2}\right).$$

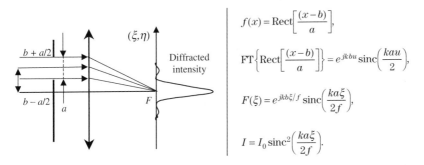

$$f(x) = \mathrm{Rect}\left[\frac{(x-b)}{a}\right],$$

$$FT\left\{\mathrm{Rect}\left[\frac{(x-b)}{a}\right]\right\} = e^{jkbu}\,\mathrm{sinc}\left(\frac{kau}{2}\right),$$

$$F(\xi) = e^{jkb\xi/f}\,\mathrm{sinc}\left(\frac{ka\xi}{2f}\right),$$

$$I = I_0\,\mathrm{sinc}^2\left(\frac{ka\xi}{2f}\right).$$

Figure 7.13. As compared to Figure 7.12, the slit has been translated parallel to itself. The diffraction pattern remains exactly the same, except that phases at point F are not the same.

The method for obtaining the Fourier transform of the function $\mathrm{Rect}(x/a)$ is indeed very similar to the calculations of formulas (7.9) and (7.10). The interest in using the Fourier method is that we can take full advantage of all the results that have been accumulated by mathematicians concerning the Fourier transform. Figures 7.13, 7.14, and 7.15 are good examples of the advantages of the method, which also justify an affirmation that we made earlier, see Figure 7.3: a diffraction pattern is all the more spread as the diffracting aperture is smaller.

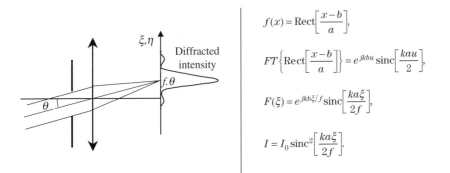

$$f(x) = \mathrm{Rect}\left[\frac{x-b}{a}\right],$$

$$FT\left\{\mathrm{Rect}\left[\frac{x-b}{a}\right]\right\} = e^{jkbu}\,\mathrm{sinc}\left[\frac{kau}{2}\right],$$

$$F(\xi) = e^{jkb\xi/f}\,\mathrm{sinc}\left[\frac{ka\xi}{2f}\right],$$

$$I = I_0\,\mathrm{sinc}^2\left[\frac{ka\xi}{2f}\right].$$

Figure 7.14. The respective positions of the slit and of the lens are the same as in Figure 7.12, but the incidence is now oblique. The diffraction pattern has the same aspect, but is now centered on the focal point associated to the direction θ.

7.2.2.2. *Several Equidistant Slits*

Two Slits

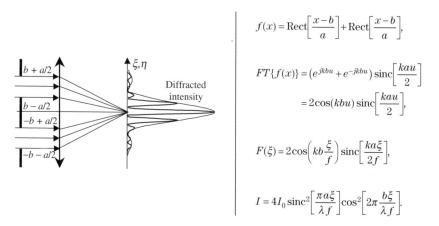

$$f(x) = \text{Rect}\left[\frac{x-b}{a}\right] + \text{Rect}\left[\frac{x-b}{a}\right],$$

$$FT\{f(x)\} = \left(e^{jkbu} + e^{-jkbu}\right)\text{sinc}\left[\frac{kau}{2}\right]$$

$$= 2\cos(kbu)\,\text{sinc}\left[\frac{kau}{2}\right],$$

$$F(\xi) = 2\cos\left(kb\frac{\xi}{f}\right)\text{sinc}\left[\frac{ka\xi}{2f}\right],$$

$$I = 4I_0\,\text{sinc}^2\left[\frac{\pi a\xi}{\lambda f}\right]\cos^2\left[2\pi\frac{b\xi}{\lambda f}\right].$$

Figure 7.15. The two slits give two diffraction patterns that have exactly the same amplitude repartitions, see Figure 7.13, but different phase repartitions. The phase difference depends on the position in the focal plane and produces an interference pattern described by a sine, superimposed on the diffraction pattern described by a sinc.

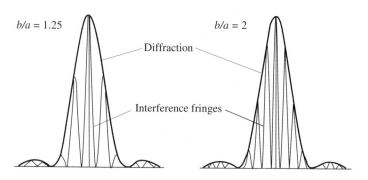

Figure 7.16. Variation of the light intensity in the focal plane of Figure 7.15. The interference fringes, analogous to Young's slit fringes, are "modulated" by diffraction fringes (see Figure 6.1).

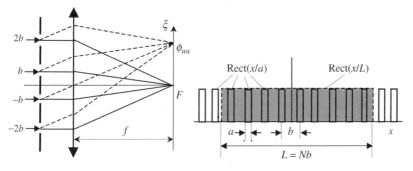

Figure 7.17. Diffraction by N identical and regularly spaced slits.

Many Slits

Before studying the diffraction pattern given by a great (possibly infinite) number of slits, we will introduce some useful mathematical functions:

- Dirac's pulse $\delta(\zeta)$, where ζ is a real dimensionless variable,

$$\delta(\zeta = 0) \to \infty,$$
$$\delta(\zeta \neq 0) = 0,$$
$$\int_{-\infty}^{\infty} \delta(\zeta)d\zeta = 1. \tag{7.11.a}$$

- Dirac's pulse occurring for a given value, n, of the variable $\delta_n(\zeta) = \delta(\zeta - n)$,

$$\delta_n(\zeta = n) = \delta(0) \to \infty,$$
$$\delta_n(\zeta \neq n) = 0,$$
$$\int_{-\infty}^{\infty} \delta_n(\zeta)d\zeta = 1. \tag{7.11.b}$$

- Product of convolution, $f(\zeta) \otimes \delta_n(\zeta)$, of a function $f(\zeta)$ by $\delta_n(\zeta)$,

$$f(\zeta) \otimes \delta_n(\zeta) = \int_{-\infty}^{\infty} f(\zeta)\delta_n(\zeta)d\psi = f(\zeta = n).$$

- Fourier transform of a product of convolution,

$$FT(f(\xi) = F(\eta)), \quad FT(g(\xi) = G(\eta)),$$
$$FT[f(\xi)g(\xi)] = F(\eta) \otimes G(\eta),$$
$$FT[f(\xi) \otimes g(\xi)] = F(\eta)G(\eta).$$

- Dirac's comb: succession of equally spaced Dirac pulses,

$$\text{Comb}(\zeta) = \sum_{n=-\infty}^{\infty} \delta_n(\zeta), \quad n \text{ is an integer.} \tag{7.11.c}$$

- Infinite comb of slits (see Figure 7.17):

$$\text{SlitComb}(x) = \text{Comb}\left(\frac{x}{b}\right) \otimes \text{Rect}\left(\frac{x}{a}\right) = \left\{\sum_{n=-\infty}^{\infty} \delta_n\left(\frac{x}{b}\right)\right\} \otimes \text{Rect}\left(\frac{x}{a}\right). \quad (7.11.\text{d})$$

The rectangle function Rect[x/a] has already been introduced in Section 7.2.2.1, formula (7.10).

- An infinite comb of slits is easily reduced to a finite comb of only N slits (see Figure 7.17), by multiplying by the function Rect[x/Nb]:

$$N/\text{SlitComb}(x) = \left\{\text{Comb}\left(\frac{x}{b}\right) \otimes \text{Rect}\left(\frac{x}{a}\right)\right\} \text{Rect}\left(\frac{x}{Nb}\right). \quad (7.11.\text{e})$$

Diffraction Pattern of a Comb of Slits

The Fourier transform is a convenient and powerful tool for studying diffraction patterns, and provides a nice illustration of the respective roles of diffraction and interference and also of multiple interference. Making the Fourier transform of the various formulas (7.11) and the products of convolution needs rather tricky mathematical manipulations, we will develop the full calculations only in the cases where the number of slits is infinite (formulas (7.11.c) and (7.11.d)).

The Fourier transform of Dirac's pulse is another Dirac pulse; in the same way, the Fourier transform of a Dirac comb is another Dirac comb. Observed in the focal plane of a lens, the Fraunhofer diffraction pattern of an infinite number of parallel equidistant infinitely narrow slits, illuminated by a monochromatic planar wave, is an infinite set of parallel and infinitely thin bright lines.

The infinitely narrow slits are now replaced by an infinite set of narrow slits (width equal to a). The diffraction pattern is obtained from the Fourier transform of formula (7.11.d),

$$\text{FT}\left[\text{Comb}\left(\frac{x}{b}\right) \otimes \text{Rect}\left(\frac{x}{a}\right)\right] = \left\{\sum_{n=-\infty}^{\infty} \delta_n(kbu)\right\} \text{sinc}\left(\frac{kau}{2}\right), \quad (7.12.\text{a})$$

$$I = I_0\left\{\sum_{n=-\infty}^{\infty} \delta_n\left(2\pi\frac{b}{\lambda}\frac{\xi}{f}\right)\right\} \text{sinc}^2\left(\pi\frac{a}{\lambda}\frac{\xi}{f}\right). \quad (7.12.\text{b})$$

The first term of (7.12.a) comes from the multiple interference of all the waves diffracted by the different apertures: constructive interference only occurs in the directions for which the waves diffracted by two neighboring slits show a phase difference exactly equal to 2π. The second term comes from the diffraction by each individual slit. Formula (7.12.b) gives the variation of the light intensity on the focal plane of the lens of Figure 7.17: a set of equidistant and infinitely thin fringes have an appreciable intensity only between the first two zeros of the *sinc-function* ($|\xi| < \lambda f/a$); the narrower the slit is, the greater is the number of fringes.

The calculation, in the case of a finite number of slits with a finite width a (formula (7.11.e)), is less trivial, because of the presence of a usual product and of a convolution product. The limited number of slits corresponds to the term $\mathrm{Rect}(x/(Nb))$ in formula (7.11.e); its Fourier transform is the *sinc-function* $\mathrm{sinc}[kNbu/2]$. In the focal plane the diffraction pattern is a comb. The light intensity of each individual tooth is given by $\mathrm{sinc}^2[\pi Nb\xi/\lambda f]$, it has a principal maximum for $\xi = 0$, surrounded by two minima equal to zero that are obtained for

$$\xi = \pm \lambda f / Nb. \tag{7.13}$$

Two Waves and Multiple Wave Interference

In the previous arrangements diffraction and interference are produced by a regular distribution of diffracting objects, the phase shift varies versus the angle and versus the position of the object. The angular repartition of the intensity of the diffracted light shows important *principal maxima* for some directions, between them *subsidiary maxima* are found, their intensity being far smaller. Between two consecutive maxima there is a minimum having an intensity equal to zero.

The directions of the principal maxima are those for which the phase shift between two neighboring objects is precisely equal to 2π. The diffracted intensity decreases all the more rapidly as the total number of diffracting objects is larger.

- In the case of only two diffracting objects (Young's slits, for example), the variation of the intensity is a sine.
- In the case of an infinite number of diffracting objects, the variation law of the intensity is a succession of Dirac pulses.
- In the case of a finite number N of diffracting objects, the variation law is a comb tooth having a width that is all the thinner as N is larger.

Resolving Power of Diffraction Grating Spectroscope

A set of regularly spaced slits is illuminated by a planar wave that has two spectral components of respective wavelengths λ_1 and λ_2. The diffraction pattern, observed on the focal plane of a lens, is made of the superposition of the two sets of fringes of two colors. Such an arrangement is a spectroscope, its ability to separate two wavelengths is all the more interesting as the diffraction peaks have a narrower angular width.

If $\Delta\lambda$ is the smallest difference of two wavelengths that can be discriminated by the spectroscope, by definition, the resolving power of the spectroscope is equal to

$$R = \frac{\lambda}{\Delta\lambda}.$$

It is considered that two wavelengths are discriminated if the distance between the two principal maxima corresponding to λ_1 and λ_2 is larger than the distance from a principal maximum to the first minimum. The positions of the principal maxima come from $\mathrm{Comb}(b\xi/\lambda f)$, and are given by

$$\xi_{\mathrm{Max},\lambda 1} = n\lambda_1 f/b, \quad \xi_{\mathrm{Max},\lambda 2} = n\lambda_2 f/b \quad \rightarrow \quad (\xi_{\mathrm{Max},\lambda 1} - \xi_{\mathrm{Max},\lambda 1}) = n(\lambda_1 - \lambda_2)f/b.$$

The distance to a principal minimum is obtained from (7.13): $\Delta\xi = \lambda f/Nb$, the resolving power is equal to

$$R = \frac{\lambda}{\Delta\lambda} = \frac{\lambda}{(\lambda_1 - \lambda_2)} = nN. \tag{7.14}$$

The total number N of diffracting slits mainly determines the resolving power.

7.2.3. Far Field Diffraction by a Circular Aperture

The circular aperture of radius a of Figure 7.18 is described by the following expression:

$$C(x, y) = 1 \quad \text{if } (x^2 + y^2) \leq a^2,$$
$$C(x, y) = 0 \quad \text{if } (x^2 + y^2) > a^2.$$

The Fourier tranform of $C(x, y)$ is a Bessel function of the first order,

$$F(u, v) = \mathrm{FT}[C(x, y)] = 2\frac{J_1(Z)}{Z} \quad \text{with} \quad \rho = \sqrt{u^2 + v^2} \quad \text{and} \quad Z = 2\pi\frac{a}{\lambda}\rho.$$

The function $J_1(Z)/Z$ plays, in the case of a circular aperture, the same role as the *sinc-function* in the case of a rectangular slit. The graph of $(J_1(Z)/Z)^2$ looks like that of a square *sinc*: a central maximum surrounded by subsidiary less intense maxima, with a minimum equal to zero between two consecutive maxima. The first minimum is obtained for $Z = 3.83$, the corresponding radius, which is the radius of the Airy disk, see Figure 7.18, is equal to

$$\rho_{\mathrm{Airy\ disk}} = f\frac{3.83}{\pi}\frac{\lambda}{2a} = 1.22\frac{\lambda}{2a}f. \tag{7.15}$$

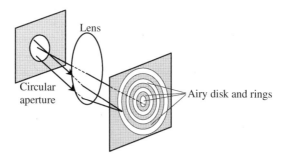

Figure 7.18. The diffraction pattern of a circular aperture, observed in the focal plane of a lens, is made of a disk surrounded by rings. The central disk is often called an "Airy disk."

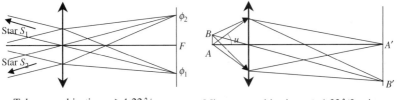

Telescope objective: $\varepsilon \geq 1.22\lambda/a$ Microscope objective: $\varepsilon \geq 1.22\lambda/2n \sin u$

Figure 7.19. Resolving power of an objective. For the two images, (ϕ_1 and ϕ_2) or (A' and B'), to be separated when observed through an eyepiece, it is considered that their distance should be greater than the radius of the first dark Airy ring.

Resolving Power of an Imaging System

Because of diffraction, the image of a point source is not a point, but an Airy disk; the images of two neighboring points are seen separately if their Airy disks don't overlap too much. It is considered that this is case if the center of one disk coincides with the first zero of the other (this approximation is called the *Rayleigh criterion*).

The distance between two points that are seen separately is the *limit of resolution* of the objective. In the case of a telescope the limit of resolution is an angle equal to $1.22\lambda/2a$ according to formula (7.15). The case of a microscope objective raises some difficulty since it involves angles that are not small, it is found that the angular limit of resolution is $1.22\lambda/2n \sin u$, where u is the greatest value of the angle that makes, with the axis, a ray entering into the objective. It is interesting to notice that, in both cases, the resolving power is all the better as the objective transmits a higher light flux to the eyepiece.

7.2.4. *Far Field Diffraction of a Rectangular Aperture*

An (a, b) rectangular aperture is represented by the product of two rectangle functions with respective arguments equal to x/a and y/b. The corresponding diffraction pattern is the product of two *sinc-functions*. In Figure 7.20 is shown the diffraction pattern of a square aperture: it is a regular arrangement of squares and rectangles having their respective centers at the nodes of a network with a step equal to λ/a. The light intensity rapidly decreases with the distance to the origin, and only the most central have been represented. At the center is a square, the side of which is equal to $2\lambda/a$.

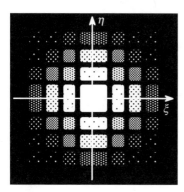

$$f(x, y) = \text{Rect}\left[\frac{x}{a}\right]\text{Rect}\left[\frac{y}{a}\right],$$

$$F(u, v) = \text{sinc}\left[\frac{\pi a}{\lambda}u\right]\text{sinc}\left[\frac{\pi a}{\lambda}v\right],$$

$$I = I_0 \text{sinc}^2\left[\frac{\pi a}{\lambda}u\right]\text{sinc}^2\left[\frac{\pi a}{\lambda}v\right].$$

Figure 7.20. Far field diffraction pattern of a square aperture (side a).

7.3. Fresnel Diffraction

The integration of the integrals in the case of Fresnel's diffraction is more complicated than in the case of Fraunhofer's diffraction. We are going to describe a method that was used by Fresnel to determine the intensity of the light that is diffracted by an opaque circular disk, at a point located on its axis. Submitted to the French Academy of Sciences in 1818, this calculation had been rejected by Poisson who objected that one consequence was that some light should exist only in the middle of the geometric shadow. Arago who participated in the Academy meeting made the experiment in his laboratory and could demonstrate experimentally the existence of the litigious bright point, bringing a clear argument in favor of the wave theory of light. The experimental arrangement is described in Figure 7.21.

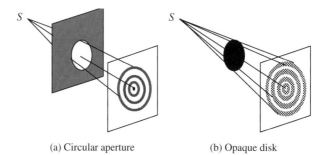

(a) Circular aperture   (b) Opaque disk

Figure 7.21. Diffraction by a circular aperture or by an opaque disk. The two diffraction patterns have the same general aspect. In the case of the opaque disk, as soon as the screen is sufficiently remote, there is always a bright dot at the center. In the case of the circular aperture the center is either bright or dark, according to the position of the screen.

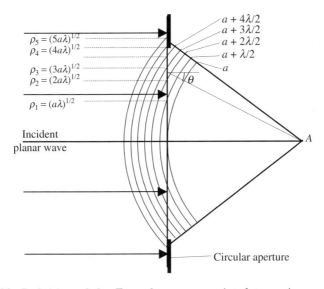

Figure 7.22. Definition of the Fresnel zones associated to a given point A. The intersection of a set of spheres, centered at point A and having radii, respectively, equal to a, $a + \lambda/2$, $a + 2\lambda/2$, $a + 3\lambda/2, \ldots$, with the plane of the aperture, are concentric circles. A Fresnel zone is delimited by two consecutive circles. The first zone is a circle, the following zones are circular rings.

Fresnel Zones

The Fresnel method of integration of the integrals is made possible because of a suitable division of the domain of integration: the elementary areas are defined in Figure 7.22. From geometric considerations it can be shown that all the Fresnel zones have equal areas $\pi a\lambda$ and that the radius of the nth zone is equal to $\rho_n = (na\lambda)^{1/2}$.

To determine the diffracted vibration sent to point A, we start by adding the elementary vibrations sent by all the points of the same zone, and then we add the contributions of the different zones. Since all the Fresnel zones have the same area, the moduli of the contributions of the different zones are almost equal. In fact, the moduli decrease slowly as the radius is increased, and this is for two reasons. First, the distance to point A increases; the second, and most important, reason is due to the fact that the amplitude of the vibration sent to point A decreases with the angle θ between the normal to the plane of the zone and the line that joins a point of the zone to point A.

The total vibration at point A can be calculated using the Huygens-Fresnel principle; it is found, but the result is quite obvious because of the method that has been chosen to divide the surface of integration, that the vibrations arriving from two consecutive rings have opposite phases. The vibration at point A is the sum of alternately positive and negative terms, let s_i be the vibration coming from the Fresnel zone labeled I, and we have

$$s = \sum_i s_i = s_1 + s_2 + s_3 + s_4 + s_5 + \cdots.$$

In the case of the circular aperture of Figure 7.21(a), the total number of Fresnel zones is either even or odd. In the first case the waves interfere destructively: the intensity at point A is almost equal to zero, the center of the diffraction pattern is dark. On the contrary, an odd number of zones gives a maximum of light at the center.

The case of an opaque disk is different since the more central Fresnel zones are not illuminated; let p be the label of the first nonobliterated zone, the total vibration is now given by

$$s = \sum_{i=p}^{\infty} s_i = s_p + s_{p+1} + s_{p+2}, \ldots,$$

$$s = \frac{s_p}{2} + \left(\frac{s_p}{2} + \frac{s_{p+1}}{2} \right) + \left(\frac{s_{p+1}}{2} + \frac{s_{p+2}}{2} \right) + \cdots. \tag{7.16}$$

To be rigorous we should consider the convergence of the sum; we will admit that the two terms in parentheses cancel and that the first term, $s_p/2$, is the only remaining one. Finally, there is always a maximum of light at the center.

Fresnel Zone Plate

Although they have been considered, for a long time, as esthetical curiosities, the Fresnel zone plates have now found interesting applications for making lenses in cases where it is difficult to substantially increase the index of refraction with regard to the surroundings, for example, in integrated optics or in X-ray imaging systems.

The principle of a zone plate is rather simple; opaque rings obliterate one Fresnel zone over two, the purpose being to keep only, in the development of formula (7.16), the terms that have the same sign.

If $a = 1\,\mathrm{m}$ and $\lambda = 0.5\,\mu\mathrm{m}$, the radius of the first zone is $0.707\,\mathrm{mm}$. A zone plate can readily be obtained by drawing, using a large scale, circles having radii varying as the square root of the successive integers. The rings between two consecutive circles are then darkened and photographed using a given magnification; the resulting slide is a zone plate. The radius of the nth ring is $\rho_n \approx \sqrt{na\lambda}$, its thickness is equal to the difference $\rho_{n+1} - \rho_n \approx \sqrt{a\lambda}/2\sqrt{n} = \rho_0/2\sqrt{n}$. It can be shown that the optical quality of the image increases with the number of zones, the transverse size (aberration) of the focal point is of the order of the thickness of the last ring. In the case of a photographic process, this thickness is fixed by the photoplate.

If a planar monochromatic wave is sent orthogonal to a Fresnel zone plate, an accumulation of light is found in the vicinity of the different points F_n, see

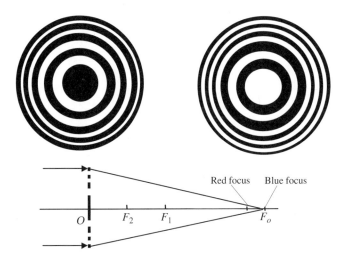

Figure 7.23. Examples of Fresnel zone plates. On the left-hand plate, the even zones are suppressed; on the right-hand plate, the odd zones have been suppressed.

Figure 7.23, that satisfy the following relation, in which n is an integer, R is the radius of the first zone, and λ is the wavelength,

$$OF_n = \frac{1}{2n+1}\frac{R^2}{\lambda}.$$

A Fresnel zone plate appears to be a lens that has several focal points. Of course the chromatic aberrations are important since the focal length is proportional to the inverse of the wavelength. If, instead of using a parallel incident beam, a spherical wave centered at some point P is sent to the plate, the transmitted light is made of several spherical waves centered at the points Q_n defined by

$$\frac{1}{OP}+\frac{1}{OQ_n}=(2n+1)\frac{R^2}{\lambda}=\frac{1}{OF_n}.$$

Number of Fresnel Zones Inside an Aperture-Limited Spherical Wave

The solid angle of a full spherical wave is equal to 4π steradians; the beams that are used in systems working in Gauss conditions are far less open. As soon as the aperture of a spherical wave is quite limited, diffraction effects are observed. A good evaluation of the effect of diffraction is provided by the notion of *Fresnel's number*. The definition of *Fresnel's number*, associated to a given spherical wave and a given circular aperture, is given in Figure 7.24 where a spherical wave is obtained by focusing a planar wave with an aberration-free lens. We consider concentric spheres centered at the focal point; the first one is tangent to the plane of the aperture and its radius is

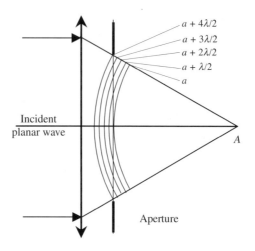

Figure 7.24. The Fresnel number of an aperture-limited spherical wave is equal to the number of Fresnel zones inside the open part of the aperture.

equal to a, the radii of the other spheres are regularly increased $\lambda/2$ by $\lambda/2$. The intersections of the different spheres with the plane of the aperture are circles. A Fresnel zone is a ring between two adjacent circles. The Fresnel number is just the total number of zones inside the aperture. The larger the Fresnel number, the smaller is the diffraction spot in the vicinity of the focus.

7.4. Diffraction Gratings

During the last decades the technology for making diffraction gratings has considerably improved, opening the way to efficient mass production methods. Diffraction gratings are now usual and relatively cheap components.

7.4.1. Diffraction–Diffusion of Light

Diffraction is observed any time that an electromagnetic wave hits some small-sized object, by small size we mean of the order of one to fifty times the wavelength. The object under consideration is often made of the collection of smaller (as compared to the wavelength) constituents that will be called *particles,* each of them participates in the diffraction process.

The light that is diffracted by the global object is the result of the interference of all the wavelets diffracted by the different particles. Although the following formal distinction is not universally used, we will distinguish two main cases:

- *Diffusion,* where the spatial distribution of the particles is at random. Typical examples are the repartition of water drops in fog, or of nitrogen and oxygen molecules in the atmosphere.

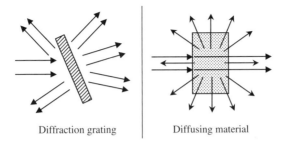

Diffraction grating | Diffusing material

Figure 7.25. Comparison between diffusion and diffraction. For a periodic arrangement, the diffracted beams are observed only along specific directions called the orders of diffraction. Diffused light is emitted in all directions.

- *Diffraction*, where the distribution of the particles is perfectly organized. The two most famous examples are the equidistant grooves of a grating and the arrangement of the atoms inside a perfect crystal.

From a practical point of view, the main difference between diffraction and diffusion is related to the directivity of the diffracted beams. No privileged direction exists in the case of diffusion, the diffused light being emitted in all directions. On the contrary, in the case of a regular arrangement of the particles there are only a few directions in which the wavelets interfere positively, these directions are called the *orders of diffraction*.

A diffraction grating is made of identical, parallel, and equidistant grooves that are considered to have an infinite length. The most important characteristic of a grating is its periodicity, which is the separation of two adjacent grooves. The directions of the diffraction orders are fixed by the periodicity.

Another characteristic of a grating is the shape of its grooves. This shape doesn't play any role in the determination of the direction of the orders, however it determinates how the incident energy is shared by the different orders. This is the reason why a significant effort is made to achieve given profiles, so that only one order will carry away most of the energy.

Working in the far field diffraction regime, a grating must be placed between two lenses, a collimating lens and then a focusing lens. As shown in Figure 7.27, two main different arrangements are used, the light being either reflected or transmitted by the grating. In fact, the transmission arrangement is handicapped by the fact that the light travels through the substrate on which the grating is deposited; consequently, this substrate should be very homogeneous with a perfectly planar and well-polished second interface, in order to avoid unwanted phase modifications.

Figure 7.26. Some usual groove profiles. The periodicity ranges from $10\,\mu m$ to $0.5\,\mu m$, i.e., 100 to 2000 grooves per millimeter.

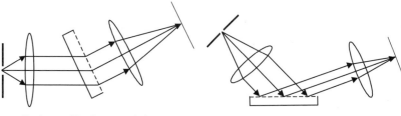

Grating working by transmission   Grating working by reflection

Figure 7.27. A grating can work either by transmission or by reflection. The second arrangement is preferred since it is not concerned with a possible lack of homogeneity of the substrate that supports the grating.

7.4.2. *The Bragg Formula*

Diffraction by a grating is a multiple interference process. The total vibration diffracted in a given direction is obtained by summing up all the waves diffracted by the different particles of the diffracting object. In the case of a grating, the wavelets diffracted by all the particles belonging to the same groove are first added, and then the contributions of the different grooves. As the total number of grooves is enormous, it's a highly multiple interference process.

The directions of constructive interference are those for which the phase difference between the contributions of two neighboring grooves is very close to a multiple of 2π; in the case of an infinite number of grooves the phase difference should be strictly equal to a multiple of 2π. This condition is often referred to as the *phase matching condition.*

The Bragg formula is a relationship between the angle of incidence and the angle of diffraction. It indicates the directions in which the different diffracted beams are emitted, but it doesn't give any indication as to how the energy is shared between them, the shape of the grooves governs the partition rule.

7.4.2.1. *The Bragg Formula in the Optical Case*

The grating will be considered here as an infinite set of parallel and equidistant infinitely thin grooves; the only parameter to be defined is the periodicity a, which is the separation of two neighboring grooves,

$$(\sin\theta_{\text{diff}} - \sin\theta_{\text{inc}}) = p\frac{\lambda}{a} \quad \text{(same orientation for the two beams),} \qquad (7.17.a)$$

$$(\sin\theta_{\text{diff}} + \sin\theta_{\text{inc}}) = p\frac{\lambda}{a} \quad \text{(opposite orientation for the two beams).} \qquad (7.17.b)$$

The orientation (7.17.b) is often more convenient.

Formulas (7.17) give, for a given value of the angle of incidence, a set of values for the angle of diffraction, labeled by the integer p. Only the values of p corresponding to absolute values of $\sin\theta_{\text{diff}}$ smaller than one should, a priori,

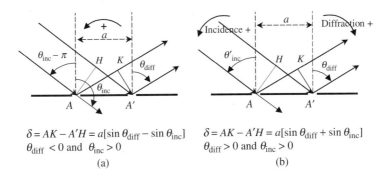

$$\delta = AK - A'H = a[\sin\theta_{\text{diff}} - \sin\theta_{\text{inc}}]$$
$$\theta_{\text{diff}} < 0 \text{ and } \theta_{\text{inc}} > 0$$
(a)

$$\delta = AK - A'H = a[\sin\theta_{\text{diff}} + \sin\theta_{\text{inc}}]$$
$$\theta_{\text{diff}} > 0 \text{ and } \theta_{\text{inc}} > 0$$
(b)

Figure 7.28. Evaluation of the difference δ of the optical lengths of the waves diffracted by two neighboring grooves. According to the authors, two different orientations are commonly used for the angles between the rays and the normal to the plane of the grating.

be considered. An interpretation of the case when the sine is greater than one will be given in Section 7.4.2.3.

The directions of the different orders of diffraction explicitly depend on the wavelength, except for the zero order, which is said to be nondispersive.

7.4.2.2. The Bragg Formula for a Three-Dimensional Crystal Lattice (Diffracting X-Rays)

The distance between atoms (angstroms) being far smaller than optical wavelengths, a crystal lattice doesn't diffract light waves; the situation is, of course, different for X-rays. By the way, gratings can also diffract X-rays; the experiment was successfully made at a time when some doubt existed about a common identity between X-rays and light waves. Let us come back to the diffraction of X-rays by a crystal; the practical interest is enormous, since it provides information about the crystal lattice. From the angular repartition of the diffracted beams and the Bragg formula, we obtain the shape and size of the crystal cell; from the comparison of the respective intensities of the different orders of diffraction, we obtain information about the chemical nature of the atoms, or groups of atoms in the cell.

When an X-ray electromagnetic wave reaches the crystal, it penetrates deep inside and each atom reemits X-rays in all directions. Because of the three-dimensional arrangement of the diffracting particles, the expression of the phase-matching condition is rather different from the case of a diffraction grating. We will proceed in two steps: first we consider the diffracting particles belonging to the same reticular plane, and then consider the phase matching between the vibrations produced by two neighboring parallel reticular planes.

We refer to Figure 7.30, the different atoms of a given reticular plane are labeled $A_1, A_2, \ldots, A_i, \ldots$, the phase-matching condition should be written for any couple of points $A_i A_k$:

$$(\sin\theta_{\text{diff}} - \sin\theta_{\text{inc}}) = p\frac{\lambda}{A_i A_k}.$$

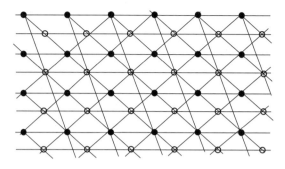

Figure 7.29. Schematic representation of some reticular planes of a two-dimensional lattice with two different atoms in each individual cell.

Due to the high value of the possible number of pairs, the only possible value is $p = 0$, which means that the angles of incidence and diffraction are equal and that each reticular plane behaves as a mirror on which X-rays are reflected according to the Snell-Descartes law for reflection. Finally, we write the phase-matching condition between the waves that are reflected by two consecutive planes. In the X-ray domain, it's more usual to refer to the direction of the beams with regard to the angle α with the reticular planes, d being the distance between two neighboring planes, the difference of the optical paths is given by

$$IA_2 + A_2J = 2d\sin\alpha \quad \rightarrow \quad 2d\sin\alpha = p\lambda \quad \text{(Bragg formula for a crystal).} \quad (7.18)$$

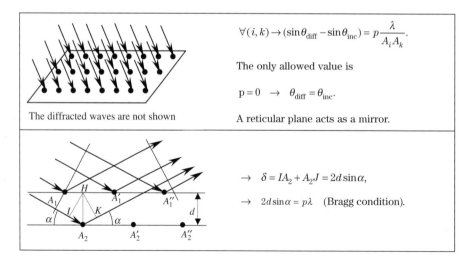

$$\forall(i,k) \rightarrow (\sin\theta_{\text{diff}} - \sin\theta_{\text{inc}}) = p\frac{\lambda}{A_iA_k}.$$

The only allowed value is

$$p = 0 \quad \rightarrow \quad \theta_{\text{diff}} = \theta_{\text{inc}}.$$

A reticular plane acts as a mirror.

The diffracted waves are not shown

$$\rightarrow \quad \delta = IA_2 + A_2J = 2d\sin\alpha,$$

$$\rightarrow \quad 2d\sin\alpha = p\lambda \quad \text{(Bragg condition).}$$

Figure 7.30. X-ray diffraction by a crystal lattice. A given incident beam will generate a diffracted beam if, and only if, its direction is properly oriented so that the Bragg condition is fulfilled.

7.4.2.3. *Numerical Applications of the Bragg Formula in the Optical Case*

Knowing the periodicity, a, of a grating, the wavelength, λ, and the angle of incidence, θ_{inc}, we want to calculate the angles of diffraction θ_{diff}. The number N of grooves per millimeter (or per inch) is often preferred to the periodicity: $N = 1/a$. We use the orientation defined by (7.13.b) (opposite orientations for the incident and diffracted beam).

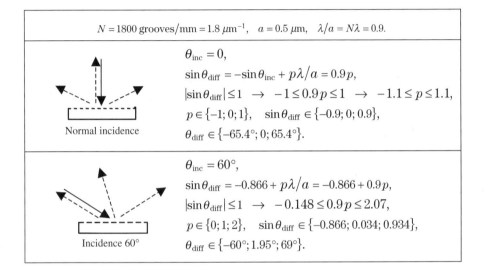

$$N = 1800 \text{ grooves/mm} = 1.8\,\mu\text{m}^{-1}, \quad a = 0.5\,\mu\text{m}, \quad \lambda/a = N\lambda = 0.9.$$

Normal incidence	$\theta_{inc} = 0,$ $\sin\theta_{diff} = -\sin\theta_{inc} + p\lambda/a = 0.9p,$ $\|\sin\theta_{diff}\| \le 1 \;\rightarrow\; -1 \le 0.9p \le 1 \;\rightarrow\; -1.1 \le p \le 1.1,$ $p \in \{-1; 0; 1\}, \quad \sin\theta_{diff} \in \{-0.9; 0; 0.9\},$ $\theta_{diff} \in \{-65.4°; 0; 65.4°\}.$
Incidence 60°	$\theta_{inc} = 60°,$ $\sin\theta_{diff} = -0.866 + p\lambda/a = -0.866 + 0.9p,$ $\|\sin\theta_{diff}\| \le 1 \;\rightarrow\; -0.148 \le 0.9p \le 2.07,$ $p \in \{0; 1; 2\}, \quad \sin\theta_{diff} \in \{-0.866; 0.034; 0.934\},$ $\theta_{diff} \in \{-60°; 1.95°; 69°\}.$

$N = 100 \text{ mm}^{-1}$; $\lambda = 0.5\,\mu\text{m}$; normal incidence, $\theta_{inc} = 0$. Calculate the total number of diffracted orders:

- $N\lambda = 0.05 \;\rightarrow\; \sin\theta_{diff} = 0.05p \;\rightarrow\; -20 \le p \le 20.$
- The total number of diffracted orders is 41. This example shows that the number of diffracted orders is all the larger as the number of grooves per unit length is smaller.

$N = 2000 \text{ mm}^{-1}$; $\lambda = 0.5\,\mu\text{m}$; $\theta_{inc} = 10°$. Calculate the total number of diffracted orders:

- $N\lambda = 1.2 \;\rightarrow\; \sin\theta_{diff} = -\sin 10° + 1.2p = -017 + 1.2p.$
- $0.69 \le p \le 0.95 \rightarrow$ the only possible value is $p = 0 \rightarrow \theta_{inc} = -\theta_{diff}$. The zeroth order is the only order allowed, the periodicity is too small and the grating is equivalent to a mirror, there is no dispersion.

7.4.2.4. *Littrow's Configuration*

Incidence conditions may be found where the diffracted beam propagates exactly in the opposite direction to the incident beam, $\theta_{inc} = -\theta_{diff}$; the grating is then said to work in the Littrow condition, which is defined by

$$\sin\theta_{Littrow} = \frac{pN\lambda}{2}. \tag{7.19}$$

For a given grating and a given order of diffraction, except the zeroth order, the angle $\theta_{Littrow}$ corresponding to the Littrow condition depends on the wavelength.

Numerical Application

- $N\lambda = 1$. There are only two possible values for p in formula (7.19):
  - $p = 0$, $\theta_{Littrow} = 0$, this case is not interesting, since the zero-order mode is not dispersive.
  - $p = 1 \rightarrow \sin\theta_{Littrow} = \frac{1}{2} \rightarrow \theta_{Littrow} = 30°$.
- $N\lambda = 0.6 \rightarrow \sin\theta_{Littrow} = 0.3p \rightarrow \sin\theta_{Littrow} \in \{0; 0.3; 0.6; 0.9\}$.
  - There are four different orders for which the Littrow condition can be satisfied.

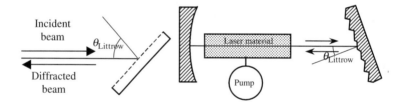

Figure 7.31. In the Littrow configuration the grating sends back the diffracted beam exactly in the opposite direction of the incident beam. As $\theta_{Littrow}$ depends on the wavelength such an arrangement is commonly used as the second mirror of a tunable laser, a rotation of the grating changes the laser frequency.

7.4.2.5. *A Diffraction Grating May Support Evanescent Waves*

A formal application of the Bragg formula may give values of $\sin\theta_{diff}$ that are larger than +1 or smaller than −1. A similar situation has already been encountered when using the Snell-Descartes law of refraction in the conditions of total internal reflection. An interpretation has been given thanks to evanescent waves. The situation here is very analogous: each time that the Bragg formula gives a sine or cosine greater than one, in absolute value, evanescent waves can be observed.

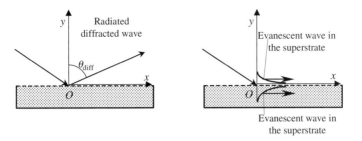

Figure 7.32. Radiated and evanescent modes of diffraction.

We consider a diffracted mode such that the sine of the angle of diffraction is equal to some real number α_p, with an absolute value greater than one,

$$\sin \theta_{\text{diff}} = \alpha_p > 1,$$

the cosine of θ_{diff} is obtained by the usual formula, $\cos^2 \theta_{\text{diff}} + \sin^2 \theta_{\text{diff}} = 1$,

$$\cos \theta_{\text{diff}} = \pm j \sqrt{\alpha_p^2 - 1} = \pm j \beta_p \quad \text{with} \quad \beta_p^2 = \alpha_p^2 - 1.$$

Let us call $V(x, y)$ the electromagnetic vibration at some point $M(x, y)$, to obtain an expression for $V(x, y)$ we go back to the usual form of a planar wave having a wave vector $\mathbf{k}$ making an angle θ_{diff} with the axis Oy, see Figure 7.32. We set

$$\mathbf{OM} = \mathbf{r} = x\mathbf{x} + y\mathbf{y}, \quad \mathbf{k} = \frac{2\pi}{\lambda}(\mathbf{x}\sin\theta_{\text{diff}} + \mathbf{y}\cos\theta_{\text{diff}}),$$

$$V(x, y) = \mathbf{A}e^{j\omega t}e^{-j\mathbf{k}_p\mathbf{r}} = \mathbf{A}e^{j\omega t}e^{-j\frac{2\pi}{\lambda}(x\sin\theta_{\text{diff}} + y\cos\theta_{\text{diff}})},$$

$$V(x, y) = \mathbf{A}e^{j\omega t}e^{-j\frac{2\pi}{\lambda}\alpha_p x}e^{-\frac{2\pi}{\lambda}\beta_p|y|}, \quad \mathbf{A} \text{ is a constant vector.}$$

$V(x, y)$ represents a wave, that is:

A progressive wave in a direction that is parallel to the grating Ox.
An evanescent wave in a direction that is orthogonal to the grating Oy.

7.4.3. Practical Considerations about Gratings

7.4.3.1. Methods for Making Gratings

Until rather recently the gratings were quite expensive; nowadays, they can be considered as relatively cheap optical components since collective methods of elaboration are now available, a large number of identical gratings being simultaneously duplicated from an initial "master grating." The grating market is an important one, thanks to applications such as spectrometers (biology, chemistry) and, more recently, to optical communications (wavelength multiplexing).

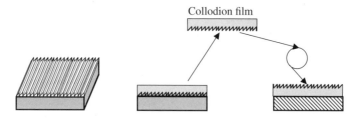

Figure 7.33. Duplication of a ruled grating. Using a diamond tool, a large number of parallel and equidistant grooves are notched on a planar glass surface, giving a master grating. A thin layer of liquid collodion is poured onto its surface. After drying, a collodion cast of the master grating is obtained, it's a solid-state film (thickness ≈ 0.1 mm) that is then glued onto a glass substrate. After metallization, an excellent replica of the initial grating is obtained.

Up to the utilization of lasers and photodyes, the gratings were obtained by engraving a glass surface with a diamond tool, the displacement of which was driven by a special ruling machine. As early as 1880, Rowland made a grating with 1700 grooves per millimeter; Michelson could produce a 30 cm long grating with 500 grooves per millimeter, which means that the diamond tool has notched 150,000 identical lines. . . . Once a first grating has been engraved, it can be replicated: Figure 7.33 shows one of the first methods that was used for duplication, it is based on the property of a varnish, called collodion.

The technology of the elaboration of gratings has largely benefited from the applied research efforts that have been devoted to the elaboration components for microelectronics, especially photolithography and cutting substrates in small pieces, using diamond saws or laser beams.

The fabrication methods rely on the polymerization of organic dyes, which can be either photosensitive (polymerization being enhanced by a suitable ultraviolet or visible radiation), or thermosensitive (polymerization being due to heating). A first grating is developed, usually 10×10 cm. Replicas are then made and cut in smaller pieces, 1×1 cm.

A dye is an organic liquid compound, made of molecules called monomers, which remains independent as long as they are not put into a situation where they polymerize, giving a solid state compound. Polymerization needs some energy that can be obtained from a light beam of a suitable color, or by a heat source of suitable temperature.

The procedure for making gratings is roughly described in Figures 7.34 and 7.35. Although there are no good reasons for that, this kind of grating is usually called a *holographic grating*.

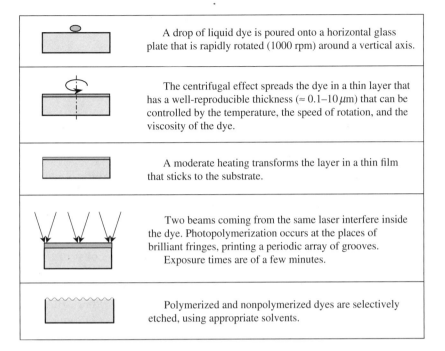

Figure 7.34. Modern techniques of fabrication of a master grating. The periodicity is determined by the angle of the two laser beams. The depth and shape of the grooves are reproducibly determined by the time of exposure and by the etching process.

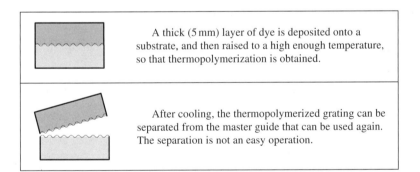

Figure 7.35. Duplication of a master grating using a thermopolymerized dye.

7.4.3.2. *Blazed Grating*

Blazed gratings are a good illustration of *why* and *how* the energy is distributed among the different orders of diffraction. The profile of a blazed grating is shown in Figure 7.36, it's a succession of parallel narrow rectangular mirrors making an angle α with the plane of the grating. The global diffracted

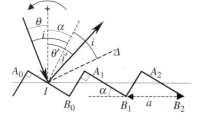

Figure 7.36. Groove profile of a blazed grating.

light is the result of the interference of the different waves diffracted by each individual mirror. The only directions in which some diffracted light can be observed are obtained from the Bragg formula.

The angular repartition of the light that is diffracted by a given mirror, A_0B_0 for example, is described by a $(\sin u/u)$ function with a maximum in the direction Δ of specular reflection on the mirror. By a suitable choice of periodicity a, angle α, angle of incidence θ, and of wavelength λ, it is possible to find conditions where Δ coincides with the direction of an order of diffraction of the grating, say the mth mode, while the directions of the other orders coincide with the zeros of the $(\sin u/u)$ function. Under such conditions the pth mode carries away all the energy

$$\cos i = \cos(\theta + \alpha) = p\frac{\lambda}{2a\sin\alpha}. \tag{7.20}$$

Numerical application: $\lambda = 0.5\,\mu m$, $N = 1000$ mm^{-1}, $a = 1\,\mu m$.

- $\alpha = 30°$. Find the angles of incidence and of refraction for the grating to be blazed in the first order: $\cos(\theta + \alpha) = 0.5 \rightarrow \theta = 0$, $\theta' = 30°$.
- Find α so that the blaze and the Littrow conditions are simultaneously fulfilled for the first order of diffraction,

$$\sin\theta = \frac{\lambda}{2a} = 0.25,$$

$$\cos(\theta + \alpha) = \frac{\lambda}{2a\sin\alpha} \quad \rightarrow \quad 2\tan\alpha\cos\theta = \frac{\lambda}{a} \quad \rightarrow \quad \alpha \approx 14.5°.$$

7.4.3.3. The Littrow Autocollimation Mounting

Figure 7.37 describes a simple and efficient spectrometer that uses a grating in the Littrow arrangement. For a given wavelength λ, the diffracted light comes back exactly in the opposite direction to the incident light. The entrance slit is placed in the focal plane of an achromatic lens; if the planes of the slit and of the grating are parallel, the autocollimated image just coincides with the slit, a few degrees tilt shifts the image in a place where it can be disposed as a photographic plate or an array of photodiodes.

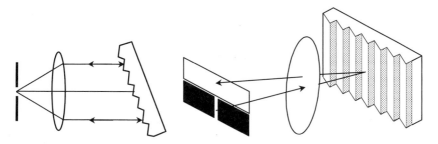

Figure 7.37. Spectrometer using the Littrow mounting.

7.4.3.4. *Monochromator*

Starting from a white source, a monochromator gives a light with a tunable wavelength λ and a narrow bandwidth. Using a grating and an optical imaging system, an entrance slit is imaged on an exit slit. The Littrow autocollimation mounting can be used, the photographic plate being replaced by a slit. The wavelength is tuned by rotating the grating around an axis in its plane. The spectral width is determined by the width of the exit slit, typically 5 to $50\,\mu m$. Chromatic aberrations should of course be avoided; this is the reason why spherical mirrors are preferred to lenses.

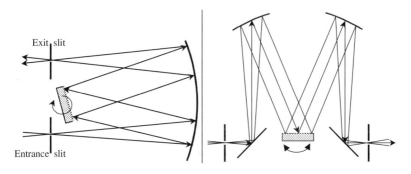

Figure 7.38. Examples of monochromators.

7.5. Holography

Denis Gabor in the middle of the twentieth century proposed the idea of holography. In the beginning, the purpose was to improve the resolving power of electron microscopes; the De Broglie waves associated to the electron beam would have been used to register the holograms that would have then been read with a monochromatic optical wave. Gabor immediately introduced the word hologram from the Greek, *holos* meaning *as a whole*, to suggest that both the amplitude and phase of the waves are registered in the photographic plate. In fact, the development of holography could only occur after the invention of lasers, the writing and reading of the hologram being made by optical waves. Denis Gabor became a Nobel Laureate in 1971.

7.5.1. *Definition of Holography*

We first consider the reason why, watching the planar picture of a landscape, we have the feeling that it is a representation of a three-dimensional space. Then we explain how different is the case of a holographic representation.

7.5.1.1. *Appreciation of the Relief on a Picture, Depth of Field*

Figure 7.39 reminds us what happens when a picture is taken of some landscape. An objective images the landscape on the photographic plate. The focusing process on some point A consists of translating the plate until it coincides exactly with the image front plate (P') associated to the object front plane (P) of point A. A landscape is not usually planar. The focusing is only valid in this couple of planes, to a point such as B is associated a tiny spot bb', instead of a dimensionless point; if the spot is small enough, we still get the impression that bb' is a good image of B. The thickness, on the inside of which must be located the points that give acceptable images, is called the depth of field of the camera.

A picture, as well as a painting made by an artist, is nothing other than a two-dimensional object; the sensation of relief, or of perspective, is purely subjective and calls up our memory.

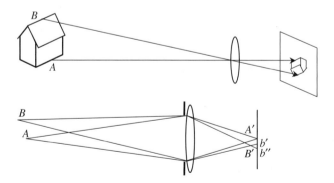

Figure 7.39. The camera is focused on the front wall of the house. The image of some point B at the top of the roof is slightly before the plate; if the tiny spot (bb'), generated by a pencil of rays coming from B, is small enough, the image of B is considered as acceptable.

7.5.1.2. *Appreciation of the Relief on a Hologram*

Holography is a technique that memorizes the phase of a wave at the instant of exposure; in fact, it's the relative phase shifts from one point to the next that are stored. The phase storage is obtained thanks to the interference of the incident wave to be recorded with a reference wave that has a well-known and well-reproducible phase repartition. The photographic plate registers interference fringes, the blackening being more intense at the place of clear fringes, where both incident and reference waves are in phase. After impression and development the plate becomes a *hologram*.

The hologram is illuminated with a reading wave that has the same phase distribution as the reference wave that was used when recording the hologram; each point of the hologram becomes a source of coherent optical wavelets. It will be shown that the repartition of the complex amplitudes along the plane of the hologram is identical with the repartition that was created by the initial wave at the level of the photographic plate of Figure 7.40(a): the Huygens-Fresnel principle says that, along the line $O_1O_2O_3O_4$, the optical fields are identical, respectively:

- In Figure 7.40(a), in the absence of the plate.
- Beyond the hologram of Figure 7.40(b), the three dimensions of the landscape have actually been reconstituted.

It is interesting to say that, if the hologram is broken in several pieces, each piece contains all the information about the landscape and can be used as the entire initial hologram.

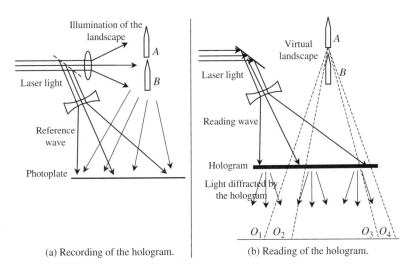

(a) Recording of the hologram.   (b) Reading of the hologram.

Figure 7.40. Recording and reading a hologram. An observer moves from O_1 to O_4. Between O_1 to O_2 the two objects are simultaneously visible; from O_2 to O_3, A is hidden by B and is not visible; from O_3 to O_4, A and B can be seen again.

7.5.2. *Principle of Holography*

Let us consider an initial wave A of complex amplitude $a_{1(x,y,z)}e^{j\varphi(x,y,z)}$. The plate where the hologram is to be recorded is taken as the xOy reference plane of coordinate $(z = 0)$; in this plane the complex amplitude is simplified as

$$a_{1(x,y)}e^{j\phi(x,y)} = a_{1(x,y,0)}e^{j\phi(x,y,0)}.$$

For the sake of simplicity, the reference wave will be a planar wave of amplitude a_2 and parallel to the plate, a_2 is supposed to be real. The amplitudes of vibration and intensity at point $P_{(x,y,0)}$ are given by

$$a_{(x,y)} = \left(a_{1(x,y)}e^{j\phi(x,y)} + a_2\right),$$

$$I_{(x,y)} = \left(a_{1(x,y)}e^{j\phi(x,y)} + a_2\right)\left(a_{1(x,y)}^{*}e^{-j\phi(x,y)} + a_2\right),$$

$$I_{(x,y)} = a_{1(x,y)}a_{1(x,y)}^{*} + a_2^2 + a_2\left(a_{1(x,y)}e^{j\phi(x,y)} + a_{1(x,y)}^{*}e^{-j\phi(x,y)}\right).$$

We admit that the conditions of exposure and of development of the photographic plate have been chosen so that the variation of the darkening varies linearly with the light intensity. Under such conditions, when the hologram is illuminated by the reference wave, the repartition of the amplitude of the Huygens-Fresnel sources is described by the following formula:

$$\alpha_{(x,y)} + \beta_{(x,y)}e^{j\phi(x,y)} + \beta_{(x,y)}^{*}e^{-j\phi(x,y)}. \tag{7.21}$$

The second term of (7.21) reproduces exactly the complex amplitude repartition of the initial wave.

7.5.3. *Examples of Holograms*

We are now going to study several examples to illustrate how the initial wave can be regenerated.

7.5.3.1. *Hologram of a Planar Wave (see Figure 7.41)*

The reference and wave to be recorded are both planar and have equal amplitudes. The intersection of the plane of the hologram with the plane of the two wave vectors is taken as the Ox axis. The interference pattern of the two waves is extremely simple; the variation of transparency follows a sine law, versus x, and is proportional to the following expression:

$$\left(1+e^{-j\frac{2\pi}{\lambda}x\sin\theta}\right)\left(1+e^{+j\frac{2\pi}{\lambda}x\sin\theta}\right) = 2 + e^{-j\frac{2\pi}{\lambda}x\sin\theta} + e^{+j\frac{2\pi}{\lambda}x\sin\theta}. \tag{7.22}$$

We recognize the three terms of the more general formula (7.16). The hologram is made parallel with equidistant lines (periodicity $a = \lambda/\sin\theta$). Illumi-

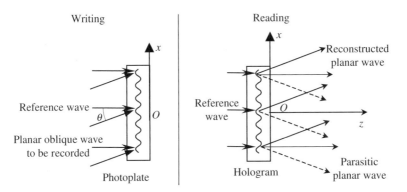

Figure 7.41. Writing and reading the hologram of a planar wave.

nated at normal incidence by the reference beam, the grating generates diffracted beams in the directions defined by the Bragg formula:

$$\sin\theta_{\text{diff}} = \sin\theta + p\frac{\lambda}{a} = (1+p)\sin\theta.$$

When the hologram is illuminated by the reference wave of Figure 7.41, its surface is covered with secondary radiating sources. The diffracted field is the Fourier transform of the amplitude repartition of formula (7.22), to each term of which is associated a parallel beam propagating in one of the following directions:

- $+\theta$, this is the reconstructed wave;
- Oz axis, corresponding to a partial transmission of the reference wave;
- $-\theta$, this is a parasitic wave.

The reason why there are only three orders of diffraction comes from the sinusoidal variation of the transparency; this effect is very analogous to the blaze effect in the case of a triangular profile.

The previous results are general: when reading a hologram, besides the reconstruction of the initial wave, parasitic waves are also obtained.

7.5.3.2. *Hologram of a Spherical Wave*

We are now going to holograph a diverging spherical wave that is supposed to have an aperture small enough to allow the same type of development that was used in Section 2.6.3. A parasitic spherical wave is also generated; an oblique reference wave efficiently separates the reconstructed wave from the parasitic wave.

7.5.3.3. *The Reference Writing and Reading Waves Are Both Orthogonal to the Hologram*

We refer to Figure 7.42; the reference wave and spherical wave have the same amplitudes a_0. ρ being the distance of a point P of the hologram to Oz and D the distance $D = A_0O$ of the center of the spherical wave to the plate, the spherical wave is described by $a_0e^{-j\pi(\rho^2/\lambda D)}$. The amplitude $a(\rho)$ and the intensity $I(\rho)$ of the optical vibration at point P are, respectively, given by

$$a(\rho) = a_0^2\left(1 + e^{-j\frac{\pi\rho^2}{\lambda D}}\right),$$

$$I(\rho) = a_0^2\left(1 + e^{-j\frac{\pi\rho^2}{\lambda D}}\right)\left(1 + e^{+j\frac{\pi\rho^2}{\lambda D}}\right) = a_0^2\left(2 + e^{+j\frac{\pi\rho^2}{\lambda D}} + e^{-j\frac{\pi\rho^2}{\lambda D}}\right), \qquad (7.23.\text{a})$$

$$I(\rho) = 2a_0^2\left(1 + \cos\frac{\pi\rho^2}{\lambda D}\right). \qquad (7.23.\text{b})$$

The hologram, as is shown by formula (7.23.b), is made of alternate clear and dark rings, their radii increasing as the square root of the successive integers. It looks like a Fresnel zone plane (see Section 7.3), except that the transparency is not equal to zero or to one, but varies sinusoidally. A Fresnel zone plate has many focal points (see Figure 7.23); because of a kind of blaze effect, in the case of a sinusoidal variation of the transparency, only two focuses are illuminated, they correspond to the points A_0 and A_0' of Figure 7.42. The spherical wave centered at point A_0 is the reconstituted wave; it corresponds to the term $e^{j\pi\rho^2/\lambda D}$ of formula (7.18.a). The wave centered at point A_0 is a parasitic wave; it should be associated to the term $e^{-j\pi\rho^2/\lambda D}$.

An observer facing the hologram has the impression that he sees a real point A_0 that would be behind the hologram. If the hologram was registered

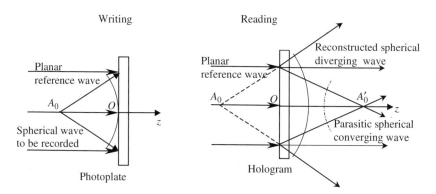

Figure 7.42. Hologram of a spherical wave with a reference wave orthogonal to the plate. Interference fringes are concentric circles.

with some extended object, the same treatment can be repeated for every point of the object. The parasitic wave also reproduces the initial object, which appears as a real object located in front of the hologram.

7.5.3.4. *The Planar Reading Reference Wave Is Oblique*

In the arrangement of Figure 7.42 the observer's eye also receives the reference wave and the parasitic wave, this drawback can be avoided by using an oblique reference wave, as is the case in Figure 7.43.

If the angle between the direction of the reference wave and the normal to the hologram, the intensity of the light vibration when recording the hologram, is equal to

$$I_{(x,y)} = a_0^2 \left(e^{-j\frac{2\pi}{\lambda}x\sin\theta} + e^{-j\frac{\pi}{\lambda D}(x^2+y^2)} \right) \left(e^{j\frac{2\pi}{\lambda}x\sin\theta} + e^{j\frac{\pi}{\lambda D}(x^2+y^2)} \right),$$

$$I_{(x,y)} = a_0^2 \left(2 + e^{-j\frac{2\pi}{\lambda}x\sin\theta} e^{j\frac{\pi}{\lambda D}(x^2+y^2)} + e^{j\frac{2\pi}{\lambda}x\sin\theta} e^{-j\frac{\pi}{\lambda D}(x^2+y^2)} \right).$$

(7.24.a)

When the hologram is illuminated by the reference wave, the repartition of the complex amplitudes, $a(\rho)$, is obtained by multiplying the amplitude of the reference wave by the transparency of the hologram, which is given by (7.24.a):

$$a(\rho) = I_{(x,y)} e^{-j\frac{2\pi}{\lambda}x\sin\theta}$$

$$= a_0^2 \left(2e^{-j\frac{2\pi}{\lambda}x\sin\theta} + e^{-j\frac{\pi}{\lambda D}(x^2+y^2)} + e^{-j\frac{4\pi}{\lambda}x\sin\theta} e^{j\frac{\pi}{\lambda D}(x^2+y^2)} \right).$$

(7.24.b)

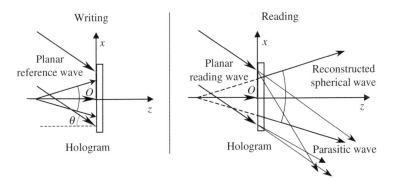

Figure 7.43. With an oblique reference wave, the reconstructed wave is well separated; the observer is not dazzled by the transmitted reference wave or by the parasitic wave.

The first term in equation (7.24.b) reconstitutes the reference wave that propagates in the direction θ. The second term reconstitutes a diverging spherical wave, which seems to have been emitted by point A_0. The last term is a spherical wave that converges at point A_0' in front of the hologram and in a direction making an angle 2θ with the normal.

7.5.3.5. *Volume Holography*

In the previous sections the holograms were planar, we are now going to use photographic emulsions that are no longer very thin. We thus consider the interference of a planar reference light wave (R) with some other wave (W) inside a volume of finite size. If the two waves are both planar, a family of parallel, alternately bright and dark, planar layers is printed inside the emulsion.

If the second wave is not a planar wave, it can always be considered as the superposition of planar waves with suitable wave vectors and intensities; each component of the planar wave decomposition of W, when interfering with the reference wave R, creates a family of parallel layers. The set of printed layers is called a *volume hologram*.

Let k_0 and k_r be the respective wave vectors of W and R that will be supposed to have the same amplitude a_0; the interference pattern is described by

$$a(r) = a_0 (e^{-jk_r r} + e^{-jk_0 r}) \quad \rightarrow \quad I(r) = 2a_0^2 [1 + \cos((k_0 - k_r)r)],$$

$$I(r) = 2a_0^2 [1 + \cos k_G r] = 2a_0^2 \left[1 + \cos \frac{2\pi}{\Lambda} u_G r \right],$$

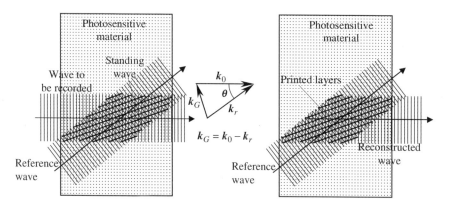

Figure 7.44. Recording and reading of a volume hologram. The Bragg reflection on the layers printed in the plate reconstructs the initial wave.

where

$k_G = (k_0 - k_r)$ is called the grating vector;
$u_G = k_G/kG$ is a unit vector parallel to one of the bisectors of the angle
 (k_0, k_r); and
$\Lambda = \lambda/(\sin \theta/2)$ is the periodicity of the printed grating.

We now examine what happens when the reference beam illuminates the hologram. The situation can be compared to the diffraction by the reticular planes of a crystal lattice; the parallel planes of a grating reflect the light if a kind of Bragg condition is satisfied. Let λ' be the wavelength of the reference beam and let α be the angle between the reference wave vector and the layers of the hologram:

$$\sin \alpha = \frac{\lambda'}{2\Lambda}. \tag{7.25}$$

The initial Gabor proposition was to make a hologram with an electron beam, and then to read it with an optical beam; λ would have been the wavelength of the De Broglie wave associated to the electrons and λ' an optical wavelength. For many reasons no realization has ever been obtained. Let us come back to the case where the writing and reading reference waves are optical waves of the same frequency (see Figure 7.44): $\lambda' = \lambda$, which implies that $\alpha = \theta/2$. The set of bright and dark layers plays the same role as the reticular planes in the diffraction by a crystal: the reading beam is selectively reflected in the direction of the second bisector of the angle (k_0, k_r), and reconstructs the initial planar beam.

Because information is stored in a three-dimensional space, volume holography is, in principle, more powerful. Several, eventually many, different holograms can be registered inside the same volume, either changing the direction or the frequency of the reference wave.

Color holograms: A volume hologram can be read by a white reference light beam, Bragg reflection occurs only for that part of the spectrum which coincides with the color that was used for recording. Using the principle of the trichromatic system (see Section 7.8), "colored holograms" can be made; three holograms are registered in the same plate with three different wavelengths, when the plate is illuminated with a white light, the initial object appears in color and in three dimensions.

7.5.3.6. *Interference Holography*

Very small deformations (μm) of large size objects can be revealed thanks to holography, for example, the stress deformations of mechanical structures such as a bridge, a shaft, or the body of a car. Two holograms are consecutively recorded in the same plate, the object being successively in the two situations to be compared. The reference beam then simultaneously illuminates the two holograms and an objective makes a projection on a screen. The image reproduces the object, but it is striped with interference fringes, the deformation remaining constant along a given fringe.

The method can work under real time conditions. A hologram of the unstressed object is first registered and the virtual image that is obtained when illuminated by the reference beam is put into exact coincidence with the object that is also illuminated. On the screen the fringes due to deformation appear as if they were painted on the object.

7.6. Diffraction and Image Processing

The Fourier transform plays a preeminent role in diffraction problems and in the formation of images. Starting from some object, described as a repartition of amplitudes $f(x, y)$ in a front plane, a lens L_1 elaborates the Fourier transform $F(X, Y)$ of $f(x, y)$, which is available in its image focal plane. A second lens L_2 then makes a second Fourier transform $f'(x', y')$ of $F(X, Y)$. $f'(x', y')$, which is available in the image focal plane of L_2, is the image of the initial object and, except for a multiplication by a "magnification coefficient," is similar to $f(x, y)$. The intermediate Fourier transform $F(X, Y)$ can be modified by placing a slide with a controlled transparency on the focal plane (Π) of L_1, (Π) is also called the Fourier plane of the arrangement.

7.6.1. *Fourier Plane*

The arrangement shown in Figure 7.45, which is analogous to that of Figure 7.11, will allow us to go deeper into the insight of the mechanism of the formation of an image by a lens. An object is placed in the focal plane (π) of a lens (focal length f). We consider an elementary area $dx\,dy$ around some point $M(x, y)$, a spherical wavelet emitted by this point is collimated by the lens into a planar wave propagating in the direction defined by the coefficients

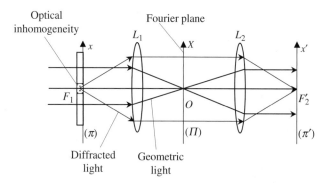

Figure 7.45. The image focal plane of L_1 coincides with the object focal plane of L_1. (π) and (π') are the antiprincipal planes of the centered system.

$u = x/f$ and y/f; its amplitude is proportional to $dx\,dy$ and is written as $f(x, y)\,dx\,dy$, $f(x, y)$ varies from one point of the object to another and is a complex number. Let us evaluate the complex amplitude $F(X, Y)$ of the optical vibration at a point $P(X, Y)$ of the second focal plane (Π) of the lens. $F(X, Y)$ results from the superposition of all the planar waves coming from the different points of the object. If M doesn't belong to the lens axis Oz, the planar wave coming from M is not parallel to Oz; the corresponding amplitude repartition is described by

$$e^{-j\frac{2\pi}{\lambda}\frac{xX+yY}{f}}f(x,y)\,dx\,dy.$$

The total amplitude at point $P(X, Y)$ is given by

$$F(X,Y) = \int_{-\infty}^{+\infty}\int_{-\infty}^{+\infty} f(x,y)e^{-j\frac{2\pi}{\lambda}\frac{xX+Yy}{f}}\,dx\,dy. \tag{7.26}$$

We have considered that the size of the lens is large enough so that the diffraction effects by the corresponding aperture are negligible, and so, the integrals are extended to infinity. Formula (7.21) shows that the amplitude distribution in the plane (Π) is the Fourier transform of the function $f(x, y)$,

$$F(X,Y) = \mathrm{FT}[f(x,y)].$$

The plane (Π) is then considered as an object playing, for lens L_2, the same role as (π) for L_1. The amplitude repartition, $f'(x', y')$, in the focal plane (π') of the second lens is the Fourier transform of $F(X, Y)$:

$$F'(x',y') = \mathrm{FT}[F(X,Y)] = \mathrm{FT}\{\mathrm{FT}[f(x,y)]\} \quad \rightarrow \quad f'(x',y') = f(-x,-y).$$

The intermediate plane (Π) is called the *Fourier plane* of the arrangement.

The previous method is nothing other than a sophisticated method for establishing that the two planes (π) and (π') are the two conjugate planes of the centered system for which the magnification is equal to -1; such planes are also called "antiprincipal planes," by analogy with the principal planes of Section 3.5.2. The method is however very interesting, since it opens the way to very powerful methods of image processing. If a plate, with a variable transparency $t(X, Y)$, is placed on the Fourier plane, the planar wave decomposition of the function $f(x, y)$ can be modified at will, as well as the appearance of the image that the second wave will reconstruct, by the inverse Fourier transform.

7.6.2. *Image Processing*

We are now going to examine different examples where optical signals are processed by modification of their planar wave decompositions. To each component of the planar wave decomposition is associated a point of the Fourier plane; if a tiny opaque obstacle is placed at this point, the corresponding component will be filtered out.

7.6.2.1. *Strioscopic Arrangement—Phase Contrast Microscopy*

We refer again to Figure 7.45, the optical inhomogeneity concerns the phase, the refractive index being slightly different from the surrounding index, this is, for example, the case of bacteria floating in water of a microscopic preparation. The amplitude repartition in the plane (π) is described by

$$a(\rho) = a_0 e^{-j\varphi(\rho)} \quad \text{with} \quad \varphi(\rho) = \frac{2\pi}{\lambda}[1 + \Delta n\delta(\rho)],$$

where ρ is the distance to the axis and $\delta(\rho)$ is a Dirac pulse. The index difference Δn between the bacteria and water is very small. Changing the phase origin and considering that $e\Delta n$ is small, as compared to the wavelength, we can write

$$\text{Plane } (\pi): \quad a(\rho) = a_0 e^{-j\frac{2\pi\Delta ne}{\lambda}\delta(\rho)} \approx a_0 \left[1 - j\frac{2\pi\Delta ne}{\lambda}\delta(\rho)\right].$$

Since the Fourier transform of a Dirac pulse is a constant and vice versa, the amplitude repartition in the Fourier plane is equal to

$$\text{Plane } (\Pi): \quad A(\rho) = a_0 \left[\delta(\rho) - j\frac{2\pi\Delta ne}{\lambda}\right]. \tag{7.27.a}$$

After a second Fourier transform, made by the second lens, we finally obtain the amplitude and intensity repartitions in the image plane,

$$\text{Plane } (\pi): \quad a'(\rho) = a_0 \left[1 - j\frac{2\pi\Delta ne}{\lambda}\delta(\rho)\right], \tag{7.27.b}$$

$$a'a'^* = a_0^2 \left[1 + \left(\frac{2\pi\Delta ne}{\lambda}\right)^2 \delta(\rho)\right]. \tag{7.27.c}$$

According to formula (7.27.c), plane (π') is homogeneously illuminated with, in the middle, a clearer spot corresponding to the presence of the bacteria. Figure 7.45 gives an interpretation of what happens, in terms of optical light rays. The planar incident wave is focused at the center of the Fourier plane and gives the term $a_0\delta(\rho)$ in formula (7.27), it then diverges and is collimated by the second lens to give a constant illumination on (π'). The second term, $-ja_0(2\pi\Delta ne/\lambda)\delta(\rho)$ in (7.27), should be associated to the light that is diffracted by the bacteria.

It's not difficult to take into account the fact that the lenses have a finite size and that the diameter of the bacteria is not equal to zero. Let R and r be the respective radii of the lens and of the bacteria. In the Fourier plane (Π) the amplitude repartition of the "geometric light" on the one hand, and of the "diffracted light" on the other are described by Airy functions $(J(\rho)/\rho)$. The spot inside which the geometric light is concentrated has a diameter equal to $1.22\lambda f/R$ and goes to zero if $R \to \infty$. On the contrary the diffracted light spreads over a large area $(1.22\lambda f/r)$.

Strioscopic Arrangement

In the case of the arrangement of Figure 7.45, the variation of intensity at the place of the image of the bacteria is very small, $\approx (2\pi\Delta ne/\lambda)^2$, and will hardly be perceived. In the spectroscopic arrangement the "geometric light" is removed by covering the focal point O with a tiny opaque stop. The plane (π') remains completely dark in the absence of bacteria, and the image of the bacteria is more easily revealed.

Phase Contrast

The strioscopic arrangement is not very luminous, and its ability to exhibit small details is quite limited by the unavoidable existence of parasitic stray light. In the phase contrast method, the tiny stop is replaced by a transparent plate, which has the same size, and a thickness such that the optical path of the geometric light is increased by an odd number of half-wavelengths, with regard to the diffracted light. Formulas (7.27.a) and (7.27.b), giving respectively, the amplitude repartitions in the Fourier plane and in the image plane, are now replaced by

$$A(\rho) = a_0\left[e^{\pm j\pi/2}\delta(\rho) - j\frac{2\pi\Delta ne}{\lambda}\right] = ja_0\left[\pm\delta(\rho) - \frac{2\pi\Delta ne}{\lambda}\right], \qquad (7.28.a)$$

$$a'(\rho) = ja_0\left[\pm 1 - \frac{2\pi\Delta ne}{\lambda}\delta(\rho)\right]. \qquad (7.28.b)$$

Neglecting second-order terms, the intensity repartition in the image plane is

$$E(\rho) = a'_{(\rho)}a'^*_{(\rho)} \cong a_0^2\left[1 \pm \frac{4\pi\Delta ne}{\lambda}\delta_{(\rho)}\right]. \qquad (7.29)$$

The intensity variation is larger, since it's now of the order of $(2\pi\Delta ne/\lambda)$. According to the fact that the geometric and diffracted beams are in phase, or in opposition, the variation is positive (clear image on a dark background) or negative (dark image on a clear background).

The phase contrast method is still handicapped by the existence of stray light. It's finally the aptitude of the observer to perceive a small intensity contrast that fixes the smallest size of detectable details. Using a plate, which at the same time, changes the phase and partially absorbs the geometric light, can accumulate the advantage of strioscopy and of phase contrast. Let us suppose that the phase difference is still $\pm\pi/2$, while the intensity is multiplied by a coefficient α smaller than unity, (7.24) then becomes

$$E(\rho) \cong a_0^2\left[\alpha^2 \pm 4\alpha\frac{\pi\Delta ne}{\lambda}\delta(\rho)\right]. \qquad (7.30)$$

The geometric light is more attenuated than the diffracted light and the contrast more favorable. With a proper choice of α, a difference of optical length as low as a few nanometers can be perceived. Phase contrast microscopy is much appreciated by biologists who generally observe transparent objects immersed inside a liquid having almost the same index of refraction. A typical example is the observation of living cells; phase contrast allows the observation in vivo, without using a coloring dye that is selectively fixed by a given component of the cell.

7.6.2.2. Optical Image Processing

Spatial Filtering

The planar waves that are focused near the center of the Fourier plane are said to be *low-frequency* spatial components. The spatial frequency is so high that the wave vector makes a greater angle with the axis, the corresponding focus being farther from the center. A tiny opaque spot, suitably placed in the Fourier plane, will filter out a given component.

A tiny opaque stop, at the center of the Fourier plane, is a *high-frequency* filter, which blocks the low-frequency component and especially the zero frequency; this is the case of strioscopy. On the contrary, a small hole is a *low-frequency* filter, transmitting only the low frequencies. With a pair of scissors and a sheet of black paper, various filters can easily be made. A horizontal slit, with broadness D, will only transmit the so-called vertical frequencies, with a bandwidth of the order of $\Delta k = D/2\lambda f$. In the same way a vertical slit only transmits the horizontal frequencies.

A function that is rapidly varying versus x and/or y, has many high-frequency components. If a high-frequency filter is used the areas, inside of which the intensity variations are smooth, will appear dark, while the zone of more rapid variation will be more brightly illuminated (see Figure 7.48).

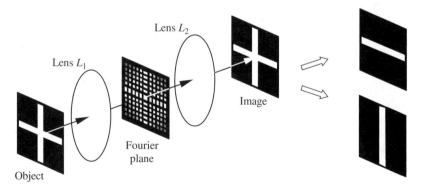

Figure 7.46. Formation of the image of a cross. At first with no filtering at all, and then with filters that select either the "horizontal" or "vertical" frequencies.

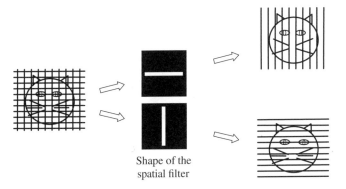

Figure 7.47. Filtering of the image of a cat seen through a square mesh grid.

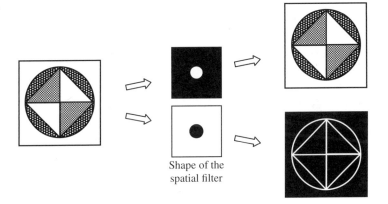

Figure 7.48. With a high-frequency filter, only the zones where the gradient of intensity is important will be illuminated. The image contour will appear as bright lines drawn on a dark background.

Analog Calculation of a Two-Dimensional Fourier Transform

The calculation of a two-dimensional Fourier transform needs a powerful computer. A simple lens achieves this operation in an analog way and can, in principle, do it in real time. This is the reason why, between 1960 and 1980, an important research effort was devoted to optically computing the Fourier transform. A drawback of this method is due to the noise that artificially introduces high-frequency components. Hopefully, or not, at the same time, the important development of microelectronics has drastically increased the storage capacity of computers and simultaneously decreased computation times, Fourier transforms are exclusively made by computers.

Computer-Assisted Image Processing

The continuous variation of the light intensity along some image can easily be sampled, using small adjacent surfaces that are called *pixels*. Each pixel is characterized by one real number (or by two real numbers if the amplitude is complex). These numbers, stored in good order in the memory of a computer, constitute a numerical image. Once the image has been replaced by a set of numbers, it's easy to imagine many operations, including the Fourier transform.

A first Fourier transform will give the repartition of complex amplitudes $A(X, Y)$ in the Fourier plane. Spatial filtering becomes very easy; it is just a multiplication by a number $\alpha(X, Y)$. A new set of numbers is obtained from which, thanks to a cathode ray tube or a printer, the new image can be visualized.

A Fourier transform followed by a spatial filtering was the first operation to be thought of. Since then, engineers have invented many other mathematical manipulations which, on one hand, are more adequately fitted to computer calculations and, on the other hand, give images of better quality. Such methods have considerably improved the numerical images that are transmitted from telescopes installed in satellites rotating around the Earth. The numerization of the image is done with a matrix of photodiode places in the focal plane of the telescope. Radio or microwaves make the transmission of the set of numbers back to Earth. This can be images of the Earth or of the stars and planets. In the last case, the main interest comes from the fact that the perturbations due to crossing the atmosphere are avoided. A difficulty with these indirect methods comes from the fact that an image is always obtained, the problem is to be sure that it has something to do with the initial object.

Fourier transform holograph

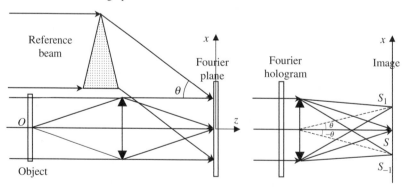

Figure 7.49. Recording and reading a Fourier hologram.

Fourier Transform Holography

Thanks to holography it is possible to store the repartition of complex amplitudes existing in a given plane. The arrangement of Figure 7.49 shows how to store the amplitude repartition of the Fourier plane; the hologram is then called a Fourier hologram. If the object is a point, the wave after the lens is a planar wave propagating parallel to Oz; the interference with the oblique reference wave gives horizontal fringes parallel to Oy: the transparency of the hologram is a sine function of x. A parallel beam illuminating the hologram is diffracted in three directions that are focused by the lens at S, S_1, and S_{-1}, the two last points can be considered as the images of the initial object.

8

Index of Refraction

The Snell-Descartes law of refraction introduces the index of refraction, n, very early in the development of Optics where, as a matter of fact, it is considered as a normalized value of the speed of propagation, V, of light in a transparent material. The normalization being made with respect to the speed of propagation, c, of electromagnetic waves in a vacuum.

In this chapter, we analyze the physical mechanisms, which indicate that V is different from c, as well as the law of variation of the index of refraction versus the frequency of the electromagnetic waves. First of all, we would like to introduce some important notions that a physicist should bear in mind when he starts to study the index of refraction:

- *Principle of Relativity*: No information can be transmitted faster than the speed of light in a vacuum; more specifically, the electric and magnetic fields that are produced at time t, by a moving electrical charge, cannot be perceived earlier than $(t + r/c)$ at a distance r from the charge.
- Vacuum is the only medium to show *no dispersion*: radio waves, as well as light waves or γ-rays, propagate exactly at the same speed c.
- The phase velocity of a sine wave in a given material depends on the frequency, the variation being more important in the vicinity of the bands of absorption of the considered material.
- Although it's not generally the case, there is no reason why the phase velocity of a wave should not be larger than c, or why the index of refraction should not be smaller than one. For example, the index of refraction of metals for X-ray waves is smaller than unity. This is not contradictory to the relativity principle, since information is carried at the group velocity, which, because of dispersion, is not equal to the phase velocity.

Chapter 8 has been reviewed by Dr. Lionel Bastard from the Institut National Polytechnique, Grenoble.

8.1. Physical Mechanisms Involved with Propagation in a Transparent Material

As far as vacuum propagation is concerned, we will make no attempt to answer the meaningless question: What is vibration? Neither shall we try to find a support to the Maxwell displacement currents that must be introduced, even in a vacuum. Maxwell's equations will be considered as a fundamental hypothesis, fully verified by many experimental consequences. Up to now, Maxwell's equations have never failed.

We will analyze the difference between the propagation in a vacuum and the propagation in a transparent material; the latter being considered as a set (C) of largely submicroscopic particles, bearing either positive charges (nuclei) or negative charges (electrons). C is very often made of subsets with equal numbers of positive and negative charges (atoms or molecules) and also of subsets in which the number of charges of one kind is only slightly larger than the number of charges of the other kind (positive or negative ions).

When a collection of atoms, such as C, receives an electromagnetic wave, an oscillating electromagnetic field, due to this wave, is superimposed on the field of forces that is responsible for the coherency of the corresponding material. It's up to the reader to imagine all the possible interactions. We will only consider the simplest one, that's to say, the action of the electric field of the wave on the charged particles, which is known as *Electric Dipole Interaction* (see equation (1.1)),

$$\frac{d^2\xi}{dt^2} + \gamma \frac{d\xi}{dt} + \omega_0^2 \xi = \frac{e}{m} f(t). \tag{8.1}$$

For the sake of simplification, C is limited by a planar interface and the electromagnetic wave is a sine planar wave with a wave vector orthogonal to the interface. The electric charges are put into a forced sine vibration and, in turn, each of them becomes the source of a spherical electromagnetic wave. The electromagnetic field, at a given point P of C, is the addition (interference) of the following fields:

- The field that would exist if C was replaced by vacuum, which is in fact the incident wave.
- The fields that are generated by all the spherical waves.

Because of the symmetry of the problem, the resulting field should have the same value at any point of a plane orthogonal to the direction of propagation. The children that hit the surface of the swimming pool of Figures 1.8 and 1.9 suggest that the superposition of the numerous waves coming from all the elements of C reconstitute a planar wave.

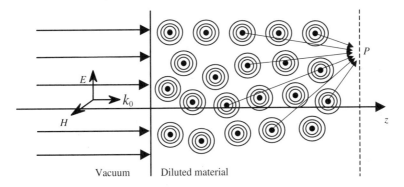

Figure 8.1. Illustration of the propagation of a wave in a material. A planar wave arrives on the planar interface of a piece of material; its electric field sets in motion the electric charges of the atoms, which become oscillating dipoles having the same frequency as the incident wave. The small circles are supposed to suggest the wave surfaces of the spherical wavelets emitted by the dipoles.

8.2. Determination of the Index of Refraction

The determination of the phase velocity of a wave propagating in a transparent material is certainly a difficult problem; at first, the equations of the motion of the charged particles should be integrated and the resulting field at a given point P should be calculated. The next problem would be to find the function written in the right-hand side of equation (8.1). Each of the Q particles of C is submitted to the action of the $Q - 1$ other particles which, at the same time, are submitted to the action of the particle under study. . . . The case of a gas is simplified by the fact that the mutual interactions between the charges are negligible compared to the action of the field of the incident wave.

8.2.1. *Case of a Diluted Material*

We consider a gas made of neutral atoms or molecules. Under the action of the electric field of the incident wave, the electrons vibrate with respect to the nucleus: the gas is then a collection of electrical dipoles, each of them having an electric dipolar momentum that varies sinusoidally versus time. Let N be the number of dipoles per unit volume.

We refer to Figure 8.2, a planar light wave crosses a thin layer of gas, and the light propagates in a vacuum. We want to obtain the expression for the electromagnetic field, $E_{P,\text{total}}$, at a point P located some distance behind the layer of gas. $E_{P,\text{total}}$ can always be considered as the addition of the field $E_{P,0}$ that exists in the absence of the layer of gas and of some extra field $E_{P,\text{slice}}$ that is due to the presence of the layer of gas. Let n and Δz be, respectively, the

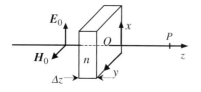

Figure 8.2. Action of a thin film of gas on an electromagnetic wave.

index of refraction and the thickness of the gas sample, we can write the following expressions:

$$E_{P,0} = E_0 e^{j\omega(t-z/c)}, \tag{8.2}$$

$$E_{P,\text{total}} = E_0 e^{j\omega[t-n\Delta z/c-(z-\Delta z)/c]} = E_0 e^{-j\omega(n-1)\Delta z/c} e^{j\omega(t-z/c)}. \tag{8.3}$$

The refractive index of a gas being very close to one, we have

$$e^{-j\omega(n-1)\Delta z/c} \approx 1 - j\omega(n-1)\Delta z/c,$$

$$E_{P,\text{total}} = E_0 e^{j\omega(t-z/c)}[1 - j(n-1)\omega\Delta z/c].$$

The field generated by the layer of gas is thus equal to

$$E_{P,\text{slice}} = -jE_0(n-1)\frac{\omega\Delta z}{c} e^{j\omega(t-z/c)}. \tag{8.4}$$

The gas is considered as a set of oscillating dipoles; let $\eta_0 = N\Delta z$ be the number of dipoles per unit of surface. The field that is created at point P is calculated in Annex 8.A and is given by formula (8.A.11):

$$E = -xj\omega\frac{N\Delta z q x_0}{2\varepsilon_0 c} e^{j\omega(t-z/c)}. \tag{8.5}$$

A comparison of the two formulas (8.4) and (8.5) gives an expression of the refractive index if we are able to relate E_0 and x_0. Because of the thinness of the slice, all the charges vibrate in phase and the function $f(t)$ in equation (8.1) is taken as $f(t) = E_0 e^{j\omega t}$; the solution is then given by

$$\xi = \frac{q}{m}\frac{1}{(\omega_0^2 - \omega^2 + j\gamma\omega)}E_0 e^{j\omega t} \quad \rightarrow \quad x_0 = \frac{q}{m}\frac{E_0}{(\omega_0^2 - \omega^2 + j\gamma\omega)}. \tag{8.6}$$

At last, we obtain the refractive index

$$n = 1 + \frac{Nq^2}{2\varepsilon_0 m(\omega_0^2 - \omega^2 + j\gamma\omega)}. \tag{8.7}$$

A physical interpretation of the index of refraction can be imagined as follows: the electric field of the incident wave polarizes the atoms and sets the electrons in a forced sine motion. The acceleration of the charges radiates a new field that is superimposed on the initial one. The two fields vibrate at the same frequency, however they are not in phase: when interfering they create, inside the material, a field that is not in phase with the incident field.

As the phase difference is proportional to the product of the propagation speed by the length covered inside the material, it seems that the light propagates with a phase velocity that is not equal to c.

8.2.2. Calculation of the Index of Refraction in a More General Case

To be able to integrate the equation, we made the assumption of a suitably diluted material so that the oscillations of the electrons were only driven by the incident wave; the mutual interaction of the dipoles was neglected. In a dense material this assumption is no longer valid, so we will come back to Maxwell's equations and make the same kind of development that was made already in Section 2.3. At that time we made no real difference between the propagation in a vacuum or in a transparent dielectric material, except that ε_0 was just replaced by ε.

8.2.2.1. Polarizability of a Material

For a more accurate description of the material inside which the propagation occurs, we come back to the definition of the Maxwell displacement vector D and we introduce the polarization vector P that is the dipolar momentum of an unitary volume of the sample. Polarization is of course originated at the atomic level; when one of its electrons is taken away from the equilibrium position by some distance ξ, an atom becomes an electric dipole with an electric dipole momentum equal to $p = q\xi$ (q is the electric charge of the electron). If N is the cubic density of atoms, we have

$$P = Np = Nq\xi. \tag{8.8}$$

In the frame of a linear interaction between the material and the light, ξ is proportional to the electric field E, the proportionality coefficient is called the *atomic polarizability* ϖ and is defined as

$$p = \varpi\varepsilon_0 E. \tag{8.9}$$

Let l be the number of electrons that, inside a given atom, are concerned with the interaction with a light wave. A simple model of a harmonic oscillator is chosen, with an eigenfrequency $\omega_{0,k}$ and a damping coefficient γ_k, k is a labeling index ($0 < k < l$),

$$m(\ddot{\xi}_k + \gamma_k \dot{\xi}_k + \omega_{0,k}^2 \xi_k) = qE_0 e^{j\omega t} \quad \rightarrow \quad \xi_k = \frac{q}{m}\frac{1}{\Delta_k(\omega)} E_0 e^{j\omega t},$$

with

$$\Delta_k(\omega) = -\omega^2 + j\gamma_k\omega + \omega_{0,k}^2. \tag{8.10}$$

The variation of the atomic polarizability versus frequency is described by

$$\varpi(\omega) = \frac{q^2}{\varepsilon_0 m} \sum_k \frac{1}{\Delta_k} = \frac{q^2}{\varepsilon_0 m} \sum_k \frac{1}{-\omega^2 + j\gamma_k \omega + \omega_{0,k}^2}. \tag{8.11}$$

Oscillator Strength

We should be reasonably worried by the previous method of establishing the equations, which is at the same time pragmatic and relaxed, since we used a very classical treatment for microscopic phenomena that fall, in principle, into the scope of Quantum Mechanics. A full quantum treatment is possible and gives very similar results, except for the introduction of a set of real coefficients, called the *oscillator strengths* f_k, which are characteristic of the atoms under consideration,

$$\varpi(\omega) = \frac{q^2}{\varepsilon_0 m} \sum_k \frac{f_k}{\Delta_k} = \frac{q^2}{\varepsilon_0 m} \sum_k \frac{f_k}{-\omega^2 + j\gamma_k \omega + \omega_{0,k}^2}, \tag{8.12.a}$$

with

$$0 < f_k < 1 \quad \text{and} \quad \sum_k f_k = 1. \tag{8.12.b}$$

Electrical Susceptibility

When the polarization results from the application of an electric field $\boldsymbol{E}$, the electrical susceptibility χ is introduced by the following formula:

$$\boldsymbol{P} = \varepsilon_0 \chi \boldsymbol{E}. \tag{8.13}$$

Going back to formulas (8.11) and (8.13) and introducing the cubic density of electrons N_k, we obtain

$$\boldsymbol{P} = \varepsilon_0 \sum_k N_k \varpi(\omega) \boldsymbol{E} = \varepsilon_0 \chi(\omega) \boldsymbol{E},$$

$$\chi(\omega) = \sum_k N_k \varpi(\omega) = \frac{q^2}{\varepsilon_0 m} \sum_k N_k \frac{f_k}{\Delta_k(\omega)}. \tag{8.14}$$

The variation of the polynomial $\Delta_k = -\omega^2 + j\gamma_k\omega + \omega_{0,k}^2$ is rather smooth, except in the close vicinity of a band of absorption, when $\omega \approx \omega_{0,k}$. The damping coefficient γ_k is much smaller than the eigenfrequency, so the roots of the polynomial are almost equal to $\omega_{0,k}$ and $\Delta_k(\omega)$ can be simplified as

$$\Delta_k(\omega) = \omega_{0,k}^2 - \omega^2 + j\gamma_k\omega = (\omega_{0,k} + \omega)(\omega_{0,k} - \omega) + j\gamma_k\omega$$

$$\approx 2\omega_{0,k}\left[(\omega_{0,k} - \omega) + j\frac{\gamma_k}{2}\right].$$

Formula (8.14) becomes

$$\chi(\omega) = \frac{q^2}{2\varepsilon_0 m} \sum_k \frac{N_k f_k}{\omega_{0,k}} \frac{1}{(\omega_{0,k} - \omega + j\gamma_k/2)},$$

setting $\chi(\omega) = \chi'(\omega) - j\chi''(\omega)$,

$$\chi'(\omega) = \frac{q^2}{2\varepsilon_0 m} \sum_k \frac{N_k f_k}{\omega_{0,k}} \frac{(\omega_{0,k} - \omega)}{\left[(\omega_{0,k} - \omega)^2 + \gamma_k^2/4\right]}, \qquad (8.15.a)$$

$$\chi''(\omega) = \frac{q^2}{2\varepsilon_0 m} \sum_k \frac{N_k f_k}{\omega_{0,k}} \frac{\gamma_k/2}{\left[(\omega_{0,k} - \omega)^2 + \gamma_k^2/4\right]}. \qquad (8.15.b)$$

Law of Refraction of the Index of Refraction

The square of the index of refraction is equal to the square of the dielectric constant ε_0. In a dielectric material, the simple relationship $\boldsymbol{D} = \varepsilon_0 \boldsymbol{E}$, that is valid in a vacuum, becomes $\boldsymbol{D} = \varepsilon_0 \boldsymbol{E} + \boldsymbol{P}$. When the existence of the polarization is the result of the action of an external electric field, we can write the following equations:

$$\boldsymbol{P} = \varepsilon_0 \chi \boldsymbol{E} \quad \rightarrow \quad \boldsymbol{D} = \varepsilon_0(1+\chi)\boldsymbol{E} = \varepsilon\boldsymbol{E} \quad \rightarrow \quad \varepsilon = \varepsilon_0(1+\chi), \qquad (8.16.a)$$

$$\varepsilon_r = \frac{\varepsilon}{\varepsilon_0} = n^2, \quad n^2 = 1 + \chi(\omega) = 1 + N\varpi(\omega), \qquad (8.16.b)$$

$$n^2 = 1 + \frac{Nq^2}{\varepsilon_0 m} \sum_k \frac{f_k}{\Delta_k(\omega)}. \qquad (8.17.a)$$

In the case of a diluted material the refractive index is of the order of one, formula (8.17.a) then takes a simpler form that coincides with (8.7),

$$n \approx 1 + \frac{\chi}{2} \quad \rightarrow \quad n' \approx 1 + \frac{\chi'}{2} \quad \text{and} \quad n'' \approx 1 + \frac{\chi''}{2}. \qquad (8.17.b)$$

8.2.2.2. *Local Field Correction (Lorentz-Lorenz or Clausius-Mossotti Formula)*

The method for establishing formula (8.15) is no longer limited by the necessity that the index of refraction is close to unity; however, it contains the implicit assumption that the incident field is the *polarizing field*, that's to say, the field that creates the dipolar oscillation. Reasonable in the case of a diluted material, this hypothesis is not true in a dense material where the field that exists in the environment of an atom, also called a *local field*, is the superposition of the applied field and of a *depolarizing field*, which corresponds to the action of all the other atoms.

The same problem is also met in Electrostatics, when it is desired to calculate the electric field inside a dielectric material. A good description may be found in a standard textbook, see, for example, *Solid State Physics* by

Charles Kittel. To evaluate the depolarizing field, we imagine that the atom is inside a small cavity dug inside the material.

In the case of an isotropic or a cubic material, the cavity can be a sphere. To obtain the local field E_{local} inside the cavity, a term equal to $P/3\varepsilon_0$ should be added to the external field E,

$$E_{local} = E + \frac{P}{3\varepsilon_0}. \tag{8.18}$$

We will do the calculation in the simpler case where there is only one electron per atom that contributes to the index of refraction,

$$P = \varepsilon_0 N \bar{\omega} E_{local},$$

$$P = \varepsilon_0 \chi E \quad \rightarrow \quad \chi = \frac{3N\bar{\omega}}{3 - N\bar{\omega}},$$

$$\varepsilon_r = \frac{\varepsilon}{\varepsilon_0} = n^2 = 1 + \chi \quad \rightarrow \quad \frac{n^2 - 1}{3} = \frac{N\bar{\omega}}{3 - N\bar{\omega}},$$

$$\frac{n^2 - 1}{n^2 + 2} = \frac{1}{3} N\bar{\omega}. \tag{8.19}$$

When more than one electron per atom contributes to the index of refraction, formula (8.19) is generalized as

$$\frac{n^2 - 1}{n^2 + 2} = \frac{1}{3} \sum_k N_k \bar{\omega}_k. \tag{8.20}$$

Formula (8.20) is called the *Lorentz-Lorenz* or also *Clausius-Mossoti formula*. Its demonstration explicitly admits a cubic environment of the atoms.

Gladstone's Law

For a gas, formula (8.20) becomes simpler and coincides with (8.7). If the gas is made of only one kind of molecule, the right-hand term is proportional to the number of molecules per unit volume, that's to say, to the density d. In this case the Lorentz-Lorenz formula is written as

$$\frac{(n - 1)}{d} = \text{constant.} \tag{8.20'}$$

Known as the *Gladstone law*, (8.20') shows that the index of refraction increases with pressure, which makes the material denser, and decreases with temperature. Previous remarks remain true in the case of a liquid or solid-state material. The index variation that can be produced by a reasonable variation of the temperature remains weak (up to three decimal places); the temperature effect is however a convenient way to finely tune the value of the refractive index.

The acoustooptic effect is an interesting consequence of the effect of pressure. The propagation of an acoustic wave inside (sonic, ultrasonic, or hypersonic) a material modulates the index of refraction at the same frequency. If a light beam and an acoustic wave, with wavelengths of the same order of magnitude, propagate simultaneously in a liquid or a solid, the light can be efficiently diffracted by the index grating formed by the acoustic wave.

8.3. The Index of Refraction Is a Complex Number

According to formulas (8.16) and (8.17) the dielectric constant and the index of refraction are complex numbers:

$$n = n' - jn'', \quad \varepsilon = \varepsilon_0 \varepsilon_r = \varepsilon' - j\varepsilon'' = \varepsilon_0(\varepsilon_r' - j\varepsilon_r''),$$
$$\varepsilon_r = n^2 \quad \rightarrow \quad \varepsilon_r' = n'^2 - n''^2 \quad \text{and} \quad \varepsilon_r'' = 2n'n''. \tag{8.21}$$

Negative signs have been introduced in the above formulas, because in most cases (sample at thermal equilibrium, for example) the imaginary parts of the index and of the dielectric constant are negative.

8.3.1. *Beer's Law*

The imaginary part of the refractive index corresponds to *attenuation*, or possibly *amplification*, of the wave during propagation. It is convenient to define a complex wave vector:

$$\boldsymbol{k} = \boldsymbol{k}' - j\boldsymbol{k}'' \quad \text{with} \quad |\boldsymbol{k}'| = n'\omega/c \quad \text{and} \quad |\boldsymbol{k}''| = n''\omega/c,$$
$$\boldsymbol{E} = \boldsymbol{E}_0 e^{j\omega(t-z/c)} = \boldsymbol{E}_0 e^{-n''\omega z/c} e^{j\omega(t-n'z/c)} = \boldsymbol{E}_0 e^{-k''z} e^{j(\omega t - k'z)}, \tag{8.22}$$

where $e^{j\omega(t-n'z/c)}$ is a complex number that keeps a constant modulus during the propagation. On the contrary $e^{-\omega n''z/c}$ is a real number that decreases if n'' is positive. The expression of the intensity is given by

$$I = I_0 e^{-2k''z} = I_0 e^{-\alpha z} \quad \text{with} \quad \alpha = 2k''. \tag{8.23}$$

The relationship (8.23) is called Beer's law of variation of the intensity during the propagation of a light beam in an absorbing material. The coefficient α is called the absorption coefficient; its variation $\alpha(\omega)$ versus frequency is the absorption spectrum of the material and is easily obtained from spectroscopic measurements.

Beer's law can also be written using the logarithm of the intensity

$$I_{db} = 10\log_{10}\left(\frac{I}{I_0}\right) = -(10\alpha\log_{10})z = -\alpha_{db}z,$$

$\alpha_{db} = 10\alpha\log_{10}(e) = 4.3\alpha$ is the "db-coefficient of absorption."

8.3.1.1. *Order of Magnitude*

- For a standard piece of glass, in the middle of the visible spectrum ($\lambda = 0.6\,\mu m$), the real part of the index is 1.5 and the absorption coefficient is of the order of some db/cm,

$$\alpha_{db} \approx 0.1\,\text{db}/\text{cm} = 10\,\text{db}/\text{m} \quad \rightarrow \quad \alpha = 10/4.3 = 2.3\,\text{m}^{-1} \quad \rightarrow \quad k'' = \alpha/2 = 1.15\,\text{m}^{-1},$$
$$n'' = k''\lambda/2\pi = 0.11.10^{-6} \quad \rightarrow \quad n = 1.5 - j0.11.10^{-6}.$$

- For silica, which is probably one of the most transparent materials in the near infrared ($1.55\,\mu m$), $\alpha_{db} \approx 0.15$ db/km $= 0.15.10^{-3}$ db/m.
- For a rather absorbing material $\alpha_{db} \approx$ some db/mm. An absorption of one decibel per wavelength is very high.

Remark

The imaginary part of the index comes from the phenomenological damping coefficient in the equation of the harmonic oscillator that describes the motion of an electron. As with any damping coefficient it corresponds to some energy dissipation: it's the reason why the wave is attenuated. Our model says nothing of what happens with the dissipated energy, the answer to this difficult question would need a more accurate description of the phenomena. The lost energy will often appear as heat, corresponding to erratic vibrations of the atoms, but it can also appear as stray light diffused around the propagation medium (see, for example, "The Reason Why the Sky Is Blue" in the Annex of Chapter 11).

8.3.2. *Index of Refraction of a Metal*

The phenomenological model of equation (8.1) corresponds to a dielectric material inside which the electrons are bound by a restoring force associated to the term $\omega_{atom}^2.\xi$. The case of a metal is different: beside bound electrons, as in the case of a dielectric, *free electrons* can move all over the sample. The phenomenological equation of the motion of a free electron is almost identical to (8.1), except that the restoring force is withdrawn,

$$m(\ddot{\xi} + \gamma_k \dot{\xi}) = qE_0 e^{j\omega t}.$$

For free electrons formula (8.17.a) is to be replaced by

$$n^2 = 1 + \frac{Nq^2}{\varepsilon_0 m} \frac{1}{(-\omega^2 + j\gamma\omega)}. \tag{8.24}$$

The index of refraction of a metal has in principle two components, however the contribution of the bound electrons is negligible as compared to the contribution of free electrons. Free electrons, by definition, can be found anywhere in the metal and they see an *average electric field*, and there is no necessity to make a local field correction.

Plasma Frequency

The dissipation of energy is due to collisions of the free electrons with the atoms. An important parameter in this case is the mean free path that is the average distance covered by an electron between two collisions; the collision time τ, which is the time necessary to cover this distance, is typically 10^{-13} s, about ten times the optical period ($\omega\tau \cong 10$). The conductivity σ of the metal is linked to the collision time and it can be established that the square of the index is given by

$$n^2 = 1 + \frac{\sigma/\varepsilon_0}{j\omega(1 + j\omega\tau)}. \tag{8.24'}$$

If $\omega\tau$ is considered to be much greater than one, (8.24') becomes

$$n^2 = 1 - \frac{\sigma}{\varepsilon_0\omega^2\tau} = 1 - \frac{\omega_p^2}{\omega^2} = 1 - \frac{\lambda^2}{\lambda_p^2}, \tag{8.25}$$

where ω_p and λ_p are, respectively, the plasma frequency and plasma wavelength of the metal; for most metals they belong to the ultraviolet or X-band domains. The optical frequencies are much smaller than the plasma frequencies, and so the square of the index of refraction is negative. As a first approximation the index of a metal is purely imaginary, which means that metals are always very absorbing for optical waves.

The modulus of the refractive index of metals has always had a high value; as a consequence, the reflection coefficient is also very high, which explains the typical sparkling appearance of nonoxidized metals. In reality, the indices of metals are complex numbers, with an imaginary part larger than the real part. At a frequency higher than the plasma frequency, in the ultraviolet or X-band, the index is real and smaller than unity.

8.3.3. *The Kramers-Krönig Formula*

When a light beam is incident on an interface separating a vacuum from some material, a reflected beam and a transmitted beam are generated. The

Table 8.1. Complex indices of some metals at $\lambda = 589.3\,\text{nm}$.

Sodium	$0.044 - j2.42$	Tin	$1.48 - j5.25$
Silver	$0.2 - j3.44$	Gold	$0.47 - j2.83$
Aluminum	$1.44 - j5.23$	Copper	$0.62 - j2.57$

material can be considered as a *system* that is submitted to an *excitation* by the incident beam and then gives a *response* that is made of the reflected and transmitted beams. As these two beams are a consequence of the forced vibrations of the dipoles of the material, the polarization of the material can also be considered as the answer to the system of the excitation of the incoming light. The existence and main properties on the optical index come from the fact that the excitation and response are linked by linear equations; these equations are, at first, the relation (8.13) between P and E, and second, the differential equation (8.1) that governs the motion of electrons.

Because of the linearity, a sine excitation produces a proportional sine response; the proportionality coefficient is a complex number that varies with the frequency. This is the reason why the permittivity, as well as the index of refraction, are dispersive complex quantities. We are now going to show that an important consequence of the principle of causality is the existence of two necessary relations between the frequency variation laws of the real and imaginary parts of the index of refraction. These relations were introduced independently by Krönig in 1926 and by Kramers in 1927; known as the Kramers-Krönig formulas, they establish relations between the dispersion laws of the real part, $\chi'(\omega)$, and the imaginary part, $\chi''(\omega)$, of the susceptibility.

The physical reason for the Kramers-Krönig formulas comes from the fact that, because of the causality principle, the response cannot happen before the excitation. In our case, the electrons cannot start to oscillate prior to the arrival of the incident beam. This mathematical demonstration is rejected in Annex 8.B; the result is given by formula (8.26):

$$\chi'(\omega) = \frac{2}{\pi} \int_0^{+\infty} \frac{\omega'\chi''(\omega')}{(\omega'^2 - \omega^2)} d\omega' \quad \text{and} \quad \chi''(\omega) = -\frac{2\omega}{\pi} \int_0^{+\infty} \frac{\chi'(\omega')}{(\omega'^2 - \omega^2)} d\omega'. \quad (8.26)$$

The examination of (8.26) immediately shows a mathematical difficulty since the denominators of the fractions to be integrated cancel if $\omega = \omega'$. A mathematical trick should be used considering the principal part of the integrals (see Annex 8.B). The interesting point is that, once the variation law $\chi'(\omega)$ has been measured, or calculated, $\chi''(\omega)$ can be obtained from (8.26), and vice versa. However, one must be aware that (8.26) is a *relationship between two functions* which has two consequences:

(i) the knowledge of the value of $\chi'(\omega_0)$ at a given frequency ω_0 doesn't allow any prediction about $\chi''(\omega_0)$;
(ii) to obtain $\chi''(\omega)$ from (8.26) requires, in principle, knowing the variations of $\chi'(\omega)$ over the whole frequency range, from zero to infinity.

8.3.4. *Analytical Expression of the Laws of Dispersion*

The index of refraction can be measured with high accuracy over all of the visible spectrum; as this knowledge is very important in many cases, various empirical formulas have been established to describe its variation versus frequency. These formulas are quite accurate and give the value of the refractive index up to five decimal places.

Cauchy's formula: This is valid for materials that have only ultraviolet absorption bands,

$$n = A + \frac{B}{\lambda^2} + \frac{C}{\lambda^4}.$$

Briot's formula: This is valid for materials with infrared and ultraviolet absorption bands,

$$n = -A'\lambda^2 + A + \frac{B}{\lambda^2} + \frac{C}{\lambda^4}.$$

The previous formulas are nothing other than series developments of the Lorentz-Lorenz formula. They concern materials that are transparent in the visible and are valid far from the absorption bands.

Normal dispersion: Inside a transparency band, the index of refraction is a decreasing function of the wavelength or, which is the same, an increasing function of the frequency. In Annex 8.B, this proposition will be established as a consequence of the Kramers-Krönig formulas. It's convenient to remember that the refractive index is higher in the blue than in the red.

$n_{blue} > n_{red}$:   a blue ray is more deviated by a prism than a red one.

A frequency domain inside of which the index increases with frequency is sometimes called a *zone of normal dispersion*.

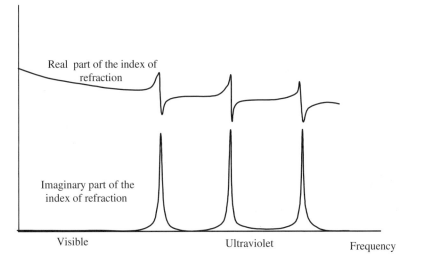

Figure 8.3. Dispersion and absorption curves of a material that is transparent in the visible and has three absorption bands in the ultraviolet.

Anomalous dispersion: When the frequency falls inside an absorption band, the index variation is no longer monotonic; one then speaks of *a zone of anomalous dispersion.*

8.4. Index of Refraction and Populations of the Energy Levels of a Transition

This section should be read in parallel with Section 9.2.3 where the absorption is related to the transitions made by an atom between the two energy levels associated to a dipolar transition (cf. formula (9.31)); the absorption coefficient α is related to the transition parameters by the following formula:

$$\alpha = \frac{c^2}{8\pi v^2 u} A f(v)(N_0 - N_1) = \frac{c^2}{8\pi v^2 u} \frac{1}{\tau_{\text{radiative}}} f(v)(N_0 - N_1). \qquad (8.27)$$

The expression of the imaginary part of the susceptibility, $\chi''(v)$, can be deduced from the absorption coefficient using the following expressions that come from equations (8.17.b) and (8.23):

$$n'' = \frac{\chi''}{2}; \quad \alpha = 2k'' = \chi'' \frac{\omega}{c},$$

$$\chi''(v) = \frac{c^3}{16\pi^2 v^3} n'^2 \frac{1}{\tau_{\text{radiative}}} f(v)(N_0 - N_1). \qquad (8.28)$$

It is left as an exercise to use the Kramers-Krönig formulas to deduce the expression of the real part, $\chi'(v)$, of the susceptibility from (8.28) in the case

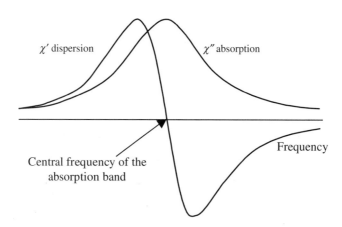

Figure 8.4. Representation of the variations χ' and χ'' versus frequency, in the case of a Lorentzian profile. For the central frequency, the absorption is maximum while χ' changes its sign.

of a Lorentzian expression of the function $f(v)$. It's easier to make the ratio of the two formulas (8.15.a) and (8.15.b):

$$\chi'(v) = \frac{2(\omega_{atom} - \omega)}{\lambda} \chi''(v),$$

$$\chi'(v) = \frac{c^3}{8\pi^2 v^3} n'^2 \frac{(\omega_{atom} - \omega)}{\gamma} f(v)(N_0 - N_1).$$

(8.29)

Annex 8.A

Electric Dipole Radiation

Any piece of material is in fact a collection of electric dipoles that can freely oscillate or that are forced to oscillate by an incoming electromagnetic wave. The purpose of this annex is to calculate the electromagnetic field radiated from an oscillating dipole.

8.A.1. Definition of an Oscillating Dipole

We refer to Figure 8.A.1 and consider a dipole made of two electric charges, $+q$ and $-q$, disposed at points Q_+ and Q_- separated by a distance d; by definition, the electric dipolar momentum $\boldsymbol{p}$ of the dipole is the product of the absolute value of the charges by the vector $\boldsymbol{a} = \boldsymbol{Q_- Q_+}$,

$$\boldsymbol{p} = q \boldsymbol{Q_+ Q_-}.$$

If $\boldsymbol{p}$ is a constant vector, the dipole is said to be static; an oscillating dipole corresponds to the case where the distance follows a sine time variation:

$$\boldsymbol{p} = \boldsymbol{p_0} \cos \omega t = \mathrm{Re}[\boldsymbol{p_0} e^{j\omega t}]. \tag{8.A.1}$$

8.A.2. Electric and Magnetic Fields Created by a Dipole

We will first successively consider the case of a static dipole and then of an oscillating dipole.

8.A.2.1. Static Dipole

The determination of the electric field created by a static dipole is a trivial electrostatic calculation: the field at point P in Figure 8.A.1 is parallel to the plane that contains point P and vector $\boldsymbol{p_0}$; using the unitary vectors $\boldsymbol{r_1}$ and $\boldsymbol{\theta_1}$, the expression of the field is

$$\boldsymbol{E} = \frac{p_0}{4\pi\varepsilon_0} \frac{1}{r^3} (2\boldsymbol{r_1} \cos\theta + \boldsymbol{\theta_1} \sin\theta_1). \tag{8.A.2}$$

Figure 8.A.1. Oscillating dipole. (r_1, θ_1, ψ_1) is a unitary direct tetrahedron; r_1 is parallel to a, θ_1 is in the plane (OP, a), ψ_1 is orthogonal to the two other vectors.

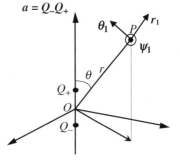

8.A.2.2. *Oscillating Dipole*

An oscillating dipole is, at the same time, a pair of charges and an electric current, since the charges are moving; and so an electric field and a magnetic field are generated. The calculation of this electromagnetic field is interesting, from a physical point of view; the Theory of Relativity is implicitly involved, since the effect of a change in the positions of the charges cannot be felt instantaneously at point P. This calculation is also trivial in Electromagnetics courses. The components of the field can be deduced from a scalar potential V and from a vector potential A,

$$V = \frac{q}{4\pi\varepsilon_0}\left(\frac{e^{j\omega(t-Q_+P/c)}}{Q_+P} - \frac{e^{j\omega(t-Q_-P/c)}}{Q_-P}\right) \approx \frac{qa}{2\varepsilon_0\lambda}\frac{1}{r}\cos\theta\left(\frac{\lambda}{2\pi r} + j\right)e^{j\omega(t-r/c)},$$

$$A = \frac{\mu_0}{4\pi r} j\omega qa e^{j\omega(t-OP/c)}(r_1\cos\theta - \theta_1\sin\theta). \tag{8.A.3}$$

$$E = \frac{\pi q a}{\varepsilon_0\lambda^2 r}e^{j\omega(t-r/c)}\left[2r_1\left(\frac{\lambda^2}{16\pi^2 r^2} + j\frac{\lambda}{2\pi r}\right)\cos\theta + \theta_1\left(\frac{\lambda^2}{16\pi^2 r^2} + j\frac{\lambda}{2\pi r} - 1\right)\sin\theta\right], \tag{8.A.4.a}$$

$$H = \frac{1}{\mu_0}\nabla \wedge A = -\psi_1\sin\theta\frac{\pi c q a}{r\lambda^2}e^{j\omega(t-r/c)}\left(j\frac{\lambda}{2\pi r} - 1\right). \tag{8.A.4.b}$$

In general, the spatial field distributions are quite complicated, they become simpler if:

- $a \ll \lambda = 2\pi c/\omega$, the distance between the charges is small as compared to the wavelength.
- The point P is far from the dipole.

If the frequency goes to zero (or if the wavelength goes to infinity), formula (8.A.4.a) becomes identical to (8.A.2). At a long distance from the dipole and keeping only the $1/r$ terms, the following formulas are obtained:

$$E = -\frac{\pi q a}{\varepsilon_0\lambda^2}\frac{e^{j\omega(t-r/c)}}{r}\sin\theta\theta_1, \tag{8.A.5}$$

$$H = -\frac{\pi c q a}{\lambda^2}\frac{e^{j\omega(t-r/c)}}{r}\sin\theta\psi_1. \tag{8.A.6}$$

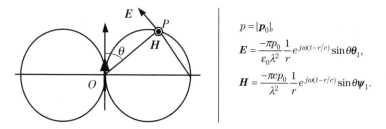

Figure 8.A.2. Diagram of emission of a dipole. The field keeps a constant modulus along a circle that is tangent to the dipole at point O. No emission in the direction of the dipole.

The expression $e^{j\omega(t-r/c)}/r$ in (8.A.5) and (8.A.6) corresponds to the propagation of a wave. (E, H, OP) is a direct trihedron. The ratio of the modulus of the electric and magnetic fields remains constant and is equal to the wave impedance

$$\frac{|E|}{|H|} = \frac{1}{\varepsilon_0 c} = \mu_0 c = \sqrt{\frac{\mu_0}{\varepsilon_0}} = 377\,\text{ohms}. \tag{8.A.7}$$

The $1/r$ law, which must be associated to the energy conservation, corresponds to a *slow diminution* of the amplitude of the wave as a function of the distance of propagation; the rate of decrease would be much faster if it were an exponential law. This is the explanation of the enormous range of the waves that are emitted by an oscillating dipole. The radio wave coming from a modest emitter (a few watts) can still be detected after having covered several thousands of kilometers; light rays coming from the very edge of the universe are seen with the naked eye. . . .

Polarization and Angular Repartition of the Radiation of a Dipole

The light radiated by a dipole is linearly polarized; at a given point P, the electric field is parallel to the plane, that is defined by P and the electric momentum p of the dipole, and is perpendicular to the line joining P to the center of the dipole.

Despite the presence of $e^{j\omega(t-r/c)}/r$ in the expression for the fields, even at a great distance, the radiation of a dipole is quite different from a spherical wave. The indicatrix of emission doesn't have any spherical symmetry, since the amplitude cancels in the direction of the dipole. From a mathematical point of view, the cancellation comes from the presence of $\sin\theta$ in formulas (8.A.5) and (8.A.6); the physical interpretation comes from the fact that an electromagnetic wave must be transversal and that, because of the cylindrical symmetry of the problem, the fields cannot have components that would be orthogonal to the dipole.

8.A.3. Power Radiated by an Oscillating Dipole

We will give two methods for calculating the power that is radiated by an oscillating dipole. The first method is fully electromagnetic and relies on the previous expressions of the fields and on the Poynting theorem. The second will use results of the Theory of Relativity: it's because of the acceleration of the moving charges that energy is radiated by the dipole.

8.A.3.1. *Calculation Using the Poynting Theorem*

The power that is radiated through a given surface Σ is equal to the flux of the Poynting vector Π through the surface,

$$\Pi = \tfrac{1}{2}\,\mathrm{Re}[\boldsymbol{E} \wedge \boldsymbol{H}^*].$$

In the above formula, the asterisk is for complex conjugation; its presence is the time to disappear ($e^{j\omega t}e^{-j\omega t}$) in the vector product; from a physical point of view this means that the power is averaged over many periods,

$$P = \iint_{\Sigma} \Pi\, d\boldsymbol{S} = \frac{\pi^2 c p_0^2}{2\varepsilon_0 \lambda^4} \iint_{\Sigma} \frac{\sin^2\theta}{r^2} r^2 \sin\theta\, d\theta\, d\varphi.$$

To evaluate the integral, Σ is a sphere with the dipole at the center,

$$P = \frac{\pi^2 c p_0^2}{2\varepsilon_0 \lambda^4} \int_0^{2\pi} d\varphi \int_0^{\pi} \sin^3\theta\, d\theta = \frac{p_0^2 \omega^4}{12\pi\varepsilon_0 c^3} = \text{constant} \quad (\forall\ R). \qquad (8.A.8)$$

- P is independent of the radius of the sphere, which ensures the conservation of energy.
- P varies as the fourth power of the frequency or, similarly, as the fourth power of the inverse of the wavelength. For a given dipolar momentum, the radiated power is $2^4 = 16$ times larger in the blue ($0.4\,\mu$m) than in the red ($0.8\,\mu$m).

8.A.3.2. *Calculation Using Relativity Considerations*

Richard Feynman gave an elegant and didactic presentation in the chapter entitled "Electromagnetic Radiation" of his book *Lectures on Physics*.

Formulas (8.A.9) give the expressions of the electric and magnetic fields that are created by a charge q moving along some trajectory Γ,

$$\boldsymbol{E} = \frac{-q}{4\pi\varepsilon_0}\left(\frac{\boldsymbol{e}_{r'}}{r'^2} + \frac{r'}{c}\frac{d}{dt}(\boldsymbol{e}_{r'}) + \frac{1}{c^2}\frac{d^2}{dt^2}(\boldsymbol{e}_{r'})\right), \qquad (8.A.9.a)$$

$$\boldsymbol{H} = \frac{-1}{\mu_0 c}\boldsymbol{e}_{r'} \wedge \boldsymbol{E}. \qquad (8.A.9.b)$$

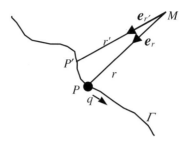

Figure 8.A.3. An electric charge q is moving along the trajectory Γ; e_r and $e_{r'}$ are the respective unit vectors of MP and MP'.

The two first terms of (8.A.9.a) are the relativistic expression of the Coulomb law for the electric field created by a charge q. *The last term is associated to the electromagnetic field that is radiated by the oscillating dipole*; to obtain its expression, we need to know the acceleration of the extremity of the unit vector $e_{r'}$. The electromagnetic field that is radiated, at a given point P and at a given time t, may be expressed as a function of the acceleration of the charge at time $(t - r/c)$.

The following important results are obtained:

A charge with a uniform rectilinear motion doesn't radiate any field.
A charge radiates electromagnetic energy, only if it is accelerated.
The total radiated power is proportional to the square of the acceleration.

Consequences

- In the case of harmonic oscillations, the acceleration is proportional to the square of the angular frequency ω, and thus the radiated power is proportional to ω^4, which agrees with formula (8.A.8).
- *Synchrotron radiation*: A charged particle orbiting along the trajectory of some accelerator (a synchrotron, for example) is permanently submitted to a centripetal acceleration and radiates an electromagnetic field, which is called synchrotron radiation. According to the working conditions of the accelerator, the radiated spectrum covers the whole electromagnetic domain, including visible, ultraviolet, and X-rays.

8.A.3.3. *Lifetime of the Excited Level of an Electric Dipole Transition*

We consider an atom that is isolated in a vacuum; an electron of this atom and the nucleus makes an electric dipole, their distance $\xi(t)$ follows the equation of a harmonic oscillator:

$$\frac{d^2\xi}{dt^2} + 2\alpha\frac{d\xi}{dt} + \omega_{\text{atom}}^2\xi = 0.$$

After having been taken away from its equilibrium position, the electron will do free oscillations, let $W(t)$ and W_0, respectively, be the values of the "mechanical energy" of the oscillator and consider that the damping is weak ($\omega_{\text{atom}} \gg \alpha$):

$$\xi(t) = \xi_0 e^{-\alpha t}\cos\omega_{\text{atom}}t,$$
$$W(t) = W_0 e^{-2\alpha t} = W_0 e^{-t/\tau_{\text{radiative}}}, \quad \text{with} \quad \tau_{\text{radiative}} = 1/2\alpha.$$

The energy $W(t)$ that is stored by the oscillator decreases as time passes; the instant power that is dissipated is given by

$$P(t) = -\frac{dW}{dt} = 2\alpha W(t),$$
$$2\alpha = \frac{1}{\tau_{\text{radiative}}} = \frac{P(t)}{W(t)},$$

where τ_{rad} is called *the radiative lifetime* of the oscillator. To obtain an expression of this time constant, we will admit that the dissipation of energy corresponds to the radiated power that has already been calculated and is given by formula (8.A.8):

$$\frac{1}{\tau_{\text{radiative}}} = \frac{1}{W}\frac{p_0^2\omega_{\text{atom}}^4}{12\pi\varepsilon_0 c^3}. \tag{8.A.10.a}$$

To obtain a value of τ_{rad} we need to know W; for no good reason, except that the result of the calculation is correct, we will replace W by $h\nu$:

$$\frac{1}{\tau_{\text{radiative}}} = \frac{p_0^2}{6h\varepsilon c^3}\omega_{\text{atom}}^3. \tag{8.A.10.b}$$

Our demonstration is not really satisfactory, since we treat in a classical way the spontaneous emission (see Chapter 9), which is typically a quantum process. Despite its impure origin, formula (8.A.10.b) is valid and verified by the experiment; a more rigorous demonstration can be given on a quantum basis. We have chosen this kind of presentation because it insists on the physical meaning of electric dipole radiation.

Order of Magnitude of the Radiative Lifetime

To do numerical applications of (8.A.10.b), we need an order of magnitude of p_0, we will make the product of the elementary charge $e = 1.6 \times 10^{-19}$ C by an order of magnitude of the diameter of an atom (1 Å $= 10^{-10}$ m) $\rightarrow$ $p_0 \approx 10^{-29}$ Cb.m. The time constant $\tau_{radiative}$ is also called the *relaxation time* of the transition.

Wavelength	1 m	1 cm	100 μm	10 μm	1 μm	0.1 μm	1 Å
Frequency	300 MHz	30 GHz	$3 \times$ 10^{12} Hz	$3 \times$ 10^{13} Hz	$3 \times$ 10^{14} Hz	$3 \times$ 10^{15} Hz	$3 \times$ 10^{18} Hz
$\tau_{radiative}$ Lifetime	10^{12} s millennium	10^6 s one day	1 s	10^{-3} s	10^{-6} s μs	10^{-9} s ns	10^{-18} s

Orders of Magnitude of the Relaxation Times of Fully Allowed Transitions, $p_0 \approx 10^{-29}$ Cb.m

The above table gives some orders of magnitude of the relaxation times. Very long (a few days or more) relaxation times are not really meaningful: they just indicate that, in such cases, the system will find other relaxation processes (collisions with other atoms or with the walls of the tank, in the case of a gas; collision with phonons in a crystal).

As far as optical transitions are concerned the relaxation times are usually in the nanosecond-microsecond range. Radiative lifetimes can be measured; experimental values, $\tau_{measured}$, are always larger than the theoretical value given by (8.A.10.b). The value f of the ratio of the calculated to the measured lifetime is smaller than one:

$$0 \le f = \frac{\tau_{radiative}}{\tau_{measured}} \le 1.$$

It can be shown that the parameter f is identical to the *oscillator strength* of formulas (8.12).

- If $f = 0$, the transition is said to be *forbidden*, the lifetime is then infinite.
- If $f = 1$, the transition is said to be *fully allowed*, the lifetime is equal to the calculated value.
- If $0 < f < 1$, the transition is said to be *partly allowed* (or *partly forbidden*).

The lifetime was calculated in the case of an isolated dipole that was radiating in the open space. Inside an atom, an electron belongs to the electron cloud, which constitutes a type of electrical screening. We can give the following analogy of a dipole that is inside a closed box. At first the walls of the box are perfect conductors, the radiation that is produced by the dipole is reflected by the walls and stored: the lifetime is infinite, but an external observer is not aware that there is an oscillating dipole inside. Let us suppose now that small holes are drilled through the walls; only a part of the radiation will be diffracted outside while the remaining part is sent back to the dipole.

A transition associated to an electron belonging to an outer shell is fully allowed, this is the case of the emission of the well-known D-light by sodium atoms (there is only one peripheral electron); the lifetime of the excited level is of a few nanoseconds.

A transition associated to an electron belonging to an inner shell is only partly allowed, such cases are often met with in rare earth atoms and transition atoms of the Mendeleev chart. The lifetimes are then measured in milliseconds ($f \approx 10^{-3}$), the excited level is then said to be *metastable* (in ancient Greek, *meta* = almost). A famous example is given by chromium ions Cr^{3+}, which have been selected to obtain the first laser emission ($0.6943 \mu m$) because of the existence of a metastable level, with a lifetime of three milliseconds.

Absorption and Stimulated Emission Are the Same Kind of Phenomena

Absorption and stimulated emission will be introduced on a quantum basis in Section 9.2, however, contrary to spontaneous emission, they can be classically interpreted. We consider a dipole oscillating at some frequency ω and receiving an electromagnetic wave of the same frequency. The dipole interacts with the electric alternating field of the wave, the difference between the phase of the field and the phase of the motion of the dipole plays a key role; it is interesting to understand what happens when $\varphi = \pi$ or when $\varphi = 0$. When the motion and the field variation are in phase, the action of the field increases the amplitude of the dipole vibrations, which in turn gains more energy; on the contrary, if φ is equal to π, the dipole loses energy.

Let us now see what happens on the field side: stimulated emission corresponds to the case of phase concordance where the two fields are mutually reinforced by positive interference; absorption corresponds to the case of opposite phases and of attenuation by destructive interference.

8.A.4. Field Radiated by a Planar Distribution of Oscillating Dipoles

Richard Feynman in his *Lectures on Physics* proposed the following calculation. We refer to Figure 8.A.4, electrical charges are spread on a plane with a surface density η that keeps a constant density η_0 inside a circle of radius R_0, and then rapidly decreases outside the circle. The distribution admits the symmetry of revolution around the axis of the circle. All the charges are vibrating in phase, with the same frequency ω and the same amplitude x_0, the direction of vibration is orthogonal to the plane. We want to calculate the electromagnetic field that is radiated at a point P located on the axis of the circle at a distance z that is far larger than R_0.

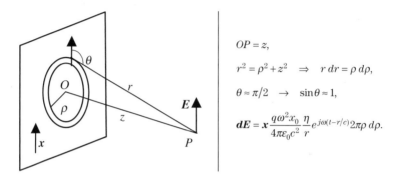

$OP = z,$

$r^2 = \rho^2 + z^2 \quad \Rightarrow \quad r\,dr = \rho\,d\rho,$

$\theta \approx \pi/2 \quad \rightarrow \quad \sin\theta \approx 1,$

$d\boldsymbol{E} = \boldsymbol{x}\dfrac{q\omega^2 x_0}{4\pi\varepsilon_0 c^2}\dfrac{\eta}{r}e^{j\omega(t-r/c)}2\pi\rho\,d\rho.$

Figure 8.A.4. The plane supports a distribution of oscillating dipoles; the vibrations are orthogonal to the plane. The elementary electric fields generated by the different dipoles are parallel to the unit vector $\boldsymbol{x}$.

The electric field created by the dipoles is given by the following integral:

$$E = \boldsymbol{x}\frac{q\omega^2 x_0}{2\varepsilon_0 c^2}e^{j\omega t}\int_z^\infty \eta e^{-j\omega r/c}dr.$$

The presence of the term $e^{-j\infty}$ is very unpleasant, this is the reason why we have chosen that η goes to zero when the distance to the center of the circle becomes infinite. The field is then given by

$$\boldsymbol{E} = -\boldsymbol{x}j\omega\frac{\eta_0 q x_0}{2\varepsilon_0 c}e^{j\omega(t-z/c)}. \tag{8.A.11}$$

Annex 8.B

The Kramers-Krönig Formula

8.B.1. Demonstration of the Kramers-Krönig Formula

The demonstration of this formula requires a rather sophisticated knowledge of mathematics and, more specifically, about Riemann and Cauchy integrals. A presentation can be found in the book *Solid State Physics* by C. Kittel, here we will just emphasize the physical articulations of the demonstration. By the way, we would like to underline the specific beauty of the argument which put the Principle of Causality on a mathematical basis and which also establishes a well-known experimental result, according to which blue light is more deviated than red light by a prism.

Let P be the polarization taken by a piece of material when submitted to an electric field E. If the field varies with time, so does the polarization. A sine variation of the field will create a sine variation of the polarization. The complex vector $\overline{P}$ is proportional to $\overline{E}$: the field plays the role of an excitation, the polarization being the response,

$$
\overline{E}e^{-j\omega t} \to \overline{P}e^{-j\omega t},
$$
$$
\overline{P}e^{j\omega t} = \chi(\omega)\overline{E}e^{j\omega t} = [\chi'(\omega) - j\chi''(\omega)]\overline{E}e^{j\omega t}.
$$

(8.B.1)

If the excitation is no longer sinusoidal, but follows some other law of variation versus time, the situation is more complicated and the Fourier transform method should be used. We will now consider the case where the excitation is a Dirac pulse,

$$
E(t) = E_0\delta(t) = \frac{E_0}{2\pi} \int\limits_{-\infty}^{+\infty} e^{j\omega t}\, d\omega,
$$
$$
P(t) = \int\limits_{-\infty}^{+\infty} P(\omega)e^{j\omega t}\, d\omega = \varepsilon_0 E_0 \int\limits_{-\infty}^{+\infty} \chi(\omega)e^{j\omega t}\, d\omega.
$$

(8.B.2)

Formula (8.B.2) is a well-known result; the harmonic susceptibility $\chi(\omega)$ (ratio response/excitation in the harmonic case), is the inverse Fourier transform of the time response to a Dirac pulse,

$$\chi(\omega) = \frac{1}{\varepsilon_0} \int\limits_{-\infty}^{+\infty} P(t) e^{-j\omega t} \, dt = \frac{1}{\varepsilon_0} \int\limits_{0}^{+\infty} P(t) e^{-j\omega t} \, dt. \tag{8.B.3}$$

Since the response cannot exist prior to excitation, $P(t)$ is identically equal to zero when $t < 0$. The domain of integration of the integral of the Fourier transform should be limited to $[0, \infty]$.

It can be established mathematically that formula (8.B.3) implies that the real and imaginary parts of the harmonic susceptibility are *Hilbert transforms* of one another. Under such conditions the function $\chi(\omega)$ has the following properties:

- The poles (values of the frequency that make the function infinite) are all above the real axis.
- $\chi(\omega) \to 0$ when $\omega \to \infty$.
- $\chi'(\omega)$ is an even function and $\chi''(\omega)$ is an odd function of the real variable ω.
-
$$\left.\begin{array}{l} \chi'(\omega) = \dfrac{2}{\pi} PP \int\limits_{0}^{+\infty} \dfrac{\omega' \chi''(\omega')}{(\omega'^2 - \omega^2)} \, d\omega', \\[4mm] \chi''(\omega) = -\dfrac{2\omega}{\pi} PP \int\limits_{0}^{+\infty} \dfrac{\chi'(\omega')}{(\omega'^2 - \omega^2)} \, d\omega', \end{array}\right] \quad \text{the Kramers-Krönig formula.}$$

The integrals raise some problems, since $\omega = \omega'$ is a pole; this is the reason why the Principal Parts (PP) have been introduced in the Kramers-Krönig formulas.

It's a good exercise to examine how the Kramers-Krönig formulas can be applied to formulas (8.15.a) and (8.15.b). As an example, consider the case of an atom with only one absorption band and derive the expression of $\chi''(\omega)$ from the expression of $\chi'(\omega)$; in other words, start from

$$\chi'(\omega) = \frac{Nq^2}{2\varepsilon_0 m \omega_0} \frac{(\omega_0 - \omega)}{\left[(\omega_0 - \omega)^2 + \gamma_k^2/4\right]},$$

and obtain

$$\chi''(\omega) = \frac{Nq^2 v}{2\varepsilon_0 m \omega_0} \frac{\gamma}{\left[(\omega_0 - \omega)^2 + \gamma^2/4\right]}.$$

8.B.2. Normal Dispersion

We consider a piece of material that is transparent inside a window limited by the two frequencies ω_1 and ω_2, the absorption being equal to zero at any frequency between ω_1 and ω_2,

$$\omega_1 < \omega < \omega_2 \quad \rightarrow \quad \chi''(\omega) = 0 \quad \rightarrow \quad \int_{\omega_1}^{\omega_2} \frac{\omega'\chi''(\omega')}{(\omega'^2 - \omega^2)} d\omega' = 0.$$

Anybody able to juggle with principal parts of integrals will easily establish that

$$\chi'(\omega) = \frac{2}{\pi} \int_0^{\omega_1} \frac{\omega'\chi''(\omega')}{(\omega'^2 - \omega^2)} d\omega' + \frac{2}{\pi} \int_{\omega_2}^{\infty} \frac{\omega'\chi''(\omega')}{(\omega'^2 - \omega^2)} d\omega',$$

$$n = n' = \sqrt{\varepsilon_r'} = \sqrt{\chi'} \quad \rightarrow \quad \frac{dn}{d\omega} = \frac{1}{2\sqrt{\chi'}} \frac{d\chi'}{d\omega},$$

$$\frac{d\chi'}{d\omega} = \frac{2}{\pi} \int_0^{\omega_1} \frac{2\omega\omega'\chi''(\omega')}{(\omega'^2 - \omega^2)^2} d\omega' + \frac{2}{\pi} \int_{\omega_2}^{\infty} \frac{2\omega\omega'\chi''(\omega')}{(\omega'^2 - \omega^2)^2} d\omega'. \tag{8.B.4}$$

The arguments of the integrals in (8.B.4) being positive, the derivatives $d\chi'/d\omega$ and $dn/d\omega$ are also positive and, inside a transparency band, the index of refraction increases with the frequency.

9

Lasers

9.1. Laser, a Feedback Oscillator

9.1.1. *Light Amplification by Stimulated Emission of Radiation*

A laser is a device which, receiving energy from a pumping source, transfers this energy to a coherent light beam.

In the above definition the word light should not be taken literally, as the wavelength of lasers very often falls outside the famous octave (0.8–$0.4\,\mu$m) inside of which the human eye can see. Lasers can emit within the deep infrared ($300\,\mu$m) as well as within the visible or the ultraviolet ($0.1\,\mu$m). There is no theoretical limitation for obtaining stimulated emission with X-rays or even Γ-rays. The family of presently existing lasers is extremely diversified and affects various branches of Physics and Chemistry. The lasers that are most often encountered are certainly semiconductor lasers, they easily emit in the near infrared (0.8–$1.6\,\mu$m) and their efficiency is exceptionally high (70%). Being produced in large quantities they are very cheap, all the more as their technology is very similar to microelectronics.

Lasers, a Brilliant Result of Basic Research

It was at the end of the 1950s that the American physicist Thomas H. Maiman obtained a burst of coherent light from of a ruby rod that had been placed along the axis of a helical flash tube, initially designed for professional film production. For many scientific or technical inventions, such as steam engines or superconductivity or even most electromagnetic effects, experimentalists or engineers had discovered and developed the phenomena before the approach of the corresponding basic research.

Chapter 9 has been reviewed by Dr. Lionel Bastard from the Institut National Polytechnique, Grenoble.

Lasers, as well as transistors ten years earlier, are however the result of the imagination of physicists (theoreticians as well as experimentalists). They have unambiguously issued from fundamental research work, with no expectation of immediate applications.

Although there were early prospects for useful applications, it has taken more than twenty years before the lasers could find a market. This long delay is probably due to the fact engineers were suddenly provided with light beams that had such new and original properties that it took some time for the new possibilities to be appreciated; at the same time, sophisticated technologies had to be invented to take full advantage of the fascinating properties of laser beams. During the two decades 1960–1980, the approach of scientists and engineers working in Optics completely changed; they had first to learn how to manage coherent and powerful light beams; they also had to cope with the fast-growing markets of Electronics and Telecommunications. Another difficulty was associated with the fact that lasers are a multidisciplinary domain, connected to many different scientific fields, among which we would like to note: Optics, Spectroscopy, Electromagnetism, Solid State and Plasma Physics, Chemistry. . . . In the same way, lasers rely on a great variety of technologies: Optics, Electronics, Precision Mechanics, Epitaxy and Thin Films Deposition, Crystal Growth, Control of Electrical Discharges. . . .

Lasers Could Have Been Invented Forty Years Earlier

The basic phenomenon for laser emission, stimulated emission of radiation by excited atoms, was introduced by Albert Einstein as early as 1917, from thermodynamical speculation on the interaction of a collection of atoms with blackbody radiation. The two main elements (synthetic ruby and a Fabry-Perot resonator) of Maiman's experiment did exist at that time; flash tubes didn't yet exist, but a flame of burning magnesium powder could have replaced them. The paradox is not that big, if we are aware that the notion of *self-oscillations* had not been discovered, as well as the notions of signal amplification and of feedback. Historically, lasers are the transposition of electrical feedback oscillators. To make a laser the following two elements are required: an amplifier working at optical frequencies and a feedback circuit.

The idea of taking advantage of stimulated emission in a material containing a population inversion for amplifying microwave or optical signals was independently and simultaneously proposed by the Americans Weber and Townes and by the Russians Basov and Prokhorov in 1953. At that time, the analogy between a tuned electrical circuit, a microwave resonator, and a Fabry-Perot resonator, had been known for a long time; the brilliant idea was just to put a piece of material with a population inversion between the two mirrors of a Fabry-Perot resonator.

9.1.2. *Laser Emission and Principle of Correspondence*

In Quantum Mechanics, the description of the state of a given system needs a set of integers that are specific to the system and called its "quantum numbers." The discovery of the quantum numbers of a system is not always straightforward; a possible solution is to start from a "classical" description and to make use of ad hoc postulates. A test of validity of the quantum numbers is to see what happens if they are given larger and larger values: the quantum description should then give results much more similar to the classical one.

The quantum description of a light beam introduces the number n of photons. A rigorous introduction of this number is of course possible. In the case of a monochromatic (frequency ν) parallel beam, the value of the number n' of photons crossing a perpendicular cross section of the beam each second, is readily obtained by dividing the power P by the individual energy of each photon $n' = P/h\nu$. If we consider that photons travel at speed c, it can be deduced that there are $n = n'/c$ photons per unit volume.

Let us try a numerical application for the case of a 1 mW laser emitting a parallel beam having a wavelength $\lambda = 0.5\,\mu m$ and a cross section of $1\,mm^2$, the result is $n \approx 8 \times 10^{10}$ photons/m^3, which is enormous: although a quantum treatment should be used, in most cases a classical method (i.e., Maxwell's equations) is quite suitable to describe a laser beam.

An X-Ray Laser Would Be Far More Dangerous

A laser such as the previous one is not at all powerful and is not difficult to build; so stimulated emission allows an easy creation of a great number of identical photons. It is possible to make far more powerful lasers, the emitted power can be as high as several terawatts (10^{12}), megawatt powers are quite common; of course this kind of laser only emits short light pulses with a duration ranging from nanoseconds (10^{-9}) to a few tens of femtoseconds (10^{-12}). One gigawatt during one nanosecond makes energy of only one joule; this energy is negligible if compared to the kinetic energy of a bullet (10 g at the speed of sound makes approximately 1 kJ). So, at the moment, laser guns mostly exist in science fiction novels; although laser weapons for blinding the enemy are in service. These rather optimistic considerations should be reconsidered, if it became possible to generate comparable densities of X or Γ photons.

Enormous laser equipments do exist for military applications and they can deliver nanosecond and megajoule pulses; if focused with lenses the corresponding beams provide, inside a volume of a hundred cubic micrometers, an energy density of the same order of magnitude as that which is obtained in a thermonuclear explosion.

9.1.3. *Feedback Oscillators*

If we exclude electromechanical devices, such as alternators or microphones, most AC electrical generators are feedback oscillators which are also called positive feedback oscillators and self-oscillating generators. Lasers belong to this type.

Among the signal processing devices, feedback oscillators are distinct in that they deliver an output signal in the absence of any input signal. To introduce the main properties of positive feedback we will consider radio oscillators using electronic components (resistors, capacitors, coils, and transistors), explanations are simpler because the frequency is low enough for the circuit size to be small as compared to the wavelength of the associated electromagnetic waves ($\lambda \approx$ meter to kilometers). Under such conditions the differences in phase between currents and voltages are well localized and solely due to capacitors and coils, it has nothing to do with wave propagation. The situation is of course very different in Optics where the wavelength is measured in micrometers.

From a very general mathematical point of view, a signal-processing device is described by some differential equation between an input signal $e(t)$, an output signal $s(t)$, and all their time derivatives. The easiest equations to be solved are linear equations; this is probably one reason why most devices are, at first sight, considered to be linear. It is then well known that the output signal of a linear system is the superposition of a free regime and of a forced regime. When conditions for self-oscillations are fulfilled, the two following properties are met, they are associated with deep mathematical difficulties:

(i) the response to some sine input signal of a well-chosen frequency is theoretically infinite; and
(ii) the amplitude of the free regime goes to infinity with time.

We will not attempt to give a general theory and will only develop two examples, one on the free regime and the other on the forced regime. We refer to the circuit of Figure 9.1(a), from an input signal $e = \hat{e}\exp(j\omega t)$ the amplifier produces an output signal $s = A\hat{e}\exp(j\omega t)$, a fraction β of the output signal is mixed with the input signal; if a signal $e_0 = \hat{e}_0\exp(j\omega t)$ is applied to the second port of the mixer we finally have the following equations:

$$e_0 = \hat{e}_0 e^{j\omega t}, \quad e = e_0 + \beta s, \quad s = Ae, \tag{9.1}$$

$$s = \hat{e}_0 \frac{A}{(1 - \beta A)} e^{j\omega t} = G\hat{e}_0 e^{j\omega t}, \quad G = \frac{A}{(1 - \beta A)}, \tag{9.2}$$

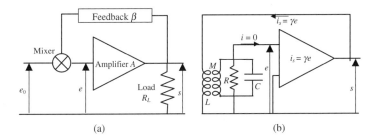

Figure 9.1. Organization of a feedback oscillator: (a) shows the principle, (b) is more realistic. The input impedance of the amplifier is infinite and the input current is equal to zero, the output generator is a current source with an electromotive current $I_s = \gamma e$ proportional to the input voltage e.

where A and β are complex numbers which depend on the angular frequency ω, according to a law that is determined by the arrangement of resistors, capacitances, and chokes, they can always be written as

$$A = A_{(\omega)}e^{j\varphi(\omega)} \quad \text{and} \quad \beta = \beta_{(\omega)}e^{j\psi(\omega)}. \tag{9.3}$$

Formula (9.2) defines the gain $G = A/(1 - \beta A)$ of an amplifier with feedback. It's well known in electronics that, playing with the feedback circuit, it is possible to change at will the value of the gain and the shape of the response curve of an amplifier. Very often a negative feedback is used to broaden the bandwidth, the price to pay being a diminution of the gain.

The values of the frequency that cancel the denominator of the gain in formula (9.2) are of course of special interest, they are obtained from the very important formula,

$$A\beta = 1. \tag{9.4}$$

Let $\omega = \omega_{osc}$ be a solution of equation (9.4), for values of the angular frequency equal to or very close to ω_{osc}, the circuit is said to be *unstable*. From an experimental point of view an output signal is obtained, even if no input signal is applied: an AC current (frequency $\approx \omega_{osc}$) exists in the load resistor R_L, the system autooscillates.

Equation (9.4) involves complex numbers and so implies two real equations:

$$A_{(\omega)}\beta_{(\omega)} = 1, \tag{9.5}$$

$$\varphi_{(\omega)} + \psi_{(\omega)} = 0 \quad \text{modulo } 2\pi. \tag{9.6}$$

Equation (9.6) shows that, to obtain autooscillation, it is necessary that the reinjected feedback signal should be in phase with the original signal; this condition is only the condition of constructive interference in Optics. The frequency, or the frequencies, of possible autooscillation are the solutions of equation (9.6).

Some initial signal e_0 will give, after amplification and feedback, a signal $A\beta e_0$, successive signals $(A\beta)^2 e_0$, $(A\beta)^3 e_0$, ... are then generated. Finally, the input and output signals will be given by

$$e = e_0 + (A\beta)e_0 + (A\beta)^2 e_0 + (A\beta)^3 e_0 + \cdots = \frac{e_0}{(1 - A\beta)}, \tag{9.7.a}$$

$$s = A\left[e_0 + (A\beta)e_0 + (A\beta)^2 e_0 + (A\beta)^3 e_0 + \cdots \right] = \frac{Ae_0}{(1 - A\beta)}. \tag{9.7.b}$$

This is another demonstration of formula (9.2), it has the interest of showing that (9.5) is in fact a *threshold condition*, since the sum of the series diverges as soon as

$$|A\beta| \geq 1. \tag{9.8}$$

When $|A\beta| = 1$ and $e_0 = 0$, s takes the indeterminate form 0/0 and the question arises as to what are the physical mechanisms which remove the indetermination and finally determine the amplitude of the oscillations produced by a self-oscillating device.

Analysis of an Autooscillator Using a Tuned RLC Circuit

We now refer to Figure 9.1(b), the feedback circuit contains an RLC resonant parallel circuit; it can be considered that the Fabry-Perot resonator of a laser has many similarities with such a circuit. The voltage e across the RLC circuit is applied at the input of an amplifier which delivers an output current $I_s = \gamma e$ proportional to e; this current goes through the primary coil of a transformer of mutual inductance M, the secondary coil is the coil of a resonant circuit. γ is considered to be independent of the frequency; it can be shown that $e(t)$ obeys the following differential equation:

$$\frac{d^2 e}{dt^2} + \frac{(1 - RM\gamma/L)}{RC} \frac{de}{dt} + \frac{1}{LC} e = 0. \tag{9.9}$$

Setting

$$\frac{1}{\tau} = \frac{(1 - RM\gamma/L)}{RC}, \quad LC\omega_0^2 = 1, \quad \text{and} \quad \omega = \sqrt{\left(1 - \frac{1}{4\omega_0^2 \tau^2} \right)},$$

we obtain

$$e(t) = \hat{e}_0 e^{-t/\tau} e^{j\omega t}. \tag{9.10}$$

It is possible to choose large enough values for M and γ so that $(1 - RM\gamma/L)$ is negative, in such conditions *the time constant τ is also negative* and the amplitude of the voltage oscillations across the resonant circuit increases exponentially versus time. According to equation (9.10) these oscillations increase indefinitely and should become infinite; for reasons that will be analyzed later this is of course not the case.

In order to understand the physical processes involved in the production of an oscillation by the circuit in Figure 9.1(b) let us take away the feedback and make $M\gamma = 0$, we then just have an ordinary RLC circuit. Let us suppose that the capacitor has been charged with an initial voltage $\hat{e}_0$, corresponding to the storage of an electrostatic energy $C\hat{e}_0^2/2$; damped sine oscillations are then produced with a frequency almost equal to $\omega_0 = (LC)^{-1/2}$. The initial energy is periodically transferred from the capacitor to the coil and vice versa, at the same time that this exchange occurs, a current exists in the resistor inside which there is heat dissipation: the total amount of energy stored in the coil and in the capacitor decreases exponentially with some time constant τ'.

We now again reestablish the feedback and the amplification and we suppose that $RM\gamma/L = 1$. In spite of the presence of the resistor R, the damping term of equation (9.9) disappears. The resonant circuit is now said to be undamped: during each period the amplifier brings to the circuit an energy which is just equal to the energy dissipated in the resistor. This energy is provided by the DC power supply of the amplifier.

If $RM\gamma/L > 1$, the energy coming from the amplifier is larger than the energy dissipated in the resistor, the amplitude of the electrical oscillations, as well as the energy dissipated in the resistor, increase exponentially. As the amplifier obviously cannot give an infinite output signal a steady state will be reached where the power in the resistor is just equal to the power that can be brought by the amplifier. This steady state is obtained after a transient state, see Figure 9.2.

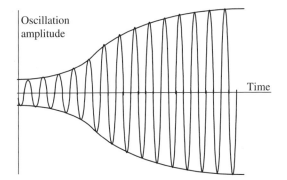

Figure 9.2. Transient state of an oscillator. As long as the amplitude of the signal remains small, linear equations are valid and give a sine oscillation with exponentially growing amplitude. Then saturation effects occur in the amplifier and a steady state is reached in which the amplitude remains constant.

We have already suspected mathematical difficulties about formulas (9.2) and (9.10) if $A\beta = 1$; the first one becoming meaningless and the second giving signals with an infinite amplitude. Our difficulties are a consequence of the fact that the circuit has been described by linear differential equations, whatever the amplitude of the signal. Let us go back to equation (9.9); this is a linear equation as long as R, L, M, and γ are perfectly constant and independent of the amplitude of the signal. If the values of the passive components (R, L, C) are certainly constant, on the contrary it's not the same for the active components of the amplifier (transistors), which will "saturate" for important values of the amplitude. As a consequence, it can be considered that a linear description is only valid for small signals when we come nearer and nearer to saturation of some parameters, A in equation (9.2) and γ in equation (9.9) become dependent on the amplitude: equations are no longer linear and their resolution needs far more than two lines. . . .

If nonlinearity is properly taken into account, the following results are obtained:

- The linear theory allows a correct determination of the threshold condition: $A\beta = 1$.
- If $A\beta$ is given a value larger than one, after a transient state, a steady state is finally reached; an almost sine periodic oscillation is produced, its amplitude is finite and all the more intense as the difference $(A\beta - 1)$ is larger.

9.1.4. Spectral Characteristics of the Ray Emitted by an Oscillator

Figure 9.3 shows a typical variation of the respective modules of the gain and feedback parameters, $A(\omega)$ and $\beta(\omega)$. They are often bell-shaped curves, to make the threshold condition $(|A(\omega)\beta(\omega)| = 1)$ possible it's necessary that the two curves overlap along a large enough area. The oscillating frequency will

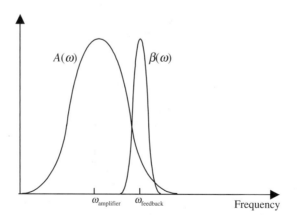

Figure 9.3. Gain and feedback coefficients versus frequency.

be intermediate between the two maxima; if they coincide their common value is the frequency of oscillation. It often happens that one curve is much narrower than the other; this curve will then define the frequency. In the example of Figure 9.1(b), the gain was represented by γ and was supposed to be independent of the frequency; the oscillation was obtained at the frequency of the resonant circuit.

An Autooscillator Starts from the Noise

In spite of its weaknesses the linear theory is quite valid for small signals, however formulas (9.2) and (9.10) still raise some problems. To give some importance to formulas (9.2) and (9.7) we had decided that $\hat{e}_0$ should be equal to zero and we had considered an autooscillator as a device giving an output signal with no input signal. Nevertheless, things are more complicated, and if we make $e_0 = 0$ in formula (9.10), $e(t)$ will remain equal to zero even if τ is negative. In reality, e_0 is never permanently equal to zero because of the *noise*. In the case of a laser the noise is the result of the spontaneous emission of radiation by excited atoms.

We now consider the arrangement in Figure 9.1(a) in which the amplifier and the feedback circuit are supposed to be tuned to the same central frequency. Let us examine what happens when the feedback coefficient is progressively increased from a small value up to the threshold value for which oscillations are obtained. In Figure 9.4 are shown the frequency spectrums of the output signals that are obtained in the absence of any input signal. Curves (1) and (2) correspond to conditions that are below the oscillation threshold: the output power remains quite small; the spectra become narrower and narrower as the threshold is approached. When the threshold is passed, even by a small amount, the phenomena change drastically: the power is suddenly increased, while the spectrum becomes very narrow.

The signal emitted by an oscillator is of course never strictly monochromatic. The theoretical determination of the spectral bandwidth $\Delta\nu_{\text{osc}}$ is an interesting, although very difficult, problem. $\Delta\nu_{\text{osc}}$ is a function of the

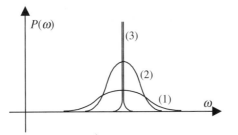

Figure 9.4. Spectra of the output power of an amplifier with positive feedback. No input signal has been applied. Curves (1) and (2) are below threshold, curve (3) is above threshold: autooscillation.

respective bandwidths, $\Delta\nu_{\text{feedback}}$ and $\Delta\nu_{\text{amplifier}}$, of the feedback circuit and of the amplifier. If the feedback circuit is more selective than the amplifier, $\Delta\nu_{\text{feedback}} \ll \Delta\nu_{\text{amplifier}}$, it can be established that

$$\Delta\nu_{\text{osc}} = \frac{2\pi h\nu}{P}\left(\Delta\nu_{\text{feedback}}\right)^2, \tag{9.11}$$

where h is Planck's constant and ν and P are, respectively, the frequency and power of the emitted oscillation.

Numerical Application

We consider the case of a 1 W laser emitting photons of 1 eV; the resonator has a "finesse" $F = 50$ and is made of two mirrors separated by a distance $d = 1$ m.

The free spectral range is (FSR) $= c/2d = 1.5 \times 10^8$ Hz; the feedback bandwidth is $\Delta\nu_{\text{feedback}} = $ (FSR)$/F = 3 \times 10^6$ Hz and, finally, $\Delta\nu_{\text{osc}} = 9 \times 10^{-6}$ Hz.

Such a spectral purity is completely meaningless, since it would require relative mechanical and thermal stabilities of the same order of magnitude for the resonator. We will just keep in mind that lasers are able to produce extremely monochromatic signals.

9.1.5. General Laser Arrangement

In Figure 9.5 is given a very general scheme of a laser: an adequate pumping device provides enough energy for a medium to become an amplifier for optical waves of suitable frequency, this medium is disposed between two parallel mirrors which permanently send the light back (feedback) into the

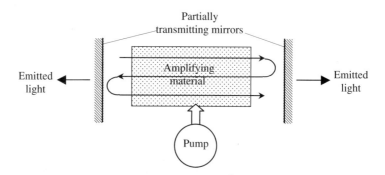

Figure 9.5. Principle of a laser. The material inside the resonator receives energy from a pumping source and stores it as a population inversion. A wave of suitable frequency travels back and forth; during each journey it receives energy from the material and loses energy at each reflection on the mirrors. When the gained and lost energies are equal the amplitude of the wave remains constant, this is the laser effect.

amplifier. Let us first consider the case of an empty resonator and let us suppose that, for some process, which is not described here, it has sent a planar wave between the two mirrors with its wave planes parallel to the plane of the mirrors. Planar waves will indefinitely travel between the mirrors; their existence corresponds to the storage of electromagnetic energy inside the resonator. If the mirrors were perfectly reflecting, the initial wave would be trapped and the amount of energy would remain constant. In the case of a partially transmitting mirror, some electromagnetic energy will escape outside and, because of these leaks, the lifetime of the stored energy is no longer infinite. It can be shown that the amplitude of the wave decays exponentially as it goes back and forth between the mirrors.

Laser Effect

Instead of an empty resonator, we now place a material that is able to amplify the waves traveling between the mirrors. If the amplification during a double transit just compensates for the losses caused by the partial reflection on the mirrors, then the system is said to be self-sustained: the resonator has become a light source which emits light through the partially transmitting mirrors. To understand the way a laser is working, it will be necessary to: (i) describe the physical reasons why a sample may amplify a wave; and (ii) understand how, and at which frequency, two parallel mirrors act as a positive feedback circuit.

9.2. Optical Amplification

9.2.1. Blackbody

Dark Red, Hot Red, or White

The story started at the end of the nineteenth century, at that time physicists were desperately chasing a law that would give quantitative indications about the radiation that is emitted by a heated body, a heated horseshoe, for example. After having identified the notion of electromagnetic radiation, it was pointed out that the power radiated from a heated body considerably increases with temperature while, at the same time, the spectral composition extended further and further toward shorter wavelengths. It was also noticed that the amount of emitted power depends on the special body under consideration, this is the reason why physicists invented an ideal object called a "blackbody." This can be thought of as a cavity almost completely isolated from the external world with which electromagnetic energy can only be exchanged through a tiny hole O, see Figure 9.6. The inner walls of the blackbody are perfectly reflecting, any ray penetrating inside will be almost indefinitely reflected, with a low probability of escaping through the aperture O. For an external observer the hole looks like a source of radiation.

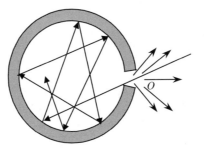

Figure 9.6. Schematic representation of a blackbody. Once it has penetrated inside the blackbody a light ray is almost indefinitely reflected on the perfectly reflecting walls, the probability that it could escape being small. A small amount of light can however escape through the small aperture O, which appears to some external observer as a point source of radiation.

At a given temperature T K the blackbody reaches an equilibrium state, the problem is then to evaluate the power that is emitted through the hole. It can be shown that the problem is completely solved as soon as we know the global electromagnetic energy density u inside the cavity; by electromagnetic energy density we mean the amount of energy inside each unit of volume of the blackbody.

The global density u has, of course, a spectral composition and we will call u_ν the associated *spectral density*: $u_\nu \, d\nu$ is the amount of energy associated to waves having a frequency falling inside the band limited by the frequencies ν and $\nu + d\nu$.

By definition we have

$$u = \int_0^\infty u_\nu \, d\nu. \qquad (9.12)$$

We now consider a blackbody having the shape of a cylinder and closed by a piston and we make an adiabatic compression of the electromagnetic radiation that is contained inside. The pressure of radiation tries to repel the piston. It can be shown that, in order that the second principle of thermodynamics should not be violated, the electromagnetic spectral density u_ν must depend on the frequency ν and the temperature T according to the law given by (9.13):

$$u_\nu = T^3 F(\nu/T), \qquad (9.13)$$

where $F(\nu/T)$ is a universal function of the variable ν/T, $F(\nu/T)$ doesn't depend either on the shape of the blackbody or on what it contains. The determination of this function had been a strong preoccupation for physicists at the end of the nineteenth century, its discovery has been the fruit of the imagination of Rayleigh and Jeans and, finally, of Planck. Rayleigh and Jeans considered the blackbody as a resonator that was ideal since its walls were perfectly reflecting. They admitted that the only waves that could exist inside such a resonator were the waves corresponding to a standing wave pattern compatible with its shape and size. Such a pattern is called a mode of the resonator. A mode is characterized by a frequency ν and a law of variation of the field amplitude versus the position inside the resonator.

We now consider a frequency domain limited by the two frequencies ν and $\nu + d\nu$, the determination of the number dg of modes with a frequency falling inside this band is not completely trivial and is described in any Thermodynamics textbook. It is obtained that dg is proportional to the volume V of the blackbody and to the spectral width

$$dg = \frac{8\pi\nu^2}{c^3} V\, d\nu \quad (c \text{ is the light speed}). \tag{9.14}$$

Rayleigh and Jeans were the first to consider the modes of a resonator as individual physical entities with two degrees of freedom. On the analogy of the kinetic theory of gases, that attributes an average individual energy equal to $kT/2$ to each microscopic object belonging to a collection at thermal equilibrium at a temperature of T K ($k = 1.37 \times 10^{-23}$ J/°K is the Boltzmann constant), each mode is given an energy equal to kT. An expression is thus obtained for the electromagnetic spectral density u_ν inside a blackbody,

$$u_\nu = dgkT \quad \rightarrow \quad u_\nu = \frac{8\pi\nu^2}{c^3} kT = \frac{8\pi k}{c^3} T^3 \frac{\nu^2}{T^2}, \tag{9.15}$$

where u_ν as given by (9.15) is in good agreement with (9.13); although this kind of measurement can never be very accurate, it can be considered that, as long as the frequency is not too high, (9.15) is in good agreement with experimental results. Unfortunately there is a severe drawback, which has been called the *ultraviolet catastrophe*: the integral of formula (9.12) diverges. A solution was found by Planck who, in 1900, had postulated that the individual energy of a mode could not vary continuously and should be an integer multiple of a quantum of energy equal to $h\nu$. Since that time the theory has been put on a more solid basis. Using Maxwell's equations and boundary conditions associated to perfectly reflecting walls, a rigorous mathematical formulation can be obtained for the modes, it can also be shown that the time dependence of the amplitude of a given mode is just the equation of an harmonic oscillator of frequency ν.

The quantification of an electromagnetic radiation then becomes an easy problem since the harmonic oscillator is extensively studied in Quantum Mechanics. The energy of a harmonic oscillator cannot have arbitrary values but must have one or another from a set of allowed values given by

$$E_n = (n + \tfrac{1}{2})h\nu. \tag{9.16}$$

The term 1/2 is the result of a theoretical refinement that will be ignored here. As a conclusion Planck considered the blackbody as a set of independent harmonic oscillators obeying the Boltzmann law, and that the number of oscillators (modes) with an energy equal to E_n are proportional to $e^{-En/kT}$. Performing a simple, although not completely straightforward calculation, Planck obtained that the average energy of a mode at temperature T is not equal to the Rayleigh-Jeans value kT, but is equal to

$$\langle u \rangle = \frac{hv}{\left(e^{hv/kT} - 1\right)}, \quad \text{blackbody formula.} \tag{9.17}$$

The angle brackets $\langle\,\rangle$ indicate an averaged value taken over the whole set of modes. An expression, now called Planck's formula, is finally obtained for u_v:

$$u_v = \frac{8\pi v^2}{c^3} \frac{hv}{\left(e^{hv/kT} - 1\right)} = \frac{8\pi h}{c^3} v^3 \frac{v}{\left(e^{hv/kT} - 1\right)}. \tag{9.18}$$

Rayleigh-Jeans (9.15) and Planck (9.18) formulas coincide, if hv goes to zero.

9.2.2. Einstein's Attempt to Establish Planck's Law

This is an attempt to find a new demonstration of the blackbody formula that Einstein had introduced, the three basic processes between the radiation and a sample of material. These processes are now universally known and, respectively, called absorption, spontaneous emission, and stimulated emission. We would like to emphasize the amazing physical simplicity of the model and perhaps, even more, the pragmatism of Einstein's treatment. He started from the fact that the function $F(v/T)$ should be universal and should not depend of the kind of blackbody under consideration, not on what is inside the blackbody which can, for example, contain a collection of atoms with two energy levels. Einstein then studied the energy exchanges between the collection of atoms and the radiation inside the blackbody and he invented a very simple model for the interaction.

In a first model only absorption and spontaneous emission is introduced, the result is a law for the electromagnetic density u_v that is not quite identical with the formula (9.18) obtained by Planck, but can converge to it when the frequency is high enough. Einstein then concluded that his model was not completely wrong and could be easily modified to obtain a good result. A third process, initially called induced emission by Einstein, is thus introduced beside the two initial ones.

9.2.2.1. Populations of the Energy Levels of an Atom

In the frame of Quantum Mechanics the absorption or emission of light are associated to a transition between two energy levels. Let us now come back to the description of a piece of material as a collection [C] of charged particles, these particles are first arranged in subsets that, for the sake of simplicity, will be called *atoms*, although they can be molecules or ions. The energy of an atom cannot be arbitrary, the atom must have one or another out of a set of allowed energies, which are called allowed energy levels or allowed states. A given atom is characterized by a ladder of "allowed levels," the rungs of which are not usually equidistant. The lowest state is the fundamental state, the others are excited states, and they are labeled as E_i. The atoms of [C] are distributed between the different levels; the number of atoms occupying a given level is the population of that level.

If N_{total} is the total number of elements of the collection and if N_i is the population of the level E_i, we have

$$\sum_i N_i = N_{\text{total}}. \tag{9.19}$$

An atom may make transitions from one level to another. To jump from a level E_i toward a higher level E_i, an atom must receive energy equal to $(E_i - E_j)$. On the contrary an atom can drop toward a less excited level and give the corresponding energy to the surrounding medium. Various processes can be involved for the exchange of energy, the most common being: absorption or emission of a photon with a suitable frequency; collision with another atom of the collection; or collision with the wall of the blackbody.

All transitions are not possible. An atom can occupy any allowed level, however it cannot jump, or drop, from a given level toward any other level: Quantum Mechanics establishes accurate rules, called selection rules, which predict which transitions are allowed or forbidden.

Inside a vessel at temperature T, the atoms of our collection exchange energy with the surrounding medium and reach a thermal equilibrium governed by the Maxwell-Boltzmann law according to which the ratio of the populations of levels are given by

$$\frac{N_{j,\text{equilibrium}}}{N_{i,\text{equilibrium}}} = e^{-\frac{(E_j - E_i)}{kT}}. \tag{9.20}$$

Order of Magnitude of Equilibrium Populations

As far as optical transitions are concerned the energy difference between the two levels is of the order of 1.5 eV, at room temperature (300 K) the thermal agitation energy kT is 25 meV; the argument of the Maxwell-Boltzmann exponential is equal to −60, which means that the ratio of the populations is almost equal to zero: in other words, all the atoms are located at ground level. The situation is quite different if the transition belongs to the radio frequency domain or even to the microwave domain, the energy difference between the energy levels is then negligible as compared to kT: the equilibrium populations are practically equal.

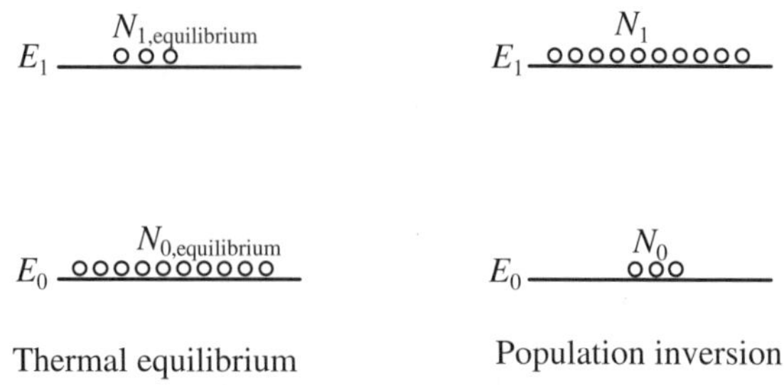

Figure 9.7. Populations of two energy levels. At thermal equilibrium the ground level is far more populated than the excited level. An inversion of population is a situation where the excited level is the most populated.

Inversion of Population

By an external action on the collection of atoms it is possible to obtain an "*out of equilibrium situation*" where the ratio between the populations is different from the equilibrium Maxwell-Boltzmann ratio. For example, it is possible to reach situations where the excited level is more populated than the ground level, the collection of atoms is then said to contain an *inversion of population*.

Relaxation Time of a Transition

Let us now consider a collection of atoms with only two energy levels E_0 and E_1, and let $N_0(0)$ and N_1, respectively, be the populations at time $t = 0$. The collection is brought into contact with a heat tank at temperature T, the populations are supposed to be different from values given by (9.20). Energy will be exchanged between the collection and the heat tank, transitions will occur between the two levels and, finally, the populations will reach their equilibrium values. The experiment shows that the time variation laws of the populations are exponential and characterized by a time constant τ,

$$N_{1(t)} = N_{1(0)}e^{-t/\tau} + N_{1,\text{equilibrium}}\left(1 - e^{-t/\tau}\right),$$
$$N_{0(t)} = N_{0(0)}e^{-t/\tau} + N_{0,\text{equilibrium}}\left(1 - e^{-t/\tau}\right),$$
$$N_{1(0)} + N_{0(0)} = N_{1(t)} + N_{0(t)} = N_{\text{total}}, \tag{9.21}$$

where τ is called the relaxation time of the transition; τ is characteristic of the transition and of the mechanisms responsible for the exchange of energy.

Radiative Transitions

Relaxation analysis is an interesting, although difficult, part of Physics. Annex 9.A describes the determination of the relaxation time in the case of a radiative transition. A transition is said to be *radiative* when the absorbed or emitted energy is of an electromagnetic kind; it will be described by an electromagnetic field in the case of a classical analysis or by photons in the case of a quantum analysis.

Transitions leading to laser action are obviously radiative transitions.

9.2.2.2. *Equilibrium of a Collection of Atoms Inside a Blackbody*

We consider a blackbody that is at thermal equilibrium at T K and that contains a collection of atoms having two energy levels E_0 and E_1; we admit that the atoms can only interact with the radiation with a frequency very close to the frequency of the atomic transition, $\nu_{\text{atom}} = (E_1 - E_0)/h$.

Einstein's First Model

In a first model Einstein only introduces absorption and spontaneous emission. The absorption allows the atoms to jump from ground level to the excited level with a probability which is all the higher as the ground level is more populated and as the electromagnetic energy density u_ν is larger: if $dN_{0,\text{abs}}$ and $dN_{1,\text{abs}}$ are the respective variations of the populations during time dt we have

$$dN_{0,\text{abs}} = -dN_{1,\text{abs}} = -Bu_\nu N_0 \, dt, \tag{9.22}$$

where B is a proportionality coefficient, the minus sign in formula (9.22) only indicates that the spontaneous emission causes the population N_0 to decrease, while the population N_1 increases by the same amount.

Spontaneous emission is introduced to allow the deexcitation of the excited level, Einstein simply considered that the number of atoms that jump downward during time dt is proportional to dt and to the population of the excited level

$$dN_{0,\text{spontaneous}} = -dN_{1,\text{spontaneous}} = +AN_1 \, dt. \tag{9.23}$$

Equilibrium is reached when upward and downward transitions are balanced and then we have

$$u_{\nu,\text{equilibrium}} = \frac{A}{B} \frac{N_{1,\text{equilibrium}}}{N_{0,\text{equilibrium}}}.$$

At equilibrium the populations are supposed to follow the law of Maxwell-Boltzmann (9.20), an expression is finally obtained for $u_{\nu,\text{equilibrium}}$:

$$u_{\nu,\text{equilibrium}} = \frac{A}{B} \frac{1}{e^{(E_1-E_0)/kT}} = \frac{A}{B} \frac{1}{e^{h\nu/kT}},$$

where A and B are usually known as the two Einstein coefficients, they have been introduced as proportionality coefficients, and they can be imposed on the following relationship:

$$\frac{A}{B} = \frac{8\pi h v^3}{c^3} = \frac{8\pi h}{\lambda^3}, \tag{9.24}$$

$$\frac{A}{B} = \frac{8\pi h v^3}{c^3} = \frac{1}{e^{hv/kT}}. \tag{9.25}$$

Although not completely identical, Planck's formula (9.18) and formula (9.25) have the same behavior if the frequency goes to infinity. Under such conditions Einstein came to the conclusion that his model should contain a part of the reality and that it could be amended to find the right formula.

Einstein's Second Model

A new interaction mechanism, called stimulated emission, is added; initially this interaction was called induced emission by Einstein. According to this interaction atoms may fall from the excited level to the ground level under the influence of the electromagnetic field. The number of atoms $dN_{1,\text{stimulated}}$ that are deexcited following this process is given by

$$dN_{1,\text{stimulated}} = -dN_{2,\text{stimulated}} = -B'u_v \, dt. \tag{9.26}$$

Taking the three processes into account and expressing the equilibrium, we obtain

$$u_{v,\text{equilibrium}} = \frac{A}{B} \frac{1}{e^{(E_1-E_0)/kT - B'/B}}. \tag{9.26'}$$

If A and B are imposed to follow (9.24) and if we make $B = B'$, (9.26') is exactly the same as Planck's formula. The fact that $B = B'$ corresponds to the fact that absorption and stimulated emission have the same physical origin; they are called *Einstein's coefficient for stimulated interaction*.

A is *Einstein's coefficient for spontaneous effect*. A is homogeneous to the inverse of time, and is the reason why it is also written as

$$A = \frac{1}{\tau_{\text{radiative}}}, \tag{9.27}$$

where $\tau_{\text{radiative}}$ is the *radiative lifetime* of the excited level.

9.2.2.3. Illustration of the Three Einstein Processes

Instead of constituting a new demonstration of the famous Planck formula, Einstein's treatment has provided a phenomenological description of the interaction of an electromagnetic field with a piece of material. The coinci-

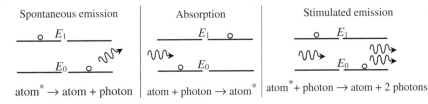

Figure 9.8. Illustration of the three Einstein basic processes; atom* means excited atom.

dence of the results of both methods is an excellent proof of the physical reality of the three basic processes that allow the collection to reach equilibrium; of course they still exist for a system out of equilibrium.

The illustration that is given in Figure 9.8 is convenient and often used. Here again phenomenology plays a more important role than rigor. The interactions are considered as "chemical reactions" between atoms and photons.

Spontaneous emission: atom* → atom + photon.

(Stimulated) absorption: atom + photon → atom.

Stimulated emission: atom* + photon → atom + 2 photons.

During the absorption process, the radiation loses energy: a photon is annihilated. The stimulated emission process creates photons that are indiscernable, with the same frequency, the same moment, and the same spin.

Role Played by the Profile of the Atomic Transition

We now consider the interaction of a monochromatic electromagnetic wave of frequency ν with a collection of atoms having two energy levels and a radiative transition of frequency ν_{atom}. Let u_ν be the electromagnetic energy density created by the presence of the wave. The phenomenological description of the interaction is governed by the following equations:

$$\frac{dN_1}{dt} = -\frac{dN_0}{dt} = -AN_1 + Bu_\nu(N_0 - N_1). \qquad (9.28.a)$$

If the energy levels are infinitely sharp an interaction will take place only if the two frequencies ν and ν_{atom} are strictly equal, to be sure that this condition is fulfilled we must introduce a Dirac distribution $\delta(\xi)$ in formula (9.28.a), where the variable ξ is related to the frequency ν by $\xi = (\nu/\nu_{atom} - 1)$. $\delta(\xi)$ is equal to zero for any value of its argument ξ, except for $\xi = 0$; $\delta(\xi = 0)$ is considered to be infinite and the integral $\int_0^\infty \delta(\xi)\,d\xi$ is by definition equal to one.

Thus equation (9.28.a) is now written as

$$\frac{dN_1}{dt} = -\frac{dN_0}{dt} = [-AN_1 + Bu_\nu(N_0 - N_1)]\delta(u). \qquad (9.28.b)$$

In fact the levels are never infinitely sharp and to any transition is associated a spectral ray with a specific profile that can easily be obtained from spectroscopic measurements.

The exact shape of the profile is determined by the physical mechanisms responsible for the broadening of the ray, it is usually a bell-shaped curve, which is maximum when $\nu = \nu_{atom}$. It can be described by a curve $f(\xi)$ and can be normalized using the condition $\int_0^\infty f(\xi)\,d\xi = 1$. Let us call $\Delta\nu_{atom}$ the broadness of the ray, if the ray is sharp enough we can write

$$\Delta\nu_{atom} \approx \alpha \frac{1}{\nu_{atom}}. \tag{9.28.c}$$

The accurate value of the proportionality coefficient α depends on the shape of the ray (Gaussian, Lorentzian, . . .) but is always of the order of one.

Finally, the phenomenological equation becomes

$$\frac{dN_1}{dt} = -\frac{dN_0}{dt} = [-AN_1 + Bu_\nu(N_0 - N_1)]f(\zeta). \tag{9.28.d}$$

We now come back to the interaction of a collection of atoms with the modes of a blackbody. In fact the atoms will interact with all the modes having a frequency falling inside the frequency band where the function $f(\xi)$ significantly differs from zero. When falling by spontaneous emission from the excited state to the ground state an atom emits a photon having a frequency falling inside the ray profile. The probability for a photon of frequency ν to be generated is proportional to the product of $f(\xi)$ by the mode density as defined by (9.14).

If the initial values of the populations are different from the Maxwell-Boltzmann equilibrium values, the system made by the collection of atoms and the blackbody is not in equilibrium. The return to equilibrium is accompanied by atomic transitions and by the emission of light rays that have the following properties:

- The emission is isotropic, the rays being emitted inside the 4π steradians of the free space.
- The emitted light is not monochromatic: its spectrum is described by the function $f(\xi)$, at least if we consider the mode density to remain roughly constant inside the frequency band $\Delta\nu_{atom}$. The spectral analysis of this light allows an experimental determination of $f(\xi)$.

There are many modes of the blackbody which have frequencies falling inside $\Delta\nu_{atom}$. Spontaneous transitions correspond to the term $AN_1 f(\xi)$ of equation (9.28.d) and produce photons that may be emitted in any of the modes near enough to the maximum of the profile. Einstein's spontaneous coefficient A is the global probability of spontaneous emission for one atom, during one second, and for a unit volume of the blackbody.

The case of stimulated emission is quite different from spontaneous emission, the two photons written on the right-hand side of the chemical reaction are *indiscernible*, and they both belong to the same mode of the blackbody, the special mode that contained the initial photon. The two photons have exactly the same frequency. The probability that a given photon of frequency ν will induce the creation of a second identical photon is proportional to $u_\nu f(\xi)$.

The directions of the momenta of spontaneous photons are at random inside the 4π steradians of the geometric space: spontaneous emission is isotropic. The photon produced by stimulated emission has the same momentum as the initial photon. Nothing very general can be said about the polarization of the spontaneous emission, on the contrary the two stimulated photons have the same polarization (or spin).

9.2.2.4. Using the Number of Photons to Write the Phenomenological Equations

It is often convenient to use the number of photons in a mode to write the phenomenological equations (9.28). Here again a rigorous treatment falls far beyond the scope of this book and a major role will be played by intuition. The number n of photons in a mode of frequency ν is proportional to the electromagnetic spectral density u_ν and equation (9.28.a) can be written as

$$\frac{dN_1}{dt} = -\frac{dN_0}{dt} = \frac{dn}{dt} = -\frac{N_1}{\tau_{\text{radiative}}} + n\frac{N_0}{\tau_{\text{radiative}}} - n\frac{N_1}{\tau_{\text{radiative}}}. \tag{9.29.a}$$

Let us call, respectively, $N_{1,\text{equilibrium}}$, $N_{0,\text{equilibrium}}$, and $n_{\text{equilibrium}}$, the equilibrium values of the populations and of the number of photons inside a blackbody. At equilibrium the time derivatives are equal to zero and we can write

$$\frac{N_{1,\text{equilibrium}}}{N_{0,\text{equilibrium}}} = \frac{n_{\text{equilibrium}}}{1 + n_{\text{equilibrium}}}. \tag{9.29.b}$$

Using the Maxwell-Boltzmann formula (9.20) we obtain the value of the number of photons at equilibrium, formula (9.29.c) is also known as the Bose-Einstein distribution,

$$\frac{N_{1,\text{equilibrium}}}{N_{0,\text{equilibrium}}} = e^{-(E_1 - E_0)/kT} = e^{-h\nu/kT},$$

$$n_{\text{equilibrium}} = \frac{1}{e^{h\nu/kT} - 1}. \tag{9.29.c}$$

9.2.2.5. The Three Einstein Processes in a Semiconductor

Of course Einstein had not considered the interaction of photons inside a semiconductor in 1917; however, the three basic processes (absorption, spontaneous, and stimulated emission) also exist in the case of a semiconductor material. A semiconductor can be considered as a collection of electrons and holes, from a statistical point of view the repartition of the holes and electrons between the various allowed energy levels is governed by a Fermi-Dirac distribution. If a piece of semiconductor is put inside a blackbody, it will be shown in Annex 9.A that the three basic processes are required to obtain a Bose-Einstein distribution for the photons and a Fermi-Dirac distribution for the electrons and holes.

A semiconductor can be considered as a single crystal inside which negative charges (electrons) and positive charges (holes) are free to move. The two kinds of charges may give the following reaction:

$$\text{electron} + \text{hole} \underset{(2)}{\overset{(1)}{\rightleftarrows}} \text{energy}.$$

where
(1) corresponds to electroluminescence and to semiconductor lasers; and
(2) corresponds to the absorption of light by a semiconductor and also to photoconduction.

The energy, which appears on the right-hand side of this kind of chemical reaction, may come from light (photons) but also from mechanical vibrations propagating along the crystal lattice (phonons). The reaction will occur:

• if a hole and an electron are simultaneously at the same place inside the crystal; and
• if energy and momentum conservation conditions are satisfied.

More details will be found in Annex 9.A. The momentum conservation laws are complicated and need the knowledge of the *shape* of the energy bands of the semiconductor, which shape is associated with the fact that the energy of an allowed level for an electron (or a hole) depends on its momentum, that's to say, on the wave vector of the associated quantum wave. For a restricted category of semiconductors, *direct band gap* semiconductors, the momentum conservation doesn't raise any problem, they are the only ones in which laser action is possible.

The most famous direct band gap semiconductors are gallium arsenide (GaAs) and indium antimonide (InSb) and also the many ternary compounds of the same family (GaAlAs, InGaAs, . . .).

Electrons and holes are not really free and are linked to the lattice. The creation, or the recombination, of an electron-hole pair occurs at the place of an atom of the lattice. In direct band gap semiconductors the moduli, p_{hole} and $p_{electron}$, of the momenta are equal, or almost equal; the conservation condition is obtained as soon as the two vectors have opposite directions, a slight difference can be transferred to the atom. On the contrary, in indirect band gap semiconductors where p_{hole} and $p_{electron}$ are quite different, a phonon is necessary to absorb the difference, the electron-hole recombination is then written as

$$\text{electron} + \text{hole} + \text{phonon} \rightarrow \text{photon}.$$

The necessary presence of a phonon at the place of the interaction strongly reduces the probability of the reaction; this is the reason why it is not possible to obtain laser action in silicon or germanium.

The three interaction processes in a direct band gap semiconductor can be summarized as:

Absorption: photon → electron + hole.
Spontaneous emission: electron + hole → photon.
Stimulated emission: electron + hole + photon → 2 photons (identical).

9.2.3. *Propagation of a Wave in a Two-Levels Material*

9.2.3.1. *Balance of the Energy Exchanges Between a Wave and a Collection of Atoms*

Let us consider a coherent light beam made of *identical-indiscernible* photons, having the same frequency and momentum. We are now going to study the interaction of photons with a collection of atoms that have only two energy levels: we will restrict ourselves to stimulated emissions and absorptions; the spontaneous emissions will be ignored. For interactions to occur, the light frequency should be close enough to the frequency of the atomic transition.

If the excited level is less populated than the ground level as, for example, is the case for thermal equilibrium, the number of absorptions is greater than the number of stimulated emissions: the number of annihilated photons exceeds the number of created photons and the intensity of the beam decreases during propagation. If, on the contrary, a situation has been achieved where the excited level is the most populated (population inversion), then the intensity increases as the light propagates and an amplification is observed: this is the laser effect (light amplification by stimulated emission of radiation).

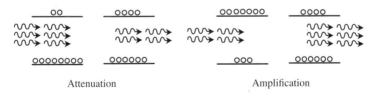

Attenuation   Amplification

Figure 9.9. Light propagation inside a two-levels material. If the higher level is the less populated, the number of absorptions is greater than the number of stimulated emissions: the beam is attenuated. In the case of population inversion, it is just the opposite and the beam is amplified.

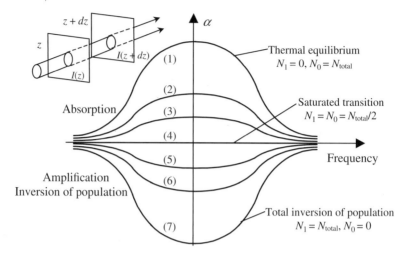

Figure 9.10. Variations of the absorption/attenuation coefficient α versus frequency for different values of the populations. The absorption coefficient α and the population inversion $(N_1 - N_0)$ have opposite signs, $\alpha < 0$ corresponds to amplification. The population inversion is maximum when all the atoms occupy the excited level.

It should be emphasized that if photons were not *bosons*, amplification of light would not be possible. It's because photons are indiscernible particles and don't obey the exclusion Pauli principle that it is possible to obtain a great number of identical photons.

Beer's law (see Chapter 8 on Index of Refraction) governs the evolution of the intensity of a light beam propagating in an absorbing or an amplifying material; we would now like to find an expression of the absorption (or amplification) coefficient α versus the populations of the levels.

We refer to Figure 9.10 and we suppose that the propagation material is made of a collection of atoms with two energy levels separated by $(E_1 - E_0)$ $= h\nu_{atom}$. Let N_0 and N_1 be the respective populations of the two levels, *expressed in numbers of atoms per unit volume*. We describe now the exchange of energy that occurs inside a slice having a cross section equal to S and a thickness equal to dz. The difference between the light intensities $I(z)$ and $I(z + dz)$ at the respective abscissas z and $z + dz$ corresponds to a variation of the energy stored inside the slice:

$$[I(z) - I(z + dz)]S = -\frac{dI}{dz} S \, dz.$$

It takes time $dt = dz/V$ for the light to cross the slice dz, using the formulas (9.22), (9.26) and introducing the absorption ray profile $f(\nu)$ we have:

- Absorption: $Bu_\nu N_0 \, f(\nu)S \, dz \, dt$.
- Stimulated emissions: $Bu_\nu N_1 \, f(\nu)S \, dz \, dt$.

When a steady state is reached the populations remain constant; remembering that each transition corresponds to an energy equal to $h\nu_{\text{atom}}$ and that the intensity I is related to the electromagnetic energy density u_ν and to the light speed V by $I = u_\nu/V$, we have

$$-\frac{dI}{dz} S\, dz\, dt = h\nu_{\text{atom}} B u_\nu [N_1 - N_0] f(\nu) S\, dz\, dt,$$

$$\frac{dI}{I} = -\frac{h\nu_{\text{atom}} B}{V} f(\nu)[N_1 - N_0]\, dz = -\alpha\, dz. \tag{9.30}$$

If we admit that the populations are independent of the abscissa z, (9.30) is nothing other than a differential expression of Beer's law:

$$I = I_0 e^{-\alpha z},$$

$$\alpha = \frac{c^3}{8\pi\nu^2 V} A f(\nu)(N_0 - N_1) = \frac{c^3}{8\pi\nu^2 V}\frac{1}{\tau_{\text{radiative}}} f(\nu)(N_0 - N_1). \tag{9.31}$$

The variation of the coefficient α versus frequency for different values of the populations is shown in Figure 9.10; we have a family of bell-shaped curves, geometrically affine to one another. The profile of the different curves is described by the same function $f(\zeta)$. The upper curve, labeled (1), corresponds to the thermal equilibrium; the following curves correspond to a collection of atoms submitted to a pumping process which is all the more intense as the level of the curve is increased. For the lowest curve a *total inversion of population* is achieved: the ground level is empty and all the atoms occupy the excited level. An interesting case is obtained when the two populations have been equalized, it is then said that *the transition is saturated*, and the coefficient α is equal to zero; the material is thus transparent whatever the frequency.

9.2.3.2. *Lifetime of a Radiative Transition*

Let us consider a situation where, by any process, the populations $N_{1(0)}$ and $N_{0(0)}$ have been given initial values that are different from the thermal equilibrium values, and let us study how the collection of atoms goes back to equilibrium. The populations are now functions of time and obey the following equations:

$$\frac{dN_0}{dt} = -\frac{dN_1}{dt} = AN_1 + B u_\nu (N_1 - N_0),$$

$$(N_1 - N_0) = N_{\text{total}} = \text{constant}, \tag{9.32.a}$$

$$\left\{ \begin{array}{l} \dfrac{dN_0}{dt} + (A + 2B u_\nu)N_0 = (A + B u_\nu)N_{\text{total}}, \\[2mm] \dfrac{dN_1}{dt} + (A + 2B u_\nu)N_1 = B u_\nu N_{\text{total}}. \end{array} \right. \tag{9.32.b}$$

Due to the presence of the product $u_\nu(N_1 - N_0)$ the solution of equations (9.32) is not straightforward. We are going to seek situations where simple solutions can be found.

Let us suppose that the pumping process is not very powerful and that the collection of atoms remains very near to thermal equilibrium at room temperature. As ν_{atomic} is an optical frequency $h\nu_{atomic}$ is considerably larger that the thermal agitation energy kT, under such conditions we have the following order of magnitude:

$$Bu_\nu << A \quad \rightarrow \quad u_\nu \approx 0,$$

$$N_1 << N_0 \quad \rightarrow \quad N_0 \approx N_{total} \quad \text{and} \quad N_1 \approx 0,$$

where u_ν can be omitted and we obtain

$$\frac{dN_0}{dt} + AN_0 = AN_{total},$$

$$\frac{dN_1}{dt} + AN_1 = 0.$$

The populations exponentially recover their equilibrium values so that, in the case of an optical transition, they are, respectively, equal to $N_0 \approx N_{total}$ and $N_1 \approx 0$. The time constant is equal to the inverse of the Einstein coefficient for spontaneous emission,

$$\tau_{radiative} = \frac{1}{A}. \tag{9.33}$$

9.2.4. Saturation of a Transition

A pumping light beam of the same frequency as the atomic transition suddenly illuminates a collection C of atoms initially at thermal equilibrium at room temperature T. At time $t = 0$ the electromagnetic density suddenly jumps from its equilibrium value $u_{\nu,equilibrium}$ to some other value u_ν which is all the larger as the pumping is more intense. For the sake of simplification we will suppose that the thickness of the collection is small enough for the beam not to be substantially attenuated during the crossing of C; as a consequence, u_ν may be considered as having the same value at any point of the collection.

The evolution of the populations is still governed by equation (9.32.b); in the beginning, the populations have their thermal equilibrium values, $N_{1,equilibrium}$ and $N_{0,equilibrium}$. Let us introduce a time constant $\theta = 1/(A + Bu_\nu)$, the evolution of the populations is given by

$$N_{0(t)} = N_{0,equilibrium} e^{-t/\theta} + N_{total} \frac{A + Bu_\nu}{A + 2Bu_\nu}(1 - e^{-t/\theta}),$$

$$N_{1(t)} = N_{1,equilibrium} e^{-t/\theta} + N_{total} \frac{Bu_\nu}{A + 2Bu_\nu}(1 - e^{-t/\theta}).$$

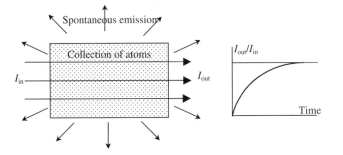

Figure 9.11. Saturation of a transition by optical pumping by an intense resonant light beam propagating inside a collection of atoms. After a transient state, during which the populations become equal, the material becomes transparent. The transmitted beam is in fact a little less intense than the incident beam due to the energy which is disseminated in the 4π steradian by spontaneous emission.

A steady state is reached where the populations are given by

$$N_{1,\text{steady}} = N_{\text{total}} \frac{A + Bu_\nu}{A + 2Bu_\nu} \xrightarrow{\text{intense pumping}} \frac{N_{\text{total}}}{2},$$

$$N_{1,\text{steady}} = N_{\text{total}} \frac{Bu_\nu}{A + 2Bu_\nu} \xrightarrow{\text{intense pumping}} \frac{N_{\text{total}}}{2}.$$

(9.34)

In the case of a powerful pumping beam, A is small compared to Bu_ν and the two levels become equally populated, the common value of the two populations being equal to half of the total number of elements of the collection of atoms: it is then said that the transition has been saturated.

It was for the sake of simplicity that the beam intensity had been supposed to remain constant during the crossing of the collection of atoms; if this is not the case the results are slightly different. During the first moments of the interaction the beam is attenuated, the energy being absorbed to make transitions that populate the excited level; as soon as the populations are equal, the material becomes transparent: the numbers of upward and downward transitions being equal. In fact the transmitted intensity is a little smaller than the incident intensity because of spontaneous transitions that emit light in all directions.

9.2.5. *Optical Pumping*

A sample is submitted to an electromagnetic pumping process when the populations of its energy levels are modified because of the interaction with an electromagnetic wave having a well-chosen frequency. We have just seen that an intense and resonant optical pumping can, at most, equalize the population of two levels, however it can never produce an inversion of population.

9.2.5.1. *Three-Levels Pumping Scheme*

In the arrangement shown in Figure 9.12 the atoms have three energy levels E_0, E_1, and E_2; the pumping is resonant for the transition between the ground level E_0 and the most excited level E_2. If the pumping power is sufficient, the respective steady state values, $N_{0,\text{steady}}$ and $N_{2,\text{steady}}$, are equal. The intermediate level is not concerned by the pumping and keeps a constant population. Finally, we can write

$$N_{0,\text{steady}} = N_{2,\text{steady}} = \frac{N_{0,\text{equilibrium}} + N_{2,\text{equilibrium}}}{2}.$$

If the intermediate level is closer to the upper level than to the ground level, $N_{2,\text{steady}} < N_{1,\text{equilibrium}}$ and an inversion of population is created between the levels E_1 and E_2. If, on the contrary, the intermediate level is closer to the ground level, this last level is depopulated and an inversion of population now exists between E_1 and E_0.

9.2.5.2. *Achieving Pumping at Optical Frequencies*

To achieve pumping according to the schemes that are proposed in Figure 9.12 we need a monochromatic pumping source with a frequency equal to the frequency of the transition between the two extreme levels.

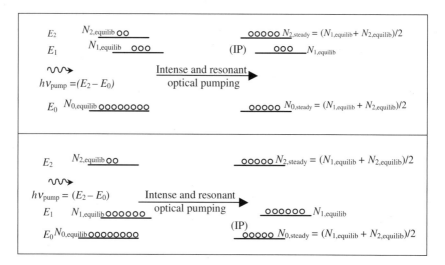

Figure 9.12. Three-levels pumping scheme. Optical pumping is resonant for the upper and lower levels and doesn't concern the intermediate level. According to the position of the intermediate level, an inversion of population (IP) is obtained either between the upper and intermediate levels or between the intermediate and ground levels.

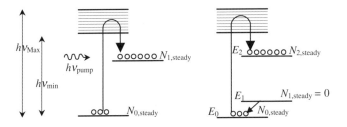

Figure 9.13. Pumping schemes using an absorption band. A fast transition occurs between the absorption band and the upper intermediate state which has a long lifetime in which the atoms accumulate. The left-hand diagram has four levels; the lower intermediate level has a very short lifetime and so is almost always empty, which makes it easier to obtain an inversion of population.

The three-levels pumping schemes were initially invented for radio or microwave devices where, thanks to frequency tunable oscillators, it is possible to concentrate an important power inside a narrow spectral band. In Optics, before the invention of lasers, powerful enough sources did not exist and the three-levels scheme was not possible; this is the reason why it is necessary to select atoms for which the upper narrow level is replaced by an absorption band (see Figure 9.13). The pumping scheme is slightly different because the sublevels of a band usually have a very short lifetime and cannot accumulate atoms; very often this band is efficiently coupled to an intermediate level on which atoms will accumulate and, if the radiative lifetime $\tau_{radiative}$ is long enough, will create an inversion with regard to the ground level.

The existence of an absorption band relaxes the necessity of a monochromatic pumping source and a white source can, in principle, be used; referring to Figure 9.13 it is seen that the useful part of the white spectrum is between the two frequencies ν_{Max} and ν_{min}. The first laser (ruby) was pumped by a flashtube designed for phototographic purposes.

Four-Levels Pumping Scheme

The previous method can be improved by using atoms that have a fourth energy level, which is denoted as E_1 on the left-hand diagram of Figure 9.13, the choice of this level obeys the following principles:

- The energy gap $(E_1 - E_0)$ being much larger than kT, the level is almost empty at equilibrium.
- The lifetime is short, so that atoms cannot accumulate population so the level remains empty.

In the three-levels method, at least one-half of the atoms should have been transferred to the upper intermediate level E_2 before reaching population inversion. This is not the case with the four-levels scheme since the first atom to arrive at level E_2 creates an inversion of population with regard with the lower intermediate level. The thresholds are considerably lower in the case of four-level lasers than in the case of three-level lasers.

Orders of Magnitude

Optical pumping is not the only method for obtaining inversions of population; it is mostly used in the case of lasers where the active medium is dense enough (solid or, more seldom, liquid) to allow an efficient coupling with the pumping source. The collection of atoms, in fact it will often be ions, is contained as impurities inside a crystalline lattice.

In the case of ruby the active elements to obtain a laser effect (wavelength $0.6943\,\mu$m) are chromium ions (Cr^{3+}) inserted into a sapphire lattice (Al_2O_3). Neodymium ions are in common use, inserted in a glass or in a crystalline matrix (YAG (Yttrium Aluminum Garnet) being very favorable), they provide a four-levels energy diagram and emit a laser light at $1.06\,\mu$m. In both cases the doping concentration is around 10^{19} to 10^{20} ions per cubic centimeter.

The radiative lifetime of the upper lever of a laser transition is an important parameter; it must be long enough to allow an accumulation of population; in other words, the transition should not be fully allowed, such a level is said to be *metastable* (in ancient Greek, *meta = almost*). According to formula (9.31) the gain per unit length is proportional to the inverse of the lifetime, which means that it should not be too long. Finally, a compromise has to be found, typical lifetimes for solid-state laser materials are of the order of milliseconds (3 ms for ruby; 200 to 800 μs, according to the host matrix, for neodymium ions).

For some special lasers, semiconductor lasers and dye lasers, for example, where very efficient pumping processes exist, fully allowed transitions can be used; extremely high gains are then obtained and the laser oscillation threshold is reached even with extremely thin pieces of material (tens to hundreds of micrometers).

Transitions from the upper band, as well as transitions from the lower intermediate state in the case of four levels, are associated to the creation or annihilation of phonons in the crystalline lattice, they are very fast (relaxation times are of the order of picoseconds).

9.2.5.3. Atomic Clocks

Cesium atomic clock: Atomic clocks are devices that deliver a periodic signal with a well-defined frequency and an extreme spectral purity. They are also a didactic illustration of electromagnetic pumping at various frequencies (optical and microwave). In the *cesium* atomic clock the frequency of a microwave oscillator, thanks to an appropriate signal processing, is tuned to be in exact coincidence with the frequency of a hyperfine transition which is

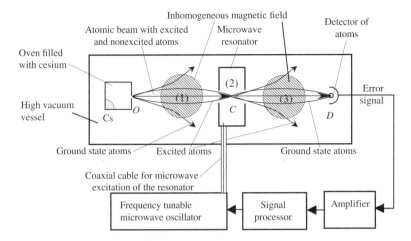

Figure 9.14. Cesium atomic clock. An atomic beam of cesium atoms is emitted from a heated furnace. The detector of atoms receives atoms if, and only if, the microwave signal frequency exactly coincides with the frequency of a transition of the *cesium* atoms. The main role of the microwave cavity is to enhance, thanks to a stationary pattern, the electromagnetic energy density of the microwave signal that pumps the collection of *cesium* atoms and equalizes the populations. The atomic beam must propagate in a high vacuum.

known to a high accuracy and is equal to 9,192,631,830 Hz. Figure 9.15 shows some of the energy levels of the atom of *cesium* that are necessary to understand the way the clock is working. As with any alkaline atom *cesium* has only one peripheral electron, the energy of the atom is different whether the spin of this electron is parallel or antiparallel to the spin of the nucleus. The energy difference is rather small and the associated frequency belongs to the microwave domain. An interesting consequence is that excited and nonexcited atoms don't behave in the same way when they propagate in an inhomogeneous static magnetic field. This property, which allows sorting one category of atoms from another, was established for the first time in a famous experiment made in 1922 by Stern and Guerlach, and gives a demonstration of the existence of the quantization of the momentum and of the existence of the spin. The experiment is well described in *Lectures on Physics* by Richard Feynman.

In the arrangement of Figure 9.14, an atomic beam of excited and nonexcited cesium atoms is emitted through the hole O of an oven containing a piece of cesium. The energy difference between the two levels of the hyperfine transition is very small and negligible as compared to the thermal agitation energy kT and, according to the Maxwell-Boltzmann law, the initial atomic beam contains the same number of the two categories of atoms.

During its propagation from O to D, the beam will cross three different regions (1), (2), and (3):

- In region (2) the static magnetic field is constant and equal to zero. Point C is inside a microwave resonator, which is excited by antenna at the exact frequency of the atomic transition.
- In region (1) the magnetic field is inhomogeneous with a repartition that has been designed in such a way that the excited atoms issued from O are focused at point C and that the ground state atoms are repelled from the ACD axis and that very few of them will reach point C.
- In region (3) the field repartition is almost the same, except that the excited atoms are now repelled from the axis, while the ground state atoms are focused onto the detector of atoms D which gives an electric output signal when hit by atoms.

When penetrating into the resonator the atomic beam contains a total inversion of population. If no microwave signal is sent into the resonator, the inversion of population remains and the gradient magnetic field in (3) will spread the atomic beam so that almost no atoms will reach the detector. If, on the contrary, the resonator is excited at a frequency equal to the atomic transition frequency, the atomic beam is submitted to a resonant pumping which equalizes the populations of the two levels and repopulates the ground level so that one-half of the initial atoms now reach the detector. The signal given by the detector is maximized when the microwave oscillator has a frequency just equal to that of the atomic transition. The microwave oscillator of Figure 9.14 is electrically controlled in such a way that its frequency is permanently equal to the atomic transition frequency. An elective signal is provided by a signal processor receiving an error signal from the atom detector.

The cesium atomic clock is an illustration of the fact that a two-levels pumping scheme cannot give population inversion; in this case, the excited level is partially depopulated by the pumping.

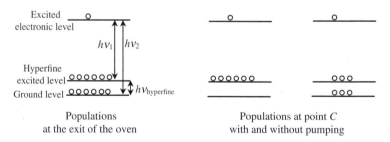

$$\nu_{\text{hyperfine}} = 9{,}192{,}631{,}830 \text{ Hz}$$

Figure 9.15. Cesium atoms levels and populations. Atoms arriving at point C are all excited. A resonant microwave pumping equalizes the populations of the hyperfine levels. ν_1 and ν_2 are optical frequencies and constitute the famous alkaline optical doublet.

Rubidium Atomic Clock

In a rubidium clock advantage is taken from the fact that there are two iso-
topic kinds of rubidium atoms. The energy levels of rubidium atoms are
described in Figure 9.16. Rb 85 atoms have no nuclear spin and, consequently,
don't have any hyperfine structure; on the contrary, Rb 87 atoms do have such
a structure. A collection of Rb 87 atoms can emit or absorb an optical doublet
of frequencies ν_1 and ν_2; a collection of Rb 85 can only emit or absorb the frequency ν_2. The lifetimes of the hyperfine levels are very long (seconds) and
mostly determined by the collisions with the walls of the cell, that's to say, by
the pressure of the vapor. The optical transitions from the upper toward the
hyperfine levels are fully allowed and correspond to a lifetime of the order of
a few nanoseconds. Atoms, when they fall from the upper level, have about
exactly the same probability to go to one of the hyperfine levels or to any
other. At thermal equilibrium, and in the absence of any pumping, the two
hyperfine levels have equal populations while the upper level is almost empty.

We refer to Figure 9.16. The lamp, which is filled with Rb 87 vapor, is a
light source emitting the doublet (ν_1, ν_2). A passive cell C_1 filled with Rb 85
vapor acts as a filter blocking ν_1 and transmitting ν_2. A second passive cell C_2,
filled with Rb 87 vapor, is placed inside a microwave resonator tuned at the
ν_3 frequency of the hyperfine transition.

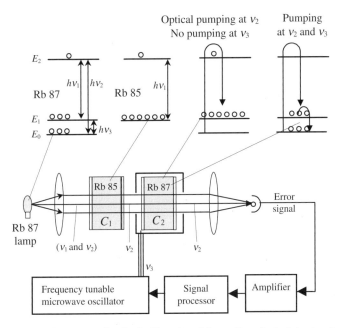

Figure 9.16. Rubidium atomic clock. The signal from the photodetector is minimum
when the microwave frequency is exactly equal to the hyperfine frequency of the
rubidium atoms.

At first we suppose that the microwave oscillator has been switched off: the atoms of cell C_2 are only submitted to an optical pumping at ν_2 which tries to make the populations of the upper level E_2 and of the ground level E_0 equal, in fact the atoms are trapped by the intermediate level E_1 on which all the atoms are stored. At the start of pumping, cell C_2 is absorbing and then becomes fully transparent at the end of a transient regime, since the ground level is now *empty*.

The microwave oscillator is now switched on and the collection is simultaneously pumped at ν_2 and ν_3. The difference from the previous case is that the pumping at ν_3 repopulates the ground level, making cell C_2 absorbing again at the optical frequency ν_2: the photodetector becomes aware of any difference occurring between the microwave and the hyperfine frequencies and elaborates an error signal to make the necessary correction.

9.3. How to Obtain an Inversion of Population

9.3.1. *Optical Pumping*

It has been seen in the previous sections how inversions of population could be obtained using optical pumping. This method is convenient mostly for dense laser materials (solids or liquids) so that the pumping light is efficiently absorbed. Referring to Figure 9.13 it is seen that in the case of a three-levels scheme the pump should be able to raise at least one-half of the total number of atoms to the upper laser level in a time shorter than its lifetime (typical value 10^{-3} s). The number of laser active centers is typically of $10^{19}/\text{cm}^3$, the energy of a transition is of the order of 1 eV; thus the energy to bring this about is 1 J/cm^3 corresponding to a power of 1 kW/cm^3. If the pumping light source is a type of blackbody, and thus is not monochromatic at all, a large fraction of its power falls outside the absorption band of the laser material: only pulsed sources can be used. Flash tubes are able to emit kilowatt pulses

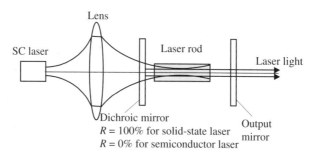

Figure 9.17. Solid-state laser pumped by a semiconductor laser. The dichroic mirror is fully transparent for the semiconductor laser and fully reflecting for the solid-state laser.

lasting a few milliseconds. With a four-levels scheme the required pumping power is lower by one order of magnitude and a continuous laser emission can be obtained using tungsten-halogen lamps.

There are two main reasons for the poor efficiency of optical pumping:

- The first and most important reason is the mismatch between the emission spectrum of the lamp and the absorption spectrum of the material.
- The second reason is a bad coupling of the pump with the laser rod.

The use of semiconductor lasers as pumping sources for solid-state lasers has been an important improvement. First of all, semiconductor lasers have an excellent efficiency for converting into light the energy coming from their electrical supply. Second, they produce a monochromatic light which can be chosen to coincide with the frequency of the absorption band. Finally, their light is spatially coherent and can be sharply focused inside a small volume.

9.3.2. *Pumping of Semiconductor Lasers*

Let us consider a sample of a semiconductor material at thermal equilibrium in darkness and at a temperature T; let $N_{\text{electron}}^{\text{equilibrium}}$ and $N_{\text{hole}}^{\text{equilibrium}}$ be the density of free carriers, the product of these two densities is a constant $K(T)$ which is determined by the material and the temperature:

$$N_{\text{electron}}^{\text{equilibrium}} N_{\text{electron}}^{\text{equilibrium}} = K(T).$$

In a semiconductor at equilibrium the number of free carriers is small, the material is very absorbing for light: each photon propagating in the semiconductor has a high probability of being absorbed thus creating an electron-hole pair. Let us suppose that, by any means, we have been able to change the carrier's densities and to give them values that are large enough so that their product is larger than $K(T)$; and let us suppose that the semiconductor sample is kept in darkness and in contact with a thermostat at temperature T: electron-hole recombination will occur until the product is again equal to $K(T)$. Had the product been smaller than $K(T)$, electron-hole pairs would have been created. ... The relaxation time, which is characteristic of this returning to equilibrium, is called the *radiative recombination time*. In a direct band gap semiconductor this time is of the same order of magnitude as the lifetime of a fully allowed radiative transition, i.e., *nanoseconds*.

The following three interaction processes are introduced:

- Absorption: photon → electron + hole.
- Spontaneous emission: electron + hole → photon.
- Stimulated emission: electron + hole + photon → 2 photons (indiscernible).

The two photons on the right-hand side of the stimulated emission reaction are indiscernible and they open the door to a potential optical amplification. When the product of the two carrier densities is greater than $K(T)$ a photon has more chance of stimulating the creation of a second identical photon rather than be annihilated; the situation is very similar to a population inversion: the semiconductor sample then behaves as an optical amplifier for a light signal having a frequency higher than $\Delta E/h$, where ΔE is the semiconductor band gap,

$$N_{\text{electron}} N_{\text{hole}} > N_{\text{electron}}^{\text{equilibrium}} N_{\text{electron}}^{\text{equilibrium}} = K(T). \qquad (9.35)$$

To obtain the situation described by equation (9.35) we can use optical pumping and create numerous electron-hole pairs by lighting the piece of semiconductor with a light wave having a frequency higher than $\Delta E/h$. A bombardment by electrons of suitable energy (cathodoluminescence) can also be used. The two previous methods are interesting and have been experimentally demonstrated, however most of the semiconductor lasers use forward biased PN junctions. Positive holes coming from the P doped region, and negative electrons coming from the N doped region, are massively injected into the junction; playing with the current allows us to give a high value to the product $N_{\text{electron}} N_{\text{hole}}$.

Figure 9.18(a) shows a very simple laser diode; real devices are far more complicated and require sophisticated techniques of epitaxy. The emitting area has the shape of a narrow and horizontal rectangle; because of diffraction the angle of divergence is much larger along the vertical direction (40°) than along the horizontal direction (10°).

Figure 9.18(b) shows a more realistic device, it is a gallium-arsenide diode. The junction is sandwiched between two layers made of an alloy, $\text{Ga}_{1-x}\text{Al}_x\text{As}$, where some atoms of gallium are replaced by aluminum; the refractive index

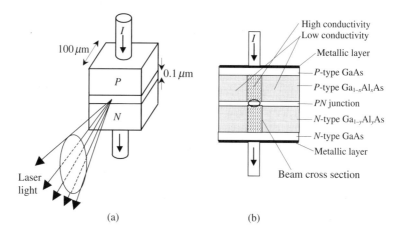

(a)   (b)

Figure 9.18. (a) represents an oversimplified laser diode. The purpose of the more realistic scheme of (b) is to confine the electric current by a control of the conductivity and to confine the light by a suitable refractive index repartition.

of the *PN* junction is higher than the indices of the two layers and so the junction is a kind of optical guide which traps the light in the region where there is a large excess of electrons and holes. To obtain laser action, high current densities are required (kA/cm^2); in order to limit the total current to an acceptable value (mA) the repartition of the electrical conductivity is designed in such a way that the current is confined to the region of the wave guide.

9.3.3. *Gas Laser Pumping*

9.3.3.1. *Principle*

In the case of a gas laser the collection of atoms (it may be ions or molecules as well) in which the laser effect is to occur is a gas; the modification of the populations of the levels is created by an electrical discharge between two electrodes immersed in the gas.

In a gas discharge electrons are emitted from a cathode and travel to the anode, meanwhile they collide with the different components of the gas mixture with which they exchange energy. By a proper choice of the physical condition of the discharge (partial pressure of the different components, current, and voltage, . . .) it is possible to tune the mean energy that is given to an atom after each collision and so to selectively populate specified energy levels, putting them in a situation of population inversion with regard to the lower levels. If the gas discharge is placed inside a Perot-Fabry resonator a laser emission is possible at a transition frequency of the atoms. Very often a device such as the one shown in Figure 9.19 is able to produce laser rays of different frequencies according to the spectral zone for which the mirrors are designed. Typical lasers of that kind are CO_2 and argon lasers.

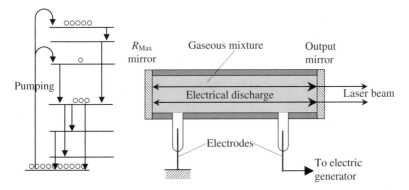

Figure 9.19. Pumping scheme using an electrical discharge in a gas. According to the setting of the various physical parameters (temperature, pressure, electric current, and voltage) the atom-electron collisions will selectively populate specified levels.

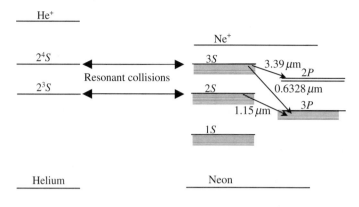

Resonant collisions   He* + Ne → He + Ne*

Figure 9.20. Energy level diagram of helium and neon atoms. Being neighbors in the Mendéleiev chart the two atoms have certain levels that are in good agreement. The existence of a greater number of levels and sublevels in the case of neon is due to the fact that this atom has more electrons (ten instead of two).

9.3.3.2. *Helium-Neon Laser*

In He-Ne lasers the laser transition occurs on a transition of the neon atoms, the helium atoms allow a selective population of certain levels of neon atoms. The gas mixture contains 90% helium and 10% neon with a total pressure of one millimeter of mercury. The electric discharge is set (few milliamps, 1000 V) so that the atoms are not ionized. The selective population of the S levels of neon atoms is a consequence of the possibility of *resonant collisions*. We consider that A and B are atoms having energy levels each having almost equal distance, in their excited state they are labeled C and B^*; it's a well-known Quantum Mechanics exercise to study the collision between excited atom A^* with atom B at the ground level: the probability is then very high that after the collision atom A is deexcited while B becomes excited (see Figure 9.20).

The energy level diagram of Figure 9.20 shows that the $2S$ levels of helium are at the same distance from the ground level as the $3S$ levels of neon, making possible *resonant collisions* between the two families of atoms. The S levels have relatively long lifetimes (microseconds) and become populated, resonant collisions then selectively populate the corresponding levels of neon, which become overpopulated with regard to the P levels. As the energy diagram of neon is quite rich, several laser transitions are possible, the most famous being the red one at $0.6328\,\mu$m.

9.3.4. *Chemical Lasers*

The products of a chemical reaction are often obtained in an excited state, the reaction is then followed by a second step during which the products go

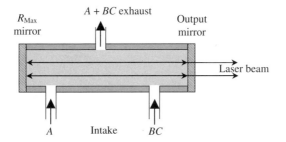

Figure 9.21. Obtaining a population inversion from a chemical reaction. The chemical reaction $A + BC \rightarrow AB + C^*$ produces the compound C in an excited state and is followed by a second reaction $C^* \rightarrow C +$ photon or $C^* +$ photon $\rightarrow C + 2$ photons.

back to the ground level and emit light, most chemical reactions used for lighting purposes belong to this type (a candle for example). Figure 9.21 describes some chemical reaction between two gaseous reactants A and BC, C is produced in an excited state and an inversion of population is immediately obtained. If the deexcitation of C occurs spontaneously light is emitted in the 4π steradian, if the reaction occurs inside a Perot-Fabry resonator stimulated emission can predominate and a laser is emitted. Chemical energy can be stored with a high volume density, this is the reason why chemical lasers are potentially very powerful and easily transportable.

9.4. The Fabry-Perot Resonator

9.4.1. *The Fabry-Perot Resonator Was Ignored for too Long*

It was the French physicists Boullouch and Fabry who, at the beginning of the twentieth century, discovered the fascinating optical properties of the arrangement of two facing parallel mirrors with excellent accuracy. For many years this arrangement was universally known as the Fabry-Perot (FP) resonator and was exclusively used for spectroscopic applications since it is a very dispersive and very luminous device. For a long time after that the FP resonator was considered as an electromagnetic resonator very analog with the resonant circuits used in radio or with microwave resonators. In Section 6.4 we have studied the FP resonator as a multiple wave interferential device, we are now going to use a new approach very near to radio and microwave methods.

9.4.2. *Transient State of a Fabry-Perot Resonator*

The FP resonator of Figure 9.22 is made of two planes and two parallel mirrors separated by a distance d. A *planar* monochromatic wave R_0, with its wave vector parallel to the axis of the FP resonator, arrives at mirror M_1 on which

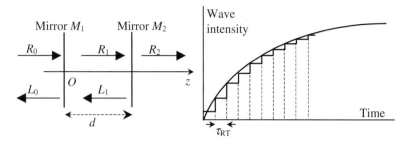

Figure 9.22. Transient state of an *FP* resonator. $\tau_{RT} = 2\,d/c$ is the time to make a round trip between the two mirrors. L_i and R_i are, respectively, waves propagating toward the left or the right side.

a transmitted wave R_1 and a reflected wave L_0 are generated. The wave R_1 propagates toward the right and arrives at the second mirror where a reflected wave L_1 and a transmitted wave R_2 are generated. There is no reason to introduce a wave L_2 since the half-space located after the second mirror extends to infinity.

We suppose that the FP resonator is initially empty of any electromagnetic energy and we take, as the time origin, the instant when wave R_0 just arrives at the first mirror. Waves R_1 and L_1 go back and forth between the two mirrors and their amplitudes reach constant values, this situation being obtained when the energy brought each second by R_0 is equal to the sum of the energies, respectively, transmitted and reflected by R_2 and L_0. Figure 9.22 describes an intuitive shape of the variation versus time of the intensities of the different waves L_0, L_1, R_1, and R_2. These intensities increase step by step, each step having a duration equal to the time $\tau_{RT} = 2\,d/c$ necessary for the light to make a double passage in the resonator. The following formulas, which give the time variations of the intensities, will be justified in the next section:

$$\left|R_i\right|^2 = \left|R_{i,\mathrm{Max}}\right|^2 (1 - e^{-t/\tau_{res}}), \quad i \in \{0, 1, 2\},$$

$$\left|L_i\right|^2 = \left|L_{i,\mathrm{Max}}\right|^2 (1 - e^{-t/\tau_{res}}), \quad i \in \{0, 1, 2\}, \tag{9.36}$$

where τ_{res} is a time constant which is characteristic of the resonator and which depends on τ_{RT}, see (9.37).

Energy Storage Inside a Fabry-Perot Resonator

After a time, which is theoretically very long but which is in fact of the order of a few times τ_{res}, according to formulas (9.36) a steady state is reached where $|R_i|^2$ and $|L_i|^2$ are constant. This steady state, which will be studied in Section 9.4.3, is the result of constructive interference of the waves L_1 and R_1 and corresponds to the storage of electromagnetic energy inside the resonator; the accumulation of energy occurs during the transient state.

Lifetime τ'_{res} of the Energy Inside the Resonator

Once the steady state has been obtained, the incident wave R_0 is suddenly switched off; the waves L_1 and R_1 keep oscillating between the two mirrors, each time they hit a mirror they lose some energy which is transmitted outside the mirrors. Let us follow the successive reflections of one of these waves; if at time t its intensity is equal to some value $I(t)$, at time $(t + \tau_{RT})$ this intensity has decreased and is equal to $\rho_1\rho_2 I(t)$ (ρ_1 and ρ_2 are the respective reflection coefficients of the two mirrors), since during a round trip the wave is reflected once from each mirror

$$I_{(t+\tau_{RT})} = \rho_1\rho_2 I_{(t)}.$$

The energy still stored in the resonator at time t is proportional to the intensity of the wave, so we can write

$$W_{(t+\tau_{RT})} = \rho_1\rho_2 W_{(t)} \quad \rightarrow \quad W_{(t+\tau_{RT})} - W_{(t)} = -(1-\rho_1\rho_2)W_{(t)},$$

$$\tau_{RT}\frac{dW_{(t)}}{dt} = -(1-\rho_1\rho_2)W_{(t)} \quad \rightarrow \quad \frac{dW_{(t)}}{W_{(t)}} = -\frac{dt}{\dfrac{\tau_{RT}}{(1-\rho_1\rho_2)}},$$

$$W_{(t)} = W_0 e^{-t/\tau'_{res}} \quad \text{with} \quad \tau'_{res} = \frac{\tau_{RT}}{(1-\rho_1\rho_2)}, \tag{9.37}$$

where τ'_{res} is the lifetime of the wave inside the resonator, the physical processes involved in emptying the resonator are identical to those involved in filling it with energy and thus we will consider that $\tau'_{res} = \tau_{res}$. The lifetime increases with the reflection coefficients and becomes infinite when the two coefficients are equal to unity.

In the case of two mirrors separated by one meter and having reflection coefficients such that $\rho_1\rho_2 = 1$, we have $\tau_{RT} = 6.7$ ns and $\tau_{res} = 134$ ns.

9.4.3. *DC Behavior of a Fabry-Perot Resonator*

We go back to Figure 9.22, we suppose that a steady state has been reached and we would like to describe it. We then have to find an electromagnetic field fulfilling Maxwell's equations and the boundary conditions on the mirrors. As we deal with plane, parallel and infinitely extended mirrors, the problem is simplified; we can choose planar waves for L_i and R_i:

$$L_i e^{-j(\omega t-kz)} \quad \text{and} \quad R_i e^{-j(\omega t-kz)},$$

where L_i and R_i are the respective complex amplitudes at $z = 0$ on the first mirror.

The boundary conditions are expressed by the following set of equations:

$$z = 0, \quad \text{mirror } M_1: \quad \begin{cases} R_1 = t_1 R_0 + r_1 L_1, \\ L_0 = r_1 R_0 + t_1 L_1, \end{cases} \tag{9.38.a}$$

$$z = d, \quad \text{mirror } M_2: \quad \begin{cases} L_1 e^{jkd} = r_2 R_1 e^{-jkd}, \\ R_2 = t_2 R_1, \end{cases} \tag{9.38.b}$$

where t_i and r_i are the transmission and reflection of the mirrors for the *amplitude* of the waves, in the case of nonabsorbing mirrors we have $|t_i|^2 + |r_i|^2 = 1$. We had previously introduced reflection coefficients, ρ_i, for the intensities, we simply have $\rho_i = |r_i|^2$. If we consider that the amplitude R_0 of the incident wave is given, the problem has four unknown L_0, L_1, R_1, and R_2 that can be obtained from the four equations (9.38), we obtain

$$\begin{cases} R_1 = R_0 t_1 A_{(\phi)}, \\ L_1 = R_0 t_1 r_2 A_{(\phi)}, \\ R_2 = R_0 t_1 t_2 A_{(\phi)}, \\ L_0 = R_0 \left(r_1 + t_1^2 r_2 e^{-j\phi} A_{(\phi)} \right), \end{cases} \tag{9.39.a}$$

with

$$A_{(\phi)} = \frac{1}{\left(1 - r_1 r_2 e^{-j\phi}\right)} \quad \text{and} \quad \phi = 2kd = \frac{4\pi d}{\lambda} = \frac{2wd}{c}, \tag{9.39.b}$$

where $A(\phi)$ is the same function that was introduced when studying the FP resonator in Chapter 6 on interference, see formula (6.17). ϕ is the phase difference corresponding to a return journey between the mirrors and is proportional to the frequency. The graph of $A(\phi)$ is a succession of sharp maxima, which are all the sharper as $r_1 r_2$ is closer to unity. The conditions for obtaining a maximum correspond to a resonance inside the resonator and are illustrated in Figure 9.23.

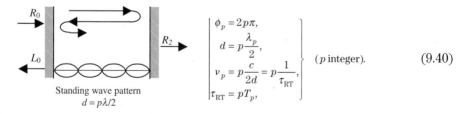

$$\begin{aligned} \phi_p &= 2p\pi, \\ d &= p\frac{\lambda_p}{2}, \\ v_p &= p\frac{c}{2d} = p\frac{1}{\tau_{RT}}, \\ \tau_{RT} &= pT_p, \end{aligned} \right\} \quad (p \text{ integer}). \tag{9.40}$$

Standing wave pattern
$d = p\lambda/2$

Figure 9.23. Illustration of the resonance condition of an FP resonator.

Let λ_p, ν_p, and T_p, respectively, be the wavelength, frequency, and period of an incident wave R_0 that fulfills the resonance condition: if τ_{RT} is a multiple of the period, the different waves are still in phase after a round trip and interfere constructively.

If the mirrors are covered with a metallic layer the phase shift at reflection is equal to π, the electric field is equal to zero on them; a node of the stationary wave pattern is located on each mirror. The separation between the mirrors is an integer number of half-wavelengths. In the case of dielectric layers, the phase shifts at reflection, ψ_1 and ψ_2, are not equal to π, so we have

$$d = \lambda_p \left(\frac{p}{2} + \frac{\psi_1 + \psi_2}{2\pi} \right),$$

usually the first term in parentheses is considerably larger than the second one.

Between the two mirrors, stationary waves are found with fixed minima (nodes) and maxima (antinodes) where the electric and magnetic fields may be considerably reinforced. If a piece of material is placed between the mirrors the electric dipolar interactions are considerably enhanced at the place of the antinodes of the electric field: at these antinodes the amplitude of the fields is equal to the amplitude of the incident wave R_0 multiplied by the quality coefficient Q.

The Fabry-Perot Resonator Is Well Adapted to Optics

The previous considerations apply whatever the frequency. They are valid in the radio or microwave domains, however, in these two cases, theoretical and practical difficulties arise; on the one hand, the wavelengths, being larger than in Optics, cannot be considered to be small as compared to the transversal size of the mirrors and planar waves are not suitable solutions. On the other hand, it's more difficult to make good reflectors with the same quality as in Optics.

Vacuum deposition techniques allow the deposition of reflecting coatings almost perfectly transparent for optical waves. Thanks to lasers the technology of thin layer deposition has been greatly improved, and any corresponding methods are carefully protected, if they are not military secrets. Given the high values of the power involved in laser beams a low value of the absorption coefficient is essential for a laser mirror.

9.4.4. The Role of Diffraction

When the transversal dimensions of the mirrors are not large enough to be considered as infinite, planar waves don't describe accurately the electromagnetic solution of the problem of a resonator made of two parallel mirrors. The exact solution resembles a planar wave that would be limited by a diaphragm, Figure 9.24 illustrates the reason why higher losses are found in the case of transversally limited mirrors: the light reflected by one mirror

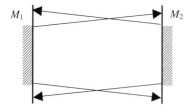

M_1 M_2

Figure 9.24. Diffraction losses. Because of diffraction, part of the light coming from one mirror covers an area which is larger than the reflecting part of the other mirror and is only partially reflected.

covers an area which is a little larger than the reflecting zone of the other mirror; even if its coatings are perfectly reflecting, the second mirror sends back an energy which is smaller than the incident energy. It can be shown that the diffraction losses are substantially decreased if the planar mirrors are replaced by spherical mirrors. Before the invention of lasers the French physicist P. Cohnes used spherical mirrors to improve the resolution of FP spectrometers. The eigenmodes of spherical resonators are made of spherical waves, instead of planar waves. The adjustment of a spherical FP resonator is easier than the adjustment of a planar FP resonator since the parallelism of the mirrors is less critical.

The theoretical description of a resonator with noninfinitely extended mirrors is not simple. We must first find special waves which will reproduce identically, reflection after reflection, on the two mirrors. Let us consider a given wave leaving mirror M_1 of Figure 9.24 and going toward mirror M_2, and let $P_0 = P_{0(x,y)}e^{j\phi_{0(x,y)}}$ be the expression of its complex amplitude on the mirror M_1. Using Huygens' wavelets, we can, in principle, calculate the figure of diffraction that is created on the other mirror and obtain the repartition of the complex amplitudes along M_2, say $P_1' = P_{1(x,y)}'e^{j\phi_{1(x,y)}}$. To calculate the wave that is reflected by M_2, P_1' is then multiplied by a complex factor $r_{2(x,y)}$, which varies with the coordinates according to the following law: $r_{2(x,y)}$ is constant inside the mirror and is equal to zero outside.

We obtain a new repartition of the complex amplitudes along mirror M_2, $P_1 = r_{2(x,y)}P_{1(x,y)}'e^{j\phi_{0(x,y)}}$, from which is obtained, using the same method, the repartition of complex amplitudes, P_2', created on M_1 by the reflected wave. P_2' is multiplied by a reflection coefficient $r_{1(x,y)}$ equal to zero outside mirror M_1 and the new figure of diffraction is now calculated after two reflections.

The iteration may be continued indefinitely; we obtain the expression of the wave after a number l of reflections, knowing its expression after $l - 1$ previous reflections. The calculations are of course not straightforward, but the result can be considered rather simple: if the initial function, $P_0 = P_{0(x,y)}e^{j\phi_{0(x,y)}}$, is selected from a special set of functions $D_{(x,y)}^{m,n}e^{j\phi_{(x,y)}^{m,n}}$, called the eigenmodes of the resonator, then P_l is simply proportional to P_{l-1},

$$P_{l(x,y)}e^{j\phi_{l(x,y)}} = KP_{l-1(x,y)}e^{j\phi_{l-1(x,y)}}, \quad K \text{ is a complex constant,} \quad (9.41)$$

where $D_{(x,y)}^{m,n}$ and $\phi_{(x,y)}^{m,n}$ are specific to the resonator under consideration: size of the mirrors, distance between them, radius of curvature if they are spherical; m and n are integers.

The case of planar mirrors of *infinite transverse extension* is *degenerated* corresponding to planar waves: $D_{(x,y)}^{mn}$ and $\phi_{(x,y)}^{mn}$ are independent of m and n as well as of the coordinates x and y, the coefficient K of formula (9.41) is equal to the product of reflection coefficients of the mirrors. In the case of perfectly reflecting mirrors of limited extension, the modulus of K is smaller than one because of the diffraction losses.

Coming back to the general cases, the eigenmodes make a family with two degrees of freedom associated to the two indices m and n. The usual form of $D_{(x,y)}^{mn} e^{j\phi_{(x,y)}^{mn}}$ is given by

$$D_{(x,y)}^{mn} e^{j\phi_{(x,y)}^{mn}} = e^{-\left(x^2+y^2\right)/w^2} H_{(x,y)}^{mn} e^{j\phi^{mn}} . \tag{9.42}$$

It is the term $e^{-(x^2+y^2)/w^2}$ that really describes the effect of the limited size of the mirrors. The modes are Gaussian beams, their waist, w, is of the order of the diameter of the mirrors. $H_{(x,y)}^{mn}$ is a real function, typical patterns are shown in Figure 9.25. The mode labeled (0, 0) is called the principal mode, $H_{(x,y)}^{00}$ and $\phi_{(x,y)}^{00}$ are independent of x and y; this mode looks very much like a planar wave. Modes for which $(m,n) \neq (0,0)$ are called transverse modes, the electromagnetic energy is confined in smaller spots inside which the phase is constant but varies from one spot to the next. The coefficient K of formula (9.41) diminishes rapidly with increasing values of m and n, since the diffraction losses increase when the size of the spot decreases.

In most cases a laser will emit on the fundamental mode of its resonator, which has the smaller losses. It may happen accidentally that the laser emission occurs on a transverse mode, the emission patterns are very esthetic and coincide with the patterns of Figure 9.25, the corresponding laser beams are more divergent because of diffraction. In commercial lasers the transverse modes are carefully avoided.

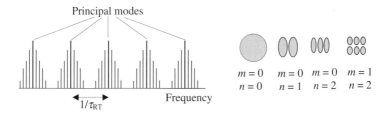

Figure 9.25. Principal and transverse modes of an FP resonator. In the left-hand diagram, a line with a height proportional to the quality coefficient represents each mode. The frequency difference between two adjacent principal modes is equal to the inverse of the round trip time. The diagram is not to scale since the frequencies of the transverse modes are closer.

9.4.5. *The Comb of Modes, Orders of Magnitude*

The eigenmodes that have been introduced in the previous sections should be considered as original and independent physical objects, which can be excited independently. From a mathematical point of view, this corresponds to the fact that they are described by *orthogonal functions*.

The frequency of a mode involves three indices, mnp, two of them, m and n, have been introduced in formula (9.42), the third one, p, is associated with the standing wave pattern and is related to formula (9.40) and is equal to the number of antinodes between the mirrors, see Figure 9.23. The frequency repartition of the modes is given in Figure 9.25; if we just consider the principal modes the diagram looks like a comb, since the frequency difference between two adjacent principal modes is constant and is equal to $1/\tau_{RT}$. Except for semiconductor lasers, which are very short, the length of a laser resonator ranging from centimeters to meters is considerably larger than an optical wavelength and p goes from 10^4 to 10^6.

In a 1 m length resonator the return time is $\tau_{RT} = 2d/c = 6.7$ ns and the frequency between adjacent modes is $(v_{p+1} - v_p) = 150$ MHz. An optical frequency being of about 10^{14}, it is seen than the different principal modes are very close.

The determination of the frequencies of the transverse modes requires knowledge of the functions ϕ^{mn}; the distance between two adjacent transverse modes is typically 10 MHz.

Quality Coefficient of a Mode

As for any resonant element it is interesting to evaluate the bandwidth and the quality coefficient Q of a mode. To do so we must first evaluate the losses; if we ignore the diffraction losses and just take into account the reflection losses, it is sufficient to know the reflection coefficients of the mirrors, ρ_1 and ρ_2. The same kind of calculation has already been made in Section 6.4.2. The quality coefficient of a 1 m long resonator, using mirrors for which $\rho_1\rho_2 = 0.95$, is about 10^9 and the bandwidth is about 1 MHz.

9.4.6. *How to Undamp a Fabry-Perot Resonator: Laser Effect*

In the presence of a weakly damped circuit, the instinctive reaction of someone who is familiar with radio frequency is to try to use it to make an autooscillator. The next requirement is the need for an amplifier working at the right frequency, and to consider an FP resonator filled with a medium like the one studied in Section 9.2.3 and then let us use the methods of Section 9.4.3. We again write the set of equations (9.38), but now the wave vector of L_1 and R_1 is a complex number, $k = k' - jk''$:

$$R_i: R_i e^{-k''z} e^{j(\omega t - k'z)}, \quad G_i: G_i e^{k''z} e^{j(\omega t + k'z)}.$$

We obtain formulas identical to (9.39) where $A_{(\phi)} = 1/(1 - r_1 r_2 e^{-j\phi})$ is replaced by

$$A'_{(\phi)} = \frac{1}{\left(1 - r_1 r_2 e^{-2k''d} e^{-j\phi}\right)} = \frac{1}{\left(1 - r_1 r_2 e^{\alpha d} e^{-j\phi}\right)}, \tag{9.43}$$

with

$$\alpha = 2k'' = -\frac{c^3}{8\pi v^2 V} \frac{1}{\tau_{\text{radiative}}} f(v)(N_0 - N_1).$$

In the case of an inversion of population, the imaginary part, $-k''$, of the wave vector is positive

$$2k'' = -\alpha = -\frac{c^3}{8\pi v^2 V} \frac{1}{\tau_{\text{radiative}}} f(v)(N_0 - N_1).$$

When ϕ is varied, the module of the complex number oscillates between the two values $1/(1 - r_1 r_2 e^{+\alpha d})$ and $1/(1 + r_1 r_2 e^{+\alpha d})$. If $r_1 r_2 e^{+\alpha d} = 1$, frequencies can be found for which the complex number $A'_{(\phi)}$ becomes infinite: the resonator is no longer damped and light is emitted.

The oscillation condition is thus given by

$$r_1 r_2 e^{-j\phi} = 1. \tag{9.44}$$

Equation (9.44) is a relation between complex numbers and implies two relations between real numbers:

$$r_1 r_2 e^{+\alpha d} = 1, \tag{9.45.a}$$

$$e^{-j\phi} = 1 \quad \rightarrow \quad \phi = p2\pi. \tag{9.45.b}$$

Equation (9.45.b) shows that laser emission is produced at the frequencies of the eigenmodes of the resonator. Relation (9.45.a) receives an easy physical explanation: after a double transit inside the resonator, the reflection losses, $r_1 r_2$, are compensated for by the amplification $e^{\alpha d}$. The condition expressed by (9.44.b) should be replaced by a threshold condition:

$$r_1 r_2 e^{+\alpha d} > 1 \quad \rightarrow \quad \alpha \geq \alpha_{\text{threshold}} = \frac{1}{d} \text{Log}_e(1/r_1 r_2). \tag{9.46.a}$$

To obtain a laser effect with given mirrors, the inversion of population should be greater than some threshold value so that $\alpha > \alpha_{\text{threshold}}$. Introducing the reflection coefficients for intensities, $\rho_1 = r_1^2$ and $\rho_2 = r_2^2$, (9.46.a) is written as

$$\alpha_{\text{threshold}} = \frac{1}{2d} \text{Log}_e(1/\rho_1 \rho_2) \approx \frac{(1 - \rho_1 \rho_2)}{2d}. \tag{9.46.b}$$

A Laser Starts from Spontaneous Emission

Formulas (9.46) are exactly equivalent to formula (9.8) that was obtained at the very beginning of this chapter for a feedback oscillator. We are now going to show how to apply the treatment that helped to solve the mathematical difficulties associated with this kind of device.

Because of pumping, an inversion of population is created, the collections of atoms being largely out of equilibrium, numerous spontaneous emissions occur. Waves are emitted in all directions with frequencies falling inside the spectral band of the function $f(\nu)$ which defines the profile of the atomic transition: among these waves, the only ones which will survive and become amplified, while propagating back and forth between the mirrors, are those which have the same properties as the eigenmodes.

The amplification is a consequence of stimulated emissions, which occur at a rate that increases with the amplitude of the wave. The more intense the laser light, the greater is the number of stimulated emissions per second and also the number of atoms falling to ground level. When the laser light becomes intense, the population inversion decreases and so does the amplification coefficient: this is the reason why the amplifier *saturates* and why the amplitude of the oscillation doesn't grow indefinitely.

9.5. Spectral Characteristics of Light Emitted by a Laser

9.5.1. *Multimode Emission*

Figure 9.26 shows the variation of the feedback coefficient of an FP resonator (comb of modes) and the variation of the optical gain of a material with an

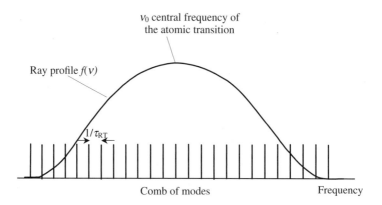

Figure 9.26. Gain and feedback coefficients versus frequency. This figure is to be compared with Figure 9.3.

inversion of population. Generally speaking, the autooscillation condition (9.44) is fulfilled for many modes, and very often for a tremendous number of modes. All the modes having a frequency near enough to the central frequency of the atomic transition will be undamped. In such a case the laser will emit several rays, which are extremely monochromatic and have very close frequencies: the laser is then said to be *multimode*.

9.5.1.1. *Random Multimode Emission*

If no special care is taken a laser will work in this way. In principle, the different modes are not coupled to one another and oscillate independently. After the pumping has been switched on, there is a transient state during which the different modes start oscillating. A given mode starts when the autooscillation condition is fulfilled, initially increasing the oscillation amplitude saturates. When a steady state is reached, the electromagnetic field can be written as $E_n \cos(2\pi v_n t - \phi_n)$, where n is the index of the mode. The signal emitted by the laser is the sum of the different oscillations of all the oscillating modes

$$E(t) = \sum_n E_n \cos(2\pi v_n t - \phi_n).$$

The phase ϕ_n is fixed by the instant when the oscillation started; usually there is no deterministic relation between ϕ_n and n. The above summation is not possible in the absence of more information about ϕ_n, however if the number of oscillating modes is large, it is considered that the global intensity is equal to the addition of the individual intensities:

$$I_n = \sum_n E_n^2.$$

9.5.1.2. *Mode Locking*

It is possible to suppress the uncertain repartition of the different phases ϕ_n. When this is the case, it is said that the different modes are *synchronized*, or that their phases are *locked* to one another.

The time variation law of the intensity of a mode-locked laser is completely different and is the result of the interference of a great number of signals whose frequencies are equally separated by the intermodal separation $c/2d = 1/\tau_{RT}$,

$$v_n = v_0 + \frac{n}{\tau_{RT}}.$$

For the sake of simplification we will consider that all the ϕ_n are equal to zero and that all the modes oscillate with the same amplitude E_0,

$$E_{(t)} = E_0 \sum_{n=0}^{N} \left[\cos 2\pi \left(\nu_0 + \frac{n}{\tau_{RT}} \right) t \right],$$

$$E_{(t)} = E_0 \, \text{Re} \left\{ e^{j2\pi\nu_0 t} \sum_{n=0}^{N} \left[e^{j2\pi \frac{n}{\tau_{RT}} t} \right] \right\}, \qquad (9.47)$$

$$\sum_{n=0}^{N} \left[e^{j2\pi \frac{n}{\tau_{AR}} t} \right] = \frac{e^{j2\pi(N+1)\frac{t}{\tau_{RT}}} - 1}{e^{j2\pi \frac{t}{\tau_{RT}}} - 1} = \frac{e^{j\pi(N+1)\frac{t}{\tau_{RT}}}}{e^{j\pi \frac{t}{\tau_{RT}}}} \frac{e^{j\pi(N+1)\frac{t}{\tau_{RT}}} - e^{-j\pi(N+1)\frac{t}{\tau_{RT}}}}{e^{j\pi \frac{t}{\tau_{RT}}} - e^{-j\pi \frac{t}{\tau_{RT}}}},$$

$$E_{(t)} = E_0 g_{(t)} \cos \left[2\pi \left(\nu_0 + \frac{N}{2\tau_{RT}} \right) t \right] \quad \text{with} \quad g_{(t)} = \frac{\sin \left[\pi(N+1) \frac{t}{\tau_{RT}} \right]}{\sin \left[\pi \frac{t}{\tau_{RT}} \right]}. \qquad (9.48)$$

The total number of modes N that have been locked is determined by the number of teeth in the comb that are contained inside the bandwidth of the amplifier, this is usually a large number (10^4–10^6). Equation (9.48) represents a sinusoidal signal having a frequency close to ν_0 and modulated by an envelope describing the function $g(t)$ of formula (9.48).

The intensity becomes equal to $I_{Max} = (N + 1)^2 E_0^2 \approx N^2 E_0^2$, everytime that t is equal to an integer multiple of τ_{RT} and decreases to zero after a duration equal to τ_{RT}/N, the intensity then oscillates between zero and E_0^2 and will grow again to I_{Max}. So, a mode-locked laser regularly emits very short and powerful pulses. The duration of an individual pulse is of the order of τ_{RT}/N, the repetition period is τ_{RT}.

An elegant physical explanation can be given for mode locking. At $t = 0$, the $N + 1$ terms of the summation of formula (9.47) are in phase and interfere constructively, the resulting amplitude is equal to $(N + 1)E_0$; as they don't oscillate exactly at the same frequency, the phase distribution spreads rapidly over 2π and their sum is almost equal to zero (it fact it oscillates between zero and E_0). The time for the amplitude to decrease from $(N + 1)E_0$ to E_0 is about equal to the round trip time divided by the number of synchronized modes. The modes will again be *in phase* when t is an integer multiple of τ_{RT}. Finally, the laser emits a succession of regularly spaced pulses. A numerical application shows that pulses as short as a picosecond, and even a tenth of a femtosecond, may be obtained. A few femtosecond light pulses have only several light periods. . . .

Fourier analysis can also obtain the previous result. According to formula (9.47) the Fourier spectrum of a mode-locked laser is a Dirac comb (separation of the teeth $1/\tau_{RT}$) multiplied by a rectangle having a length equal to N/τ_{RT}. To obtain the time law of variation we make a Fourier transform. If N were

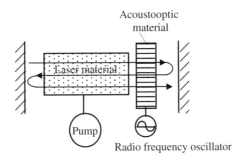

Figure 9.27. Synchronization of the modes using an acoustooptic effect.

infinite, the amplitude would also be a Dirac comb with a tooth separation equal to $(1/\tau_{RT})^{-1} = \tau_{RT}$. If there is only a finite number of teeth, the Fourier transform is again a comb, each tooth being the Fourier transform of a rectangle, that's to say, a *sinc* function having its first zero at time τ_{RT}/N as given by (9.48).

There are many methods of synchronizing the modes of a laser, they, more or less, find a way to break the randomness of the modes, their dates of birth. In the experimental set-up of Figure 9.27, use is made of the acoustooptic effect in a well-chosen crystal placed inside the laser cavity. When a transparent material supports an acoustic wave of frequency ν_{acoust} its index of refraction is modulated at the same frequency; the experiment shows, and the theory verifies (see the Brillouin effect), that a light wave of frequency ν_{opt} propagating in such a medium is partly converted into two new light waves of respective frequencies $(\nu_{opt} \pm \nu_{acoust})$. If the frequency of the oscillator that generates the acoustic wave frequency is tuned to be just equal to the intermodal separation, $\nu_{acoust} = (\nu_p + 1 - \nu_p) = 1/\tau_{RT}$, the optical waves obtained by the acoustooptic effect from one initial mode belong to the family of modes of the FP resonator. The successive appearance of the modes is no longer at random.

9.5.2. *Single Mode Emission*

As opposed to the multimode case, a laser may be a single mode and emit only one frequency. If the bandwidth of the optical amplifier is narrow (a few gigahertz), the FP resonator can be given a short enough length (10 cm), so that only one tooth of the comb of modes is inside the gain curve: only one mode will be able to oscillate, see Figure 9.28. A fine-tuning of the distance between the mirrors is necessary to put the mode frequency in good coincidence with the central frequency of the gain curve; this is obtained thanks to a piezoelectric cell, see Figure 9.29. The DC voltage across the piezo is adjusted so that the output signal is maximized; the laser frequency is exactly equal to the frequency of the atomic transition.

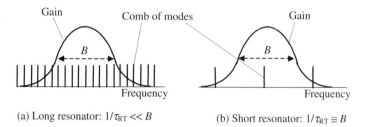

(a) Long resonator: $1/\tau_{RT} \ll B$ (b) Short resonator: $1/\tau_{RT} \cong B$

Figure 9.28. Relative disposition of the gain curve and of the comb of modes for two different sizes of the resonator. (a) is well adapted to a multimode emission. In the case of (b) the intermodal separation is large, only one mode falls inside the bandwidth of the gain curve.

If the gain curve is broad, thousands of gigahertz, a single mode emission is difficult to achieve by the previous method; in this case a thin FP etalon (thickness e) is inserted inside the main resonator. This arrangement is shown in Figure 9.30, the incidence θ on the etalon is oblique and tunable. The beams having a frequency equal to an eigenfrequency of the etalon are well transmitted, the others being rejected in a direction making an angle 2θ with the axis. Two combs of modes are now involved: one for the laser resonator and one for the etalon; their respective intermodal separations are quite different; laser emission can only occur at frequencies that belong to the two combs. The etalon is narrow, so its modes are quite distant if its index of refraction is equal to n; the mode separation, which is equal to $c/2ne\cos\theta$, is finely tunable by playing with θ: One frequency of the etalon can be put in coincidence with a frequency of the laser resonator.

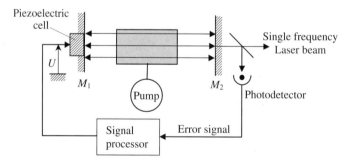

Figure 9.29. Single frequency laser. The FP resonator is short and only one mode falls inside the gain curve. A piezoelectric crystal attached to one of the mirrors finely tunes the mirror separation. An electronic device permanently adjusts the voltage U to maximize the laser intensity.

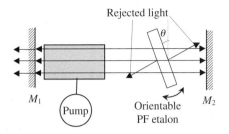

Figure 9.30. Mode selection using a FP etalon. Laser emission can only occur at a frequency which belongs to the two combs of modes of the etalon and of the laser resonator. A rotation of the etalon finely shifts the frequencies of the associated comb.

Frequency Tunable Lasers

The invention of frequency tunable lasers has considerably modified the art of optical spectroscopy. Figure 9.31 shows two possible arrangements. A prism is placed between the amplifying material and the second mirror; let us consider a light ray IJ propagating orthogonal to mirror M_1 and having a wavelength λ, after refraction by the prism it gives the light ray KL. If mirror M_2 is disposed perpendicular to KL, the reflected beam propagates along the same direction and then indefinitely goes back and forth between the mirrors: a laser emission is possible if the autooscillation condition is fulfilled. As the deviation by a prism changes with the wavelength, a rotation of the prism tunes the wavelength, the wavelength for which laser emission is obtained.

The second arrangement of Figure 9.31 uses a diffraction grating in the Littrow arrangement; for a given angle of incidence on the prism we can find a wavelength for which the diffracted beam propagates in the opposite direction to the incident beam. If the grating is suitably blazed the diffracted beam D doesn't exist. Rotating the grating allows the tuning of the laser frequency.

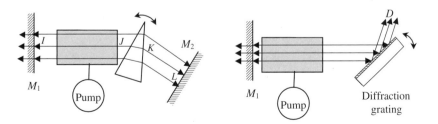

Figure 9.31. Dispersive optical components are used to tune the frequency of a laser.

9.6. Laser Transient Effects

9.6.1. *Laser Rate Equations*

Our purpose is now to describe the time evolution of the light emitted by a laser. The laser is supposed to work only on one mode.

We use the set of equations (9.29) to which must be added the description of the pumping and of the optical losses of the resonator. The losses are "useful losses" since they represent the light that is emitted by the laser. Let us call dn_{loss} the number of photons that leave the resonator during the time dt,

$$dn_{loss} = -\frac{n}{\tau_{res}} dt, \quad dn_{loss} = -\frac{n}{\tau_{res}} dt. \tag{9.49.a}$$

The power emitted by the laser is given by formula (9.49.b):

$$P_{laser} = -hv \frac{dn_{loss}}{dt} = \frac{n}{\tau_{res}} hv. \tag{9.49.b}$$

We suppose that the lower level of the laser transition is the ground level, the pumping becomes completely inefficient when this level is empty, so the pumping term in the equation is written as $W_p N_0$,

$$\begin{cases} \dfrac{dN_0}{dt} = \dfrac{N_1}{\tau_{rad}} + \dfrac{n(N_1 - N_0)}{\tau_{rad}} - W_P N_0, \\[2mm] \dfrac{dN_1}{dt} = -\dfrac{N_1}{\tau_{rad}} - \dfrac{n(N_1 - N_0)}{\tau_{rad}} + W_P N_0, \\[2mm] \dfrac{dn}{dt} = -\dfrac{N_1}{\tau_{rad}} - \dfrac{n}{\tau_{res}} + \dfrac{n(N_1 - N_0)}{\tau_{rad}}, \end{cases} \tag{9.50}$$

$$N_{total} = (N_1 + N_0) \quad \text{total number of atoms,}$$
$$N = (N_1 - N_0) \quad \text{inversion of population,}$$

$$N_1 = \frac{N_{total} + N}{2}, \quad N_0 = \frac{N_{total} - N}{2}.$$

Equations (9.50) become

$$\frac{dn}{dt} + n\left(\frac{1}{\tau_{res}} - \frac{N}{\tau_{rad}}\right) = \frac{N + N_{total}}{2\tau_{rad}}, \tag{9.51}$$

$$\frac{dN}{dt} = 2\frac{dN_1}{dt} = -\frac{N_{total} + N}{\tau_{rad}} - \frac{2nN}{\tau_{rad}} + W_P(N_{total} - N). \tag{9.52}$$

Equations (9.50), (9.51), and (9.52) are called the *laser rate equations*; they need two variables η and n, the variation of which has to be calculated. Their solution is not immediate because of the presence of the *stimulated term* nN/τ_{rad} and can only be made by numerical integration.

Initial Conditions

At time $t = 0$, the pump is suddenly switched on and the pumping term jumps from zero to W_P and then remains constant. The initial conditions are:

- $N_{0(0)} = N_{total}$.
- $N_{(0)} = -N_{total}$, initial inversion of population.
- $n_{(0)} = 0$, initial number of photons in the mode.

The following scenario can summarize the result of the numerical integration:

- Because of the pumping, the inversion of population, initially negative, starts growing and at a certain time becomes positive.
- The collection of atoms is then an optical amplifier; however, the inversion of population is not sufficient for the gain to compensate for the losses. The autooscillation can start only when the inversion of population reaches the threshold value $N_{threshold}$.
- The excited level becoming more and more populated, spontaneous emission creates more and more photons. The number n of photons remains small as long as $N < N_{threshold}$. Once the threshold is passed, n suddenly increases faster and, in a few nanoseconds, goes from almost nothing to an enormous value.
- The number of stimulated emissions per second is now important and the inversion of population decreases to keep a value close to the threshold value.

Threshold Population Inversion

If the inversion of population is considered as constant, the solution of equation (9.51) is

$$n = \frac{N + N_{total}}{2\tau_{rad}/\tau}(e^{+t/\tau} - 1) \quad \text{with} \quad \frac{1}{\tau} = \frac{N}{\tau_{rad}} - \frac{1}{\tau_{res}}. \tag{9.53}$$

The condition that the number of photons increases is given simply by

$$\frac{1}{\tau} = \left(\frac{N}{\tau_{rad}} - \frac{1}{\tau_{res}}\right) > 0 \quad \rightarrow \quad N_{threshold} = \frac{\tau_{rad}}{\tau_{rés}}.$$

In formula (9.53), the term $(N + N_{total})/(2\tau_{rad}/\tau) = N_1/(\tau_{rad}/\tau)$ represents the spontaneous emission, which clearly demonstrates that the laser oscillation starts from spontaneous emission.

Maximum Value of the Inversion of Population

Let us suppose that the collection of atoms has not been placed inside a resonator or, which is equivalent, that one mirror is hidden. No autooscillation is to occur, we have $1/\tau_{\text{res}} = 0$ and $n = 0$ in equation (9.52). The inversion of population is then given by

$$N = -N_{\text{total}}e^{-t/\Theta} + N_{\text{Max}}(1-e^{-t/\Theta}), \tag{9.54}$$

with

$$N_{\text{Max}} = N_{\text{total}}\frac{W_P - 1/\tau_{\text{rad}}}{W_P + 1/\tau_{\text{rad}}} \quad \text{and} \quad \Theta = \frac{1}{W_P + 1/\tau_{\text{rad}}},$$

where in N_{Max} is the maximum value that can be obtained with a given pumping represented by W_P; this value is reached with a time constant equal to Θ_0. In the presence of a resonator it can be considered that formula (9.54) remains almost valid until the instant Θ_0 when the laser oscillation starts, which instant is close to the instant when the inversion of population becomes positive.

Θ_0 is given by

$$\Theta_0 = \Theta \operatorname{Log}_e\left(\frac{1 + W_P\tau_{\text{rad}}}{2W_P\tau_{\text{rad}}}\right).$$

We arrive at a time t that is larger than Θ_0, the spontaneous emission is now negligible as compared to the spontaneous emission and the associate term, N_1/τ_{rad}, can be neglected in equations (9.50) and (9.51).

We introduce the reduced variables

$$\theta = \frac{t}{\tau}, \quad y = n\frac{\tau}{\tau_{\text{rad}}}, \quad x = N\frac{\tau}{\tau_{\text{rad}}},$$

$$x_{\text{Max}} = N_{\text{Max}}\frac{\tau}{\tau_{\text{rad}}}, \quad x_{\text{threshold}} = N_{\text{threshold}}\frac{\tau}{\tau_{\text{rad}}}, \tag{9.55}$$

$$\frac{dx}{d\theta} = x_{\text{Max}} - x - 2xy, \quad \frac{dy}{d\theta} = y(x_{\text{threshold}} - x).$$

Cancellation of the time derivative in (9.55) gives the steady state values of x and y,

$$x_{\text{steady}} = x_{\text{threshold}} \quad \text{and} \quad y_{\text{steady}} = \frac{x_{\text{Max}} - x_{\text{threshold}}}{2x_{\text{threshold}}}.$$

Once the steady state is reached, the population inversion remains constant and equal to the threshold value. The permanent value of the number of photons is all the higher as the maximum possible inversion of the population is larger than the threshold. A rigorous integration of the rate equation being impossible, we are going to see what happens if the pumping is not too strong and that x and y remain close to their steady state values. We introduce the auxiliary functions $p(t)$ and $q(t)$,

$$x = x_{\text{steady}}[1 + p(t)] \quad \text{and} \quad y = y_{\text{steady}}[1 + q(t)],$$

$$K = \frac{x_{\text{Max}}}{x_{\text{threshold}}},$$

$$\frac{d^2 p}{d\theta^2} + K \frac{dp}{d\theta} + (K - 1)x_{\text{threshold}} p = 0. \tag{9.56}$$

The solutions of (9.56) are damped oscillations: $y = Ae^{-\gamma t}\cos\omega_m t$, with

$$\omega_m = \frac{1}{2\tau}\sqrt{4(K-1)x_{\text{threshold}} - K^2} \approx \frac{1}{\tau}\sqrt{(K-1)x_{\text{threshold}}} \quad \text{and} \quad \gamma = K/2\tau,$$

where ω_m is a good indication of the order of magnitude of the frequency of the oscillations of relaxation that are often emitted by solid-state lasers and that will be described in the next section.

9.6.2. *Oscillation of Relaxation*

To study what happens in the case of a strong pumping rate we use a hydraulic analogy in which the populations are assimilated to the respective levels in two tanks.

We refer to Figure 9.32. Water is pumped from a lower tank to another situated above. A siphon can empty the upper tank if the level of the water is higher than a threshold level, which can be reached only if the pump is able

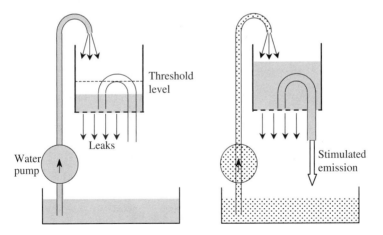

Figure 9.32. Hydraulic analogy of a laser. The two tanks are, respectively, the ground and excited levels. The leaks represent spontaneous emission. The syphon is for stimulated emission.

to compensate for the leaks. When this is so, the upper tank fills until the syphon starts operating.

If the rate of filling by the pump and the rate of emptying by the syphon are comparable, the level in the upper tank remains just above the threshold and the syphon continuously pours water into the lower tank.

If the syphon has a large diameter, the upper tank is suddenly emptied, a large amount of water being lowered; we must then wait until the upper tank is again filled before a second quantity of water is lowered: a relaxation process then takes place.

The results of a numerical solution of the rate equations are in good agreement with the hydraulic analogy. According to this model, relaxation pulses are periodically emitted and their intensity decreases from one pulse to the next.

The agreement with experimental results is reasonable but not excellent. A solid state laser emits short (microsecond) and intense (10 to 100 kW) relaxation light pulses that don't have the nice regularity of the theoretical pulses. They are erratically emitted, the average time separation between two successive pulses being of the order of 1 μs.

Several explanations can be given for this experimental discrepancy. The rate equations implicitly suppose a monomode laser; in fact, many modes are simultaneously emitted and they compete in sharing the inversion of population. Coming back to the hydraulic analogy, to simulate the competition between the different modes, many different syphons should be introduced. ... The erratic character of the relaxation light pulse comes from the fact that, because of the nonlinear term nN, the solution of equations (9.51) and (9.52) may have a *chaotic behavior*.

Let us conclude by saying that the relaxation emission of pulses is not safe for a solid state laser and is often accompanied by a deterioration of the laser material.

9.6.3. Q-Switching

One of the mirrors of the laser is hidden during the start of the pumping, meanwhile the quality coefficient Q of the cavity is kept at a low value, which allows the inversion of population to reach a high value, almost equal to the value N_{Max} that is calculated in formula (9.54). As a result, an important amount of energy (J/cm^3) is stored in the collection of atoms.

If a high value is again given to Q (this is the reason for the expression *Q-switch*), the laser will immediately start oscillating. Coming back to the hydraulic analogy, the syphon is first closed allowing the accumulation of water in the upper tank, which is then rapidly emptied after reopening.

As far as lasers are concerned, the time necessary to restore a high value to the quality coefficient must be shorter than the relaxation period of the laser (i.e., 1 μs); techniques for fast optical switching will be described later.

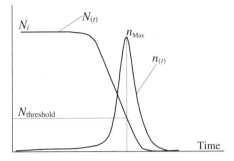

Figure 9.33. Variation of the inversion of population and of the number of photons in a Q-switched laser. The initial population inversion is far greater than the threshold value.

The experiment shows that a powerful pulse (megawatt to gigawatt) is emitted with duration in the nanosecond range; such a pulse is commonly called a *giant pulse*.

Let us come back to the rate equations and integrate equations (9.51) and (9.52) using the following initial conditions: time $t = 0$ is the instant when Q takes a high value. The laser has not yet started oscillating and the number of photons $n(0)$ is equal to zero. The initial value of the inversion of population has a high value $N(0) = N_i$. Spontaneous emission and pumping are negligible during the emission of the giant pulse. Changing the time unit by setting $t' = t/\tau_{res}$, we can write

$$\begin{cases} \dfrac{dN}{dt'} = -2\dfrac{nN}{N_{threshold}}, \\ \dfrac{dn}{dt'} = n\left(\dfrac{N}{N_{threshold}} - 1\right), \end{cases} \qquad N_{threshold} = \dfrac{\tau_{rad}}{\tau_{res}},$$

$$\dfrac{dn}{dN} = -\dfrac{1}{2}\left(1 - \dfrac{N_{threshold}}{N}\right) \quad \rightarrow \quad n_{(N)} = \dfrac{N_i - N}{2} + \dfrac{N_{threshold}}{2}\log_e\left(\dfrac{N}{N_i}\right).$$

The number of photons is maximum at the instant when the population inversion is just equal to its threshold value and is then equal to

$$n_{Max} = \dfrac{N_i - N_{threshold}}{2} + \dfrac{N_{threshold}}{2}\log_e\left(\dfrac{N_{threshold}}{N_i}\right).$$

According to (9.49), the peak power of the pulse is equal to

$$P_{Max} = h\nu\,\dfrac{n_{Max}}{\tau_{res}}.$$

N_{fiina} being the inversion of population at the end of the pulse, the total number of atomic transitions during the pulse is simply equal to $N_i - N_{fiina}$; the global energy W of the pulse is then easily deduced. If the pulse is assimilated to a square pulse we obtain an order of magnitude of its duration θ;

$$W = h\nu \frac{N_i - N_{\text{final}}}{2} \quad \rightarrow \quad \theta = \frac{W}{P_{\text{Max}}} = \tau_{\text{res}} \frac{(N_i - N_{\text{final}})}{2n_{\text{Max}}},$$

where N_{fiina} is obtained by making $n = 0$ and solving the transcendental equation

$$\log_e\left(\frac{N_i}{N_{\text{final}}}\right) = \frac{N_i - N_{\text{final}}}{N_{\text{threshold}}}.$$

Numerical application: Volume of the laser crystal $1\,\text{cm}^3$; doping concentration $N_{\text{total}} = 10^{20}$ atoms/cm^3; photon energy $1.5\,\text{eV}$.
  $N_i = 0.1N_{\text{total}}$; $N_{\text{threshold}} = 0.01N_{\text{total}}$; $N_{\text{fiina}} = 0$; $\tau_{\text{res}} = 10\,\text{ns}$. $n_{\text{Max}} = 0.2210^{19}$;
$W = 1.2\,\text{J}$; $P_{\text{Max}} = 0.5310^8\,\text{W}$; $\theta = 2.2610^{-8}\,\text{s}$.

Q-Switching Devices

Figure 9.34 shows two devices used for Q-switching a solid-state laser. The rotating prism has been largely used and is now obsolete. A Porro prism rotates at high speed about an axis that is orthogonal to the edge and parallel to the hypotenuse. The rays can oscillate inside the optical cavity when the angle between the edge of the prism and the plane of the mirror is smaller than the diffraction angle of divergence θ of a beam limited by a diaphragm having the same diameter D as the laser rod ($\theta = \lambda/D \approx 10^{-3}\,rad$). The speed of rotation should be greater than θ divided by the relaxation period of the laser, which makes radian/microsecond $\approx$ ten thousand rounds per minute.
  Electrooptic Q-switching was initially obtained with a Kerr cell, but Pockels' cells using the electrooptic effect in a crystal are far better since they

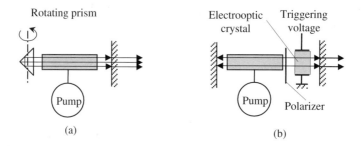

(a)   (b)

Fig. 9.34. The laser in (a) can oscillate only when the edge of the rotating prism is parallel to the mirror with an accuracy better than λ/D (diameter of the rod divided by the wavelength). The neutral lines of the electrooptic crystal of (b) are oriented at $45°$ with the polarizer; if the voltage is such that it behaves as a $\lambda/4$ plate, the reflected light is polarized at $90°$ with the polarizer and is blocked, the switch becomes transparent again when the voltage drops to a value for which the crystal is a λ plate.

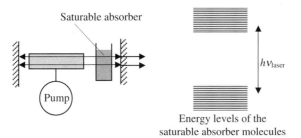

Figure 9.35. Passive Q-switch using a saturable absorber. A cell, filled with a solution of molecules having an absorption band at the laser frequency, is placed inside the laser cavity. Initially opaque, the cell becomes transparent after the spontaneous emission has saturated the transition.

need a lower voltage (one hundred instead of several thousand volts). A polarizer and an electrooptic crystal are inserted between the laser material and one of the mirrors. Two different voltages, respectively, $V_{\lambda/4}$ and V_{λ}, are used to bias the crystal. When the voltage is equal to $V_{\lambda/4}$, the crystal is equivalent to a $\lambda/4$ wave plate with its neutral axis at $45°$ with the polarizer, for a double passage of the rays the crystal behaves as a $\lambda/2$ wave plate. For V_{λ} the crystal is a λ wave plate. The voltage being equal to $V_{\lambda/4}$, let us consider a beam that is transmitted by the polarizer and then comes back to it, after one reflection and two passages in the crystal, its polarization, which is obtained by considering a symmetry with regard to the neutral lines of the crystal, is orthogonal to the direction of the polarizer and the light is blocked by the polarizer. The light is fully transmitted for V_{λ}. Q-switching is achieved by maintaining a voltage equal to $V_{\lambda/4}$ during the beginning of the pumping, and commuting to V_{λ} when the inversion of population is maximum.

The Q-switching device of Figure 9.35 is completely passive and takes advantage of the saturation of an atomic transition. A cell, filled with a solution of molecules having an absorption band including the laser frequency, is placed between the laser material and one of the mirrors. Initially, the cell is opaque and the laser material doesn't see the mirror; as the pumping continues the light emitted by spontaneous emission saturates the transition and the cell becomes transparent; the concentration of the solution is chosen so that the time to reach transparency is equal to the time that is necessary for the inversion of population to become maximum, a giant pulse is then emitted.

9.7. Originality of Laser Light

The main originality of a laser beam is to have two coherencies, spatial and temporal. Spatial coherency is synonymous with directivity, temporal coherency with monochromaticity. These two properties come directly from

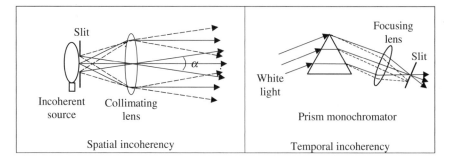

Figure 9.36. Illustration of spatial and temporal incoherencies. A collimator delivers an almost parallel beam. The narrower slit is, the smaller is the divergence of the beam. When the slit is closed the divergence is only limited by the diameter of the lens, but the intensity is then equal to zero. . . . In the same way, the beam delivered by the collimator is all the more monochromatic as the slit is narrow; a perfectly monochromatic light is obtained when the width of the slit goes to zero, the intensity also goes to zero. . . .

the stimulated emission, which produces photons in the same mode of the resonator. It's because of the properties of *autooscillation* that the optical characteristics (frequency and shape of the wave surfaces) of the emitted wave fit exactly with the resonator.

Directivity: The wave surfaces of a laser beam are perfectly defined and don't vary erratically from time to time; their shape is determined by the geometrical characteristics of the mirrors. If the output mirror is plane, the rays are orthogonal to its surface, the difference is only determined by diffraction and the angle is equal to $1.22\lambda/D$ (λ is the wavelength and D the diameter of the mirror). With the usual values of λ and D, the divergence is of the order of a milliradian. If the mirror is spherical, the laser wave is spherical and of course divergent, however a stigmatic optical arrangement can always transform it in a parallel beam, the divergence of which is limited by diffraction.

$1.22\lambda/D$ is a theoretical limit, the divergence of beams delivered by real lasers is always a little larger. When the divergence is almost equal to this value, the laser is said to be *diffraction limited*, which is a criterion of high optical quality.

Monochromaticity: The time variation law of the electromagnetic field of the laser beam is fully deterministic. In the case of a single-mode laser the signal is in principle sinusoidal, which means perfectly monochromatic. The residual bandwidth of a laser is estimated in Annex 9.B and is incredibly narrow; the real value is mostly due to the mechanical and thermal instabilities of the resonator.

Annex 9.A

Light-Semiconductor Interaction

9.A.1. Energy Levels in a Semiconductor

A semiconductor can be considered as a collection of negative charges (electrons) and of positive charges (holes) free to move inside a crystal. Because of Quantum Mechanics, the energy of the electrons and holes cannot take any value and should be equal to some allowed values. Figure 9.A.1 shows the diagram of the energy levels that are allowed for electrons and holes in a semiconductor: energy levels gather in *energy bands*.

The upper band is called the conduction band and corresponds to electrons; the lower band is called the valence band and corresponds to holes. The repartition of the electrons and holes among the different levels is fixed by the temperature and follows the Fermi-Dirac statistic; it is described by a Fermi function, see Figure 9.A.1. This is to be compared with the case of atoms and molecules that followed a Maxwell-Boltzmann distribution, see formula (9.20) in Section 9.2.2.1.

We now intend to show that, if a piece of sample of semiconductor is placed inside a blackbody, the three basic Einstein processes are necessary to ensure that:

- The photons are distributed according to a Bose-Einstein distribution.
- The electrons and holes are distributed according to a Fermi-Dirac distribution.

The interaction with light is considered as a "chemical reaction" which is illustrated in Figure 9.A.2. The energy at the right-hand side of the reaction may correspond to not only light (photons) but also to mechanical vibrations of the lattice of the semiconductor crystal. If this reaction is to occur, the two following conditions should be met:

- An electron and a hole are simultaneously at the same place.
- Energy and momentum conservation conditions are satisfied.

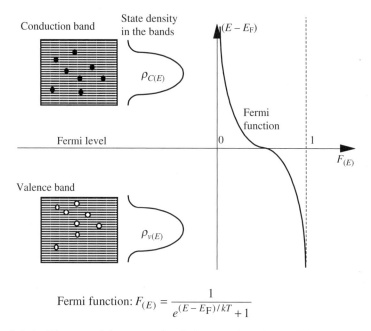

Fermi function: $F_{(E)} = \dfrac{1}{e^{(E-E_F)/kT}+1}$

Figure 9.A.1. Diagram of the energy levels in a semiconductor. The energy is plotted vertically. $\rho_{C(E)}$ and $\rho_{V(E)}$ are the densities of the allowed states of energy. The Fermi function indicates which levels are occupied; E_F is the energy of the Fermi level.

The energy conservation is very simple:

$$E_{\text{electron}} - E_{\text{hole}} = h\nu.$$

The conservation of the momentum is less easy to understand. It should be considered that the energy of electrons and holes depends on their respective momenta or, which is the same, on the wave vectors of the associated

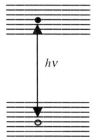

$$\text{electron} + \text{hole} \underset{2}{\overset{1}{\rightleftarrows}} \text{energy}.$$

(1) Corresponds to the light emission by electroluminescence and to lasers.
(2) Corresponds to the absorption of light by a semiconductor and to photoconduction.

Figure 9.A.2. Creation and recombination of pairs of electrons and holes.

quantum waves. For a certain category of semiconductors, *direct band gap semiconductors*, the conservation of the momentum doesn't raise very difficult problems. Typical direct band gap semiconductors are GaAs and InSb and their derivatives. The momenta of holes and electrons have almost equal moduli and the conservation is obtained if they have opposite directions, a motion of an atom of the lattice can compensate for a possible discrepancy.

For *indirect band gap semiconductors* the presence of a photon is required, its main role being to provide, or absorb, a small momentum discrepancy. The reaction is then written

$$\text{electron} + \text{hole} + \text{phonon} \rightarrow \text{photon}.$$

9.A.2. Spontaneous and Stimulated Effects in a Semiconductor

We will restrict ourselves to direct band gap semiconductors; the three basic mechanisms are the following:

Absorption (stimulated):   photon → electron + hole.
Spontaneous emission: electron + hole → photon.
Stimulated emission: electron + hole + photon → 2 photons (identical).

We introduce the densities of allowed states in the two bands, $\rho_{C(E)}$ and $\rho_{V(E)}$: the number of states having an energy between E and $E + dE$ are, respectively, equal to $\rho_{C(E)} \, dE$ for the conduction band and to $\rho_{V(E)} \, dE$ for the valence band.

At thermal equilibrium, the probability that a level E is occupied by an electron is given by a Fermi function $P_{\text{electron}(E)} = F_{(E)}$, the probability of a hole being the complementary probability $P_{\text{hole}(E)} = (1 - F_{(E)})$. Of course the probabilities are different if the semiconductor is out of equilibrium. The numbers, dN_{electron} and dN_{hole}, of electrons and holes having their energy between E and $E + dE$ are, respectively, given by the following formulas:

Electrons in the conduction band: $dN_{C,\text{electron}} = \rho_{C(E)} P_{\text{electron}(E)} \, dE.$
Electrons in the valence band: $dN_{V,\text{electron}} = \rho_{V(E)} P_{\text{electron}(E)} \, dE.$
Holes in the conduction band: $dN_{C,\text{hole}} = \rho_{C(E)} P_{\text{hole}(E)} \, dE.$
Electrons in the valence band: $dN_{V,\text{hole}} = \rho_{V(E)} P_{\text{hole}(E)} \, dE.$

The above *chemical* reactions couple together two levels separated by $h\nu$. A recombination of an electron and a hole will happen if level E is occupied by a hole, while level $E + h\nu$ is occupied by an electron. In the same way, an electron-hole pair is generated if the same respective levels are simultaneously empty.

Let n be the number of photons and let dn_{spon}, dn_{absorp}, and dn_{stimul} be the number of photons that are, respectively, created or absorbed during time dt, by spontaneous emission, absorption, or stimulated emission. If u_ν is the electromagnetic energy density at frequency ν, and A and B are Einstein coefficients, we can write

$$dn_{\text{spon}} = A\,dt \int P_{\text{hole}(E)} \rho_{V(E)} P_{\text{electron}(E+h\nu)} \rho_{C(E+h\nu)}\,dE,$$

$$dn_{\text{absorp}} = Bu_\nu\,dt \int \{(1-P_{\text{hole}(E)})\rho_{V(E)}(1-P_{\text{electron}(E+h\nu)})\rho_{C(E+h\nu)}\}\,dE, \qquad (9.A.1)$$

$$dn_{\text{stimul}} = Bu_\nu\,dt \int P_{\text{hole}(E)} \rho_{V(E)} P_{\text{electron}(E+h\nu)} \rho_{C(E+h\nu)}\,dE. \qquad (9.A.2)$$

At thermal equilibrium, the time derivatives cancel. The fact that, if the probabilities $P_{\text{hole}(E)}$ and $P_{\text{electron}(E+h\nu)}$ are Fermi-Dirac functions, the blackbody formula is obtained for u_ν will be considered as a justification of our description,

$$\int \rho_{C(E+h\nu)}\rho_{V(E)}\{AF_{(E+h\nu)}(1-F_{(E)}) + \cdots$$
$$+ Bu_\nu[F_{(E+h\nu)}(1-F_{(E)}) - F_{(E)}(1-F_{(E+h\nu)})]\}\,dE = 0,$$

$$u_\nu = \frac{A}{B}\,\frac{F_{(E+h\nu)}(1-F_{(E)})}{F_{(E)}(1-F_{(E+h\nu)}) - F_{(E+h\nu)}(1-F_{(E)})}.$$

Dividing the numerator and denominator by $F_{(E+h\nu)}F_{(E)}$, we obtain

$$u_\nu = \frac{A}{B}\,\frac{\dfrac{1}{F_{(E)}} - 1}{\left(\dfrac{1}{F_{(E+h\nu)}} - 1\right) - \left(F\dfrac{1}{f_{(E)}} - 1\right)} = \frac{A}{B}\,\frac{e^{(E-E_F)/kT}}{e^{(E-E_F+h\nu)/kT} - e^{(E-E_F)/kT}},$$

$$u_\nu = \frac{A}{B}\,\frac{1}{e^{h\nu/kT} - 1} \qquad \text{which is the expected formula.}$$

9.A.3. The Bernard-Duraffourg Formula, Inversion of Population

If electron-hole pairs have been generated in excess, the numbers of electrons and holes are larger than the equilibrium values. To recover equilibrium, the semiconductor will follow a relaxation process with a time constant equal to the inverse, $1/A$, of the Einstein coefficient for spontaneous emission. $1/A$ is also called the recombination time of electrons and holes τ_{recomb} which is of the order of a few nanoseconds in a direct band gap semiconductor and is much longer in the case of an indirect band gap.

An efficient way of creating electrons and holes is to use a forward biased PN junction. Figure 9.A.3 illustrates the phenomenon that occurs in such a case. When the electric current is equal to zero, there are only a few electrons in the conduction band and a few holes in the valence band. The forward bias

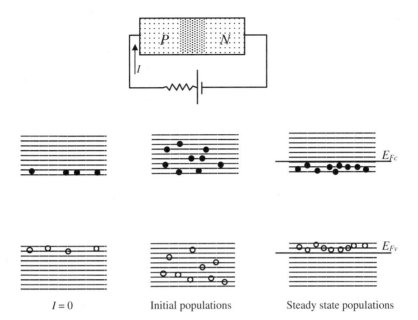

Figure 9.A.3. Populations of the conduction and valence bands of a forward biased *PN* junction. Hollow circles are holes and dark circles are electrons.

repels the electrons of the *P* region and the holes of the *N* region toward the junction area inside which they will remain during a time equal to τ_{recomb}. The equilibrium recovery is made in two steps:

- During the first period (picoseconds) the excess carriers rapidly reach thermal equilibrium between the sublevels of the band: the electrons gather at the bottom of the conduction band, while the holes gather at the top of the valence band, as illustrated in the right-hand diagram of Figure 9.A.3.
- During the second period (nanoseconds) electrons and holes recombine.

To describe the carriers' repartition among the sublevels it is convenient to introduce the "quasi-Fermi levels" E_{Fc} and E_{Fv}, the probabilities of occupation are given by the following formulas:

$$\text{Conduction band: } P_{C,\text{electron}} = \frac{1}{e^{(E-E_{Fv})/kT}+1}, \qquad (9.\text{A}.3)$$

$$\text{Valence band: } \quad P_{V,\text{electron}} = \frac{1}{e^{(E-E_{Fv})/kT}+1},$$

$$P_{V,\text{hole}} = (1 - P_{V,\text{electron}}) = 1 - \frac{1}{e^{(E-E_{Fv})/kT}+1}. \qquad (9.\text{A}.4)$$

To obtain the amplification condition, we come back to formulas (9.A.1) and (9.A.2) and we replace the occupation probabilities by (9.A.3) and (9.A.4).

Amplification is obtained when the stimulated emissions overcome the number of absorptions. It is found that $[(dn/dt)_{\text{stimul}} - (dn/dt)_{\text{absorp}}]$ is proportional to $(1 - e^{(E_{Fv} - E_{Fc} + h\nu)/kT})$. The semiconductor can amplify optical waves whose frequency fulfills the double inequality, which is called the Bernard-Durrafourg equation, because of the names of the two French physicists who established it for the first time,

$$E_C - E_V < h\nu < E_{FC} - E_{FV}.$$

Annex 9.B

Spectral Width of a Laser Oscillation

The oscillation emitted by a monofrequency laser is very monochromatic, but it has, of course, a spectral width $\Delta\nu_{\mathrm{osc}}$. The laser oscillator is supposed to have reached a steady state and emits a beam of constant intensity. We will only consider the case when the inversion of population has a high value while the lower laser level remains unpopulated. n is the number of photons inside the resonator, N_2 is the population of the upper level, and τ_{res} is the lifetime of the electromagnetic energy when the resonator is empty.

The number of photons that are lost, or gained, by the mode during time dt is given by

$$dn_{\mathrm{spon}} = AN_2\, dt, \quad \text{spontaneous photons,}$$

$$dn_{\mathrm{stimul}} = AnN_2\, dt, \text{ stimulated photons,}$$

$$dn_{\mathrm{loss}} = -\frac{nh\nu}{\tau_{\mathrm{res}}}dt, \quad \text{photons leaving the resonator through the mirrors,}$$

$$\frac{dn}{dt} + n\left(\frac{1}{\tau_{\mathrm{res}}} - AN_2\right) = AN_2. \tag{9.B.1}$$

For the steady state we have

$$n_{\mathrm{steady}} = \frac{AN_{2,\mathrm{steady}}}{1/\tau_{\mathrm{res}} - AN_{2,\mathrm{steady}}} \quad \rightarrow \quad AN_{2,\mathrm{steady}} = \frac{n_{\mathrm{steady}}}{n_{\mathrm{steady}} + 1}1/\tau_{\mathrm{res}}.$$

The population $N_{2,\mathrm{steady}}$ cannot be greater than $(1/A\tau_{\mathrm{res}})$ since n_{steady} cannot be infinite; in the case of low losses, it can however be almost equal to this limit. The emitted power P is given by

$$P = \frac{h\omega}{2\pi} = h\nu\frac{n_{\mathrm{steady}}}{\tau_{\mathrm{res}}} \quad \rightarrow \quad n_{\mathrm{steady}} = \frac{P\tau_{\mathrm{res}}}{h\nu}.$$

From a physical point of view, the residual spectral width of the laser light is a consequence of the losses of the resonator. Let us consider the case of an empty resonator (with no amplifying material inside) using the same mirrors as the laser and suppose that at time $t = 0$ the number of photons inside is n_0, the evolution of the number n of photons follows a first-order differential equation

$$\frac{dn}{dt} + \frac{n}{\tau_{\mathrm{res}}} = 0. \qquad (9.B.2)$$

The empty resonator is a filter whose bandwidth $\Delta\omega_{\mathrm{res}}$ is given by

$$\Delta\omega_{\mathrm{res}} = \frac{\omega_{\mathrm{res}}}{Q_{\mathrm{res}}} = \frac{1}{\tau_{\mathrm{res}}} \quad \rightarrow \quad \Delta\nu_{\mathrm{res}} = \frac{1}{2\pi}\frac{1}{\tau_{\mathrm{res}}}.$$

A comparison of (9.B.1) and (9.B.2) suggests that a laser behaves as a resonator in which the losses are not completely compensated for by the gain and whose time constant τ' is given by

$$\frac{1}{\tau'} = \left(\frac{1}{\tau_{\mathrm{res}}} - AN_{2,\mathrm{steady}} \right).$$

The laser spectral width $\Delta\nu_{\mathrm{osc}}$ is finally equal to $\Delta\nu_{\mathrm{osc}} = 2\pi(h\nu/P)(\Delta\nu_{\mathrm{res}})^2$. A numerical application of this formula has been given in Section 9.1.4.

10

Nonlinear Optics

A problem often encountered in Physics is to find the response $r(t)$ of a system to some excitation $e(t)$ that is time dependent. In many cases the relationship (excitation $\leftrightarrow$ response) is linear, the simplest case being proportionality. If the excitation is harmonic, which means sinusoidal time variation, the linear response is also harmonic, with the same frequency. In many cases, the linear approach is a simplification of more general systems that are said to be *nonlinear*. The response to a harmonic signal (frequency ω) is still periodic with the same frequency, but is no longer harmonic: the frequency spectrum has many more components, among which the double frequency (2ω) component is often the most important.

As a light beam propagates inside a piece of material (see Annex 8.B of the chapter on the Index of Refraction), the electric field of the incident wave can be considered as an excitation, while the polarization taken by the material is the response. As long as the light intensity is not very powerful, the interaction remains linear and the light that is transmitted has only one component at the same frequency. The nonlinear aspect of the light/material interaction could not be revealed before the appearance of the powerful beams delivered by lasers. The first demonstration was made in 1961: the incident *red light* (0.6943 μm) of a ruby laser, after propagation inside a quartz crystal, gave birth to an *ultraviolet beam* (0.34715 μm). A spectacular and useful application is the generation of green light (0.553 μm) from the near infrared light (1.06 μm) of a neodymium laser.

If the excitation is a sine function of frequency ω, so is the response. $e(t)$ and $r(t)$ vibrate at the same frequency but don't have, in general, the same phase; this is because a linear interaction is described by a linear relation between the functions $e(t)$ and $r(t)$ and all their derivatives,

$$\sum_n a_{n(\omega)} \frac{d^n r}{dt^n} = \sum_p b_{p(\omega)} \frac{d^p e}{dt^p}, \tag{10.1}$$

Chapter 10 has been reviewed by Dr. Ludovic Brassé from Teemphotonics.

with $e(t)$ and $r(t)$ being considered as the real part of the two complex numbers $\overline{E}$ and $\overline{R}$, we have

$$\overline{R} = \overline{E}\,\frac{\sum_p b_p (j\omega)^p}{\sum_n a_n (j\omega)^n} = \overline{A}(\omega)\overline{E}. \tag{10.2}$$

Equation (10.2) clearly shows that the response is proportional to the excitation. Any linear system has the following properties:

- The proportionality coefficient: $\overline{A}(\omega)$ is independent of the amplitude of the signal. In the case of Optics the index of refraction doesn't depend on the light intensity.
- Principle of superposition: If r_1 is the response to e_1, and r_2 the respective response to e_2; when e_1 and e_2 are simultaneously applied, the response is $(r_1 + r_2)$.
- If the excitation is purely sinusoidal, the response is also purely sinusoidal of the same frequency.
- If the excitation has several components ω_1, ω_2, ω_3, the response has no other components. The moduli of the different proportionality coefficients $\overline{A}(\omega_1)$, $\overline{A}(\omega_2)$, $\overline{A}(\omega_3)$ are not equal. Some of them can be equal to zero: the corresponding component is then filtered out.

A nonlinear process doesn't follow the preceding rules. It's not possible to give a general formulation for nonlinear interactions. To exhibit the properties that are useful for nonlinear optics, we will restrict ourselves to the simple example where the (excitation ↔ response) relation is not a differential equation and has a polynomial form

$$r(t) = a_1 e_{(t)} + a_2 e_{(t)}^2 + a_3 e_{(t)}^3 + \cdots. \tag{10.3}$$

The first term is the linear part of the response; the higher-order terms are the nonlinear part. If the excitation $e(t)$ is small, the higher-order terms are negligible. The most important consequence of (10.3) is the appearance of new frequencies in the Fourier spectrum of the answer. Let us first consider the case of a monochromatic excitation and keep only the two first terms of (10.3):

$$e_{(t)} = \hat{e}\cos\omega t,$$

$$r(t) = a_1 \hat{e}\cos\omega t + a_2 \hat{e}^2 \cos^2\omega t = a_1 \hat{e}\cos\omega t + \frac{a_2 \hat{e}^2}{2} + \frac{a_2 \hat{e}^2}{2}\cos 2\omega t. \tag{10.4}$$

In the Fourier spectrum, in addition to a term at the frequency of the excitation that comes from the linear part of (10.3), are found:

- A DC term (frequency equal to zero): $a_2 \hat{e}^2/2$.
- A double frequency term, also called a *second harmonic*, $(a_2 \hat{e}^2/2)\cos 2\omega t$; the expression Second Harmonic Generation (SHG) is commonly used.

We now consider the case of an excitation with two frequencies ω_1 and ω_2:

$$e_{(t)} = \hat{e}_1 \cos \omega_1 t + \hat{e}_2 \cos \omega_2 t,$$

$$r(t) = a_1 \hat{e}_1 \cos \omega_1 t + a_2 \hat{e}_2 \cos \omega_2 t + \frac{a_2^2}{2}[\hat{e}_1^2 + \hat{e}_1^2] + \cdots$$

$$+ \frac{a_2^2}{2}[\hat{e}_1^2 \cos 2\omega_1 t + \hat{e}_2^2 \cos 2\omega_2 t] + a_2 \hat{e}_1 \hat{e}_2 [\cos(\omega_1 - \omega_2)t + \cos(\omega_1 + \omega_2)t].$$

The principle of superposition doesn't apply and the Fourier spectrum is considerably enriched.

10.1. Microscopic Interpretation of a Nonlinear Optical Interaction

Light propagation in a transparent material is accompanied by the vibrations of electrons bound to the atoms, whose vibrations are forced by the electric field of the light wave. As a result of the fact that the electrons are taken away from their equilibrium positions by the action of the electric field (excitation), an electric polarization (response) is created. The global action of the nucleus and of the other electrons of an atom corresponds to a potential $V(r)$, as a first approximation this potential is parabolic and creates a restoring force that is proportional to the distance from the equilibrium position, and whose proportionality is responsible for the linear character of the interaction. As suggested by Figure 10.1, the parabolic potential is a first approximation, and the real potential should be represented by a serial Taylor development.

As long as the amplitude of the forced oscillations keeps a small amplitude, the parabolic approximation is accurate enough. If the incident field is more intense, the motion of the electrons has three components: one at the incident frequency, another at the double frequency, and the last one is a DC component.

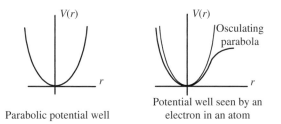

Parabolic potential well   Potential well seen by an electron in an atom

Figure 10.1. A bound electron is trapped inside a potential well. In the limit of a linear interaction this well is parabolic and the electron is submitted to a restoring force proportional to the distance of its equilibrium position.

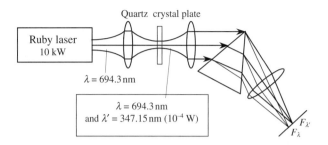

Figure 10.2(a). First SHG experiment by Maker, Terhune, and Savage (1961). The conversion efficiency is very low.

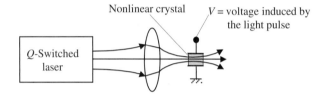

Figure 10.2(b). Optical rectification. The two sides of the nonlinear crystal are coated with a metal layer. When the short and intense light pulse of a Q-switched laser travels across the crystal, a voltage is generated.

It will be shown that if the environment of the electron is centrosymmetric, which happens in cubics and in isotropic materials, the development of equation (10.3) has no second-order term, the first higher term being the third one. Because of the absence of a component at 2ω in the Fourier spectrum of $\cos^3 \omega t$, the second harmonic generation is then impossible.

10.2. Phase Matching Condition

Nonlinear optics is dominated by the important notion of *phase matching* that will be introduced on a phenomenological basis in the case of second harmonic generation. Because of the presence of the incident wave, each atom of the nonlinear crystal is made an electric dipole, which radiates electromagnetic wavelets at two frequencies, ω and 2ω. The amplitude of the wavelets at the fundamental frequency ω is much larger than the amplitude at the double frequency. At a given point P, the field is the result of the interference of all wavelets. We first consider the wavelets of frequency ω; the value of the phase velocity V_ω is, *by construction*, such that the wavelets interfere constructively (see Chapter 8 on Index of Refraction). The situation is different for the wavelets at frequency 2ω, since they propagate at $V_{2\omega}$, but are emitted by oscillating dipoles that have a phase repartition corresponding to a propagation speed equal to V_ω. Because of the law of normal dispersion the two speeds are not equal ($V_\omega < V_{2\omega}$): the interference between the double frequency wavelets will not normally be constructive, this is the reason why the conversion efficiency is very small in the experiment described in Figure 10.2(a). More favorable conditions can be found where the interference is constructive at both frequencies; they constitute the *phase matching condition*.

We refer to Figure 10.3, the crystal is very transparent and only a small amount of energy is transferred from the fundamental beam to the second harmonic beam: the amplitude of the fundamental beam remains roughly constant along the crystal and can be written as

$$E_\omega(t) = A_0 \cos\left[\omega\left(t - \frac{z}{V_\omega}\right)\right].$$

We first evaluate the amplitude $E_{2\omega}(Z)$ of the second harmonic field at point P of abscissa Z. This field is created by all the dipoles between 0 and Z. The contribution, $dE_{2\omega}$ of the dipoles of the slice z, $z + dz$ is proportional to the square of the amplitude of the field at the fundamental frequency, the time of propagation from abscissa z to abscissa Z is equal to $(Z - z)/V_{2\omega}$; it seems natural to write

$$dE_{2\omega} = KA_0^2\, dz \cos\left[2\omega\left(t - \frac{z}{V_\omega} - \frac{(Z-z)}{V_{2\omega}}\right)\right],$$

$$dE_{2\omega} = KA_0^2\, dz \cos[2\omega t - z(k_{2\omega} - k_\omega) - k_{2\omega}Z], \qquad (10.5)$$

$$E_{(t,e)}^{2\omega} = K\int_0^e A_0^2 \cos(z\Delta k + 2\omega t - k_{2\omega}e)\,dz,$$

$$E_{(t,e)}^{2\omega} = 2KA_0^2\, \frac{\sin\left(\dfrac{\Delta k e}{2}\right)}{\Delta k} \cos\left(2\omega t - \frac{k_{2\omega} + k_\omega}{2}e\right),$$

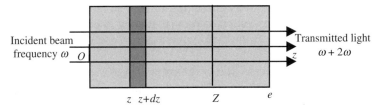

Figure 10.3. A planar wave propagates inside a nonlinear crystal, parallel to the Oz axis; the transmitted light is due to the interference of wavelets emitted by dipoles oscillating at ω and 2ω.

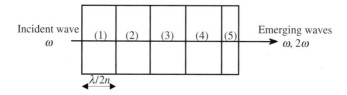

Figure 10.4. Second harmonic generation using a thick plate. The crystal may be decomposed as pairs of slices with a thickness of $\lambda/2\Delta n$. The SHG signals that are, respectively, generated in the first and second halves of a given slice interfere destructively. Only the very last section contributes to the generation of second harmonic light in the emerging beam.

where $\Delta k = k_{2\omega} - 2k_\omega$ is called the *phase mismatch* between the two waves. The light intensity at the double frequency is given by

$$I_{2\omega} = K^2 e^2 A_0^4 \frac{\sin^2\left(\dfrac{\Delta ke}{2}\right)}{\left(\dfrac{\Delta ke}{2}\right)^2}. \tag{10.6}$$

The intensity of the second harmonic is a periodic function of the thickness e of the crystal plate and is maximized for well-chosen values of the thickness

$$\frac{\Delta ke_{max}}{2} = (2p+1)\frac{\pi}{2} \quad \rightarrow \quad e_{max} = \frac{\lambda}{2(n_{2\omega} - n_\omega)} + p\frac{\lambda}{(n_{2\omega} - n_\omega)},$$

where n_ω and $n_{2\omega}$ are, respectively, the indices of refraction at the fundamental and second harmonic frequencies.

A better efficiency of the generation of the second harmonic generation is obtained when the phase mismatch, Δk, is made equal to zero; according to (10.6), the intensity of double frequency light is then equal to $k^2 e^2 A_0^4$ and it continuously increases with the thickness. The cancellation of the phase mismatch implies the equality of the two indices n_ω and $n_{2\omega}$, which, because of dispersion, is impossible if the two beams have the same polarization. The trick is to use beams with orthogonal polarizations, and to find suitable conditions where the ordinary index at one frequency is equal to the extraordinary index at the other frequency, which is not forbidden by the law of dispersion.

In Figure 10.5 are shown the ordinary and extraordinary sheets of the index surface of a negative uniaxial crystal, respectively, at the frequencies ω and 2ω. The ordinary layer at ω and the extraordinary layer at 2ω intersect at the four points P_1, P_2, P_3, and P_4. The conditions for an efficient $\omega/2\omega$ conversion are obtained when the incident wave (ω) is ordinary polarized and propagates along the direction OP_1. The wavelets that are generated at the

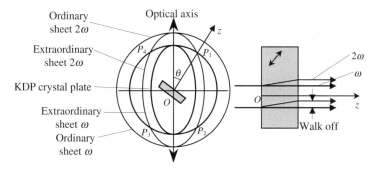

Figure 10.5. Determination of the phase matching condition. Along OP_1, an ordinary frequency ω propagates at the same speed as an extraordinary wave of frequency 2ω. The faces of the crystal plate are cut perpendicular to OP_1, so that the incidence is normal. Inside the crystal, the wave vectors at ω and 2ω are parallel, the rays don't exactly coincide, the walk-off is however small.

frequency 2ω propagate at the same phase velocity as the incident wave. Under such conditions, an important part of the energy is transferred from the fundamental beam to the second harmonic beam; equations (10.5) and (10.6) are no longer valid since the fundamental beam is now depleted as it propagates.

Let us now calculate the angle θ between the optical axis and the direction OP_1 for which the two indices $n_{o,\omega}$ and $n_{e,2\omega}(\theta)$ are equal. Whatever the direction of propagation, the ordinary wave sees the same index $n_{o,\omega}$; the index $n_{e,2\omega}(\theta)$ of the extraordinary wave, at frequency 2ω, is given by

$$\frac{1}{n_{e,2\omega}^2(\theta)} = \frac{\cos^2\theta}{n_{o,2\omega}^2} + \frac{\sin^2\theta}{n_{e,2\omega}^2},$$

$$n_{o,\omega} = n_{e,2\omega}(\theta) \quad \rightarrow \quad \sin^2\theta = \frac{1/n_{o,\omega}^2 - 1/n_{o,2\omega}^2}{1/n_{e,2\omega}^2 - 1/n_{o,2\omega}^2}.$$

In the case of KDP, the indices have the following values: $n_{e(\lambda=694\,\text{nm})} = 1.465$, $n_{o(\lambda=694\,\text{nm})} = 1.505$, $n_{e(\lambda=347\,\text{nm})} = 1.487$, $n_{o(\lambda=347\,\text{nm})} = 1.534$; the angle θ is equal to $51°$. As it's more convenient to work at normal incidence, the end faces of the crystal are cut perpendicular to the direction OP_1. The ω and 2ω wave vectors are parallel, but not the light rays, see Figure 10.5. Frequency doublers are now very common, especially for doubling the $1.06\,\mu\text{m}$ light of neodymium lasers into a green $0.53\,\mu\text{m}$ light. The most popular crystal being KDP (potassium diphosphate). Lithium niobate (LiNbO_3) and "banana" ($\text{Ba}_2\text{NaNb}_5\text{O}_{15}$) have important nonlinear coefficients; in this case, the values of the ordinary and extraordinary indices vary appreciably with temperature, and it is possible to find a temperature at which the ordinary index of one color is equal to the extraordinary index of the other color.

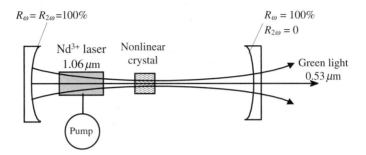

Figure 10.6. Efficient second harmonic generation. The conversion efficiency increases with the square of the intensity of the fundamental beam; this is why the nonlinear crystal is placed inside a laser cavity. One mirror is fully reflecting at the fundamental (ω) frequency and the harmonic (2ω) frequency, the other one is dichroic, fully reflecting at ω and perfectly transparent at 2ω. The only losses from the laser resonator are due to harmonic conversion.

Quantum Interpretation of Nonlinear Effects

Nonlinear effects belong to the category of multiphoton processes, harmonic generation being a two-photon process that can be described by the following reaction:

$$\text{photon }(h\nu) + \text{photon }(h\nu) \rightarrow \text{photon }(h2\nu).$$

The probability of such an event happening is maximum and, in fact, is different from zero, only if two specific conditions are satisfied: energy conservation and momentum conservation. Energy conservation is to be associated with the frequency doubling and momentum conservation to phase matching. In the case of photons, of frequency ν and traveling along parallel directions, we have

$$\text{Energy conservation: } \quad h\nu + h\nu = h2\nu,$$

$$\text{Momentum conservation: } \quad n_1 \frac{h\nu}{c} + n_1 \frac{h\nu}{c} = n_2 \frac{h2\nu}{c}.$$

10.3. Nonlinear Polarization

The fact that the permittivity of dielectric material is different from that of a vacuum has already been interpreted in terms of polarization, the polarization being a consequence of the motion of the electrons that is forced by the electric field. As long as the motion of an electron is described by a linear differential equation with constant coefficients, the abscissa of an electron, within a possible phase difference, is proportional to the electric field. This is the reason why a permittivity, independent of the field amplitude, can be introduced. The situation is different when the potential is no longer parabolic and the restoring force is not proportional to the distance from equilibrium.

10.3.1. *Motion of an Electron in a Nonparabolic Potential*

Let us write the equation of the motion of an electron submitted to an electric field $f(t)$ and placed in a potential $V(\xi)$ that will be developed as

$$V(\xi) = -m(\tfrac{1}{2}\omega^2_{\text{atom}}\xi^2 + \tfrac{1}{3}D\xi^3),$$

$$\frac{d^2\xi}{dt^2} + \gamma\frac{d\xi}{dt} = \frac{e}{m}f(t) - \frac{1}{m}\frac{dV}{d\xi},$$

$$\frac{d^2\xi}{dt^2} + \gamma\frac{d\xi}{dt} + \omega^2_{\text{atom}}\xi + D\xi^2 + \cdots = \frac{e}{m}f(t). \tag{10.7}$$

Inside a crystal the situation is more complicated, since the displacement, the forces, and the electric field are related by tensors. . . . To understand what happens we will limit ourselves to scalar relations. Another complication comes from the fact that the utilization of the complex numbers, to obtain the response of a system to a sine excitation, is only valid because equation (10.1) is a linear equation with constant coefficients.

The function $f(t)$ that represents the electric field is a sine, but we must be very careful when using the imaginary notation and always treat it with real functions and write $f(t)$ as

$$f(t) = E_\omega \cos\omega t = \frac{E_\omega}{2}(e^{j\omega t} + e^{-j\omega t}),$$

To obtain a solution for equation (10.7), we will follow an iterative approach. As a first step we will omit the x^2 term and obtain

$$\xi_\omega(t) = \frac{e}{2m}E_\omega\left(\frac{e^{j\omega t}}{\Delta_{(\omega)}} + \frac{e^{-j\omega t}}{\Delta_{(-\omega)}}\right) = \frac{e}{2m}\left(\frac{e^{j\omega t}}{\Delta_{(\omega)}} + \text{cc}\right), \tag{10.8}$$

$$\Delta_{(\omega)} = \omega^2_{\text{atom}} - \omega^2 + j\gamma\omega.$$

The term "cc" in (10.8) indicates that the "complex conjugate" of the term already written should automatically be added.

The value of $\xi(t)$ is taken from (10.8) and was introduced in (10.7):

$$\xi(t) = \xi_\omega(t) + \xi_{2\omega}(t) \quad \text{with} \quad \xi_{2\omega}(t) = -\frac{e^2}{m^2}(E^2_\omega)^2\frac{D}{\Delta^2_\omega\Delta_{2\omega}} + \text{cc}.$$

The next step of the iteration would be to introduce $\xi_{2\omega}(t)$ in (10.7) and then determine the 4ω component, which will be considered as negligible.

10.3.2. *Polarization of a Second-Order Nonlinear Susceptibility*

To the term $\xi_\omega(t)$ corresponds, for the polarization, a term of frequency ω, which, after multiplication by the number N of atoms per volume unit, gives the *linear susceptibility* χ_ω:

$$P_\omega(t) = Nex_\omega(t) + \text{cc} = \varepsilon_0 \chi_\omega E_\omega e^{j\omega t} + \text{cc},$$

$$\chi_\omega = \frac{Ne^2}{m\varepsilon_0} \frac{1}{\Delta_{(\omega)}}, \quad \text{this relation is very similar to (8.14).}$$

In the same way, the polarization at 2ω can be written as

$$P_{2\omega}(t) = \tfrac{1}{2}[P_{2\omega} e^{j2\omega t} + \text{cc}],$$

$$P_{2\omega} = d_{2\omega} E_\omega^2 \quad \text{with} \quad d_{2\omega} = \frac{DNe^3}{2m^2} \frac{1}{\Delta_\omega^2 \Delta_{2\omega}},$$

where $d_{2\omega}$ is called the *second-order nonlinear susceptibility* at the pair of frequencies $\omega/2\omega$, and we can introduce the linear susceptibilities at the respective frequencies $\omega/2\omega$, and

$$d_{2\omega} = \varepsilon_0^3 \frac{mDN}{2N^2 e^3} \chi_\omega^2 \chi_{2\omega}.$$

Generalization of the Notion of Susceptibility

Let P be the polarization of the material under the influence of an electric field E, a Taylor serial development can always be established

$$P = \varepsilon_0 (\chi E + \chi^{(2)} E^2 + \chi^{(3)} E^3 + \cdots), \tag{10.9}$$

where $\chi^{(2)}$ and $\chi^{(3)}$ are called the second- and third-order susceptibilities. As in the previous section, another second-order susceptibility is also introduced by $\varepsilon_0 \chi^{(2)} = 2d$. It's also convenient to consider two parts in the expression of polarization: a linear part P_L and a nonlinear part P_{NL}:

$$P_L = \varepsilon_0 \chi E \quad \text{and} \quad P_{NL} = \varepsilon_0 (\chi^{(2)} E^2 + \chi^{(3)} E^3 + \cdots). \tag{10.10}$$

We refer to Figure 10.7, a nonlinear material receives two different beams having two different frequencies ω' and ω'', if we consider only the "two-photons interactions" the emergent light beam will have the following frequencies: ω', ω'', $2\omega'$, $2\omega'$, $2\omega''$, $\omega' \pm \omega''$; the polarization is then

$$P = \varepsilon_0 \chi_{\omega'} E_{\omega'} + \varepsilon_0 \chi_{\omega''} E_{\omega''} + d_{2\omega'} E_{\omega'}^2 + d_{2\omega''} E_{\omega''}^2 + \cdots$$
$$+ 2d_{\omega'\omega''} E_{\omega'} E_{\omega''}^* + 2d_{\omega'+\omega''} E_{\omega'} E_{\omega''} + \text{cc}.$$

Because the light vibration is a vector, formula (10.10) is not sufficient, and the different terms are in fact tensors. . . . As an example, we will write

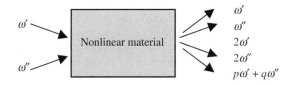

Figure 10.7. When a nonlinear material receives two different frequencies, ω' and ω'', a large number of new frequencies, $p\omega' + q\omega''$, may be generated.

the relation that gives the component P_x of the polarization along the Ox axis and in the case of a second harmonic generation,

$$P_x^{2\omega} = d_{xxx}^{2\omega} E_x^\omega E_x^\omega + d_{xyy}^{2\omega} E_y^\omega E_y^\omega + d_{xzz}^{2\omega} E_z^\omega E_z^\omega + \cdots$$
$$+ 2d_{xzy}^{2\omega} E_z^\omega E_y^\omega + 2d_{xzx}^{2\omega} E_z^\omega E_x^\omega + 2d_{xxy}^{2\omega} E_x^\omega E_y^\omega + \text{cc}. \qquad (10.11)$$

Kleinman Relation

The tensor that connects the nonlinear polarization $\boldsymbol{P}_{2\omega}$ to the nine terms $(E_x^2,\ E_y^2,\ E_z^2,\ E_xE_y,\ E_yE_x,\ E_yE_z,\ E_zE_y,\ E_zE_x,\ E_xE_z)$ is a (6×6) tensor.

A permutation of E_j and E_k, having no physical consequence, $d_{ijk} = d_{ikj}$, the tensor has only 18 independent elements. Formula (10.12) give the general aspect of this tensor, when the following contraction rules are used:

$$x = 1, \quad y = 2, \quad z = 3, \quad xx = 1, \quad yy = 2, \quad zz = 3, \quad yz = zy = 4,$$
$$xz = zx = 5, \quad xy = yx = 6,$$

$$\begin{pmatrix} P_x \\ P_y \\ P_z \end{pmatrix} = \begin{pmatrix} d_{11} & d_{12} & d_{13} & d_{14} & d_{15} & d_{16} \\ d_{21} & d_{22} & d_{23} & d_{24} & d_{25} & d_{26} \\ d_{31} & d_{32} & d_{33} & d_{34} & d_{35} & d_{36} \end{pmatrix} \begin{pmatrix} E_x^2 \\ E_y^2 \\ E_z^2 \\ 2E_zE_y \\ 2E_zE_x \\ 2E_xE_y \end{pmatrix}. \qquad (10.12)$$

Kleinman found good reasons to show that the three indices can be written in any order, and that it still only contains ten coefficients:

$$\begin{pmatrix} d_{11} & d_{12} & d_{13} & d_{14} & d_{15} & d_{16} \\ d_{16} & d_{22} & d_{23} & d_{24} & d_{14} & d_{12} \\ d_{15} & d_{24} & d_{33} & d_{23} & d_{13} & d_{14} \end{pmatrix}.$$

In the end, if the symmetry of the crystal is taken into consideration and if the referential is suitably chosen, most of the coefficients are equal to zero;

Table 10.1. Values of some second-order nonlinear coefficients of the most usual crystals. The fundamental and SHG wavelengths are of the order of $1\,\mu m$ and $0.5\,\mu m$.

Crystal	10^{-23} (mks units)
KDP	$d_{14} = 0.50$
Quartz	$d_{11} = 0.41$
GaP	$d_{14} = 90$
GaAs	$d_{14} = 120$
LiNbO$_3$	$d_{31} = 5.3$
	$d_{22} = 2.5$
Ba$_2$NaNb$_5$O$_{15}$	$d_{33} = 11.5$
	$d_{32} = 8$
LiIO$_3$	$d_{31} = 7$

for example, in the case of a KDP crystal, the matrix of the tensor is written as

$$\begin{pmatrix} 0 & 0 & 0 & d_{14} & 0 & 0 \\ 0 & 0 & 0 & 0 & d_{14} & 0 \\ 0 & 0 & 0 & 0 & 0 & d_{14} \end{pmatrix} \rightarrow \begin{array}{l} P_x = 2d_{14}E_zE_y, \\ P_y = 2d_{14}E_zE_x, \\ P_z = 2d_{14}E_xE_y. \end{array}$$

There is no second-order nonlinear effect in a centrosymmetric material.

A centrosymmetric material has a center of inversion, which means that if E_j^ω and E_k^ω are changed into $-E_j^\omega$ and $-E_k^\omega$, the value of the polarization remains unchanged; we write the expression of the polarization as

$$P_1^{2\omega} = \sum_{j,k=x,y,z} d_{ijk}^{2\omega} E_j^\omega E_k^\omega.$$

It is easily seen that the matrix $(d_{ijk}^{2\omega})$ is identically equal to zero, which establishes the proposition. Of course third-order nonlinear effects are possible.

10.4. Equations of Propagation in a Nonlinear Material

When the nonlinear polarization is introduced into Maxwell's equations, the wave equation takes a different form (see formula (10.13)), since the second member is no longer equal to zero and contains terms that vibrate at 2ω.

In the case of a linear material illuminated by a monochromatic beam of frequency ω on the boundary, the field is identically equal to zero at the frequency 2ω and nothing happens at this frequency. On the contrary, in the case of a nonlinear material, a wave is created inside the sample.

The term $\partial^2 P_{NL}/\partial t^2$ is often called a *source term*.

Maxwell's equations	Wave equation
$\nabla \wedge E = -\mu_0 \dfrac{\partial H}{\partial t},$	
$\nabla \wedge H = +\varepsilon \dfrac{\partial E}{\partial t},$	$\nabla^2 E - \varepsilon \mu_0 \dfrac{\partial^2 E}{\partial t^2} = \mu_0 \dfrac{\partial^2 P_{NL}}{\partial t^2}.$ (10.13)
$\nabla D = 0,$	In a linear material P_{NL} is equal to zero.
$\mu_0 \nabla H = 0,$	
$D = \varepsilon E = \varepsilon_0 E + P_L + P_{NL}.$	

10.4.1. *Planar Waves Coupled by a Second-Order Nonlinear Effect*

We consider three planar waves propagating in a nonlinear transparent material, parallel to the Oz axis; the respective frequency and wave vector moduli are (ω_1, k_1), (ω_2, k_2), (ω_3, k_3),

$$E^{\omega_1}_{(z,t)} = \tfrac{1}{2}\left[E^{\omega_1}_{(z)} e^{j(\omega_1 t - k_1 z)} + cc\right],$$

$$E^{\omega_2}_{(z,t)} = \tfrac{1}{2}\left[E^{\omega_2}_{(z)} e^{j(\omega_2 t - k_2 z)} + cc\right],$$

$$E^{\omega_3}_{(z,t)} = \tfrac{1}{2}\left[E^{\omega_3}_{(z)} e^{j(\omega_3 t - k_3 z)} + cc\right].$$

Many frequencies may be created, and we suppose that the phase matching condition is obtained only if $\omega_3 = (\omega_1 + \omega_2)$. The only source terms to be considered are then

$$\mu_0 \frac{\partial^2}{\partial t^2}\left[E^{\omega_1}_{(z)} E^{\omega_2}_{(z)} e^{j[(\omega_1+\omega_2)t - (k_1+k_2)z]}\right] + cc,$$

$$\mu_0 \frac{\partial^2}{\partial t^2}\left[E^{\omega_3}_{(z)} E^{\omega_2 *}_{(z)} e^{j[(\omega_3-\omega_2)t - (k_3-k_2)z]}\right] + cc,$$

$$\mu_0 \frac{\partial^2}{\partial t^2}\left[E^{\omega_1}_{(z)} E^{\omega_3 *}_{(z)} e^{j[(\omega_1-\omega_3)t - (k_1-k_3)z]}\right] + cc.$$

The problem is to find $E^{\omega_1}_{(z)}$, $E^{\omega_2}_{(z)}$, $E^{\omega_3}_{(z)}$. In this kind of calculation, it is always assumed that the functions vary slowly with z, which means that their variations, and the variations of their derivatives, are very small when the distance is increased by one wavelength and is expressed by

$$\left|\frac{d^2 E^{\omega_i}_{(z)}}{dz^2}\right| << k_i \left|\frac{dE^{\omega_i}_{(z)}}{dz}\right|. \tag{10.14}$$

We write the equation for $E^{\omega_3}_{(z)}$; the other equations are obtained by permutation

$$\nabla^2 E^{\omega_1}_{(z,t)} = \frac{1}{2}\frac{\partial^2}{\partial z^2}\left[E^{\omega_1}_{(z)}e^{j(\omega_1 t - k_1 z)} + \text{cc}\right] \cong -\frac{1}{2}\left[k_1^2 E^{\omega_1}_{(z)} + 2jk_1\frac{\partial E^{\omega_1}_{(z)}}{\partial z}\right]e^{j(\omega_1 t - k_1 z)} + \text{cc},$$

$$\nabla^2 E^{\omega_1}_{(z,t)} - \varepsilon\mu_0\frac{\partial^2 E^{\omega_1}_{(z,t)}}{\partial t^2} = \frac{\mu_0 d}{2}\frac{\partial^2}{\partial t^2}\left[E^{\omega_3}_{(z)}E^{\omega_2^*}_{(z)}e^{j[(\omega_3-\omega_2)t-(k_3-k_2)z]}\right] + \text{cc}.$$

We make $(\omega_3 - \omega_2) = \omega_1$ and $\partial/\partial t = j\omega_i$:

$$-\frac{1}{2}\left[k_1^2 E^{\omega_1}_{(z)} + 2jk_1\frac{\partial E^{\omega_1}_{(z)}}{\partial z}\right]e^{j(\omega_1 t - k_1 z)} + \text{cc} + \frac{\omega_1^2\varepsilon\mu_0}{2}\left[E^{\omega_1}_{(z)}e^{j(\omega_1 t - k_1 z)} + \text{cc}\right]$$

$$= -\frac{\omega_1^2\varepsilon\mu_0}{2}d\left[E^{\omega_3}_{(z)}E^{\omega_2^*}_{(z)}e^{j[\omega_1 t - (k_3-k_2)z]} + \text{cc}\right],$$

since $\omega_i^2\varepsilon_i\mu_0 = k_i^2$ the formulas get simpler

$$\frac{\partial E^{\omega_1}_{(z)}}{\partial z} = -j\frac{\omega_1}{2}d\sqrt{\frac{\mu_0}{\varepsilon_1}}E^{\omega_3}_{(z)}E^{\omega_2^*}_{(z)}e^{-j(k_3-k_2-k_1)z}. \tag{10.15.a}$$

We have similar formulas at the other two frequencies:

$$\frac{\partial E^{\omega_2^*}_{(z)}}{\partial z} = j\frac{\omega_2}{2}d\sqrt{\frac{\mu_0}{\varepsilon_2}}E^{\omega_1}_{(z)}E^{\omega_3^*}_{(z)}e^{-j(k_1-k_3+k_2)z}, \tag{10.15.b}$$

$$\frac{\partial E^{\omega_3}_{(z)}}{\partial z} = -j\frac{\omega_3}{2}d\sqrt{\frac{\mu_0}{\varepsilon_3}}E^{\omega_1}_{(z)}E^{\omega_2}_{(z)}e^{-j(k_1+k_2-k_3)z}. \tag{10.15.c}$$

Equations (10.15) describe the interactions of three waves, the frequencies of which are linked by the relation $\omega_3 = (\omega_1 + \omega_2)$. The phase mismatch that was already intuitively introduced is included in the term $k_3 - k_1 - k_2$. To cancel the phase mismatch, the three waves must have different polarizations; the nonlinear coefficient is then equal to the corresponding element, d_{ij}, of the nonlinear susceptibility tensor.

10.4.2. *Second Harmonic Generation*

In the case of a second harmonic generation we have $\omega_1 = \omega_2 = \omega$ and $\omega_3 = 2\omega$. We will make the assumption that the fundamental signal ω is not depleted and that $\left|E^\omega_{(z)}\right|^2$ keeps a constant value. The incident intensity at 2ω is equal to zero, $E^{2\omega}_{(0)} = 0$. The set of equations (10.15) reduces to only one equation

$$\frac{dE^{2\omega}_{(z)}}{dz} = -j\omega d\sqrt{\frac{\mu_0}{\varepsilon_3}}\left[E^\omega_{(z)}\right]^2 e^{j(\Delta k)z},$$

$$\frac{\partial E^{2\omega}_{(z)}}{\partial z} = -j\frac{\omega}{2}d\sqrt{\frac{\mu_0}{\varepsilon}}\left(E^\omega_{(z)}\right)^2 e^{j\Delta kz} \quad \text{with} \quad \Delta k = k_3 - 2k_2 = k_{2\omega} - k_\omega. \quad (10.16)$$

The integration of the first-order linear equation (10.16) is straightforward and gives the second harmonic intensity on the end face of the crystal, at $z = \delta$,

$$E^{2\omega}_{(z)} = -j\frac{\omega}{2}d\sqrt{\frac{\mu_0}{\varepsilon}}\left(E^\omega\right)^2\frac{e^{j\Delta kz}}{j\Delta k} + \text{constant},$$

$$E^{2\omega}_{(\delta)} = -\frac{\omega}{2}d\sqrt{\frac{\mu_0}{\varepsilon}}\left(E^\omega\right)^2\frac{e^{j\Delta kz} - 1}{\Delta k},$$

$$E^{2\omega}_{(\delta)}E^{2\omega*}_{(\delta)} = 2\frac{\mu_0\omega^2 d^2}{\varepsilon_3}\left|E^\omega\right|^2\frac{\sin^2(\Delta k\delta/2)}{\Delta k^2} = \frac{\mu_0\omega^2 d^2}{2\varepsilon_3}\delta^2 \mathrm{sinc}^2\left(\frac{\Delta k\delta}{2}\right).$$

To obtain the light intensity $I_{2\omega}$ at 2ω, we must use the wave impedance $Z = \sqrt{\varepsilon/\mu}$,

$$I_{2\omega} = \frac{1}{2Z}E_{2\omega}E^*_{2\omega} = \frac{1}{2Z}\left|E_{2\omega}\right|^2 = \frac{1}{2}\sqrt{\frac{\varepsilon_3}{\mu_0}}\left|E_{2\omega}\right|^2 = \frac{n_3}{2}\left|E_{2\omega}\right|^2. \quad (10.17)$$

When the phase matching condition is satisfied, the cardinal sine is maximum, since $\Delta k = 0$. The refraction indices have almost the same value, and the best parameter to be considered for comparing two different nonlinear materials is the ratio of the nonlinear susceptibility to the third power of the index of refraction.

Conversion Efficiency When the Depletion of
the Incident Beam Is Not Neglected

We are now going to again integrate the set of equations (10.15), but we will not suppose that the incident beam is not attenuated as it propagates and generates a beam at the double frequency. To simplify the equations we will consider that the phase matching condition is fulfilled ($\Delta k = 0$); we set

$$A_{i(z)} = E_{(z)}^{\omega_i}\sqrt{\frac{n_i}{\omega_i}} \quad \text{and} \quad \kappa = \frac{d}{2}\sqrt{\frac{\mu_0}{\varepsilon_0}\frac{\omega_1\omega_2\omega_3}{n_1 n_2 n_3}},$$

$$\frac{dA_{1(z)}}{dz} = -j\kappa A_{3(z)}A_{1(z)}^*, \tag{10.18.a}$$

$$\frac{dA_{3(z)}}{dz} = -j\kappa A_{1(z)}^2,$$

$$\frac{d}{dt}\left(A_{3(z)}A_{3(z)}^*\right) = \frac{d}{dt}\left(A_{1(z)}A_{1(z)}^*\right) = j\kappa\left(A_{3(z)}^*A_{1(z)}^2 - A_{3(z)}A_{1(z)}^{2*}\right). \tag{10.18.b}$$

Equation (10.17.b) is the expression of the energy conservation.

As boundary conditions we chose $A_{3(z=0)} = A_{1(z=0)} = A_0$, we suppose that A_0 is a real number, and we set $A_{3(z)}' = -jA_{3(z)}$, and equation (10.18.a) becomes

$$\frac{dA_{1(z)}}{dz} = -\kappa A_{3(z)}' A_{1(z)},$$

$$\frac{dA_{3(z)}'}{dz} = \kappa A_{1(z)}^2, \tag{10.19.a}$$

$$A_{3(z)}'^2 + A_{1(z)}^2 = \kappa A_0^2 = \text{constant}, \tag{10.19.b}$$

$$\frac{d}{dz}A_{3(z)}' = \kappa\left(A_0^2 - A_{3(z)}'^2\right), \tag{10.19.c}$$

$$A_{3(z)}' = A_0 \tanh(\kappa A_0 z),$$

where $A_{3(z)}' \to A_0$ if $z \to \infty$, which means that, in the case of phase matching, the conversion efficiency is 100% if the crystal is long enough.

10.4.3. *Optical Parametric Amplification*

Parametric phenomena are well known in electronics where varactors (capacitor with a capacitance $C(V)$ that varies with the voltage V) are used to make parametric amplifiers and parametric oscillators. In such capacitors the charge Q is not proportional to the voltage, but is a nonlinear function of V. This technique can be transposed in nonlinear optics.

In a parametric device there are always three signals, with three different frequencies ω_1, ω_2, and ω_3 ($\omega_3 = \omega_1 + \omega_2$). The terminology has been defined for electronic devices and adopted in Optics. The signal which has the highest frequency is called the *pump* and is usually more intense, and brings some energy that contributes to the amplification of the two other beams that are, respectively, called the *signal* and the *idler*; but there is no rule for defining which is which.

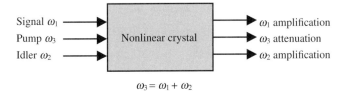

$$\omega_3 = \omega_1 + \omega_2$$

Figure 10.8. Principle of a parametric interaction. Three beams of respective frequencies ω_1, ω_2, ω_3 propagate simultaneously in a nonlinear crystal. The highest frequency beam is called the pumping beam; energy is transferred from the pump, which is attenuated, to the two other beams that are amplified.

We come back to the three equations (16.15) and we consider that the phase matching condition is satisfied,

$$\frac{dA_1}{dz} = -j\kappa A_3 A_2^*,$$

$$\frac{dA_2^*}{dz} = +j\kappa A_1 A_3^*, \qquad (10.20)$$

$$\frac{dA_3}{dz} = -j\kappa A_1 A_2.$$

A linear combination of equations (10.20) gives

$$2\left(A_3^* \frac{dA_3}{dz} + A_3 \frac{dA_3^*}{dz}\right) = \left(A_1^* \frac{dA_1}{dz} + A_1 \frac{dA_1^*}{dz}\right) + \left(A_2^* \frac{dA_2}{dz} + A_2 \frac{dA_2^*}{dz}\right).$$

We will only consider the case when the pump is undepleted and keeps a constant amplitude $A_3(0)$ that is supposed to be real. The system is now reduced to only two equations:

$$\frac{dA_1}{dz} = -j\kappa A_3(0)A_2^* = -j\gamma A_2^*,$$

$$\frac{dA_2^*}{dz} = +j\kappa A_1 A_3(0) = +j\kappa A_1, \qquad \text{with} \quad \gamma = \kappa A_3(0) = \kappa A_3^*(0),$$

$$\frac{d^2 A_1}{dz^2} - \gamma^2 A_1 = 0, \qquad A_{1(z)} = A_{1(0)} \cosh \gamma z - jA_{2(0)}^* \sinh \gamma z,$$

$$\rightarrow \qquad \qquad (10.21.\text{a})$$

$$\frac{d^2 A_2^*}{dz^2} - \gamma^2 A_2^* = 0, \qquad A_{2(z)}^* = A_{2(0)}^* \cosh \gamma z + jA_{1(0)} \sinh \gamma z.$$

We suppose that there is no incident beam at the idler frequency $A_{2(0)}^* = 0$:

$$A_{1(z)} = A_{1(0)} \cosh \gamma z \quad \rightarrow \quad \frac{A_{1(Z)} A_{1(Z)}^*}{A_{1(0)} A_{1(0)}^*} \cong \frac{e^{2\gamma z}}{4} \quad \text{if } z \rightarrow \infty. \qquad (10.21.\text{b})$$

The fact that the amplified signal becomes infinite with z is a consequence of the restrictive hypothesis of an undepleted pump.

Calculation of the Parametric Gain in the Case of Phase Mismatch

If the mismatch term Δk is kept, equations (10.20) become

$$\frac{dA_1}{dz} = -j\kappa A_3 A_2^* e^{-j\Delta kz},$$

$$\frac{dA_2^*}{dz} = +j\kappa A_1 A_3^* e^{+j\Delta kz},$$

$$\frac{d^2 A_1}{dz^2} + j\Delta k \frac{dA_1}{dz} - \gamma^2 A_1 = 0,$$

$$\frac{d^2 A_2^*}{dz^2} + j\Delta k \frac{dA_2^*}{dz} - \gamma^2 A_2^* = 0,$$

$$(10.22)$$

$$A_{1(z)} = \left[A_{1(0)}\left(\cosh \beta z + j \frac{\Delta k}{2\beta} \sinh \beta z \right) - j \frac{\gamma}{\beta} A_{2(0)}^* \sinh \beta z \right] e^{-j\frac{\Delta kz}{2}},$$

$$A_{2(z)}^* = \left[A_{2(0)}^*\left(\cosh \beta z - j \frac{\Delta k}{2\beta} \sinh \beta z \right) + j \frac{\gamma}{\beta} A_{1(0)} \sinh \beta z \right] e^{+j\frac{\Delta kz}{2}}, \quad (10.23)$$

with $\beta^2 = \gamma^2 - \Delta K^2/4$.

In the case of phase mismatch, the condition for having gain is $2\gamma > \Delta k$.

Numerical application: Let us evaluate the gain coefficient γ of a lithium niobate crystal pumped by a green pump $(0.53\,\mu\text{m})$ having an intensity of 10 MW/cm^2:

$$\omega_1 \cong \omega_2 \cong 10^{15} \text{ rad/s}, \quad n_1 \cong n_2 \cong n_3 \cong 2.2, \quad d = d_{14} = 5 \times 10^{-23},$$
$$\text{(mks units)}$$

$$\gamma = \kappa E_{3(0)} = \frac{d}{2}\sqrt{\frac{\mu_0}{\varepsilon_0} \frac{\omega_1\omega_2\omega_3}{n_1 n_2 n_3}} E_{3(0)} \cong 1 \text{ cm}^{-1}. \quad (10.24)$$

10.4.4. *Parametric Oscillator*

A gain of $1\,\text{cm}^{-1}$ is of the same order of magnitude as the gain in a solid-state laser, Nd-YAG for example, and an autooscillation may be obtained if the nonlinear crystal is placed inside a resonator having a high enough quality coefficient.

The condition of oscillation is easily obtained from the length l of the crystal and the product; as the phase condition is satisfied for only one direction of propagation the condition of oscillation is given by $R_1 R_2 e^{\gamma l} = 1$; R_1 and R_2 are the reflection coefficients of the two mirrors of the resonator and l is the length of the crystal. A pumping power of about $50\,\text{W/cm}^2$ is needed with a lithium niobate crystal and $R_1 R_2 = 98\%$.

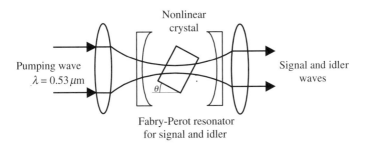

Figure 10.9. Parametric oscillator. The mirrors of the resonator are totally reflecting for the signal and idler, and transparent for the pump. Signal and idler frequencies can be tuned by tilting the angle θ.

Interpretation of the Parametric Interaction in Terms of Photons

A parametric interaction can be described by the *reaction*:

$$\text{photon } (\omega_3) \rightarrow \text{photon signal } (\omega_1) + \text{photon idler } (\omega_2). \qquad (10.25)$$

The energy conservation on one hand, and the momentum conservation on the other, give two equations that determine the signal and idler frequencies, if the frequency of the pump is known:

- Energy conservation: $\omega_1 + \omega_2 = \omega_3$. $\qquad\qquad\qquad\qquad\qquad$ (10.26.a)
- Momentum conservation: $n_1\omega_1 + n_2\omega_2 = n_3\omega_3$ $\qquad\qquad\qquad$ (10.26.b)
  (in the case when the three beams have parallel wave vectors).

The quantum theory of parametric interactions is completely beyond the scope of this book; the theory that was shown in Section 10.3.3 is a classical theory corresponding to a *stimulated parametric interaction*, beside which there is also a *spontaneous interaction*. Let n_1, n_2, and n_3 be the respective numbers of photons at the three frequencies (beware possible confusion between the index of refraction and the number of photons), and the time variations obey the following equations:

$$\frac{dn_1}{dt} = \frac{dn_2}{dt} = -\frac{dn_3}{dt} = K[n_3(n_1+1)(n_2+1) - n_1 n_2(n_3+1)]. \qquad (10.27)$$

To justify equation (10.27) we can say that, at thermal equilibrium ($d/dt = 0$), the numbers of photons are consistent with the Bose-Einstein equation (see Section 9.2),

$$\frac{dn_1}{dt} = \frac{dn_2}{dt} = -\frac{dn_3}{dt} = 0 \;\rightarrow\; \left[\frac{(n_3+1)}{n_3}\right]_{\text{equilibrium}} = \left[\frac{(n_1+1)}{n_1}\right]_{\text{equilibrium}} \left[\frac{(n_2+1)}{n_2}\right]_{\text{equilibrium}},$$

$$\left[\frac{(n_i+1)}{n_i}\right]_{\text{equilibrium}} = e^{\frac{h\nu_i}{kT}}, \quad \text{Bose-Einstein equation,}$$

$$e^{\frac{h\nu_3}{kT}} = e^{\frac{h\nu_1}{kT}} e^{\frac{h\nu_2}{kT}} = e^{\frac{h(\nu_1+\nu_2)}{kT}} \quad \text{which is satisfied since } \nu_3 = \nu_1 + \nu_2.$$

The unit terms in the brackets of equation (10.27) represent the spontaneous parametric interaction; if they were absent a parametric oscillator could not start: if we make $n_1(0) = n_2(0)$ and $n_3(0) \neq 0$, we obtain

$$\frac{dn_1}{dt} = \frac{dn_2}{dt} = -\frac{dn_3}{dt} = Kn_3.$$

10.4.5. Parametric Frequency Conversion

The parametric frequency conversion in Optics is comparable to the heterodyne methods that are currently used in Electronics. A heterodyne device is a device that receives a first carrying wave (frequency ν_1^{carrier}) modulated by some signal $f(t)$, and delivers a new carrying wave (frequency ν_2^{carrier}) modulated by the same signal $f(t)$. Since new frequencies are introduced, heterodyning is obviously a nonlinear process. A good example is described in Figure 10.10. The beam of a CO_2 laser (wavelength $\lambda_1 = 10.6\,\mu\text{m}$), that carries a signal $f(t)$, is mixed with the light of an Nd-YAG laser (wavelength $\lambda_2 = 1.058\,\mu\text{m}$) in a nonlinear crystal. The light emerging from the crystal is modulated by $f(t)$, the new carrying wavelength is now equal to $\lambda_3 = \lambda_1\lambda_2/(\lambda_1 \pm \lambda_2)$, $0.96\,\mu\text{m}$ or $1.17\,\mu\text{m}$. The $\pm$ sign is determined by the phase-matching conditions.

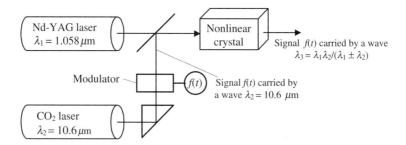

Figure 10.10. The signal $f(t)$ is first carried by a far infrared beam and then by a near infrared beam, after an up-conversion of the frequency in a nonlinear crystal.

The Manley-Rowe Equation

There are two possibilities of achieving a frequency conversion:

- A photon of frequency ν_3 is split into two photons of frequencies ν_2 and ν_1.
- Two photons ν_1 and ν_2 are mixed and generate a photon ν_3.

We consider three waves of frequencies ν_1, ν_2, and ν_3 that interact in a nonlinear crystal with $\nu_1 + \nu_2 = \nu_3$. During the interaction they exchange energy, the highest frequency wave loses some energy P_3, while the two other beams, respectively, gain P_1 and P_2. Any time that a photon ν_3 is created (or annihilated) a photon v_1 and a photon v_2 are simultaneously annihilated (or created), the result is a relation that is called the Manley-Rowe relation:

$$\frac{P_3}{v_3} = \frac{P_1}{v_1} = \frac{P_2}{v_2} = \frac{P_1 + P_2}{v_1 + v_2}. \tag{10.28}$$

10.4.6. *Interpretation of the Electrooptic Effect*

The electrooptic effect, also called the Pockels effect, has already been introduced on the occasion of induced birefringence (cf. Section 5.7.4.2), and it was described how the application of a DC electric field modifies the index of refraction of a crystal. We are now going to show that this effect can be considered as a nonlinear effect, considering the DC field as having a frequency equal to zero.

A DC electric field is applied to a nonlinear crystal inside of which is also sent an optical wave of frequency ω. The polarization of the crystal has a linear part P_L and a nonlinear part P_{NL}. P_L has a zero-frequency component to which is associated the static permittivity $\varepsilon(0) = \varepsilon_0$, and a component at ω, to which corresponds a permittivity $\varepsilon(\omega)$ and an index of refraction $n(\omega) = \sqrt{\varepsilon(\omega)/\varepsilon_0}$. Because of dispersion, $\varepsilon(0)$ and $\varepsilon(\omega)$ are often quite different. The nonlinear part of the polarization now has a DC field and a 2ω term. The electric field in the nonlinear material and the polarization are given by the following equations:

$$E(t) = E_{(0)} + E_{(\omega)}e^{j\omega t} + cc,$$

$$P_{(t)} = P_{L(t)} + P_{NL(t)},$$

$$P_{(t)} = \varepsilon_0 \chi_{(0)} E_{(0)} + \varepsilon_0 \chi_{(\omega)} E_{(\omega)}e^{j\omega t} + 2d\left[E_{(0)} + E_{(\omega)}e^{j\omega t}\right]^2 + cc, \tag{10.29.a}$$

$$P_{\omega(t)} = \left[\varepsilon_0 \chi_{(\omega)} + 4dE_{(0)}\right]E_{(\omega)}e^{j\omega t} + cc, \tag{10.29.b}$$

where $\varepsilon_0 \chi_{(\omega)}$ comes from the linear interaction and $4dE_{(0)}$ originates from the nonlinear interaction; in fact it comes from the double product of the development of the square brackets and is responsible for the presence of a term that vibrates at ω and contributes to the index of refraction at this frequency. $n + \Delta n$ and n being, respectively, the indices with and without a DC field, we have

$$(n + \Delta n)^2 \cong n^2 + 2n\Delta n = 1 + \chi_{(\omega)} + \frac{4dE_{(0)}}{\varepsilon_0},$$

$$\tag{10.30}$$

$$n^2 = 1 + \chi_{(\omega)} \quad \text{and} \quad \Delta n = \frac{4dE_{(0)}}{2n\varepsilon_0}.$$

Unfortunately, a difficulty is hidden behind equations (10.29). The nonlinear coefficient d is obtained from equations (10.8.b) and (10.10.b) and depends on the frequency and surely takes a different value if the frequency is equal either to zero or to ω. The same kind of difficulty was also encountered when the question was to give a general formulation of Maxwell's equations. In Section 2.3.1.2 the difficulty was discarded by considering fields with a narrow spectrum; the same trick cannot be used here, since we have, at the same time, DC and optical signals. Ignoring this difficulty, we have obtained an interesting interpretation of the Pockels effect; the calculated values of the different coefficients are, however, in good agreement with the experimental values.

10.5. Third-Order Nonlinear Phenomena

Third-order nonlinear effects come from the product of $\chi^{(3)}$ by three optical fields having, respectively, three different frequencies ω_1, ω_2, and ω_3; in the nonlinear polarization are found expressions such as

$$\cos\omega_1 t \cos\omega_2 t \cos\omega_3 t = \tfrac{1}{8}\left(e^{j\omega_1 t} + e^{-j\omega_1 t}\right)\left(e^{j\omega_2 t} + e^{-j\omega_2 t}\right)\left(e^{j\omega_3 t} + e^{-j\omega_3 t}\right).$$

The Fourier spectrum is made of frequencies chosen from among the various combinations of $\pm\omega_1 \pm \omega_2 \pm \omega_3$:

- $\omega_1 = \omega_2 = \omega_3 = \omega$ and $(\omega_1 + \omega_2 + \omega_3) = 3\omega$ →   frequency tripling.
- $\omega_1 = \omega_2 = \omega_3 = \omega$ and $(\omega_1 + \omega_2 - \omega_3) = \omega$ →   Kerr effect (see Section 10.4.1).

In the most general case three different fields, $E_i^{\omega_1}$, $E_j^{\omega_2}$, $E_k^{\omega_3}$, are combined to give a fourth field, $E_l^{\omega_4}$; such interactions are also called *four waves interactions*. The notation should provide many indications about the different frequencies and polarizations and so the letters are heavily loaded with indices. The nonlinear polarization is written as

$$P_{i,\text{NL}}^{\omega_4} = \chi_{ijk}^{\omega_1\omega_2\omega_3\omega_4} E_i^{\omega_1} E_j^{\omega_2} E_k^{\omega_3}. \tag{10.31}$$

The lower indices (*ijkl*) indicate on which axis (Ox, Oy, Oz) the field is projected, the upper indices indicate the frequencies. The frequency ω_4 is taken among $\pm\omega_1 \pm \omega_2 \pm \omega_3$.

Third-harmonic generation is easily obtained experimentally, and is observed even in centrosymmetric materials, and is often used to obtain coherent ultraviolet beams.

10.5.1. *The Kerr Effect*

The Kerr effect was already encountered on the occasion of induced birefringence; we distinguished two types of the Kerr effect, the DC Kerr effect and the optical Kerr effect. The first one corresponds to the application of a DC electric field which, in fact, is often a field that varies slowly at the optical scale; it is interpreted as a nonlinear third-order effect, with:

- $\omega_1 = \omega_2 = 0$ (DC field).
- $\omega_3 = \omega_4 = \omega$ (optical field).

The polarization has two components at ω, one of linear origin and the other of nonlinear origin. The index of refraction which comes from the polarization at ω now has two components:

- The first component, which is the most important, comes from the linear part of the polarization, and is equal to the value of the index in the absence of any DC field.
- The second, which has a nonlinear origin, is much weaker. Because of the tensor character of equation (10.31), the material becomes anisotropic.

10.5.1.1. *The Optical Kerr Effect*

In the case of the optical Kerr effect the following set of frequencies is selected: $(\omega_1 = \omega_2 = \omega)$, $(\omega_3 = -\omega)$, $(\omega_4 = 2\omega - \omega = \omega)$, all these fields have the same frequency. The nonlinear polarization is given by

$$P_{i,\mathrm{NL}}^{\omega} = \chi_{iii}^{\omega}[E_i^{\omega}E_i^{\omega}E_i^{-\omega} + E_i^{\omega}E_i^{-\omega}E_i^{\omega} + E_i^{-\omega}E_i^{\omega}E_i^{\omega}]e^{j\omega t} + \mathrm{cc},$$

$$P_{i,\mathrm{NL}}^{\omega} = 3\chi_{(\omega)}^{(3)}|E_i^{\omega}|E_i^{\omega}e^{j\omega t} + \mathrm{cc} \quad \text{with} \quad \chi_{iii}^{\omega} = \chi_{(\omega)}^{(3)}.$$

The modification of the susceptibility at ω and the expression of the index of refraction are then given by the following expressions, where I is the light intensity and Z is the wave impedance:

$$\varepsilon_0 \Delta\chi = \frac{P_{\mathrm{NL}}^{\omega}}{E^{\omega}} = 3\chi_{(\omega)}^{(3)}|E^{\omega}|^2 = 6\chi_{(\omega)}^{(3)}ZI,$$

$$n^2 = 1 + \chi \quad \rightarrow \quad 2n\Delta n = \Delta\chi = \frac{6}{\varepsilon_0}\chi_{(\omega)}^{(3)}ZI,$$

$$n(I) = n + n_2 I \quad \text{with} \quad n_2 = \frac{3\chi_{(\omega)}^{(3)}}{\varepsilon_0 n}. \tag{10.32}$$

Because of the optical Kerr effect, the index of refraction is a linear function of the light intensity. The coefficient n_2 is positive or negative; the absolute value is very small; a very high power (kW-MW/cm^2) is necessary to demonstrate the Kerr effect. Equation (10.32) is too simple, because it doesn't take into account the induced birefringence. Under the action of the electric field of the light wave the material becomes uniaxial, with the optic axis parallel to the field, the two principal indices are given by

$$n_o = n - \Delta n, \quad n_e = n + \Delta n, \quad \text{with} \quad \Delta n = n_2 I.$$

One of the most famous materials for the Kerr effect is carbon disulfide (CS_2) for which $n_2 = +3 \times 10^{-12}$ mm^2/W. In the case of silica (SiO_2), $n_2 = +3 \times 10^{-14}$ mm^2/W. In some glasses, doped with crystallites of lead sulphide (PbS), n_2 can be as high as $n_2 = 10^{-7}$ mm^2/W.

10.5.1.2. *Self-Phase Modulation*

The index of refraction, that's to say the phase velocity, of a Kerr-effect material depends on the intensity of the light; as a consequence, the phase of the light wave is modulated by the intensity. After propagation over a distance L the phase shift is:

- $\phi = n\omega L/c$ if the intensity is low.
- $\phi = (n + n_2 I)\omega L/c$ for a higher intensity.

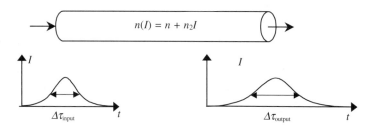

Figure 10.11. The index of refraction is larger when the intensity is maximum. The middle part of the pulse propagates slower than the front and the tail of the pulse, as a result the duration of the pulse increases during propagation.

Another consequence is the augmentation of the duration of a light pulse as it propagates inside a material where the index varies with the intensity, as is shown in Figure 10.11. The case of the propagation of a light pulse is more complicated, because of the time variation of the intensity. The instant phase of a sine signal is proportional to the time, the proportionality coefficient being the frequency $\phi(t) = \omega t$. The phase of a sine wave can be written as $\phi(t, z) = (\omega_0 t - nz/c)$ and the frequency can be defined as the partial time derivative of the argument. If the index of refraction is now a function of the intensity, and the intensity a function of time, the phase at some point z is given by $\phi(t, z) = [\omega_0 t - (n + n_2 I_{(t)})z/c]$, an instant frequency can be defined as the time partial derivative of the argument

$$\omega = \frac{\partial \phi}{\partial t} = \omega_0 + n_2 z \frac{dI}{dt}. \tag{10.33}$$

The equation shows that the frequency of the optical wave that carries the pulse varies during propagation. This effect is known as chirping, from a word that describes the modulated song of a bird. It can be shown that, if n_2 is positive, the front of the output signal has as a lower frequency than the tail, see Figure 10.14.

10.5.1.3. *Self-focusing*

We refer to Figure 10.12. A parallel light beam arrives at a transparent material; the intensity is not constant over the cross section and is maximum at the center where, because of the optical Kerr effect, the index of refraction is also maximum. The material becomes a gradient index lens, which is converging if n_2 is positive and focuses the beam: the cross section decreases while the intensity increases, in other words, the focusing effect is more and more important as the light propagates. The cross section would go to zero if the diffraction didn't have an opposite action. The effect of diffraction is noticeable only when the size of the cross section becomes of the order of the wavelength. In the end, the initial beam is divided into tiny filaments (with a diameter of the order of a few micrometers) inside which the intensity may be extremely high.

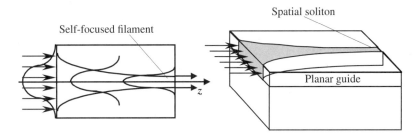

Figure 10.12. A beam with an inhomogeneous intensity repartition creates a gradient of index of refraction, which has the same effect as a lens. If the lens is converging the intensity along the axis is increased, the focusing effect becomes increasingly important, and the cross section of the beam decreases. When the cross section is of the order of the wavelength, diffraction has the opposite action. As a result, tiny filaments are produced with a diameter of a few micrometers and a very high intensity.

These phenomena are known as *self-focusing* or *self-trapping*; they are only observed with the light of powerful lasers. From a theoretical point of view they are very complicated, since the equation of Helmholtz is no longer linear and contains terms that are proportional to the square of the modulus of the electric field of the light wave. The mathematical behavior of the solution is *unstable* and often *chaotic*. As the transversal repartition of intensity across their sections is never very homogeneous, high-power laser beams are transformed in many very brilliant filaments when they travel inside a Kerr material. When the support of propagation has only two dimensions, a planar waveguide for example, experimental conditions may be found where the solution is stable; the filaments are then called *spatial solitons*. If n_2 is positive, the filaments are bright and called white spatial solitons; if n_2 is negative, dark spatial solitons are produced.

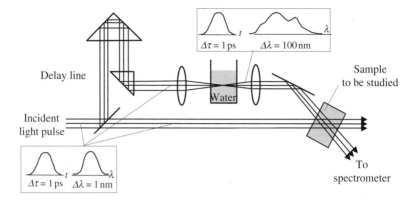

Figure 10.13. Time-resolved absorption spectroscopy experiment. This experiment is also an elegant illustration of the inequality $\Delta\nu\Delta\tau \geq 1$.

From a practical point of view self-focusing is often considered as a drawback which deteriorates the optical quality of powerful laser beams. Because of the extremely high electric field existing in the filaments, irreversible damage may be created inside the transparent materials.

In the case of a laser working in a multimode regime, the Fourier spectrum has many spectral components, the intense power enhances a large variety of nonlinear effects (frequency beat, Raman and Brillouin scattering, . . .), and the final result is a new light beam with a broad almost continuous spectrum. An elegant application is described in Figure 10.13. A very short (1 ps) and rather monochromatic (1 nm) powerful laser pulse is divided into two parts. One beam is focused on a cell filled with water; the transmitted pulse has about the same duration and a larger spectrum (100 nm). The two beams intersect inside a sample; the first beam is used to probe the sample when it is submitted to the illumination of the second beam. Thanks to an optical delay line, the absorption spectrum of the sample is exactly analyzed during the illumination of the sample, or at any time before or after.

10.5.1.4. *Optical Solitons*

Optical solitons were invented before spatial solitons; in both cases there are two antagonistic mechanisms: one is linear, the other comes from a nonlinear optical interaction. The linear processes are, respectively, the dispersion, for optical solitons, and the diffraction, for spatial solitons. The nonlinear processes are automodulation and self-focusing. The linear process tends to decrease the light intensity either by extending the pulse over a longer period (dispersion) or by expanding the beam inside a larger area (diffraction). Automodulation and self-focusing have exactly the opposite effect, and tend to decrease the intensity. Because of these two antagonistic actions, equilibrium is sometimes reached where the size and shape of the beam, in one case, or the duration and profile of the pulse, in the other case, are constant during propagation.

Optical solitons are thus light pulses that keep a constant duration and profile, whatever the distances over which they propagate. The only practical case where optical solitons are produced is in silica optical fibers; the permanency of the properties of optical solitons has been observed over 1000 km . . . . We will not attempt to write equations that are really complicated, and will only consider the mechanisms that are illustrated in Figure 10.14.

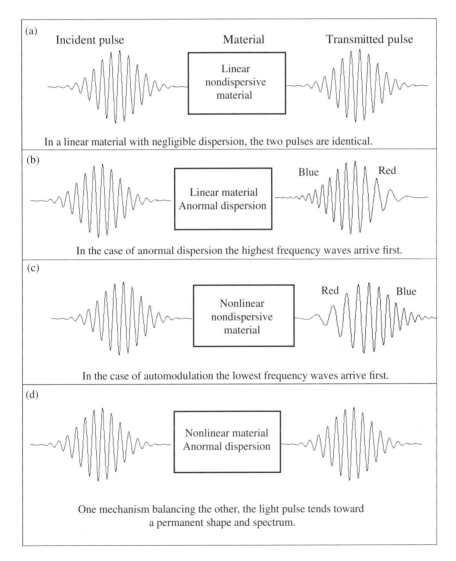

Figure 10.14. Qualitative explanation of the propagation of solitons.

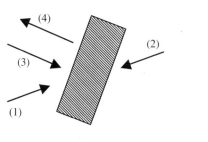

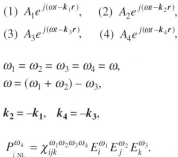

(1) $A_1 e^{j(\omega t - k_1 r)}$,   (2) $A_2 e^{j(\omega t - k_2 r)}$,

(3) $A_3 e^{j(\omega t - k_3 r)}$,   (4) $A_4 e^{j(\omega t - k_4 r)}$,

$\omega_1 = \omega_2 = \omega_3 = \omega_4 = \omega$,

$\omega = (\omega_1 + \omega_2) - \omega_3$,

$k_2 = -k_1$, $k_4 = -k_3$,

$$P^{\omega_4}_{i,NL} = \chi^{\omega_1 \omega_2 \omega_3 \omega_4}_{ijk} E^{\omega_1}_i E^{\omega_2}_j E^{\omega_3}_k.$$

Figure 10.15. Principle of phase conjugation. The nonlinear interaction of (1), (2), and (3) generates a polarization repartition at ω which, in turn, creates a fourth beam propagation in the opposite direction of (3).

White and dark solitons: The explanation of Figure 10.15 is valid for anormal dispersion. In the case of normal dispersion, the equilibrium of the two mechanisms is still possible and produces "dark solitons," which correspond to a brief extinction of the intensity of a permanent beam.

Electromagnetic waves are the only domain where solitons may be observed, they also exist in acoustic waves. Lord Rayleigh said that he observed the propagation of a soliton along a channel and that, riding his horse, he was able to follow it for several kilometers.

10.5.2. *Phase Conjugation*

Phase conjugation is a *four-waves interaction*, which is often considered as *degenerated*, because the frequencies of the three waves are equal. In Figure 10.16 is shown the general arrangement of what is also called a *four-waves*

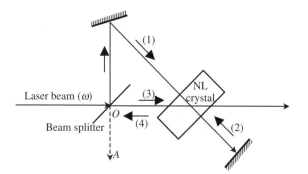

Figure 10.16. Experimental demonstration of phase conjugation. In order to be coherent and to have exactly the same frequency, beams (1), (2), and (3) are obtained from the same laser beam. The existence of light along OA establishes the existence of phase conjugation.

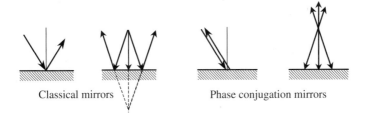

Classical mirrors   Phase conjugation mirrors

Figure 10.17. Comparison between a classical and a self-conjugation mirror. In a classical reflection, the tangential component of the wave vector is conserved and the normal component is reversed. For a phase conjugation mirror both components are reversed.

mixing experiment. Inside a nonlinear material, in which χ^3 is especially high, are sent three waves labeled (1), (2), and (3); the source terms of equation (10.13) generate a fourth wave, labeled (4). Waves (1) and (2) propagate in opposite directions and are more powerful than (3).

For the sake of simplicity, we will ignore the polarization of the waves and drop the indices i, j, k in equation (10.31). The three planar waves (1), (2), and (3), and the nonlinear polarizations that they generate, are written as

$$E_{1(r,t)} = A_1 e^{j(\omega t - k_2 r)}, \quad E_{2(r,t)} = A_1 e^{j(\omega t + k_1 r)},$$

$$E_{3(r,t)} = A_3 e^{j(\omega t - k_3 r)}, \quad E_{3(r,t)}^* = A_3^* e^{j(-\omega t + k_3 r)},$$

$$P_{NL}^{\omega + \omega - \omega} = K A_1 A_2 A_3^* e^{j\omega t} e^{jk_1 r} e^{-jk_1 r} e^{jk_3 r}, \quad K \text{ proportionality coefficient.} \quad (10.34)$$

To the polarization of equation (10.34) is associated a planar wave, propagating in the opposite direction to the wave labeled (3). Everything happens as if this last wave was reflected on the nonlinear crystal, which behaves as a kind of mirror. From a physical point of view the two beams, (1) and (2), produce, inside the crystal, a standing wave pattern on which beam (3) is reflected.

Figure 10.17 is a comparison of the ways that light rays are reflected, either on a classical mirror or on a phase conjugation mirror. In the last case, the reflected ray follows the same path as the incident ray, but in the opposite direction. The image of a real object is a virtual object, symmetric with respect to the mirror plane. In the case of phase conjugation, the image and the object are exactly superimposed.

11

Raman-Brillouin-Rayleigh Diffusion

11.1. Raman, Brillouin, Rayleigh, and Mie Scattering

As early as 1923, Einstein generalized the notion of stimulated and spontaneous emission to a new kind of light scattering process, in which the frequency of the scattered light is shifted with regard to the incident light. It was only after their experimental demonstration that these new effects were given a name, the spontaneous Raman effect in 1928 and the Brillouin effect in 1932. It was not until 1963, and with the availability of powerful and monochromatic laser beams, that the existence of the stimulated Raman effect was experimentally demonstrated.

In the case of stimulated interactions many photons are emitted in the same mode (see Section 9.2.1) and are coherent. It is possible to obtain a laser action without the necessity of an inversion of population.

Raman and Brillouin diffusion of light is based on the electric dipolar interaction of an incident light beam of frequency ω with the molecular electric dipoles of a transparent material. The same model that allowed the calculation of the index of refraction (8.1) can describe the situation. In this model the electric field of the light wave creates oscillating dipoles vibrating at the same frequency ω as the incident wave; the light that is transmitted by a piece of material is the result of the interference of the incident beam with the electromagnetic field radiated by the dipoles. In the case of the Raman and Brillouin interactions, the oscillation frequency ω_μ of the dipoles is different from the incident frequency ω. The frequency shift $(\Omega_\mu = \omega - \omega_\mu)$, which can be positive or negative, is always small as compared to ω and is specific to the molecular dipole.

As new frequencies are generated, the interaction between a light wave and a piece of material is a nonlinear interaction.

The distinction between the Raman and Brillouin diffusions is not always clear-cut; it is based on the fact that the vibrations of the different dipoles of the transparent material can either be independent or coupled to one another.

- The Raman effect corresponds to the case when the vibrations are independent, which mainly occurs for gases or liquids.
- The Brillouin effect corresponds to the case when the molecular dipoles are coupled by mechanical waves (acoustic waves if the frequency is low enough) propagating inside the transparent material.

Because of the electric dipolar origin of the scattering process the efficiency varies, in both cases, as the inverse of a fourth of the wavelength ($1/\lambda^4$).

The Brillouin diffusion effect is more general than the Raman diffusion effect and can be observed in all materials, with an efficiency that is much higher in dense materials (liquid or solid) than in dilute material (gas). Raman diffusion is present only for specific molecules that are said to be *Raman active*, or to show *Raman activity*.

Importance of the Homogeneity of the Medium Supporting the Propagation

We come back to the usual propagation of a light beam in a transparent material and consider a planar light wave (ω, k) traveling in a transparent material considered as a collection of electric dipoles that interact with the oscillating electric field of the wave. Each dipole becomes a secondary source emitting a spherical wavelet of frequency ω. At a given point P (see Figure 11.1) which can be either inside or outside the transparent material, will arrive all the wavelets that have been emitted by all the different points; the electromagnetic field at P results from the interference of all these wavelets. The same theory was developed in Chapter 8; here we just want to emphasize what happens when the medium is not perfectly homogeneous. It's only in the case when the homogeneity is perfect that the different spherical wavelets reconstruct a *planar wave* with a wave vector parallel to the wave

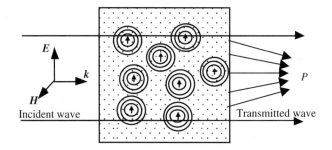

Figure 11.1. A transparent material is made of a set of electric dipoles. Set in vibration by the electric field of an incident light wave, the dipoles emit spherical wavelets. The electromagnetic field at some point P is the result of the interference of the wavelets. For the sake of symmetry, the transmitted wave that is reconstructed from an incident planar wave (of infinite transverse extension) should also be planar.

vector of the incident wave. The existence of nonhomogeneous domains breaks the symmetry, the reconstructed transmitted wave is no longer planar: the Fourier planar wave decomposition shows a wave in different directions.

If the medium is not homogeneous, the situation is different: some light is scattered in directions other than the incident wave vector. The intensity of the diffused light may be important and, for example, is responsible for the blue color of a clear sky, as well as for the milky appearance of clouds. The properties of the diffused light are determined by the size of the domains inside which the material is not homogeneous:

- *Rayleigh diffusion*: The domains are much smaller than the wavelength. The efficiency of the diffusion process varies as $1/\lambda^4$ is a function of the wavelength.
- *Mie diffusion*: The dimension of the domains is comparable to the wavelength. The efficiency is almost independent of the wavelength.

11.2. Experimental Introduction to the Raman Effect

11.2.1. *Description of the Original Raman Experiment*

For his original experiment Raman used benzene (C_6H_6). The light source, (nowadays we would say the pumping light), was a mercury vapor discharge followed by a filter isolating only one ray of frequency ν_{pump}. If no special care was taken, the diffused light appeared to be monochromatic and to have the same frequency as the pumping light; in fact the diffusion occurred on the dust particles and on the bubbles contained in the liquid. If the benzene was carefully filtered and distilled several times, the diffused light became less intense allowing the appearance of two families of rays, the frequencies of which are different from the pumping frequency, some frequencies being higher and others being lower than ν_{pump}.

The two families of rays are now, respectively, called the Stokes and the anti-Stokes rays:

- *Stokes rays*: Their frequencies, ν_S, ν'_S, ν''_S, ..., are lower than the pumping frequency.
- *Anti-Stokes rays*: Their frequencies, ν_{AS}, ν'_{AS}, ν''_{AS}, ..., are higher than the pumping frequency.
- The anti-Stokes rays are *less intense* than the Stokes rays.

The absolute values of the Stokes and anti-Stokes frequency shifts are *equal* and specific to the Raman active molecules (benzene in the original Raman experiment),

$$\begin{cases} \nu_{AS} - \nu_{pump} = \nu_{pump} - \nu_S = \nu_{molec}, \\ \nu'_{AS} - \nu_{pump} = \nu_{pump} - \nu'_S = \nu'_{molec}, \\ \nu''_{AS} - \nu_{pump} = \nu_{pump} - \nu''_S = \nu''_{molec}. \end{cases} \tag{11.1}$$

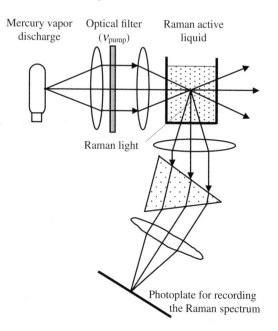

Figure 11.2. Original Raman arrangement. A monochromatic light beam illuminates a cell, filled with a Raman active material. Part of the Raman light, which is spontaneously emitted in the 4π steradian, is collected by a lens and analyzed by a dispersing prism. To avoid the dazzling effect of the pumping light, the diffused light is observed in an orthogonal direction. The Raman light is extremely weak; the exposure time was several weeks.

11.2.2. *Raman Spectroscopy*

Independently of any theoretical interpretation, Raman diffusion is a very interesting tool for chemical analysis, since a Raman spectrum is a type of *signature* of the diffusing molecules. Furthermore, it provides information about the chemical bonds between the atoms of a molecule.

Using a laser, instead of a spectral lamp, considerably shortens the time of exposition of the photoplates, which, in most Raman spectrometers, are now replaced by an array of photodiodes.

Raman Effect Microprobe

The Raman microprobe is a transposition of the electron scanning microprobe in which a beam of monokinetic electrons is sharply focused on the surface of material to be analyzed. Under the impact of the electrons the atoms of the target emit X-rays. The identity of the atoms is revealed by a spectral analysis of the emitted X-rays.

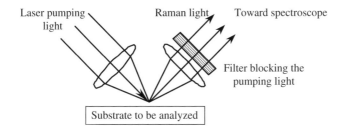

Figure 11.3. Scanning Raman microprobe. A laser beam is sharply focused on the surface; the focal point becomes a light source emitting, by the spontaneous Raman effect, the Stokes and anti-Stokes frequencies.

In a Raman microprobe, a laser beam is sharply focused on the surface of a Raman active sample, the molecules of which emit Stokes and anti-Stokes rays. The local chemical composition is deduced from a spectroscopic analysis of the Raman light. A mechanical scanning of the target provides cartography of the chemical composition.

Lidar (Light Detection and Ranging) is the apparatus that inspired Radar (Radio Detection and Ranging). Short and powerful light pulses coming from a Q-switched laser are sent into the atmosphere where they interact with the molecules they meet, mostly nitrogen, oxygen, and water, and also the inevitable impurities. These molecules reemit some light by a spontaneous Raman effect, each of them having a specific spectrum. Part of the Raman back-scattered light is collected by the telescope and analyzed by a spectroscope which gives an electric signal, from which the position (by measuring the flight time) and the identity (by spectral identification) are measured.

11.2.3. *Stimulated Raman Effect, the Woodbury and*
Ng Experiment

The diffusion of light in the original Raman experiment corresponds to a spontaneous Raman effect. In 1923 Einstein introduced the possibility of a stimulated effect, an experimental demonstration was made by chance in 1963, at the early stages of laser Q-switching. The two physicists, Woodbury and Ng, used a Kerr cell filled with mono-nitrobenzene ($C_6H_5NO_2$) to obtain giant pulses from a ruby laser (see Figure 11.6 and also Figure 9.34) and were surprised to find that the laser light contained *two* spectral components: beside the usual component at 694.3 nm of a ruby laser another component was observed at 765.8 nm which, having nothing to do with any frequency of the ruby spectrum, was rapidly identified as a Raman-Stokes frequency of $C_6H_5NO_2$.

As many aromatic compounds, mono-nitrobenzene is well known to show Raman activity; the Raman shift is equal to $1340 \, cm^{-1}$ and corresponds to a

Table 11.1. Chart of some Raman frequencies.

Frequency, cm^{-1}	Type of chemical bond	Chemical compounds
445–550	S—S	aliphatic disulfide
490–522	C—I	aliphatic iodide
510–594	C—Br	aliphatic bromide
570–650	C—Cl	aliphatic chloride
600–700	C—SH	mercaptans
630–705	C—S	aliphatic sulfide
700–1100	C—C	aliphatic carbon-carbon bond
750–850	benzene ring	paraderivatives of benzene
884–899	5 carbon ring	monosubstituted cyclopentane
939–1005	4 carbon ring	cyclobutane and derivates
990–1050	benzene ring	benzene and mono-, bi-, tri-substituted benzene
1020–1075	C—O—C	aliphatic compound
1085–1125	C—OH	aliphatic alcohol
1120–1130	C=C=O	aliphatic compound
118–1207	3 carbon ring	cyclopropane and derivativees
1190	SO_2	aliphatic compound
1216–1230	—S=O	aliphatic compound
1340	NO_2	aromatic compound
1380	NO_2	aliphatic compound
1610–1640	N=O	aliphatic compound
1620–1680	C=C	aliphatic compound
1630	C=N	aliphatic compound
1654–1670	C=N	aliphatic compound
1650–1820	C=O	aliphatic compound
1695–1715	C=O	aliphatic compound
1974–2260	triple bond C≡C	aliphatic compound
2150–2245	triple bond C≡N	nitriles
2800–3000	C—H	aliphatic compound
3000–3200	C—H	aliphatic compound
3150–3650	O—H	aromatic compound
3300–3400	N—H	aliphatic compound
4160	H—H	gaseous hydrogen

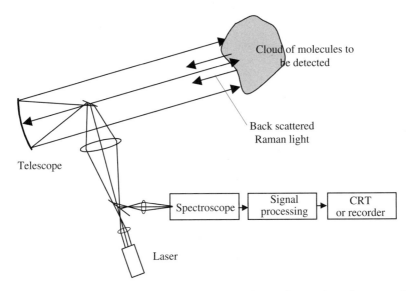

Figure 11.4. Lidar scheme. The light of a powerful laser is sent into the atmosphere via a telescope, which also collects the spontaneous Raman light emitted by molecules located several miles away.

vibration of the radical NO_2. If we make $1/\lambda_1 - 1/\lambda_2 = 1340\,nm$ and $\lambda_1 = 694.3$ nm, we obtain $\lambda_2 = 765.8\,nm$. The monochromatic and well-collimated characteristics of the beam emitted at this frequency clearly demonstrate the stimulated origin of the new beam.

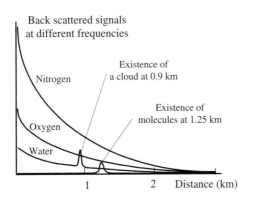

Figure 11.5. Typical curves obtained with a Lidar. At frequencies corresponding to the molecules that are always met in the atmosphere (nitrogen, oxygen, water, . . .) the return signal decreases as the inverse of the square of the distance.

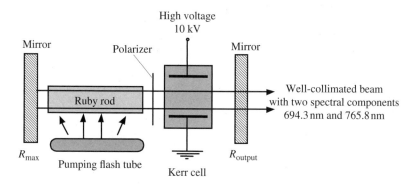

Figure 11.6. Q-switching of a ruby laser with a Kerr cell filled with nitrobenzene. The electromagnetic density is so high inside the laser cavity that the stimulated Raman effect becomes very important.

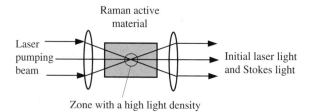

Figure 11.7. Stimulated Raman interaction is all the more efficiently produced when the pumping energy density is higher. Focusing the pumping beam increases the density inside the focal volume. The coherent Stokes light that is generated comes from the focal point of the first lens and is then collimated by the second lens.

*Other Experimental Arrangements for Obtaining
the Stimulated Raman Effect*

An important research effort was produced during the 1965–1975 decade to develop Raman lasers, which had the interest of enriching the number of available coherent laser frequencies. Efficient materials for the Raman effect are often made of polar molecules and, consequently, the threshold for obtaining self-focusing is quite low. It is now known that self-focusing often has a *chaotic behavior,* that is the reason why Raman lasers appeared impossible to tame. Meanwhile, dye lasers and optical parametric oscillators proved to be more convenient, and the interest in Raman lasers rapidly decreased. There

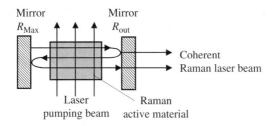

Figure 11.8. Stokes photons are emitted in any direction. In this case the photons that are emitted perpendicular to the mirrors are stored: because of the cumulative effect of stimulated emission a coherent beam is emitted.

are two cases where self-focusing effects can be ignored:

- *High-pressure gaseous material*: The small value of the number of molecules per cm^3 makes it difficult to run a stimulated Raman effect in a gas; however, very nice experiments have been reported in high-pressure hydrogen.
- *Single-mode silica fibers*: See Section 11.3.2.

11.3. Theoretical Analysis of the Raman Effect

Raman interaction can be either spontaneous or stimulated. As in the case of the usual interaction of radiation with a collection of two-level atoms, the stimulated effect can receive either a classical or a quantum interpretation. We will first describe the classical model of interaction and then describe the Einstein theory for introducing spontaneous and stimulated effects.

11.3.1. *The Classical Model of Raman Interaction*

An atom is made of a nucleus surrounded by electrons: most of the electrons are bound to the nucleus. A molecule is made of atoms that are linked together by chemical bonds, which correspond to most external electrons that are not bound to a specific nucleus and make a cloud ensuring the chemical stability of the arrangement. A molecule is thus a set of heavy positively charged nuclei and of light negatively charged electrons. All these particles can vibrate; a harmonic oscillator, which has a specific eigenfrequency, represents each vibration:

- The eigenfrequencies of nuclei belong to the infrared ($\approx 1000\,cm^{-1}$); they are designed by $\nu_{molec,i}$, where i is used to label the atom inside the molecule.
- The eigenfrequencies of electrons are much higher and belong to the ultra-violet part of the spectrum ($\approx 25{,}000\,cm^{-1}$), they are designed by $\nu_{electron,k}$, where k is used to label the electron inside the atom.

When a molecule is submitted to the action of an optical wave, the nuclei are too heavy to follow and vibrate with negligible amplitude; on the contrary, the electron will vibrate more easily. The two kinds of motion (nucleus or electron) can either be completely independent, or they can be coupled together. In the last case, part of the excitation of the electrons is transferred to the nuclei, which are set in vibration: of course, the nuclei will vibrate at one of their eigenfrequencies $\nu_{\text{molec},i}$. The ith mode of vibration of the molecule is then said to be *Raman active*.

If a Raman active molecule is illuminated by a light beam of frequency $\omega = 2\pi\nu$, its electrons vibrate at the same frequency, while the nuclei vibrate at $\nu_{\text{molec},i}$. Adopting a phenomenological point of view, we consider that the electric dipolar momentum of the molecule varies according to the following law:

$$\boldsymbol{\mu} = \alpha \boldsymbol{e} \cos \omega t \cos \omega_{\text{molec},i} t = \boldsymbol{e} \frac{\alpha}{2} [\cos(\omega - \omega_{\text{molec},i})t + \cos(\omega + \omega_{\text{molec},i})t], \quad (11.2)$$

where α is a proportionality coefficient and $\boldsymbol{e}$ is a unit vector.

11.3.1.1. *Raman Diffusion Is a Nonlinear Process*

Formula (11.2) clearly shows the nonlinear aspect of Raman interaction. The two frequencies ($\nu \pm \nu_{\text{molec},i}$) come from the corresponding terms in the expression of the electric dipolar momentum. This model gives a nice interpretation of the Raman diffusion; unfortunately, the Stokes and anti-Stokes rays should have the same intensity, which is in contradiction to the experimental observations. Einstein's theory anticipates different amplitudes for the two rays, the ratio of which is in good agreement with the experimental results.

11.3.2. *Einstein's Theory, Raman Effect*

11.3.2.1. *Real and Virtual Energy Levels*

By absorption of a photon of frequency ν, a molecule reaches the level $E' = (E_0 + h\nu)$ if initially on level E_0, and level E'' if initially on level E_1.

Since they are not *allowed levels* of the molecules, E' and E'' are called *virtual levels*. The molecule cannot stay for long on these energy levels and makes almost immediately a transition toward one of the two allowed states, E_0 or E_1:

- If the molecule makes the following trip:

$$E_0 \rightarrow E' = (E_0 + h\nu) \rightarrow E_1,$$

  a photon of frequency ν is first annihilated and then a photon of frequency $\nu_{\text{Stokes}} = (\nu - \nu_{\text{molec}})$ is created. This is *Raman-Stokes diffusion*.

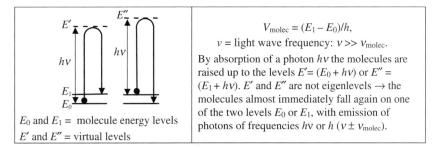

E_0 and E_1 = molecule energy levels
E' and E'' = virtual levels

$V_{molec} = (E_1 - E_0)/h,$
v = light wave frequency: $v \gg v_{molec}$.
By absorption of a photon hv the molecules are raised up to the levels $E' = (E_0 + hv)$ or $E'' = (E_1 + hv)$. E' and E'' are not eigenlevels → the molecules almost immediately fall again on one of the two levels E_0 or E_1, with emission of photons of frequencies hv or $h(v \pm v_{molec})$.

Figure 11.9. Real and virtual energy levels of a molecule.

- If the molecule makes the following trip:

$$E_1 \to E' = (E_1 + hv) \to E_0,$$

a photon of frequency v is first annihilated and then a photon of frequency $v_{anti\text{-}Stokes} = (v + v_{molec})$ is created. This is *Raman-anti-Stokes diffusion*.
- If the molecules make one of the following trips:

$$E_0 \to E' = (E_0 + hv) \to E_0 \quad \text{or} \quad E_1 \to E'' = (E_1 + hv) \to E_1,$$

a photon v is first annihilated and then a new photon of frequency v is created again. This is *Rayleigh diffusion*.

We now consider the interaction of a collection of molecules at thermal equilibrium between the two states E_0 and E_1 with a pumping beam: the numbers of Stokes and anti-Stokes diffusions are, respectively, proportional to the populations of the two levels. The lower level being more populated, the number of Stokes photons exceeds the number of anti-Stokes photons. The ratio of the two intensities is obtained from the Maxwell-Boltzmann ratio of populations,

$$\frac{I_{anti\text{-}Stokes}}{I_{Stokes}} = e^{-(E_1 - E_0)/kT} = e^{-hv_{molec}/kT}. \tag{11.3}$$

11.3.2.2. *Equilibrium in a Blackbody*

After having invented this new mechanism for the interaction of an electromagnetic radiation with a collection of molecules, Einstein could not resist the temptation of enclosing a sample of some Raman active material inside a blackbody. The problem is now to describe the coupling mechanisms between the following objects:

- Mode (1) of the blackbody with n_1 photons of frequency v_1.
- Mode (2) of the blackbody with n_2 photons of frequency $v_2 = v_1 - v_{molec}$.
- Mode (3) of the blackbody with n_3 photons of frequency $v_3 = v_1 + v_{molec}$.

- Collection of molecules with two energy levels E_1 and E_0 of respective populations N_1 and N_0.

We will consider independently the equilibrium of the collection of molecules with modes (1) and (2) on one hand and with modes (1) and (3) on the other hand. The interaction mechanisms are phenomenological and find their justification in the fact that they give the right statistical distributions for the two kinds of photons and for the molecules. Following Einstein we will use the electromagnetic energy densities u_1 and u_2 at the respective frequencies ν_1 and ν_2 and we introduce two phenomenological coefficients A and B.

We first consider the case of the Stokes interaction, $\nu_2 = (\nu_1 - \nu_{\text{molec}}) < \nu_1$, and introduce the spontaneous and stimulated processes:

- Spontaneous effect: $\left(\dfrac{dN_1}{dt}\right)_{\text{spon}} = -\left(\dfrac{dN_0}{dt}\right)_{\text{spon}} = AN_0 u_1 - AN_1 u_2.$

- Stimulated effect: $\left(\dfrac{dN_1}{dt}\right)_{\text{stimul}} = -\left(\dfrac{dN_0}{dt}\right)_{\text{stimul}} = BN_0 u_1 u_2 - BN_1 u_1 u_2.$

- Cumulative action of the two effects:

$$\frac{dN_1}{dt} = -\frac{dN_0}{dt} = A(N_0 u_1 - N_1 u_2) + Bu_1 u_2 (N_0 - N_1). \tag{11.4}$$

At equilibrium the populations remain constant:

$$\frac{d}{dt} = 0 \quad \rightarrow \quad \frac{N_1}{N_0} = \frac{Au_1 + Bu_1 u_2}{Au_2 + Bu_1 u_2} = \frac{\dfrac{1}{u_1} + \dfrac{B}{A}}{\dfrac{1}{u_2} + \dfrac{B}{A}}.$$

Inside a blackbody at thermal equilibrium, u_1 and u_2 are given by

$$u_1 = \frac{8\pi h\nu_1^3}{c^3}\frac{1}{e^{h\nu_1/kT} - 1} \quad \text{and} \quad u_2 = \frac{8\pi h\nu_2^3}{c^3}\frac{1}{e^{h\nu_2/kT} - 1}.$$

As the two frequencies ν_1 and ν_2 are almost equal we can write

$$\frac{A}{B}\frac{8\pi h\nu_1^3}{c^3} \cong \frac{A}{B}\frac{8\pi h\nu_2^3}{c^3} = 1, \quad \text{formulas identical to (9.25),}$$

$$\frac{N_1}{N_0} = \frac{(e^{h\nu_1/kT} - 1) + \dfrac{B}{A}\dfrac{8\pi h\nu_1^3}{c^3}}{(e^{h\nu_2/kT} - 1) + \dfrac{B}{A}\dfrac{8\pi h\nu_2^3}{c^3}} = \frac{e^{h\nu_1/kT}}{e^{h\nu_2/kT}} = e^{-h\nu_{\text{molec}}/kT}.$$

Formula (11.3) gives exactly the expected value for the ratio of the populations, which can be considered as proof of the validity of the Einstein model and justification for the existence of spontaneous and stimulated Raman processes.

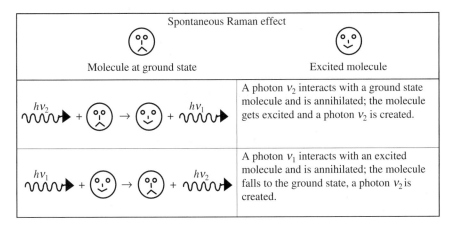

Figure 11.10. Illustration of Raman spontaneous interactions.

Instead of using the electromagnetic energy density, u_ν, we can also introduce the number of photons. Let n_1 and n_2 be the respective numbers of photons at the frequencies ν_1 and ν_1, the phenomenological equations become

$$\frac{dn_2}{dt} = -\frac{dn_1}{dt} = -\frac{dN_1}{dt} = \frac{dN_0}{dt},$$

$$\frac{dn_2}{dt} = -\frac{dn_1}{dt} = K\Big[\underbrace{(n_1 N_0 - n_2 N_1)}_{\text{spontaneous}} + \underbrace{(n_1 n_2 N_0 - n_1 n_2 N_1)}_{\text{stimulated}}\Big], \qquad (11.5)$$

$$\frac{dn_2}{dt} = -\frac{dn_1}{dt} = K[n_1(n_2+1)N_0 - n_2(n_1+1)N_1],$$

At equilbrium:
$$\left(\frac{N_1}{N_0}\right)_{\text{equilibrium}} = \left[\frac{n_1(n_2+1)}{n_2(n_1+1)}\right]_{\text{equilibrium}}. \qquad (11.6)$$

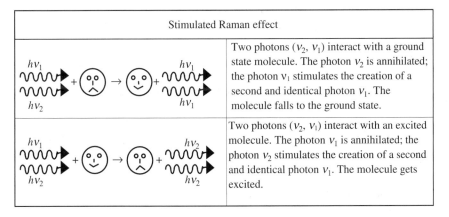

Figure 11.11. Illustration of Raman stimulated interactions.

If we make $n_1 = 1/(e^{h\nu_1/kT} - 1)$ and $n_1 = 1/(e^{h\nu_1/kT} - 1)$ we obtain

$$\left(\frac{N_1}{N_0}\right)_{\text{equilibrium}} = \left[\frac{1+1/n_2}{1+1/n_1}\right]_{\text{equilibrium}} = e^{h(\nu_2-\nu_1)/kT} = e^{h\nu_{\text{molec}}/kT}.$$

Very similar equations can be written for the anti-Stokes ray,

$$-\frac{dN_1}{dt} = +\frac{dN_0}{dt} = A(N_1 u_1 - N_0 u_3) + B u_1 u_3 (N_0 - N_1), \tag{11.7}$$

$$\frac{dn_3}{dt} = -\frac{dn_1}{dt} = -\frac{dN_1}{dt} = \frac{dN_0}{dt},$$

$$\frac{dn_3}{dt} = -\frac{dn_1}{dt} = K\left[\underbrace{(n_1 N_1 - n_3 N_0)}_{\text{spontaneous}} + \underbrace{(n_1 n_3 N_1 - n_1 n_3 N_0)}_{\text{stimulated}}\right], \tag{11.8}$$

$$\frac{dn_3}{dt} = -\frac{dn_1}{dt} = K[n_1(n_3+1)N_1 - n_3(n_1+1)N_0],$$

$$\left(\frac{N_1}{N_0}\right)_{\text{equilibrium}} = \left[\frac{n_3(n_1+1)}{n_1(n_3+1)}\right]_{\text{equilibrium}} = \left[\frac{1+1/n_1}{1+1/n_3}\right]_{\text{equilibrium}} = e^{h\nu_{\text{molecule}}/kT}.$$

Remarks

(a) In the parentheses such as $(n_i + 1)$, the term n_i corresponds to the stimulated effects (emission or absorption), while the term +1 comes from the spontaneous emission.

(b) Although n_i is a number of photons, its value is not necessarily an integer, since it is the *expected value* of the number of photons over a large number of identical systems placed in similar situations. This is the reason why we will often speak of numbers of photons that are *small* compared to unity.

11.3.2.3. *Interpretation of the Raman Initial Experiment*

The conditions of the initial Raman experiment of Figure 11.2 are not strictly the conditions of a blackbody at thermal equilibrium, since the collection of benzene molecules is illuminated by a pumping beam of frequency ν_1. It can be considered that:

• The collection of benzene molecules is very close to thermal equilibrium: $N_1/N_0 \cong e^{-h(E_1 - E_0)/kT}$.
• The number of Stokes photons n_2 is equal to zero.

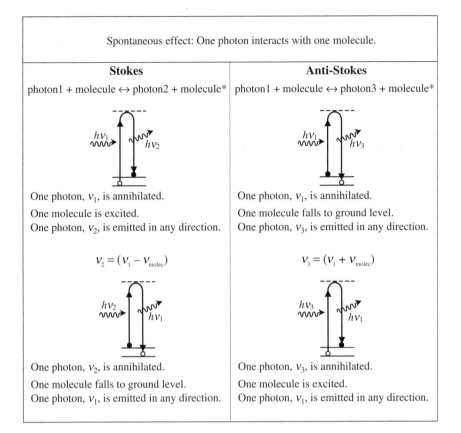

Figure 11.12(a). Spontaneous Raman transitions.

- The number of anti-Stokes photons n_3 is equal to zero.
- The number of pump photons n_1 is far greater than one.

Under such conditions equations (11.5) and (11.8) reduce to

$$\frac{dn_2}{dt} = -\frac{dn_1}{dt} = An_1N_0 \quad \text{for the Stokes ray,}$$

$$\frac{dn_3}{dt} = -\frac{dn_1}{dt} = An_1N_0 \quad \text{for the anti-Stokes ray.}$$

Spontaneous photons haven't any specific direction and are emitted in the 4π steradians. The Raman light intensity is proportional to the intensity of the pumping beam n_1. The Stokes light is proportional to the population of the ground level N_0; the anti-Stokes light is proportional to the population of the excited level N_1.

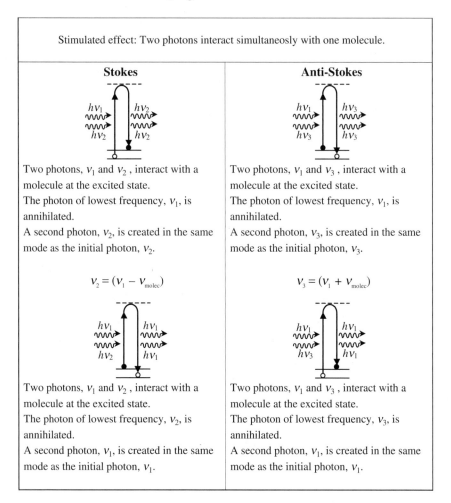

Figure 11.12(b). Stimulated Raman transitions.

11.3.2.4. *Raman Laser*

We refer to Figures 11.13 and 11.14, a coherent and powerful light beam (ν_1) is sent into a collection of Raman active molecules. This pumping beam is parallel to the Oz axis. The Raman material extends from the abscissa $z = 0$ to $z = L$. A lot of identical Stokes photons ($\nu_2 = \nu_1 - \nu_{molec}$) are produced by stimulated Raman emission and make a parallel beam and a monochromatic beam, which have the properties of a laser beam and will be designated as "the Raman laser beam." We admit that the two beams are collinear and propagate at the same speed V, we have the following expressions where $n_1(z)$

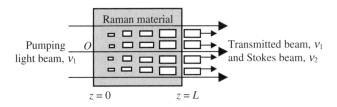

Figure 11.13. Emission of a coherent beam at the Stokes frequency. The size of the small rectangles is proportional to the number of the Stokes photons at abscissa z.

and $n_2(z)$ are the respective photon densities:

$$\frac{d}{dz} = \frac{1}{V}\frac{d}{dt},$$

$$\frac{dn_2}{dz} = \frac{1}{V}\frac{dn_2}{dt} = -\frac{dn_1}{dz} = \frac{A}{V}[n_1(n_2+1)N_0 - n_2(n_1+1)N_1], \qquad (11.9.a)$$

$$n_1(z) + n_2(z) = n_1(0) = \text{constant.} \qquad (11.9.b)$$

Relation (11.9.b) is only a consequence of the fact that, each time a Stokes photon ν_2 is created, a photon ν_1 of the pump is annihilated; it's also an expression of the energy conservation. We consider that the populations of the molecular levels are constant and keep their Maxwell-Boltzmann equilibrium values.

The integration of equation (11.9.a) is not easy because of the existence of the term $n_1(z)n_2(z)$. We first look at what happens at the entrance of the sample ($z \cong 0$); there are very few Stokes photons, most of them are emitted spontaneously in all directions (4π steradians) and constitute the *noise* from which the Raman laser will start. For $z \cong 0$, the number of Stokes photons belonging to the Raman laser beam is only a small part of the spontaneous photons; we assume that $n_2(z)$ linearly increases with the abscissa,

$$n_2(z) = Kn_1(0)N_0 z.$$

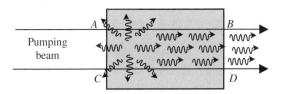

Figure 11.14. A pumping beam is sent into a cell filled with a Raman active material. The Stokes photons are produced by spontaneous and stimulated effects. The Stokes photons that are emitted in the same direction as the pumping beam interact over a longer distance and have a better opportunity to stimulate other interactions: as a consequence, a coherent beam is generated at the Stokes frequency, its geometrical characteristics are identical to those of the pumping beam.

This linear relation is valid as long as the number of photons remains much smaller than one. If the sample is short enough, this condition is valid at any point inside the sample: only a spontaneous Raman effect is observed. For longer samples, it may happen that $n_2(z)$ becomes greater than one. Let ζ be the abscissa for which $n_2(\zeta) = 1$, for $z > \zeta$ the numbers of photons $n_1(z)$ and $n_2(z)$ are both greater than one; equation (11.9.a) simplifies as

$$\frac{dn_2}{dz} = -\frac{dn_1}{dz} = \frac{A}{V} n_1 n_2 (N_0 - N_1) = D n_1 n_2, \qquad (11.10.a)$$

$$D = \frac{A}{V} n_1 n_2 (N_0 - N_1). \qquad (11.10.b)$$

The derivative of $n_2(z)$ is positive and proportional to $n_2(z)$, which implies an exponential growth. This fast variation cannot last forever, the limitation comes from relation (11.9.b): as $n_2(z)$ increases, $n_1(z)$ and, correlatively, the constant D of equation (11.10.b) decreases. The integration of equation (11.10.a) is now possible:

$$\frac{dn_2}{dz} = D n_1 n_2 = D n_2 (n_{1(0)} - n_2),$$

$$\frac{dn_2}{n_2 (n_{1(0)} - n_2)} = \frac{dn_2}{n_{1(0)}} \left(\frac{1}{n_2} + \frac{1}{n_{1(0)} - n_2} \right) = D \, dz,$$

$$\text{Log} \frac{n_2}{n_{1(0)} - n_2} = D n_{1(0)} (z - a), \quad \text{an integration constant,}$$

$$n_2(z) = n_{1(0)} \frac{e^{D n_{1(0)}(z-a)}}{1 + e^{D n_{1(0)}(z-a)}}. \qquad (11.11)$$

From equation (11.11) we see that $n_2(z)$ becomes equal to $n_1(0)$ as $z \to \infty$: all the energy is transferred from the pumping beam to the Raman beam.

The arrangements of Figures 11.15 and 11.14 are basically the same, except that in the first one a lens is used to obtain a higher energy density in the focal zone, which lowers the global power required for the stimulated regime to become predominant.

We refer again to Figure 11.14 where an incident cylindrical beam propagates inside a Raman active material. Once it has been emitted, a spontaneous Stokes photon may stimulate the creation of identical photons, this is only possible as long as this photon remains inside the volume that is illuminated by the pumping beam, which is bounded by $ABCD$. The photons that are emitted perpendicular to the direction of propagation rapidly escape from this volume and produce only a few stimulated emissions. On the contrary, a photon, which by chance has been emitted in the direction of the pumping beam, will stimulate the creation of many identical phonons.

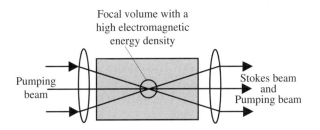

Figure 11.15. A lens is used to increase the local energy density. A coherent Stokes beam, having the same geometric characteristics as the incident beam, is emitted in the focal zone and then recollimated by a second lens.

11.3.2.5. *Raman Laser Oscillator*

We refer to Figure 11.16(a). Only one beam is sent into the Raman active material. The laws of variation of the numbers of pump photons, $n_1(z)$, and of the Stokes photons, $n_2(z)$, have been given by formulas (11.11) and are plotted in the figure. At the beginning the interaction is weak: the variation of $n_1(z)$ and $n_2(z)$ is very slow and mostly due to spontaneous emission. After a distance $L_{threshold}$, which is called the *threshold length*, stimulated emissions suddenly predominate and coherent Stokes photons are massively produced: we then speak of *Raman laser emission*. It is important to notice that this laser effect doesn't require any inversion of population.

For a given material $L_{threshold}$ decreases with the pumping power. Laser action is obtained when the length of the cell is greater than the threshold length. Conversely, inside a given cell, there is a threshold power above which laser action is observed. In fact it's not the global power that matters, but

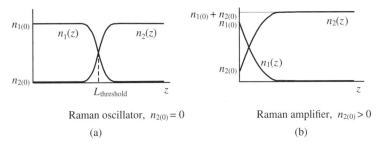

Figure 11.16. Variation of the number of pump and Stokes photons as a function of the length of propagation. In the case of (a), there is no input signal at the Stokes frequency: above a distance $L_{threshold}$, stimulated interactions predominate, corresponding to a new kind of laser oscillator. In the case of (b), two beams are simultaneously sent into the Raman active material; $n_{2(0)}$ is not equal to zero, stimulated interactions immediately predominate: the Stokes beam is amplified.

the electromagnetic energy density inside the pumping beam, which can be increased by focusing.

Laser Action Is not Possible at the Anti-Stokes Frequency

For the anti-Stokes ray, equation (11.10.a) becomes

$$\frac{dn_3}{dz} = -\frac{dn_1}{dz} = -\frac{A}{c}n_1n_3(N_0 - N_1) = -Dn_1n_2, \qquad (11.12)$$

with $D = (A/c)(N_0 - N_1)$. The derivative of the number of anti-Stokes photons, dn_3/dz, is negative making any exponential growth impossible.

Forward and Backward Raman Diffusion

When writing equations (11.9) and (11.10), we considered that the spontaneous Stokes photons emitted in the same direction could interact over a long distance with the pumping beam. In fact this is also true for the photons that are emitted in the opposite direction. A coherent beam is also emitted at the Stokes frequency and propagates in the opposite direction to the pumping beam.

11.3.2.6. *Higher-Order Raman Diffusion*

We now come back to the experimental arrangements of Figures 11.14 and 11.15 and we suppose that the cell is longer than $L_{\text{threshold}}$: the coherent pumping beam, ν_1, is converted into a new coherent beam of frequency $\nu_2' = (\nu_2 - \nu_{\text{molecule}})$, which can in turn act as a pumping beam and generate a second coherent beam of frequency $\nu_2' = (\nu_2 - \nu_{\text{molecule}}) = (\nu_1 - 2\nu_{\text{molecule}})$. If the initial pumping beam is powerful enough a third coherent beam of frequency $\nu_2'' = (\nu_1 - 3\nu_{\text{molecule}})$ will be generated. Finally, the spectrum of the light emitted by the cell may be very rich. The different beams are, respectively, called first-, second-, and third-order, ... Raman beams.

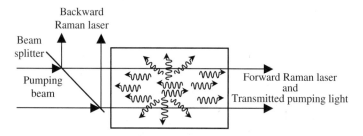

Figure 11.17. Raman diffusion may occur in any direction. The Stokes photons that are emitted in the opposite direction to the pumping beam also remain in an illuminated zone, which makes a laser action possible.

Complication Due to Self-Focusing

Most Raman active molecules are also polar molecules, which means that self-focusing is easily observed in the corresponding liquids. The energy density being very high inside an autofocused filament, the stimulated Raman threshold is more easily reached; unfortunately, self-focusing is often chaotic, making it almost impossible to tame the Raman lasers. Because of the high pumping density a large variety of nonlinear effects is obtained, the result being complex combinations of the initial laser frequency and of the various Raman frequencies, a very rich and almost continuous spectrum is obtained (see Figure 10.13).

11.3.2.7. *Raman Effect in Optical Fibers*

The SiO_2 molecule shows a Raman active transition at $N = 250\,\mathrm{cm}^{-1}$. Single-mode optical fibers are a very favorable medium for achieving the stimulated Raman effect because extremely long samples are available and, more important, there is no self-focusing. There are two main arrangements:

- Optical oscillator providing a coherent source at the Stokes frequency.
- Optical amplifier.

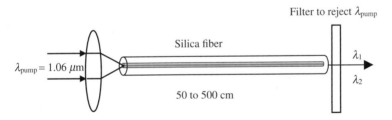

Figure 11.18. The Raman laser oscillator. The light of an Nd-YAG laser is focused in a monomode fiber. According to the pumping power, one or several wavelengths (λ_1, λ_2, ...) are obtained, corresponding to the different Raman orders ($1/\lambda_1 = 1/\lambda_{pump} - N$, $1/\lambda_2 = 1/\lambda_{pump} - 2N$, ...).

In the arrangement of Figure 11.18, we must wait until the stimulated emission has produced enough Stokes photons before the stimulated interaction predominates. In the case of Figure 11.19, another procedure is followed: two coherent beams, of respective frequencies ν_{pump} and $\nu_{signal} = (\nu_{pump} - \nu_{molecule})$, are simultaneously sent into the fiber. This situation is described in Figure 11.16(b): the stimulated interactions immediately predominate and the energy of the highest frequency beam (pumping beam) is transferred to the lowest frequency beam, which is amplified.

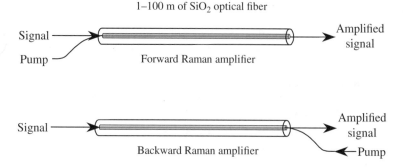

Figure 11.19. Raman optical amplifiers. A low optical signal is mixed with the light of a powerful laser. The difference between laser and signal frequencies is equal to the Raman frequency of SiO_2.

11.4. Brillouin Diffusion

Brillouin diffusion of light is very similar to Raman diffusion:

- Nonlinear process involving the emission of new optical frequencies.
- Existence of spontaneous and stimulated effects.

There are however important differences, which are listed below:

- The frequency shifts are much smaller in the Brillouin case ($\approx cm^{-1}$) than in the Raman case (thousands of cm^{-1}).
- The Brillouin effect occurs in dense materials (liquids or solids). Any elementary volume, even if it's small compared to the wavelength, will always contain many atoms or molecules that diffuse the light. Because of the close proximity of those radiating elements, the vibrations that are diffused are coherent and interfere.

Because of the small value of the frequency shift, very monochromatic light sources are required to study the Brillouin effect. For the same reasons, the liquids should be perfectly free of bubbles and dust and the solid samples free of impurities, dislocations, or other imperfections.

11.4.1. *A Homogenous Dense Material Doesn't Diffuse Any Light*

To evaluate the light that is diffused out of a macroscopic sample of material, we must integrate the contributions of the different elements. We are going to show that this integral is equal to zero if the material is perfectly homogeneous, and that the existence of diffused light originates from the existence of irregularities. We refer to Figure 11.20, a parallel monochromatic beam propagates in a transparent material, and we want to determine the light that is diffused in a direction that makes an angle θ with the incident beam.

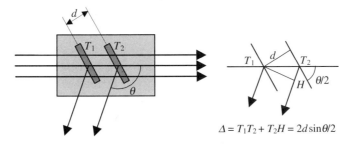

Figure 11.20. Brillouin diffusion of light. The diffusing volume is divided into parallel slices that are thin compared to the wavelength.

The diffusing volume is divided into thin parallel slices oriented parallel to the bisector of the angle θ; all the molecules of a given slice, T_1 for example, send in this direction elementary contributions that all have the same phase (this is easily understood by virtually replacing the slice by a mirror). We now consider a second slice T_2 parallel to T_1 and separated by a distance d; the optical path difference Δ and the phase difference φ, and between the respective contributions of the two slices, respectively, are given by

$$\Delta = 2d \sin\frac{\theta}{2} \quad \text{and} \quad \varphi = 2\pi\frac{\Delta}{\lambda}. \tag{11.13}$$

A given elementary slice can always be paired with another one located at a distance d such that $\varphi = \pi$; the two contributions then have opposite phases; if the material is perfectly homogeneous, they have equal amplitude and, finally, they mutually cancel by destructive interference. In other words, there is no diffusion in the direction θ; except if $\theta = 0$, which corresponds to the propagation of the incident beam. The previous demonstration is no longer valid in the following two cases:

- The sample is very thin and reduces to a thin film. In this case, the slice reflects the light as if it were a mirror, except that the reflected intensity varies with the wavelength (think of antireflection coatings).
- The material is not homogeneous; the amplitude of the contributions of the two paired slices don't exactly cancel by interference. The light diffused from a dense material is due to the existence of irregularities.

11.4.2. *Diffusion of Light by a Crystal*

Let us consider a crystal sample of "good quality"; this may contain two kinds of heterogeneities that contribute to light diffusion:

(a) Permanent irregularities due to crystal imperfections: interstitial atoms, vacancies, and dislocations of the network.

(b) Transient irregularities due to thermal agitation, to which are associated *density fluctuations* that can, in turn, be subdivided into two
categories:

Entropy fluctuations (in fact, temperature fluctuations).
Acoustic vibrations propagating erratically inside the crystal.

We will only be concerned with (b) type heterogeneities.

Rayleigh Diffusion

The entropy fluctuations are accompanied by fast density fluctuations, occurring erratically with no correlation from one point to the next. The light diffusion occurs at constant frequency, except for a broadening of the spectrum.
This kind of diffusion is referred to as *Rayleigh diffusion;* its efficiency is
proportional to the absolute temperature and cancels at 0 K.

Brillouin Diffusion

In a solid sample, because of thermal agitation, a given atom vibrates around
an equilibrium state; the motions of two adjacent atoms are not independent
and are coupled together, which corresponds to the propagation of acoustic
waves. To these waves, Quantum Mechanics associates particles that are
called *phonons.*

At a given temperature T K, acoustic waves in any direction run across
any solid sample; their frequency spectrum goes from low frequencies up to
very high frequencies (possibly infrared). A modulation of the pressure inside
the material and, consequently, a modulation of the density and of the index
of refraction should be associated to any acoustic wave propagating inside a
transparent sample. This optical heterogeneity produces the diffusion of a
light beam propagating through the sample.

11.4.3. *Classical Theory of the Brillouin Effect*

We consider a transparent material that supports a parallel optical beam of
frequency ν_{opt} and a parallel acoustic wave of frequency ν_{acoust}. Because of the
acoustic wave the atoms are set in vibration at this last frequency; the electric field of the optical vibration that is emitted by an atom is phenomeno-
logically written as

$$E_{diffused} = K \cos 2\pi\nu_{acoust} t \, \cos(2\pi\nu_{opt} t - \varphi)$$

$$= \frac{K}{2}[\sin 2\pi(\nu_{acoust} + \nu_{opt})t + \sin(2\pi(\nu_{opt} - \nu_{acoust})t - \varphi)], \quad (11.14)$$

where K is a proportionality coefficient and φ is a phase shift that depends
on the point where the diffusion is made.

According to formula (11.14), the diffusion is accompanied by a frequency shift. Two new components are obtained:

- The Stokes ray of frequency: $\nu_{Stokes} = \nu_{opt} - \nu_{acoust}$.   (11.15.a)
- The anti-Stokes ray of frequency: $\nu_{anti\text{-}Stokes} = \nu_{opt} + \nu_{acoust}$.   (11.15.b)

The lowest frequency is called the Stokes ray; the highest frequency is called the anti-Stokes ray. The previous theory, which is a classical theory, indicates that the two rays have the same intensity. The result of a quantum theory that will be given in the next section indicates that the anti-Stokes ray is less intense; however, owing to the small value of acoustic frequencies, the Stokes and anti-Stokes rays have almost the same intensity.

11.4.4. *Quantum Theory of the Brillouin Effect*

Acoustic vibrations that may exist inside a crystal can of course be quantized, the associated particles are called *phonons*, which are bosons and follow the Bose-Einstein statistics.

The Brillouin effect can be written as a chemical reaction between phonons and photons:

Brillouin diffusion (Stokes)

Incident photon → diffused photon + phonon,

$$h\nu_1 \quad = \quad h\nu_2 \quad + h\nu_{phonon}.$$

An incident photon (ν_1) is annihilated.
A diffused photon (ν_2) and a phonon (ν_{phonon}) are created.
The diffused light has a lower frequency than the incident light,

$$\nu_2 = (\nu_2 - \nu_{phonon}).$$

Brillouin diffusion (anti-Stokes)

Incident photon + phonon → diffused photon,

$$h\nu_1 \quad = h\nu_{phonon} \quad = h\nu_3.$$

An incident photon (ν_1) and a phonon (ν_{phonon}) are annihilated.
A diffused photon (ν_3) is created.
The diffused light has a higher frequency than the incident light,

$$\nu_3 = (\nu_1 + \nu_{phonon}).$$

As we did for the Raman effect, we are going to show that if we want the Bose-Einstein statistics to be simultaneously satisfied for both the photons and the phonons, it is necessary to introduce spontaneous and stimulated effects.

We will consider the interaction of the following three items:

- A mode of radiation of frequency ν_1 with n_1 photons.
- A mode of radiation of frequency ν_2 with n_2 photons.
- A mode of vibration of frequency ν_{phonon} with N phonons.

The different interactions are stated in Table 11.2. Following a phenomenological approach, the rate equations are

$$
\begin{aligned}
\frac{dn_2}{dt} = \frac{dN}{dt} = -\frac{dn_1}{dt} \\
= A[n_1 - n_2 N + n_1 n_2 - n_1 N] \\
= A[n_1 - n_2 N + n_1 n_2 - n_1 N + n_1 n_2 N - n_1 n_2 N] \\
= A[n_1(n_2+1)(N+1) - (n_1+1)n_2 N].
\end{aligned}
\tag{11.16}
$$

At thermal equilibrium the time derivatives are equal to zero, which implies that

$$
\frac{n_1}{n_1+1} = \frac{n_2}{n_2+1}\frac{N}{N+1}.
$$

The above equation is satisfied and the different numbers of photons and phonons, respectively, are given by

$$
n_1 = \frac{1}{e^{h\nu_1/kT}-1}, \quad n_2 = \frac{1}{e^{h\nu_2/kT}-1}, \quad N = \frac{1}{e^{h\nu_{acoust}/kT}-1},
$$

with $\nu_1 = (\nu_2 + \nu_{phonon})$.

11.4.5. The Brillouin Doublet

11.4.5.1. Classical Approach

Let us consider the arrangement of Figure 11.21; a planar monochromatic wave of frequency ν_{optic} propagates in a solid-state transparent material. Diffused light is, a priori, emitted in all directions and with different frequencies; however, we are going to show that the light that is diffused in a given direction, making an angle θ with the incident direction, has a well-defined frequency, $\nu'_{optic}(\theta)$, and that this frequency is given by formula (11.17), which is called the *Brillouin doublet formula*.

We return to the demonstration that allowed us to establish that a perfectly homogeneous material cannot diffuse any light (see Figure 11.20). Among the numerous thermal acoustic waves that run across the sample, we consider one that has its wave planes parallel to the two slices T_1 and T_2, let

Table 11.2. Recapitulation of the different Brillouin interactions.

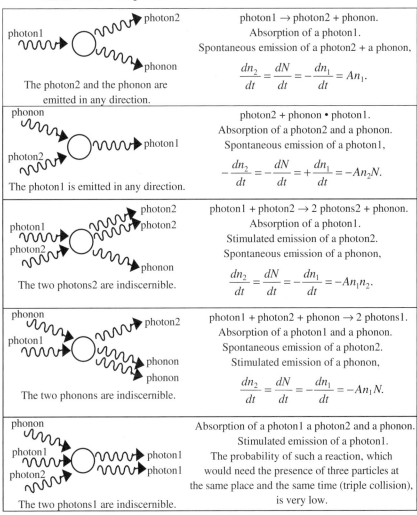

The photon2 and the phonon are emitted in any direction.	photon1 → photon2 + phonon. Absorption of a photon1. Spontaneous emission of a photon2 + a phonon, $$\frac{dn_2}{dt} = \frac{dN}{dt} = -\frac{dn_1}{dt} = An_1.$$
The photon1 is emitted in any direction.	photon2 + phonon • photon1. Absorption of a photon2 and a phonon. Spontaneous emission of a photon1, $$-\frac{dn_2}{dt} = -\frac{dN}{dt} = +\frac{dn_1}{dt} = -An_2N.$$
The two photons2 are indiscernible.	photon1 + photon2 → 2 photons2 + phonon. Absorption of a photon1. Stimulated emission of a photon2. Spontaneous emission of a phonon, $$\frac{dn_2}{dt} = \frac{dN}{dt} = -\frac{dn_1}{dt} = -An_1n_2.$$
The two phonons are indiscernible.	photon1 + photon2 + phonon → 2 photons1. Absorption of a photon1 and a phonon. Spontaneous emission of a photon2. Stimulated emission of a phonon, $$\frac{dn_2}{dt} = \frac{dN}{dt} = -\frac{dn_1}{dt} = -An_1N.$$
The two photons1 are indiscernible.	Absorption of a photon1 a photon2 and a phonon. Stimulated emission of a photon1. The probability of such a reaction, which would need the presence of three particles at the same place and the same time (triple collision), is very low.

λ_{acoust} be the wavelength. The amplitudes of the light that is diffused from the two slices are no longer equal and they don't exactly cancel by interference. The difference between the two diffused signals is maximum when the acoustic pressure is, respectively, maximum on one slice and minimum on the other. The distance d between the two slices is then equal to $\lambda_{\text{acoust}}/2$. As the slices have been paired in such a way that $\Delta = \lambda_{\text{opt}}/2$, the maximum of diffused light is obtained when

$$\Delta = 2d \sin\frac{\theta}{2} = \lambda_{\text{acoust}} \sin\frac{\theta}{2} = \frac{\lambda_{\text{opt}}}{2} \quad \rightarrow \quad \lambda_{\text{opt}} = 2\lambda_{\text{acoust}} \sin\frac{\theta}{2}.$$

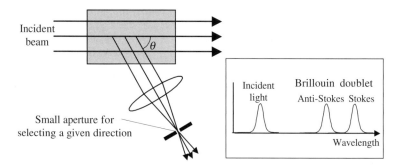

Figure 11.21. Experimental arrangement for studying the Brillouin doublet. The collecting lens and the small aperture select the light that is diffused in a given direction and send it into a spectroscope.

Let us introduce the respective speeds of propagation of the optical waves, V_{opt}, and of the acoustic waves, V_{acoust}, and the frequencies ν_{opt} and ν_{acoust}:

$$\nu_{acoust} = 2\nu_{opt} \frac{V_{acoust}}{V_{opt}} \sin\frac{\theta}{2},$$

$$\nu_{Stokes} = \nu_{opt}\left(1 - 2\frac{V_{acoust}}{V_{opt}}\sin\frac{\theta}{2}\right) \quad \text{and} \quad \nu_{anti\text{-}Stokes} = \nu_{opt}\left(1 + 2\frac{V_{acoust}}{V_{opt}}\sin\frac{\theta}{2}\right).$$

Finally, we see that the light diffused in direction θ has two spectral components, which are defined by

$$\nu = \nu_{opt}\left(1 \pm 2\frac{V_{acoust}}{V_{opt}}\sin\frac{\theta}{2}\right), \quad \text{the Brillouin doublet formula.} \quad (11.17)$$

11.4.5.2. *Quantum Approach*

The emission of Stokes phonons can be described by the reaction:

$$\text{photon1} \rightarrow \text{photon2} + \text{phonon}.$$

The probability of this interaction is significantly different from zero only if the following conservation conditions are satisfied:

- Energy conservation: $h\nu_1 = (h\nu_2 + h\nu_{acoust}) \rightarrow \nu_{Stokes} = (\nu_1 - \nu_{acoust})$.
- Momentum conservation.

The momentum conservation will provide an alternative demonstration of the Brillouin formula. $\mathbf{k}_1$, $\mathbf{k}_2$, and $\mathbf{K}$ are the respective momenta of photon1, photon2, and of the phonon; k_1, k_2, and K are their moduli. We consider that the two optical waves propagate at the same speed V_{opt}. The momentum conservation implies that it should be possible to draw a triangle with the three vectors.

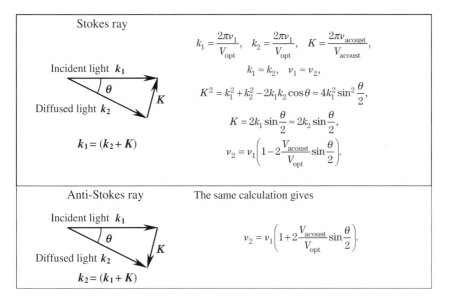

Figure 11.22. Demonstration of the Brillouin doublet formula using the conservation of momentum.

11.4.6. *Diffraction of Light by Ultrasonic Waves*

In 1921 Brillouin predicted that a liquid traversed by an ultrasonic wave of very short wavelength should behave like a diffraction grating. The first experimental demonstration was made in 1932 by the Americans Debye and Sears and by the Frenchmen Lucas and Biquard who opened a new domain called *acoustooptics*. These experiments have now found many useful applications in optical processing and for Q-switching lasers. They have been transposed in *integrated optics* where guided optic waves interact with acoustic surface waves.

Let us again examine the rate equation (11.16) in the case of the two situations described by Figures 11.21 and 11.23.

11.4.6.1. *The Brillouin Doublet Experiment (Figure 11.21)*

The initial conditions are: $n_2(0) = 0$, $N(0) = 0$, and $n_1(0) \neq 0$ ($n_1(0) \gg 1$).

For $t \approx 0$, $dn_2/dt = dN/dt = -dn_1/dt = An_1 \rightarrow$ the intensity of the diffused light is proportional to the intensity of the incident beam: photons2 and phonons are spontaneously created.

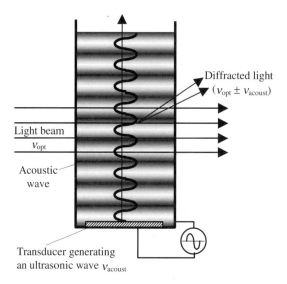

Figure 11.23. Diffraction of light by ultrasonic waves. A transducer generates a hypersonic wave of frequency ν_{acoust} in a liquid, which is traversed by compression waves that create a stratification of the density. At a given time the liquid is made of a succession of layers of alternately higher and lower refraction index and thus behaves as a diffraction grating.

11.4.6.2. Diffraction by an Ultrasonic Wave (Figure 11.23)

$$n_2(0) = 0, \quad N(0) \gg 1, \quad \text{and} \quad n_1(0) \gg 1.$$

For $t \approx 0$:

$$\frac{dn_2}{dt} = \frac{dN}{dt} = -\frac{dn_1}{dt} = An_1(N+1) \cong An_1N.$$

Thanks to the transducer the number of phonons is very high: $N(0)$ is much larger than one, the diffraction of light by an ultrasonic wave is more efficient than diffusion by thermal phonons.

11.4.6.3. Production of Phonons by Mixing Two Light Beams of Different Frequencies (Figure 11.24)

Let us send two light beams of frequencies ν_1 and ν_2 into a transparent material. The initial conditions are now $n_1(0) \gg 1$, $n_2(0) \gg 0$, and $N(0) = 0$; for $t \approx 0$, equation (11.16) then gives $dn_2/dt = dN/dt = -dn_1/dt = An_1n_2$. The term An_1n_2 may be given a high value, the two derivatives dn_2/dt and dN/dt are then positive and also have high values, which means that phonons are massively generated, the corresponding energy being borrowed from the highest fre-

Figure 11.24. Interference pattern of two light beams. Because of electrostriction the index of refraction is increased at the location of the bright fringes. The light from one beam is exactly diffracted in the direction of the other beam. If the two frequencies are equal the fringes are immobile, while they move vertically if they are different.

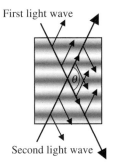

First light wave

θ

Second light wave

quency optical beam. This generation of acoustic waves can be understood as a beat between the two optical beams.

Electrostriction: The presence of an electric field E in a liquid (or crystal) gives rise to an electrostrictive strain which is analogous to a hydrostatic pressure P. P is proportional to the squared amplitude of the electric field $P = \gamma E^2$ (γ is the electrostriction constant of the material). Electrostriction is a nonlinear process. A static electric field generates a static pressure. A time-varying electric field is capable of driving acoustic waves in the material: however, the frequency should not be too high; if not, the pressure takes a constant value, which is proportional to the squared value of the amplitude of the oscillating electric field. This is the case of a light wave, which creates a static pressure proportional to the light intensity.

The superposition of two optical waves inside a piece of material creates an interference field. The optical electric field is maximum at the bright fringes and minimum at the dark fringes. Because of electrostriction, the interference field creates a periodic stratification of matter. The pressure, density, and index of refraction are maximum along the bright fringes and minimum along the dark fringes. Finally, the interference pattern prints a phase grating inside the material.

Figure 11.24 shows the interference pattern of two planar waves propagating in two different directions. The interference fringes are planes that are parallel to the bisector plane of the dihedron made by the wave planes at an angle equal to θ. Two different cases are considered. In the first case, the two waves have the same frequency and an interference pattern is fixed. In the second case, their frequencies, ν_1 and ν_2, are slightly different, and the interference fringes move in a direction that is perpendicular to their planes.

11.4.6.4. *Interference of Two Beams of the Same Frequency*

Electric field of the first beam: $E_1 = A\cos(2\pi\nu t - \mathbf{k}_1\mathbf{r})$.
Electric field of the second beam: $E_2 = A\cos(2\pi\nu t - \mathbf{k}_2\mathbf{r})$.

The electrostriction pressure is equal to the time-averaged value of $\gamma(E_1 + E_2)^2$:

$$\overline{(E_1 + E_2)^2} = A^2 + 2A^2\overline{\cos(2\pi vt - \mathbf{k}_1\mathbf{r})\cos(2\pi vt - \mathbf{k}_2\mathbf{r})},$$

$$\overline{(E_1 + E_2)^2} = A^2\left[1 + \overline{\cos\left(2\pi vt - \frac{(\mathbf{k}_1 + \mathbf{k}_2)\mathbf{r}}{2}\right)} + \cos\left(\frac{(\mathbf{k}_1 - \mathbf{k}_2)\mathbf{r}}{2}\right)\right]$$

$$= A^2\left[1 + \cos\left(\frac{(\mathbf{k}_1 - \mathbf{k}_2)\mathbf{r}}{2}\right)\right],$$

$$P = \gamma A^2\left[1 + \cos\left(\frac{(\mathbf{k}_1 - \mathbf{k}_2)\mathbf{r}}{2}\right)\right]. \qquad (11.18.a)$$

The pressure is spatially modulated but the fringes are immobile.

11.4.6.5. *Interference of Two Beams of Different Frequencies*

The calculation is exactly the same with

$$E_1 = A\cos(2\pi v_1 t - \mathbf{k}_1\mathbf{r}) \quad \text{and} \quad E_2 = A\cos(2\pi v_2 t - \mathbf{k}_2\mathbf{r}),$$

$$P = \gamma A^2\left[1 + \cos\left(2\pi\frac{(v_1 - v_2)t}{2} - \frac{(\mathbf{k}_1 - \mathbf{k}_2)\mathbf{r}}{2}\right)\right]. \qquad (11.18.b)$$

If the frequency difference ($v_1 - v_2$) is small enough and corresponds to possible acoustic or hypersonic waves, the time-averaged value of (11.18.b) is different from zero. This formula describes fringes that propagate in the direction of the difference of the wave vectors ($\mathbf{k}_1 - \mathbf{k}_2$). The speed of propagation is equal to

$$V_{\text{fringes}} = 2\pi\frac{(v_1 - v_2)}{(k_1 - k_2)}. \qquad (11.19)$$

We suppose that v_1 is the highest frequency and we call B_1 the corresponding beam; B_2 is the other beam, frequency v_2. The theoretical analysis is not straightforward, however the results are quite simple.

- Each beam diffracts light that is exactly emitted in the direction of the other beam.
- Because the diffraction occurs on a moving target, the Doppler effect changes the frequency. The frequency of light that is diffracted from one beam is equal to the frequency of light from the other beam.
- The light that is diffracted from B_1 (highest frequency) is *in phase* with B_2. Reciprocally the light that is diffracted from B_2 is in opposition to the phase with B_1. The two optical beams thus exchange energy: *the lowest frequency beam is amplified*, while the highest energy beam is attenuated.

• *Mechanical vibrations are generated*: The energy received by B_2 is slightly smaller than the energy lost by B_1; the difference is used to generate a mechanical vibration of frequency $(\nu_1 - \nu_2)$.

There is a resonance effect if the speed of propagation of the fringes is equal to the speed of propagation of an acoustic wave of frequency $(\nu_1 - \nu_2)$. The amplification of the beam B_2 is then maximized, as is the power of the acoustic wave.

Let I_1 and I_2 be the intensities of the two optical beams before interaction, and let I_{acoust} be the intensity of the generated acoustic wave. During the interaction the respective variations of intensities of the beams are ΔI_1 and ΔI_2, the theory establishes that

$$\frac{\Delta I_1}{\nu_1} = \frac{\Delta I_2}{\nu_2} = \frac{I_{\text{acoust}}}{(\nu_1 - \nu_2)} = \frac{I_{\text{acoust}}}{\nu_{\text{acoust}}}. \tag{11.20}$$

Equation (11.20) is a Manley-Rowe relation (see formula (10.28)) and is more easily obtained from the reaction

$$\text{photon1} \rightarrow \text{photon2} + \text{phonon}.$$

Annex 11.A

Diffusion of Light by a Scattered Medium

The Earth Is a Blue Planet

Seen from space, the Earth appears as a nice blue balloon on which white clouds are often painted. The atmosphere looks like a dome, the color of which varies from the well-known blue to red, according to the time of day. The coloration of the atmosphere is one of the most elegant proofs of electric dipolar radiation.

The atmosphere and clouds work in the same way as a diffuser placed around an electric bulb to ensure a better repartition of the light. Under the action of the electric field of the light coming from the Sun, the electrons of the molecules of the atmosphere are set in vibration and become light sources. By the way, we may notice that the planets that are surrounded by an atmosphere are the only ones to have a clear sky, which is not the case for the Moon.

The diffusion of light occurs differently according to whether it's obtained from independent molecules (mostly nitrogen, oxygen, and water vapor) or from clusters of molecules (water droplets, tiny crystals of ice, dust, aerosols, . . .). Independent molecules are isolated and separated by distances that are greater than an optical wavelength; since the light of the Sun is not coherent, the different wavelets have random phases and cannot interfere, our eyes will just add their intensities.

Diffusion by droplets or solid particles is different; first, it is more efficient because the number of diffusing particles per unit volume is much higher and, second, the wavelets emitted by different molecules of the same particle are coherent and interfere. Of course, the two signals coming from two different particles are not coherent. The diffusion by particles with a dimension of the order of the wavelength is called "Mie diffusion" from the name of the physicist who studied it for the first time. This is a problem of Electromagnetism that is easily formulated but doesn't have a straightforward solution, since the diffusing object is neither very large nor very small as compared to the wavelength. The calculation is exhaustively described in the book *Prin-*

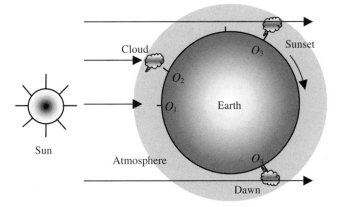

Figure 11.A.1. The white light coming from the Sun is partly diffused in the 4π steradians, either by the molecules of oxygen and nitrogen in the atmosphere ($1/\lambda^4$ Rayleigh diffusion) or by the droplets of water clouds (Mie diffusion), which is almost independent of the frequency. Seen from points O_1 or O_2 the sky is blue and the clouds are white. At points O_3 and O_4, the sky, as well as the clouds, are colored red. Seen from space the Earth appears as a blue sphere on which are often painted white clouds.

ciples of Optics by Born and Wolf. The main result is that the efficiency of light scattering is roughly constant over all of the visible spectrum; this is the reason why clouds appear to be white in sunlight.

Let us come back to the diffusion by independent gaseous molecules; as this is the result of a dipolar radiation, the efficiency varies as $1/\lambda^4$, and it is sixteen times more efficient in the blue ($0.4\,\mu$m) than in the red ($0.8\,\mu$m).

A Red Sky at Night Is the Shepherd's Delight

At points O_1 and O_2 in Figure 11.A.1, the rays coming from the Sun have traversed a rather small distance in the atmosphere: the sky is blue and the clouds are white. The situation is different for points O_3 and O_4, which correspond to sunset and dawn: before arriving there the rays had to cover a large distance in space: the blue components of the sunlight have been progressively attenuated, the sky is now red, as well as an eventual cloud. If, before arriving at O_3 and O_4, the rays crossed important clouds, in which the diffusion is more or less achromatic, the sky has almost no coloration.

We can understand the popular British dictum "a red sky at night is the shepherd's delight": a nice red sunset sky witnesses the absence of clouds in the western part of the country. As in Europe, the wind often blows from the west; this constitutes a favorable indication of good weather in the near future.

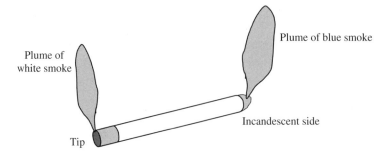

Figure 11.A.2. Light diffusion by the smoke of a cigarette. At the incandescent end the smoke is made of very thin solid particles that follow the hot gases, the diffused light obeys the $1/\lambda^4$ Rayleigh diffusion law and has a bluish color. At the tip end the smoke is white, because of Mie diffusion on the condensation droplets of water produced by tobacco combustion.

The Light That Is Diffused by the Sky Is Partly, or Possibly, Totally Polarized

The amplitude of the electromagnetic field that is radiated by a dipole depends upon the direction of observation; equal to zero in the direction of the dipole, this amplitude is maximum in a perpendicular direction. We refer to Figure 11.A.3; a molecule is placed at the origin O and is illuminated by a ray of natural light that propagates parallel to the Oz axis. The molecule acquires an electric momentum that has a random orientation in the xOy plane. For a person observing the field that is emitted in the Oz direction, the radiation is not polarized; on the contrary, the light that is emitted in the Oy (or Ox) direction is linearly polarized. In the same way, light coming from a blue sky is linearly polarized, if observed in a direction orthogonal to the Sun's rays.

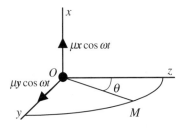

Figure 11.A.3. The amplitude of the field that is radiated by the dipole $\mu x \cos \omega t$ at point M is independent of θ. On the contrary, the field radiated by the dipole $\mu y \cos \omega t$ varies as $\cos^2 \theta$, it cancels for $\theta = 90°$.

Absorption of Light by the Atmosphere

During its propagation inside a diffusing material a light beam is attenuated. Let $I(z)$ and $I(z + dz)$ be the values of the intensity at the respective abscissas z and $z + dz$. The variation dI of the intensity is proportional to I, and we

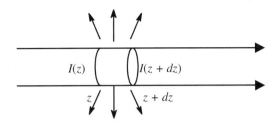

Figure 11.A.4. Attenuation of a light beam because of the diffusion by the molecules contained inside a slice of thickness dz.

have $dI = -kI\,dz$. The calculation of the proportionality coefficient k is given in the French book *Optique* by Bruhat and Kastler, the result is

$$k = \frac{32}{3}\pi^3 \frac{(n-1)}{N} \frac{1}{\lambda^4}, \qquad (11.A.1)$$

where n is the index of refraction and N is the number of molecules per unit volume.

After propagation over a distance z, the light intensity is given by

$$I(z) = I(0)e^{-kz},$$

$$\left(\frac{I(z)}{I(0)}\right)_{db} = 10\log_{10}\left(\frac{I(z)}{I(0)}\right) = 10\log_{10}(e^{-kz}) = -4.3\,kz. \qquad (11.A.2)$$

Numerical Application

Air under normal conditions: $n = 1.0003$; $N = 6 \times 10^{23}$ molecules for 22.4 liters. At $\lambda = 0.4\,\mu m \rightarrow k_{0.4} = 4 \times 10^{-5}\,m^{-1}$, which makes 0.17 db/km. It is interesting to compare the atmosphere attenuation to the attenuation of the best optical fibers (0.16 db/km at 1.55 μm): $k_{1.55} = 0.17(0.4/1.55)^4 = 7.5 \times 10^{-4}$ db/km.

It is interesting to calculate the attenuation of the Sun's rays before they reach the surface of the Earth. As the total number of molecules only determines the attenuation, it's enough to evaluate the attenuation of a cylinder of air that would have the same mass as a cylinder of the same cross section, of length 76 cm, and filled with mercury. The result is 27%.

12

Guided Optics

The capacity of a wave for transporting signals is all the more important, as the frequency is higher. What is needed if a wave is to be used as a carrier wave? First of all, coherent and powerful enough sources are necessary, then a support for the propagation of the wave is needed, with a low attenuation. Finally, one should be able to "hang to the wave" the information to be carried; in other words, to modulate the wave. Since 1960, with the appearance of lasers, coherent light sources were available. However, ten more years were necessary before it became possible to start thinking of optical telecommunications. 1970 was the important year, it was at that time that engineers became aware of the fact that silica capillary tubes made of silica SiO_2 were amazingly transparent for light waves and could thus carry light waves over very long distances, compatible with telecommunications requirements. 1970 was also the year when a laser emission was obtained from a semiconductor at room temperature.

After some time of propagation, a few kilometers in the case of radio waves propagating along metallic wires and one hundred kilometers in the case of optical fibers, the signal vanishes and must be reamplified. In 1990 a new breakthrough happened with the invention of optical amplifiers using erbium-doped silica and working at $1.55 \mu m$. Prior to this date the only possibility was to use electronic amplifiers; at each step of regeneration the optical signal was detected into an electronic signal that was amplified (with a severe bandwidth limitation) and then used to modulate a new laser diode.

Chapter 12 has been reviewed by Pierre Labeye, Physicist at the CEA (Commissariat à l'Energie Atomique).

12.1. Introduction

The Transportation Capacity of a Light Wave Is Really Enormous

The frequency band occupied by a telephone conversation is of the order of a few kilohertz: with a frequency of 10^{15} Hz, a light wave is able to carry easily 10^{10} to 10^{12} simultaneous telephone calls, which is larger than the total number of worldwide telephone calls at any given time. This limit is theoretical and raises difficult problems of modulation, of multiplexing and demultiplexing. At present the rate of transportation of a single fiber is 100 gigabits (ten million telephone calls).

Open space is the first propagation medium to think about, it has several drawbacks. The Earth being spherical and light propagating along straight lines, the range is limited by the horizon (250 km if the emitter is raised to a height of 100 m). Even on a clear day the atmosphere is not completely transparent and transmits light rays poorly in the case of rain or fog, intersatellite communications don't have this problem. Coming back to Earth, optical communications had been developed only after the introduction of optical fibers; the attenuation of the early fibers was 40 db/km (at 0.8 μm), it has now been reduced to 0.17 db/km (at 1.55 μm). This low value is to be compared to the attenuation of the best coaxial radio and microwave cables (100 db/km).

Why Dielectric Optical Guides?

In a coaxial cable, or along a pair of conducting wires, the guiding effect is due to the interaction of an electromagnetic wave with a metal, that's to say with the free electrons of the metal. In the case of an optical fiber, the propagation is made inside a dielectric material. In both cases the attenuation comes from the imperfection of the material and is described by the fact that the conductivity σ of the metal, or the permittivity ε of the dielectric material, are complex numbers, characterized "loss angles," ϕ_{metal} or ϕ_{dielec}.

$$\sigma = \sigma' - j\sigma'' = \sigma_{\text{metal}}e^{-j\phi_{\text{metal}}} \quad \text{and} \quad \varepsilon = \varepsilon' - j\varepsilon'' = \varepsilon_{\text{dielect}}e^{-j\phi_{\text{dielec}}}. \quad (12.1)$$

It's because a metal is not a single crystal but an agglomerate of microcrystals that the losses are higher in a metal than in an amorphous dielectric, which has no boundaries between grains, as is the case in a polycrystal: ϕ_{metal} is much larger than ϕ_{dielec}.

Total Internal Reflection

Total internal reflection occurs when a light beam is reflected on an interface between two transparent materials, the index of the first material being higher than the second one, and when the angle of incidence is larger than a critical

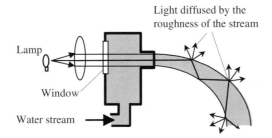

Figure 12.1. Illuminated fountain. Sent along the axis of the out-going stream of a fountain and totally reflected on the boundary, a bundle of light rays are trapped inside the jet. The reflecting surface is chaotic, which allows a small part of the light to escape, giving the impression of a liminous pipe.

angle. Under total internal reflection the cosine of the angle of refraction, that is obtained by a direct application of the Snell-Descartes law for refraction, is greater than one. Consequently (see Section 4.3.4), the sine is purely imaginary and the Fresnel law shows than the reflection coefficient is a complex number, with a modulus equal to one (see Figure 4.18). The refracted wave is evanescent and doesn't carry any energy. The incident and reflected waves have the same modulus, but their difference in phase varies with angle of incidence and is not the same for TE or TM waves.

Definition of Guided Propagation

Guided optics is defined with reference to the usual optics, which is also called three-dimensional optics or free-space optics. In the latter case, the obstacles met by the light rays are always made of surfaces that are, more or less, orthogonal to the direction of propagation; there are no boundary conditions due to interfaces that would be parallel to the rays. The volume inside which electromagnetic energy is present, a priori, extends indefinitely in a direction perpendicular to propagation.

In guided optics, on the contrary, longitudinal boundary conditions are imposed and the energy is confined inside a more restricted volume. As a first approximation, it can be considered that guided optic devices are one-dimensional (fiber) or two-dimensional objects. Using an ambiguous language, we will say that the one dimension (planar guides) or two dimensions (fibers) of the propagating space are small compared to the third dimension, which is often used as the propagation axis.

Brief Description of an Optical Fiber

Guided optics may be the occasion of sophisticated and often tedious mathematical developments, our purpose, here, is to give some orders of magni-

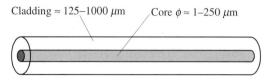

Figure 12.2. Scheme of an optical fiber.

tude of the objects of the components about which those calculations will be done. As far as fibers are concerned we will mostly consider silica fibers, which are exclusively used in telecommunications. Other important fibers also exist, especially plastic fibers.

We refer to Figure 12.2; the length (km) of an optical fiber is far larger than the diameter. The cladding is made of pure silica and has a constant index of refraction n_0. The core is made of silica, suitably doped, to increase the index up to a value n_{core} which is slightly higher than n_0. Inside the core, the index is constant, or not; it often varies with the distance to the axis

$n_{core} = n_0 + \Delta n f(r)$, Δn is usually very small; $\Delta n / n_0 \approx 10^{-3}$ is a typical value.

The domain inside of which $f(r)$ is significantly different from zero defines the diameter of the core. The following classification is useful:

- *Step index fiber:* $f(r) = 1$ if $-d/2 \le r \le d/2$, elsewhere $f(r) = 0$.
- *Gradient index fiber:* $f(r)$ has usually one maximum inside the core, sometimes two, $f(r)$ is a dimensionless function; its maximum value is normalized to one.
- *Single mode fiber:* The core diameter is small: a few wavelengths (5–10 μm).
- *Multimode fiber:* The core diameter is large: many wavelengths (150–200 μm).

The conditions for a fiber to be single mode are fixed by the wavelength, by the index variation, and by the core diameter and will be given more accurately later.

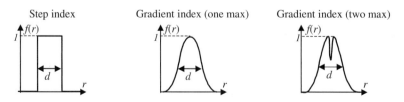

Figure 12.3. Several usual index profiles.

Integrated Optics

The expression *integrated optics* was invented in the early 1970s, by analogy with integrated electronics; many of the integrated optical technologies are inspired from microelectronics. Examples of integrated optical components are described in Figure 12.4, they are mainly made of narrow (μm) and rather long (mm) stripe guides deposited on top of a planar substrate, or embedded inside it.

A layer made of a transparent material is deposited on a transparent substrate with a lower index. As in the case of optical fibers the index variation

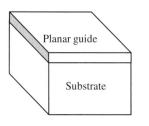 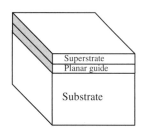

The planar wave guide is on top of the substrate.

Buried guide: the guide is sandwiched between two materials of lower index.

Figure 12.4(a). Planar optical guides for integrated optics.

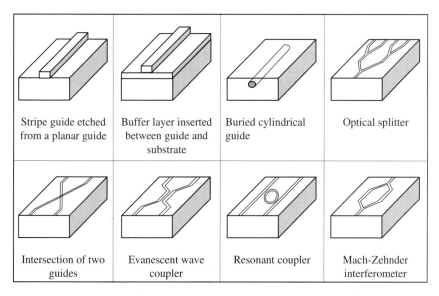

| Stripe guide etched from a planar guide | Buffer layer inserted between guide and substrate | Buried cylindrical guide | Optical splitter |
| Intersection of two guides | Evanescent wave coupler | Resonant coupler | Mach-Zehnder interferometer |

Figure 12.4(b). Some usual objects of integrated optics.

versus depth may be a step or a gradient. The index variation can be very small (10^{-3}) or more important (0.1 to 2), 0.05 is the usual value. The width and depth are of the same order of magnitude as the diameter of the core of a fiber. Because of the difficulty of making quality high optical thick layers, it's almost impossible obtain highly multimode planar guides. The losses of integrated guides are still quite low (0.1 db/cm). The losses are much larger than in the case of fibers, but propagation lengths are usually limited to the centimeter range.

12.2. Propagation in a Step Index Planar Guide

From a theoretical point of view, a step index planar guide is the simplest arrangement in guided optics; we will use it as an introduction to the behavior of guided waves.

12.2.1. *Simple Theory Using Light Rays*

A dielectric slice is sandwiched between a substrate and a superstrate that are considered as infinitely thick, the index repartition is indicated in Figure 12.5. The extension of the three elements is infinite in any direction parallel to the interfaces.

Guidance Conditions

Surprisingly enough, a theory using light rays and geometric optics gives good results, although the transversal dimensions are of the order of the wavelength.

A guide works as the illuminated fountain of Figure 12.1, and a first condition is that total internal reflection occurs on both interfaces; the angle of incidence, i, should overcome the two critical angles

$$i \geq \text{Sup}[\text{Arcsin}(n_1/n_2); \text{Arcsin}(n_3/n_2)]. \tag{12.2}$$

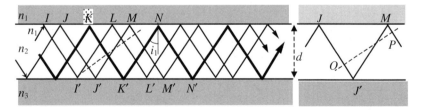

Figure 12.5. Zigzag of light rays alternately reflected by the upper and lower interfaces. A ray is guided if the angular conditions of total reflection are satisfied *and* if the phase difference between points O and P is a multiple of 2π.

A second condition is added to (12.2); the different rays, such as (JJ'), (KK'), (LL'), (MM') (see Figure 12.5), belong to the same planar wave, the wave planes of which are orthogonal to the previous rays: the phase differences along a path such as $(OJ'MP)$ must be equal, modulo 2π. The phase difference has two origins:

- $\phi_{\text{propag}} = 2\pi n_2/\lambda(OJ' + J'M + MP)$ coming from the propagation.
- ϕ_{21} and ϕ_{23} coming from the reflection on the interfaces, which is obtained from Fresnel's formula:

$$\frac{2\pi n_2}{\lambda}(OJ' + J'M + MP) + \phi_{21} + \phi_{23} = p2\pi,$$

$$\frac{2\pi n_2 d\cos i}{\lambda} + \phi_{21} + \phi_{23} = p2\pi \quad \rightarrow \quad i \in \{i_1, i_2, i_3, \ldots, i_p\}. \tag{12.3}$$

Guided Modes, Leaky Modes

Equations (12.3) (there is one equation for each value of the integer p), allow the determination of a finite set of allowed values of the angle of incidence $(i_1, i_2, \ldots, i_p)$. For those angles, the waves interfere constructively as they bounce from one interface to the other. If the two relations (12.2) and (12.3) are satisfied, the electromagnetic energy is trapped inside the guide.

To each allowed angle is associated what is called a *mode of propagation*. We are now going to examine the repartition of the electromagnetic energy inside the guide; it comes from the interference between two families of oblique waves propagating upward and downward.

- Along a direction perpendicular to the interfaces, a standing wave pattern is observed. The amplitude of the minima is equal to zero, since the two

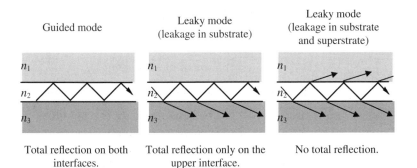

Guided mode	Leaky mode (leakage in substrate)	Leaky mode (leakage in substrate and superstrate)
Total reflection on both interfaces.	Total reflection only on the upper interface.	No total reflection.

Figure 12.6. Guided mode and leaky modes.

interfering waves have the same amplitude. The field on the interfaces is not equal to zero, corresponding to the presence of evanescent fields in the substrate and superstrate.

- Along a direction parallel to the interface a progressive wave is observed. Its wave vector is parallel to the interfaces, in a direction that is fixed by the excitation conditions.

If the conditions for total reflection are not fulfilled, rays are refracted in the substrate and superstrate, inside of which the evanescent waves are replaced by progressive waves that carry energy outside the guide. An interference pattern still exists in the guide; we then speak of *leaky modes*.

TE Modes, TM Modes

As they are reflected on a planar interface, only the TE or TM polarized beams keep their initial polarization. As a consequence, the polarization modes are also linearly polarized. Fresnel's laws are not exactly identical for TE or TM polarizations and formula (12.3) defines two sets of allowed angles, corresponding to two families of modes, the TE and TM modes. Fresnels' equations are transcendental and can only receive graphical (or numerically computed) solutions, the angles, $i_{p,\text{TE}}$ and $i_{p,\text{TM}}$, associated to the same integer p are not usually very different.

Monomode Guides, Multimode Guides

When solving equation (12.3), we can either choose a value for the thickness d of the guide and vary the wavelength λ or, on the contrary, attribute a given value to the wavelength and vary the thickness. Given a step-index guide (n_1, n_2, n_3, and d) and a wavelength λ, the following possibilities can be met:

- Equation (12.3) has no solution, whatever the value of p. This will happen if the wavelength is too large as compared to the thickness of the guide. If λ is progressively decreased, the equation will have one, two, or many solutions. There is a wavelength $(\lambda_{\text{cut-off}})_0$ (and an associated angular frequency $\omega_{\text{cut-off}} = 2\pi c/\lambda_{\text{cut-off}}$) called the *cut-off wavelength* (cut-off frequency) of the guide which is such that:

$$\lambda > (\lambda_{\text{cut-off}})_0 \rightarrow \text{the set of modes is empty,}$$

$$\lambda < (\lambda_{\text{cut-off}})_0 \rightarrow \text{the set of modes has at least one element.}$$

- *Monomode guide*: Equation (12.3) has one and only one solution.
- *Multimode guide*: Equation (12.3) has many solutions. To a given guide is associated a set of cut-off wavelengths $\{(\lambda_{\text{cut-off}})_0, (\lambda_{\text{cut-off}})_1, (\lambda_{\text{cut-off}})_2, \ldots\}$:

$$\lambda < (\lambda_{\text{cut-off}})_p \rightarrow \text{the set of modes has } (p-1) \text{ elements,}$$
$$p \text{ is called the cut-off wavelength of the } p\text{th mode.}$$

The cut-off wavelengths are different for TE and TM modes, however they are very close for a given order. A guide supporting only one TE mode and one TM mode is often considered to be a monomode guide.

Order of Magnitude of the Angle of Incidence
of the Different Modes

Fundamental mode: The zeroth-order mode is also called the fundamental mode, or principal mode. The rays are at grazing incidence, the angle of incidence is very close to $\pi/2$, and the propagation is almost parallel to the boundary interfaces. The phase velocity, which will be calculated in the next section, is practically equal to the speed of propagation, c/n_2, of light waves propagating in an open space filled with a material of index n_2.

Highest-order mode: The angle of incidence decreases with the mode order, the highest possible value of p is obtained when the conditions for total reflection are no longer satisfied; let $i_{p,\mathrm{Max}}$ be the corresponding angle of incidence. A thick guide is highly multimode; the number of allowed values of p is large. For the highest-order mode we have $n_2 \sin i_{p,\mathrm{Max}} \approx n_3 \sin \pi/2 = n_3$, it can then be shown that the phase velocity is equal to $V_3 = c/n_3$, which is the speed of propagation in an open space filled with a material of index n_3.

12.2.2. *Electromagnetic Approach of a Step Index Guide*

We consider again the step index guide of Figure 12.5 and we would like to obtain in a more rigorous way the previous results, equation (12.3) for example, and also obtain the repartition of the electromagnetic field. We will use the fact that, for given boundary conditions, the solution of Maxwell's equations is unique. The geometric space is divided into three subspaces, superstrate-guide-substrate, which are in contact along the two interfaces. In each subspace, the solution is supposed to be a planar wave that is of the following form:

$$A_i e^{j\omega t} e^{-jk_i r}, \quad i \in \{1, 2, 3\}. \tag{12.4}$$

Planar waves are solutions of Maxwell's equations, the problem is now to find suitable values for the amplitudes A_1, A_2, A_3, and for the three wave vectors k_1, k_2, k_3. We choose the following referential: the plane yOz is parallel to the interfaces and in the middle of the guiding layer, Ox is orthogonal to yOz. Because of the symmetry of the problem, which is invariant in any translation parallel to the interfaces, the three wave vectors are parallel to xOz and have only two components, k_{ix} and k_{iz}. The moduli of the wave vectors of planar waves propagating in open spaces filled with transparent media of respective indices $n_0 = 1$, n_1, n_2, and n_3 are equal to $k_0 = \omega/c$, $k_1 = n_1 k_0$, $k_2 = n_2 k_0$, and $k_3 = n_1 k_0$. The Pythagoras theorem indicates that

$$
\begin{aligned}
k_{1x}^2 + k_{1z}^2 &= n_1^2 = n_1^2 k_0^2, \\
k_{2x}^2 + k_{2z}^2 &= n_2^2 = n_2^2 k_0^2, \\
k_{3x}^2 + k_{3z}^2 &= n_3^2 = n_3^2 k_0^2.
\end{aligned}
\tag{12.5.a}
$$

Analytical Expressions of the Guided Modes

Invoking again the translation invariance parallel to Oz, we see that the Oz components of the three wave vectors must have a common value, which will be called k_z. This condition is also known as the *phase matching condition* since the progressive wave in the guide and the progressive part of the waves in the substrate and superstrate travel at the same speed in the Oz direction,

$$
k_{1z} = k_{2z} = k_{3z} = k_z, \quad \text{phase matching condition.}
\tag{12.5.b}
$$

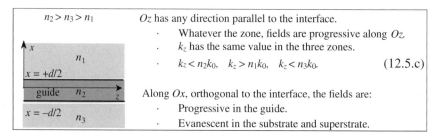

$n_2 > n_3 > n_1$

Oz has any direction parallel to the interface.
· Whatever the zone, fields are progressive along Oz.
· k_z has the same value in the three zones.
· $k_z < n_2 k_0, \quad k_z > n_1 k_0, \quad k_z < n_3 k_0.$ (12.5.c)

Along Ox, orthogonal to the interface, the fields are:
· Progressive in the guide.
· Evanescent in the substrate and superstrate.

Figure 12.7. Structure of a guided mode.

In order to have a progressive wave in the region of the guide, it's necessary that the x and z components of the wave vector are both real and, therefore, because of (12.5.a), $k_z < k_2 = n_2 k_0$. On the contrary, the fields must be evanescent in the Ox direction and $k_z > k_1 = n_1 k_0$ and $k_z > k_2 = n_2 k_0$ so that k_{1x} and k_{2x} are purely imaginary; let us introduce two real numbers, α_{1x} and α_{3x}, by the following formulas:

$$
\begin{aligned}
k_{1x} &= \sqrt{\frac{n_1^2 \omega^2}{c^2} - k_z^2} = \sqrt{n_1^2 k_0^2 - k_z^2} = j\alpha_{1x} \;\Rightarrow\; \alpha_{1x} = \sqrt{k_z^2 - n_1^2 k_0^2}, \\
k_{3x} &= \sqrt{\frac{n_3^2 \omega^2}{c^2} - k_z^2} = \sqrt{n_3^2 k_0^2 - k_z^2} = j\alpha_{3x} \;\Rightarrow\; \alpha_{3x} = \sqrt{k_z^2 - n_3^2 k_0^2}.
\end{aligned}
\tag{12.5.d}
$$

The modes are linearly polarized, either TE or TM. The disposition of the different vectors is recalled in Figure 12.8.

TE wave

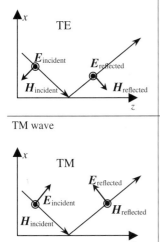

Electric field parallel to the interface and parallel to the plane of incidence: $E = E_y\,y$.

Magnetic field parallel to the interface and parallel to the plane of incidence:

$$H = H_x\,x + H_z\,z, \quad H_z = \frac{j}{\omega\mu_0}\frac{\partial E_y}{\partial x}. \tag{12.6a}$$

TM wave

Magnetic field parallel to the interface and orthogonal to the plane of incidence: $H = H_y\,y$.

Electric field parallel to the plane of incidence:

$$E = E_x\,x + E_z\,z, \quad E_z = \frac{-j}{\omega\varepsilon}\frac{\partial H_y}{\partial x}. \tag{12.6b}$$

Figure 12.8. Respective positions of the electric and magnetic fields for TE and TM waves.

TE Modes

Explicit expressions of the fields in the three regions can be obtained by writing the continuity of the tangential components on both sides of each interface:

$x > d/2$ superstrate:
$$E_{y(x,z)} = A_1 e^{-\alpha_{1x}x} e^{-jk_z z},$$
$$H_{z(x,z)} = \frac{-j}{\omega\mu_0} A_1 e^{-\alpha_{1x}x} e^{-jk_z z}, \tag{12.7.a}$$

$-d/2 \le x \le +d/2$ guide:
$$E_{y(x,z)} = A_2 \cos(k_{2x}x + \psi) e^{-jk_z z},$$
$$H_{z(x,z)} = \frac{-jk_{2x}}{\omega\mu_0} A_2 \sin(k_{2x}x + \psi) e^{-jk_z z}, \tag{12.7.b}$$

$x < -d/2 < 0$ substrate:
$$E_{y(x,z)} = A_3 e^{-\alpha_{3x}x} e^{-jk_z z},$$
$$H_{z(x,z)} = \frac{j\alpha_{3x}}{\omega\mu_0} A_3 e^{+\alpha_{3x}x} e^{-jk_z z}. \tag{12.7.c}$$

Formulas (12.6) are not quite identical to formulas (12.4) that were initially proposed for the fields; however, they can be made identical, using the complex expressions of the sine and cosine.

The problem is now to find suitable values for the amplitudes of the fields and for the components of the wave vector along the Ox axis. k_z is first given some value that is compatible with the inequality expressed by formula (12.5.c) of Figure 12.7. We then write that the tangential components of the fields have, respectively, the same value on both sides of the interfaces located at $\pm d/2$,

$$x = +d/2: \quad \begin{aligned} E_{\tan} = E_y &\Rightarrow A_1 e^{-\alpha_{1x}d/2} = A_2 \cos\left(k_{2x}\frac{d}{2}+\psi\right), \\ H_{\tan} = H_z &\Rightarrow A_1 e^{-\alpha_{1x}d/2} = \frac{k_{2x}}{\alpha_{1x}} A_2 \sin\left(k_{2x}\frac{d}{2}+\psi\right), \end{aligned} \quad (12.8.a)$$

$$x = -d/2: \quad \begin{aligned} E_{\tan} = E_y &\Rightarrow A_3 e^{-\alpha_{3x}d/2} = A_2 \cos\left(k_{2x}\frac{d}{2}-\psi\right), \\ H_{\tan} = H_z &\Rightarrow A_3 e^{-\alpha_{3x}d/2} = \frac{k_{2x}}{\alpha_{3x}} A_2 \sin\left(k_{2x}\frac{d}{2}-\psi\right). \end{aligned} \quad (12.8.b)$$

To determine the three quantities A_1, A_2, and A_3 we obtain a set of four equations (12.8) that are linear and homogeneous. As there are more equations than unknown quantities, a solution doesn't always exist. The condition required for having a solution is called the *guiding condition*, which is obtained when the determinant of the linear system is made equal to zero. When a solution exists, there is an indetermination: only the amplitude is known, within a multiplication coefficient. From a physical point of view, this constant is obtained from the total power carried by the guided mode.

Guiding Condition

The cancellation of the determinant leads to

$$\tan(k_{2x}d/2+\psi) = \frac{\alpha_{1x}}{k_{2x}} \quad \text{and} \quad \tan(k_{2x}d/2-\psi) = \frac{\alpha_{3x}}{k_{3x}}, \quad (12.9)$$

$$\phi_1^{TE} = 2\arctan\left(\frac{\alpha_{1x}}{k_{2x}}\right) \quad \text{and} \quad \phi_3^{TE} = 2\arctan\left(\frac{\alpha_{3x}}{k_{2x}}\right), \quad (12.10)$$

$$k_{2x}\frac{d}{2}+\psi = \frac{\phi_1^{TE}}{2}+n\pi \quad \text{and} \quad k_{2x}\frac{d}{2}-\psi = \frac{\phi_3^{TE}}{2}+m\pi, \quad (12.11)$$

$$2k_{2x}d - \phi_1^{TE} - \phi_3^{TE} = 2p\pi, \quad p \text{ is an integer.} \quad (12.12)$$

Formula (12.12) is identical to formula (12.3) that was established from geometrical considerations. The electromagnetic approach is however more powerful, since the maps of the field repartition for each mode are also obtained,

$$x > d/2: \quad \begin{cases} E_y^{\text{TE}}(x,z) = A_2 \cos\left(k_{2x}\dfrac{d}{2} + \psi\right)e^{-\alpha_{1x}(x-d/2)}e^{-jk_zz}, \\[4mm] H_z^{\text{TE}}(x,z) = \dfrac{-j\alpha_{1x}}{\omega\mu_0}A_2 \cos\left(k_{2x}\dfrac{d}{2} + \psi\right)e^{-\alpha_{1x}(x-d/2)}e^{-jk_zz}, \end{cases} \quad (12.13.\text{a})$$

$$|x| \le d/2: \quad \begin{cases} E_y^{\text{TE}}(x,z) = A_2 \cos(k_{2x}x + \psi)e^{-jk_zz}, \\[4mm] H_z^{\text{TE}}(x,z) = \dfrac{-jk_{2x}}{\omega\mu_0}A_2 \cos\left(k_{2x}\dfrac{d}{2} + \psi\right)e^{-jk_zz}, \end{cases} \quad (12.13.\text{b})$$

$$x < d/2: \quad \begin{cases} E_y^{\text{TE}}(x,z) = A_2 \cos\left(k_{2x}\dfrac{d}{2} + \psi\right)e^{\alpha_{3x}(x-d/2)}e^{-jk_zz}, \\[4mm] H_z^{\text{TE}}(x,z) = \dfrac{j\alpha_{3x}}{\omega\mu_0}A_2 \cos\left(k_{2x}\dfrac{d}{2} + \psi\right)e^{\alpha_{3x}(x-d/2)}e^{-jk_zz}. \end{cases} \quad (12.13.\text{c})$$

TM Modes

The method for obtaining the TM modes is exactly the same as for the TE modes; the results area is a bit more complicated because, by opposition to the magnetic permeability μ_0, the dielectric constant is not the same in the three regions. Formula (12.12) is replaced by

$$2k_{2x}d - \phi_1^{\text{TM}} - \phi_3^{\text{TM}} = 2p\pi. \quad (12.14)$$

In fact, the TM expressions for the magnetic field are identical to the TE expressions of the electric field in the TE case, the TM electric field being more complicated.

The fields that are obtained from formulas (12.13) just represent possible forms of the field in this structure. The real field is determined by the input signal, at $z = 0$, especially the amplitude A_2. The more an external source shape is similar to a mode field shape at the input interface, the more power of the source is transmitted to and carried by that mode in the guide.

Field Distribution of a Guided Mode

To each value of p, equation (12.12) associates a value of k_{2x}, which depends on the size d of the guide and on the frequency ω. Once k_{2x} is known, $k_z = k_{1z} = k_{2z} = k_{3z}$ is also known from (12.5.a) and then k_{2x}, α_{1x}, and α_{3x}. For each value of p, equations (12.13) give a specific distribution of the field. In Figure 12.9 are represented the field distributions of the fundamental mode and the third mode of a given guide, for four different frequencies; two frequencies are, respectively, close to the cut-off frequencies of the fundamental and first-order modes, the other two being quite different.

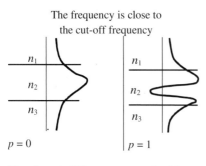

The frequency is close to the cut-off frequency

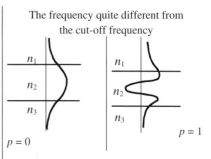

The frequency quite different from the cut-off frequency

Near the cut-off frequency, a noticeable part of the energy is in the superstrate and the substrate: the mode is weakly guided.

For a frequency further away from the cut-off, most of the energy is inside the guide. Guidance is more efficient.

Figure 12.9. Examples of field distributions of the fundamental and first guided modes.

Dispersion Curves of the Modes

The propagation constant k_z of a given mode depends on the frequency. A relation between the frequency and the modulus of a wave vector is generally called a relation of dispersion, the associated graph being a dispersion curve. To a given guide is associated a set of dispersion curves, one curve per mode. Figure 12.10 shows a typical set of dispersion curves. The dotted lines are the

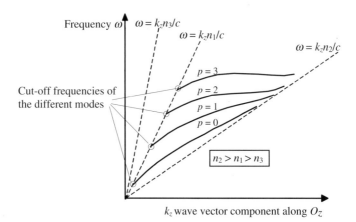

Figure 12.10. Set of dispersion curves of a planar guide. The dotted lines are the dispersion of light in an open space filled, respectively, with materials of indices n_1, n_2, and n_3.

dispersion curves of the light in an open space filled with transparent materials of respective indices n_1, n_2, and n_3; as the variation of the index with the frequency has been omitted they are straight lines of equation $\omega = k_z n_i/c$. The index n_3 of the substrate being the larger, the set of dispersion curves of the guide is between the dispersion curves of the material of the guide and of the superstrate. The dispersion curve of a mode is asymptotic to the dispersion curve of the material of the guide; it starts from a point located on the dispersion curve of the substrate, the frequency of which is the cut-off frequency of the mode.

12.2.3. *Excitation of a Planar Wave Guide—Numerical Aperture*

The previous analysis was only preoccupied by the field repartition and not by the way it could be excited. It's clear from Figure 12.11(a) that oblique planar waves can only excite leaky modes and cannot excite guided modes.

A guide can be excited through a lateral side, see Figure 12.11. For total internal reflection to occur on the interfaces, the angle of incidence θ on the input face must be smaller than some value, called the numerical aperture

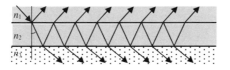

Figure 12.11(a). An oblique light ray, after refraction, makes an angle with the normal to the interfaces, which is smaller than the critical angle and cannot be totally reflected. Only leaky modes can be excited.

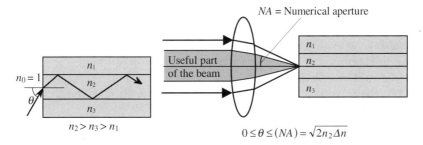

$$0 \leq \theta \leq (NA) = \sqrt{2 n_2 \Delta n}$$

Figure 12.11(b). Numerical aperture of a guide. A planar guide is never really infinite and has a lateral side through which the guide can be infinite. The numerical aperture (NA) is the maximum angle of incidence that still ensures total internal reflection on the interfaces.

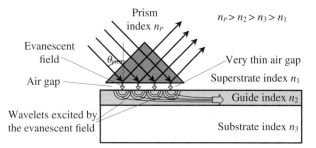

Figure 12.12. Prism coupler. A light beam is totally reflected on the hypotenuse side of the prism, which is very close (less than $\lambda/4$) to the guide. An evanescent field penetrates inside the guide that is excited by an optical tunnel effect.

(NA) of the guide. Let L_{12} and L_{23} be the respective critical angles on the two interfaces. Usually $L_{23} = \text{Arcsin}(n_3/n_2)$ is the smaller and

$$n_0 \sin(NA) = n_2 \sin\left(\frac{\pi}{2} - L_{23}\right) = \sqrt{n_2^2 - n_3^2}, \tag{12.15.a}$$

$$(NA) = \text{Arcsin}\left(\frac{\sqrt{n_2^2 - n_3^2}}{n_0}\right) \approx \text{Arcsin}\left(\frac{\sqrt{2n_2\Delta n}}{n_0}\right),$$

$$(NA) \approx \sqrt{2n_2\Delta n}, \quad \text{if } \Delta n \text{ is small.} \tag{12.15.b}$$

12.2.3.1. *Prism Coupler*

A guide is a type of closed box for a guided mode and, in principle, it's not possible to enter into a closed box; except if a kind of *tunnel effect* can be imagined. The field of a guided mode is evanescent in the superstrate; to excite the guide, one must think of an arrangement that would create an evanescent field along the upper interface of a planar guide. Such an arrangement is proposed in Figure 12.12, which is called a *prism coupler*.

We refer to Figure 12.12, the hypotenuse of an isosceles and rectangular prism is disposed very close to the surface of a planar guide, the air gap being smaller than a quarter of the wavelength. The refractive index n_{prism} of the prism is higher than the refractive index n_2 of the guiding layer. A planar wave propagates inside the prism and is totally reflected from the hypotenuse. Below the hypotenuse, the field is evanescent in a direction orthogonal to the interface and progressive parallel to the interface. Let $K_{z,\text{prism}} = (2\pi/\lambda)n_{\text{prism}} \sin r_{\text{prism}}$ be the propagation constant along the interface (r_{prism} is the angle of incidence inside the prism).

12.2.3.2. *M-Lines Method of Determination of the Propagation Constants of the Guided Modes*

The evanescent field due to total reflection penetrates inside the guide through the very thin air gap and sets in vibration the electrons of the guide which, in turn, generate spherical wavelets that can possibly be guided. The wavelets are not in phase, but have a phase repartition that is represented by $e^{-jk_{z,\text{prism}}z}$; if this phase repartition fits the phase repartition of a guided mode the wavelets interfere positively and the guided mode is excited. Let i_{mp} and $k_m = (2\pi/\lambda)n_2 \sin i_m$ be, respectively, the angle of incidence of the zigzag path (see Figure 12.5) and the propagation constant of the mth mode; the condition of constructive interference is nothing other than the Snell-Descartes law:

$$k_{z,\text{prism}} = k_m \quad \rightarrow \quad n_{\text{prism}} \sin\theta_{\text{prism}} = n_2 \sin i_m. \tag{12.16}$$

When relation (12.16) is satisfied, the corresponding rays are no longer totally reflected, part of the energy being coupled inside the guide; this occurs only for some specific angles of incidence. In the arrangement of Figure 12.13, a lens focuses light on the hypotenuse of the prism, most of the rays are totally reflected; the rays that contribute to the excitation of guided modes are only partly reflected. On a sheet of paper receiving the reflected beam can be seen an illuminated circular area with darker lines that correspond to the different modes of the guide. The directions of the dark lines, that are usually called *m-lines*, can be measured accurately providing a method of measurement of the propagation constants of the guided modes.

The Bragg equation for diffraction gratings may give angles with a sine (or a cosine) greater than one, under such conditions evanescent waves are produced; they can be used to couple a planar wave to a guide. This method of excitation of a guide, which has many useful applications, is described in Figure 12.14.

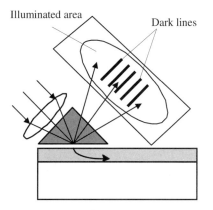

Figure 12.13. *m*-lines determination of the propagation constants of the guided modes. An incident ray that contributes to the excitation of a mode is only partly reflected. To each guided mode is associated a darker line, also called an *m*-line.

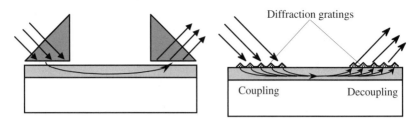

Figure 12.14. Prism and grating coupler.

12.2.4. *Generalization to Any Type of Guide*

The example of a step index planar guide was chosen because of its simplicity. We are now going to generalize the different notions that have been introduced and, more specifically, the notion of the *mode of propagation*. The basic parameters are:

- Frequency ω, the associated vacuum wavelength $\lambda = 2\pi c/\omega$, and the vacuum wave vector $k_0 = \omega/c$.
- Optogeometric characteristics of the guides (index repartition, shape, and size of the cross section).

Local modes: Modes of propagation can be introduced *only* for devices that are invariant in a translation parallel to some axis, which requires an infinite length. Real systems are not that long; the concept remains valid if the length is long compared to the wavelength. We will often deal with guides having optogeometric properties that vary smoothly with distance; examples are given in Figure 12.15. The notion of a local mode is then introduced: the local modes are the modes of a guide which would have an infinite length and optogeometric properties identical to those of the guide at the place under consideration.

Cut-off frequency: At a given frequency is associated a set of guided modes, each of them being referenced by a labeling coefficient p. If the frequency is too low and lower than a frequency $(\omega_{\text{cut-off}})_0$, the set of modes is empty. As the frequency is increased the set of modes has more and more

Figure 12.15. Examples of guides that are not invariant in a translation: the optogeometric properties (diameter, index of refraction, radius of curvature, . . .) vary smoothly from point to point and can be considered to remain constant along a section that is much larger than λ.

elements. According to the frequency, a guide will support no mode, be monomode, weakly multimode, and largely multimode. Each mode has a specific cut-off frequency.

Cut-off wavelength: The vacuum wavelength can be used instead of the frequency. A wave can be guided if the wavelength is smaller than the cut-off wavelength. Using a metaphor, it can be said that it's not possible to enter a large wavelength inside a small guide.

Propagation speed of a guide mode: At a given frequency, the electromagnetic analysis of a guide gives the set of constants of propagation of the different modes $k_{z,p}$. The ratio $V_p = k_{z,p}/\omega$ is the phase velocity of the mode, it should be noted that V_p is larger than the phase velocity in an open space filled with a material having the index as the guide.

Effective index of a mode: The ratio $n_p = V_p/c$ is the effective refractive index of the mode. Three indices are usually involved in a guiding structure, we have the following relations:

$$n_{\text{superstrate}} < n_{\text{substrate}} < n_{\text{guide}} \quad \rightarrow \quad n_{\text{substrate}} < n_p < n_{\text{guide}}.$$

In the case of an optical fiber, $n_{\text{superstrate}} = n_{\text{substrate}} = n_{\text{cladding}}$ and $n_{\text{guide}} = n_{\text{core}}$.

Dispersion: The phase velocities of the modes and, consequently, the effective indices vary with the frequency. A guide is characterized by its sets of dispersion curves. There are two categories of dispersion curves. In the first one the frequency is plotted versus the modulus of the wave vector, in the second the effective index is plotted versus frequency.

The phase velocity concerns sinusoidal waves. If the wave is modulated, the spectrum has more than one component; the group velocity, which is then introduced, may be obtained from the dispersion curves.

The dispersion of a guided mode has two origins. One comes from the geometry while the second comes from the dispersion of the material, that's to say, the variation of refractive indices with the frequency.

Structure of a guided mode: The electromagnetic description of a mode involves two main terms:

- A propagation term $e^{-jk_{z,p}z}$, where $k_{z,p}$ is the propagation constant of the pth mode.
- A description of the electromagnetic field as a function of the transverse coordinates (x, y).

From a mathematical point of view, the functions $f(x, y)$ that describe the field have the property of being *orthogonal*. We will not speculate about this property, however, we would like to insist on an important consequence. Let us suppose that one mode, and only one mode, has been excited, during the propagation the electromagnetic energy will remain confined in this mode, and will not be coupled to the other modes. In reality this orthogonal character is rather theoretical and, because of unavoidable irregularities along the guide, the different modes are always more or less coupled to one another.

Numerical Aperture: This notion was introduced by formula (12.15) and is very important for any kind of guide. Taking typical values for the indices, $n = 1.50$ and $\Delta n = 0.01$, we obtain $(NA) = \sqrt{2n\Delta n} = 0.17$.

12.3. Optical Fibers

Optical fibers are basically made of a core and a cladding. At a given frequency and according to the diameter of its core, a fiber is monomode or multimode. The index of the core, which is higher than the index of the cladding, can be constant (step index profile) or vary with the distance to the axis (graded index profile). Very often the index difference is weak, and most fibers are said to be weakly guiding.

12.3.1. *Gradient Index Fibers*

In the case of graded index multimode fibers, where the core diameter is much larger than the wavelength, a ray analysis is possible and often useful. In Figure 12.16 is shown the scheme of an optical fiber and a convenient analytical expression of the index profile. Very often $\Delta \ll 1$ and $p = 2$, so that $n(r)$ can be simplified as

$$n(r) = n_1\left(1 - \Delta\frac{r^2}{a^2}\right) = n_1\left(1 - \frac{r^2}{2\rho^2}\right) \quad \text{with} \quad \rho = \frac{a}{\sqrt{2\Delta}}.$$

Using the Snell-Descartes laws, it can be shown that the rays, inside the fibers and within Gauss conditions, follow two linear differential equations,

$$\frac{d^2x}{dz^2} + \frac{x}{\rho^2} = 0, \quad \rightarrow \quad x = A\cos\left(\frac{z}{\rho} + \psi\right),$$

$$\frac{d^2y}{dz^2} + \frac{y}{\rho^2} = 0, \quad \rightarrow \quad y = A'\cos\left(\frac{z}{\rho} + \psi'\right),$$

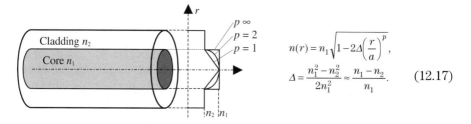

$$n(r) = n_1\sqrt{1 - 2\Delta\left(\frac{r}{a}\right)^p},$$

$$\Delta = \frac{n_1^2 - n_2^2}{2n_1^2} \approx \frac{n_1 - n_2}{n_1}. \qquad (12.17)$$

Figure 12.16. Representation of an optical fiber. a is a parameter that is homogeneous to a length and indicates the order of magnitude of the core diameter. p has no dimension parameter and it varies from 1 (triangular profile) to infinity (step profile). For weakly guiding fibers $\Delta \ll 1$.

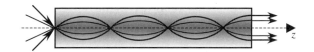

Figure 12.17. Ray propagation inside a fiber. An incident ray contained in a meridian plane will remain in this plane. Nonmeridian rays follow helical trajectories winding around the Oz axis.

where A, ψ, A', and ψ' are constants of integration, they are obtained from the injection conditions. The maximum angle between an incident light ray and the Oz axis is obtained by expressing that the constants A (or A') are at most equal to the diameter of the fiber, this maximum angle defines the *numerical aperture* (NA) of the fiber.

12.3.2. *Modal Analysis of Graded Index Fibers*

The tremendous development of optical telecommunications has generated the development of a great variety of methods to describe the electromagnetic field in fibers. It's completely out of the question to give an exhaustive list, our ambition is to give general ideas that are supposed to help the reader in his/her comprehension of calculations that are always tricky and specialized. We will start by studying step index fibers that are equivalent to the step index planar guides. Their modal analysis is not conceptually more complicated than the case of planar guides, except that, because of the cylindrical symmetry, the trigonometric functions are replaced by Bessel functions; there are also a family of orthogonal functions, unfortunately less familiar to most people than sine or cosine.

Let us consider a step index fiber, the core and cladding indices are, respectively, equal to n_1 and n_2, the radius of the core is equal to a, and the radius of the cladding is infinite. The Helmholtz equation takes two different expressions in the core or in the cladding:

$$\text{in the core:} \quad r < a: \quad \frac{\partial^2 U}{\partial r^2} + \frac{1}{r}\frac{\partial U}{\partial r} + \frac{1}{r^2}\frac{\partial^2 U}{\partial \phi^2} + \frac{\partial^2 U}{\partial z^2} + n_1^2 k_0^2 U = 0, \quad (12.18.a)$$

$$\text{in the cladding:} \quad r > a: \quad \frac{\partial^2 U}{\partial r^2} + \frac{1}{r}\frac{\partial U}{\partial r} + \frac{1}{r^2}\frac{\partial^2 U}{\partial \phi^2} + \frac{\partial^2 U}{\partial z^2} + n_2^2 k_0^2 U = 0, \quad (12.18.b)$$

where $k_0 = \omega/c = 2\pi/\lambda_0$ is the wave vector modulus in a vacuum and U represents any component of the electromagnetic field.

The variables are separated by setting

$$U(r, \phi, z) = u(r)e^{-jl\phi}e^{-jk_z z}, \quad l \in \{0, 1, 2, 3, \dots\}, \quad (12.19)$$

$$\frac{d^2 u}{dr^2} + \frac{1}{r}\frac{du}{dr} + \left(n_2^2 k_0^2 - k_z^2 - \frac{l^2}{r^2}\right)u = 0. \quad (12.20)$$

We introduce two constants of propagation in a direction orthogonal to Oz:

$$k_T^2 = n_1^2 k_0^2 - k_z^2, \quad \text{inside the core,} \tag{12.21.a}$$

$$\gamma^2 = k_z^2 - n_2^2 k_0^2, \quad \text{inside the cladding.} \tag{12.21.b}$$

The differential equation (12.20) is now written as

$$\frac{d^2 u}{dr^2} + \frac{1}{r} \frac{du}{dr} + \left(k_T^2 - \frac{l^2}{r^2} \right) u = 0, \quad \text{inside the core } r < a, \tag{12.22.a}$$

$$\frac{d^2 u}{dr^2} + \frac{1}{r} \frac{du}{dr} - \left(\gamma^2 + \frac{l^2}{r^2} \right) u = 0, \quad \text{inside the cladding } r > a. \tag{12.22.b}$$

Using the usual notations, the solution of equation (12.22.a) is a Bessel function $J_l(\xi)$, while the solution of (12.22.b) is a modified Bessel function $K_l(\eta)$,

$$r < a \quad \rightarrow \quad u(r) = A J_l(k_T r) \quad \text{in the core,} \tag{12.23.a}$$

$$r > a \quad \rightarrow \quad u(r) = B K_l(k_T r) \quad \text{in the cladding,} \tag{12.23.b}$$

where A and B are proportionality coefficients.

Developments of Bessel functions show that the solutions are very similar to the case of the step index planar guide, since

$$J_l(x) \approx \sqrt{\frac{2}{\pi x}} \cos \left[x - (2l+1)\frac{\pi}{4} \right] \quad \text{(if } x \to 0\text{),}$$

$$K_l(x) \approx \sqrt{\frac{\pi}{2x}} \left(1 + \frac{4l^2 - 1}{8x} \right) e^{-x} \quad \text{(if } x \to \infty\text{).}$$

The choice of Bessel functions ensures that Maxwell's equations (in fact the Helmholtz equation) will be satisfied; writing the boundary conditions along the only surface of discontinuity, which is located at $r = a$, will give the constants A and B of formulas (12.23). The expression of the continuity of E_z and H_z on both sides of the interface provides a first equation; only one equation is obtained, since the ratio H_z/E_z is equal to the wave impedance. To obtain a second equation, we go back to Maxwell's equations ($j\omega\varepsilon E = \Delta \wedge H$ and $-j\omega\mu_0 H = \Delta \wedge E$) and write the continuity of the components E_ϕ and H_ϕ.

Using Maxwell's equations and equations (12.21), a dispersion curve is obtained for each value of the integer l: k_z is expressed as a function of the integer l, the frequency ω. The mathematical expressions are not simple and need a computer; they are represented by a diagram identical to the diagram of Figure 12.10.

12.3.3. *Total Number of Guided Modes*

For a mode to be a guided mode it's necessary that the coefficients k_T and γ introduced by formulas (12.21) are real numbers, which implies a condition

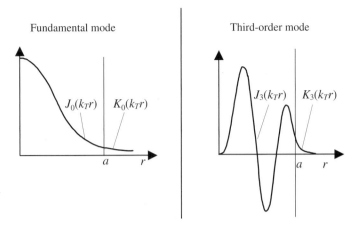

Figure 12.18. Examples of the field repartition of two modes of a step index fiber.

on $k_0 = \omega/c$. As was the case for a planar wave guide, an optical fiber can only guide a definite number of modes, $N_{\text{Max}}(\omega, l)$, of frequency ω. Each mode is characterized by a constant along Oz, a cut-off frequency $\omega_{l,\text{cut-off}}$, and a map of the electromagnetic field.

The number $N_{\text{Max}}(\omega, l)$ depends on the optogeometric characteristics of the fiber (n_1, n_2, index profile). The number of modes that really propagate along the fiber is at most equal to $N_{\text{Max}}(\omega, l)$, but is mostly fixed by the source to which the fiber is coupled. The excitation of a mode is all the more efficient as the field repartition created by the source on the entrance face of the fiber is similar to the field repartition of the mode under consideration.

The V Parameter

To evaluate the number of modes, it is convenient to introduce a new parameter, which is called the *V parameter* (or the *guide normalized frequency*), and is defined by

$$V^2 = (k_T^2 + \gamma^2)a^2 \quad \rightarrow \quad V^2 = (n_1^2 - n_0^2)k_0^2 a^2.$$

We introduce the numerical aperture: $(NA) \approx \sqrt{n_1^2 - n_2^2}$:

$$V = k_0 a(NA) = 2\pi\frac{a}{\lambda}(NA) = \frac{\omega c}{a}(NA). \tag{12.24}$$

The main interest of the *V parameter* is to allow an easy evaluation of the number M of modes. If M is big enough, it can be shown that

$$M = \alpha V^2, \tag{12.25}$$

where α is a coefficient of the order of one, it is equal to $4/\pi^2 \approx 1/2$ for a step index multimode fiber, and to 0.25 for a parabolic gradient index fiber.

12.3.4. *Effective Index and Phase Velocity of Guided Modes*

The propagation along the fiber is described by $e^{j(\omega t - \beta_q z)}$; the integer q is smaller than M. For the profiles that are described by equation (12.17), it can be shown that

$$\beta_q = n_1 k_0 \left[1 - \left(\frac{q}{M} \right)^{p/p+2} \Delta \right].$$

The phase velocity and the corresponding effective index are easily obtained

$$V_q^{\text{phase}} = \frac{c}{1 - \left(\dfrac{q}{M} \right)^{p/p+2} \Delta} \quad \text{and} \quad n_q = \frac{c}{V_q^{\text{phase}}} = n_1 \left[1 - \left(\frac{q}{M} \right)^{p/p+2} \Delta \right].$$

Group Velocity

The propagation constant is not proportional to the frequency, the phase velocity then varies with the frequency, and a group velocity V_q^{group} should be introduced for each mode, it can be shown that

$$V_q^{\text{group}} = \frac{c}{n_1} \left[1 - \Delta \frac{p-2}{p+2} \left(\frac{q}{M} \right)^p \right].$$

For step index fibers ($p = \infty$): $V_q^{\text{group}} = \dfrac{c}{n_1} \left[1 - \Delta \dfrac{q}{M} \right].$

For a parabolic gradient index fiber ($p = 2$): $V_q^{\text{group}} = \dfrac{c}{n_1} \left[1 - \Delta^2 \dfrac{q}{2M} \right].$

The dispersion is less important in the case of a parabolic profile.

Flight Time of a Light Pulse in an Optical Fiber

The carrying capacity of a fiber can be evaluated from the highest number of elementary pulses that can be transmitted each second. A source emits at a high frequency, called the clock frequency f_{clock}, a succession of very short pulses (Dirac pulses) that are coupled to the fiber. At the output, see Figure 12.20, a succession of light pulses is observed; the repetition frequency is still f_{clock}, but the pulse duration is no longer equal to zero and is equal to $\Delta\theta = (n_{\text{core}} - n_{\text{cladding}})L/c$. If $\Delta\theta$ is longer than $1/f_{\text{clock}}$, the pulses overlap and cannot be discriminated,

$$(n_{\text{core}} - n_{\text{cladding}}) = 0.01 \quad \rightarrow \quad \Delta\theta = 33\,\text{ns/km}.$$

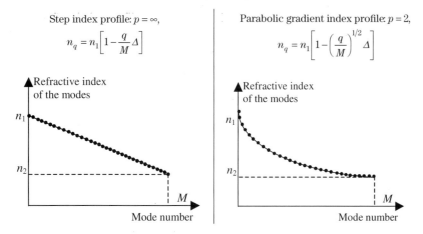

Step index profile: $p = \infty$,

$$n_q = n_1\left[1 - \frac{q}{M}\Delta\right]$$

Parabolic gradient index profile: $p = 2$,

$$n_q = n_1\left[1 - \left(\frac{q}{M}\right)^{1/2}\Delta\right]$$

Figure 12.19. Variation of the effective index as a function of the mode number. The index of the lowest-order modes is equal to the index in the middle of the core; for higher-order modes, the index is very close to the index of the cladding.

This limitation doesn't apply to the case of a monomode fiber; if there is only one mode, there is only one speed of propagation. This is true, except that we must now pay attention to another phenomenon that we had previously ignored and which is due to the dispersion of the material of which the fiber is made.

The spectrum of pulse extends over a frequency band that is all the more broad, as it is short. Because of the side bands, a light wave, modulated by a short pulse, cannot be considered as monochromatic. The different components of its Fourier spectrum don't propagate at the same speed: if the dispersion of the material is normal, the lowest frequencies travel slower than the highest frequencies. Here again, the pulse duration is increased after propagation in the fiber.

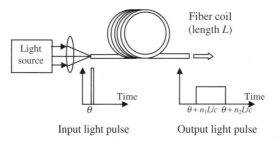

Figure 12.20. A very short light pulse is coupled, at time θ, inside a multimode fiber. We consider that all the modes are excited with the same efficiency. The fastest modes arrive at the end of the fiber at time $(\theta + n_{\text{cladding}}L/c)$, the slowest arrive later, at time $(\theta + n_{\text{core}}L/c)$. The duration is increased by $\Delta\theta = n_{\text{core}} - n_{\text{cladding}})L/c$.

To treat this problem properly, we start from the variation law of the pulse and obtain, by Fourier transform, the spectrum. The Fourier spectrum of the modulated light wave is deduced and the propagation of each component along the fiber is then studied using the proper value of the index of refraction. The spectrum of the emerging signal is calculated and, by inverse Fourier transform, the time variation law is finally obtained. During the calculation, the group velocity is introduced; it depends on the dispersion curve of the mode and also on the group velocity of light waves propagating in an open space filled with the same material as the fiber (usually silica). The group velocity is calculated in the last chapter of this book, see formula (13.19):

$$V_1^{group} = \frac{c}{n_1}\left(1 + \frac{\lambda}{n_1}\frac{dn_1}{d\lambda}\right).$$

We consider a light pulse of central wavelength λ, occupying a spectral width $\Delta\lambda$. It can be shown that, after having propagated over a distance L, the duration of the pulse is increased of $\Delta\theta$, which is given by

$$\Delta\theta = -\frac{L}{c}\frac{d^2n_1}{d\lambda^2}\lambda\Delta\lambda. \tag{12.26}$$

It's not surprising that $\Delta\theta$ is proportional to $\Delta\lambda$. In the case of a silica fiber carrying a 1.55 μm wave, a numerical application of (12.26) indicates that, for the pulses that are commonly used in optical telecommunications, $\Delta\theta$ is measured in picoseconds per kilometer.

In the case of silica, the experiment shows that the second derivative of the index of refraction, $d^2n_1/d\lambda^2$, cancels for $\lambda = 1.3\,\mu$m. For a carrier wave of this wavelength, the effect of the material dispersion vanishes; the duration increase then comes only from the dispersion of the mode and is very small.

Annex 12.A

Splitters and Couplers

In guided optics the electromagnetic energy is confined inside guides that have a very small cross section; a rather difficult problem arises if it is desired to transfer this energy to one or several other guides. It will be shown in Annex 12.C how fibers can be efficiently coupled to one another; the purpose of this annex is first to study how one guide, fiber, or integrated guide, can be coupled to one or two guides.

12.A.1. The Y-Junction

A signal can be introduced into a Y-junction, either by the "one-arm" side as in Figure 12.A.1(a), or by one of the two channels of the "two-arms" side, as in Figure 12.A.1(c). An accurate analysis is necessary, involving all the modes of propagation, including the leaky modes. In the case of single-mode guides and a symmetric Y-junction, the results are simple and can be described as follows:

- In the case of Figure 12.A.1(a), and if we omit the losses, the emerging signals are equal to one-half of the incident signal.
- In the case of Figure 12.A.1(c), a unit signal is introduced in the upper arm, no signal being introduced in the lower arm. The transmitted signal is only one-half of the incident signal; the other part goes to the cladding. In that case, a Y-junction has at least 3 db losses.
- In the case of Figure 12.A.1(b), a second Y-junction, very similar to the first one, has been placed as shown in the figure. When they arrive at the node of the second Y, the two signals are in phase and interfere constructively in the emerging arm: the transmitted signal is equal to one mode and no light is transferred to the leaky modes.

In conclusion, Y-junctions, since they are easy to make, are efficient devices for splitting a signal into 2, 4, 8, ... parts, but they cannot be used to mix independent signals.

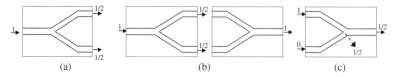

Figure 12.A.1. Division of a signal by a Y-junction.

12.A.2. Distributed Couplers

Two single-mode identical guides are disposed parallel and in close vicinity; the guided modes are coupled by their evanescent parts. In Figure 12.A.2(a), the upper guide is excited from the outside; the percentage of energy in each output port can be tuned by changing the length of the guides. The coupling is efficient when the two guides are close enough.

This kind of coupling is said to be *directional*; the coupling is made in the direction of propagation of the light: the light that is injected into arm (1), goes to arm (3) and/or arm (4), but doesn't go to arm (2). In fact, because of unavoidable imperfections, a small part will go to arm (2); this part is however very small and less than 60 db.

From a technological point of view this technique is easy to use in integrated optics; it can also be made available for optical fibers. Two or more fibers are twisted together and then heated with a flame or with a laser, so that the temperature almost reaches the melting point of silica; simultaneously, the fibers are gently pulled apart. As a result the cores of the different fibers get closer and a coupling effect is obtained.

Phase Matching Condition

A directional coupler works in the same way as a coupling prism. The second guide is in a situation that is often met in electromagnetism: a medium M is excited at points that are distributed over a distance that is made of a large number of wavelengths. Starting from those points, wavelets propagate at some velocity, which is characteristic of the medium M. The initial phases of

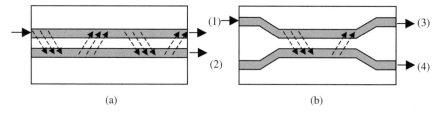

Figure 12.A.2. Distributed couplers. Two monomode guides are put in close vicinity; the two guides periodically exchange optical energy.

the wavelets are governed by some external mechanism, the propagation in the first guide in our case.

The field, at a given point of M, is the result of the interference of many wavelets. As is always the case for highly multimode interference, the field will be different from zero, if and only if the condition of constructive interference is satisfied. It is then easy to understand that the coupling will occur if the guided waves propagate at the same speed inside the two guides: the wave vectors of two modes must be equal, which is obviously the case if the guides are identical.

A Distributed Coupler Is Reciprocal

The coupling length, which is defined in Figure 12.A.3, rapidly increases with the distance between the guides. The energy exchange is total when the two wave vectors are equal; if not, the exchange rate is less than 100%, and goes to zero when the wave vectors are very different. The coupling is reciprocal, which means that the exchange rate from one port to another, for example from (1) to (3), is equal to the exchange rate in the opposite way, from (3) to (1). A directional coupler is characterized by a coupling length L_c, which is the propagation distance necessary for the signal to be completely transferred from one guide to the other; L_c is a function of the distance between the two guides and of the wavelength. A coupler can be designed so that the coupling length at λ_1 is equal to one-half of the coupling length at λ_2; such an arrangement is shown in Figure 12.A.4 and is currently used for Wavelength Division Multiplexing (WDM).

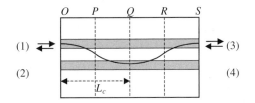

Figure 12.A.3. Illustration of the energy exchanges between the two guides. At the levels of the two planes P and R, the intensities are the same in the two guides. At the level of Q, all the energy is transferred to the second guide. $L_c = OG$ is called the characteristic length, or the coupling length.

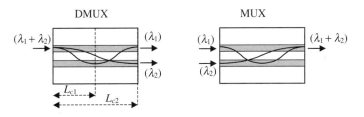

Figure 12.A.4. Wavelength Division Multiplexing (WDM).

12.A.3. Coupled Mode Theory

12.A.3.1. *Normalization of the Guided Modes*

In the description of the electromagnetic field of a guided mode, see, for example, formulas (12.13.a, b, c), there is a proportionality coefficient A_2, in many cases it doesn't matter knowing exactly this coefficient that is determined by the power carried by the mode. For an accurate definition of the coupling coefficient between the two guides of a directional coupler, it is necessary to have a definition of the proportionality coefficients of the modes. The electric and magnetic fields will be written as

$$\boldsymbol{E}_n(x, y, z, t) = A\boldsymbol{E}_n(x, y)e^{-jk_z z}e^{j\omega t},$$
$$\boldsymbol{H}_n(x, y, z, t) = A\boldsymbol{H}_n(x, y)e^{-jk_z z}e^{j\omega t}. \tag{12.A.1.a}$$

The two vectors $\boldsymbol{E}_n(x, y)$ and $\boldsymbol{H}_n(x, y)$ constitute the representation of the mode labeled n; their analytical expressions are chosen so that the boundary conditions are satisfied. The electromagnetic power W_n, which crosses a surface orthogonal to the direction of propagation Oz, is obtained from the Poynting vector $\boldsymbol{\varPi}_n$:

$$\boldsymbol{\varPi}_n = \boldsymbol{E}_n \wedge \boldsymbol{H}_n^* \quad \rightarrow \quad W_n = \frac{1}{2}\int_{-\infty}^{\infty}\int_{-\infty}^{\infty}(\boldsymbol{\varPi}_n \boldsymbol{z})\,dx\,dy.$$

A mode is said to be *normalized* when W_n is equal to one. If the normalized fields are noted as $\boldsymbol{e}_n(x, y)$ and $\boldsymbol{h}_n(x, y)$, when the power is equal to W_n, the fields are equal to

$$\boldsymbol{E}_n(x, y, z, t) = \sqrt{W_n}\,\boldsymbol{e}_n(x, y)e^{-jk_z z}e^{j\omega t},$$
$$\boldsymbol{H}_n(x, y, z, t) = \sqrt{W_n}\,\boldsymbol{h}_n(x, y)e^{-jk_z z}e^{j\omega t}, \tag{12.A.1.b}$$

where $\boldsymbol{e}_n$ and $\boldsymbol{h}_n$ are of course orthogonal and, with no loss of generality, the electric fields and the magnetic fields can be, respectively, oriented parallel to Oy and Ox. Using Maxwell's equations, it can be shown that

$$h_{nx} = -\frac{k_z}{\omega\mu_0}e_{ny} \quad \rightarrow \quad \boldsymbol{e}_n \wedge \boldsymbol{h}_n = e_{ny}h_{nx} = \frac{\omega\mu_0}{k_z}e_n^2 = \frac{k_z}{\omega\mu_0},$$

$$\int_{-\infty}^{\infty}\int_{-\infty}^{\infty}(e_n^2)\,dx\,dy = \frac{2\omega\mu_0}{k_z} \quad \text{and} \quad \int_{-\infty}^{\infty}\int_{-\infty}^{\infty}(h_n^2)\,dx\,dy = \frac{k_z}{2\omega\mu_0}. \tag{12.A.1.c}$$

Lorentz Reciprocity Theorem

Starting from the modal structure of a given device and using the Lorentz reciprocity theorem, a perturbation method gives indications about another device that has almost similar, although not strictly identical, optogeometric characteristics. The demonstration is not pleasant, however the result is simple.

We consider two lossless and nonmagnetic ($\mu = \mu_0$) systems that are described by two different repartitions, $\varepsilon(x, y, z)$ and $\varepsilon'(x, y, z)$, of the permittivity. [$E(x, y, z)$, $H(x, y, z)$], on the one hand, and [$E'(x, y, z)$, $H'(x, y, z)$], on the other, being two fields that satisfy Maxwell's equations and the respective boundary conditions in each system, we have the following equations:

$$\nabla \wedge E = -j\omega\mu_0 H, \quad \nabla \wedge E' = -j\omega\mu_0 H',$$
$$\nabla \wedge H = j\omega\varepsilon E, \quad \nabla \wedge H' = j\omega\varepsilon E',$$

$$(E^* \nabla \wedge H' - H'\nabla \wedge E^*) - (H^* \nabla \wedge E' - E'\nabla \wedge H^*) = +j\omega(\varepsilon' - \varepsilon)E^*E'.$$

Making use of the vector identity $\nabla(A \wedge B) = B\nabla A - A\nabla B$, we obtain

$$\nabla(E^* \wedge H' + E' \wedge H^*) = -j\omega(\varepsilon' - \varepsilon)E^*E'.$$

If we now consider a volume V enclosed inside a surface S and make use of the following theorem:

$$\iiint_V \nabla A \, dv = \iint_S A \, dS,$$

we obtain

$$\iint_S (E^* \wedge H' + E' \wedge H^*) \, dS = -j\omega\iiint_V (\varepsilon' - \varepsilon)E^*E' \, dv. \qquad (12.A.2)$$

Equation (12.A.2), which is called the *Lorentz reciprocity theorem*, is valid for any system; we are now going to use it for a system that is invariant in a translation parallel to the Oz axis. The volume V is shown in Figure 12.A.5, it's a cylindrical box with thickness Δz and a large radius R. For small values of Δz, we can reduce the volume integral to a surface integral,

$$\iiint_V (\varepsilon' - \varepsilon)E^* E' \, dv \cong \Delta z\iint_S (\varepsilon' - \varepsilon)E^*E' \, dx \, dy.$$

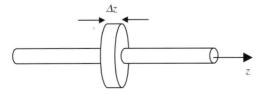

Figure 12.A.5. Application of the Lorentz reciprocity theorem to a cylindrical box.

The flux of some vector A across a surface S is made of several contributions:

- The flux across the lateral surface; this flux is considered as equal to zero if the field goes to zero for large values of the radius, which is the case for a guided mode.
- The two fluxes across the side faces, which have opposite signs:

$$\iint_S (A_z(x, y, z + \Delta z) - A_z(x, y, z)) \, dx \, dy = \Delta z \iint_S \frac{\partial}{\partial z}(A_z(x, y, z)) \, dx \, dy.$$

Coming back to (12.A.2) we can write

$$\iint_S \frac{\partial}{\partial z}[E^* \wedge H' + E' \wedge H^*]_z \, dx \, dy = -j\omega \iint_S (\varepsilon' - \varepsilon)E^* \, E' \, dx \, dy.$$

Finally, the z component of the vector products are only functions of the transverse components of the vectors

$$\iint_S \frac{\partial}{\partial z}[E_t^* \wedge H_t' + E_t' \wedge H_t^*]_z \, dS = -j\omega \iint_S (\varepsilon' - \varepsilon)E^* \, E' \, dS. \quad (12.A.3.a)$$

The Modes of a Dielectric Wave Guide Are Orthogonal

We are now going to use formula (12.A.3) to show that the modes of the dielectric guide are orthogonal. We consider the electromagnetic fields of two different modes, respectively labeled n and n'. As the guiding structure is the same for the two modes, the difference $(\varepsilon' - \varepsilon)$ is identically equal to zero. The z derivative is equivalent to a multiplication by $-jk_{nz}$ for one mode and by $-jk_{n'z}$ for the other:

$$(k_{nz} - k_{n'z}) \iint_S \frac{\partial}{\partial z}[E_{nt}^* \wedge H_{n't}' + E_{n't}' \wedge H_{nt}^*]_z \, dS = 0,$$

$$(12.A.3.b)$$

$$n \neq n' \quad \rightarrow \quad (k_{nz} - k_{n'z}) \neq 0 \quad \rightarrow \quad \iint_S \frac{\partial}{\partial z}[E_{nt}^* \wedge H_{n't}' + E_{n't}' \wedge H_{nt}^*]_z \, dS = 0.$$

Equation (12.A.3.b) is called the mode orthogonality relation. Using the Kronecker notation ($\delta_{nn} = 1$ and $\delta_{nn'} = 0$ if $n \neq n'$) we can write

$$\int_{-\infty}^{\infty} \int_{-\infty}^{\infty} (e_{ny} e_{n'y}) \, dx \, dy = \frac{2\omega\mu_0}{k_z} \delta_{nn'},$$

$$\int_{-\infty}^{\infty} \int_{-\infty}^{\infty} (h_{nx} h_{n'x}) \, dx \, dy = \frac{k_z}{2\omega\mu_0} \delta_{nn'}. \quad (12.A.3.c)$$

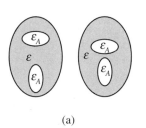

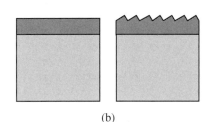

(a)

(b)

Two cylindrical guides are embedded in a cladding; on the left side the separation is smaller.

The first system is made of a planar guide deposited on a substrate. A grating has been etched on top of the guide.

Figure 12.A.6. Examples of systems with almost similar optogeometric characteristics.

Coupled Mode Equations

In this section the electromagnetic properties of a given system (Σ') will be deduced from the properties of another system (Σ), which has almost the same optogeometric characteristics.

Let us write the normalized expressions of the electric and magnetic fields of the nth mode of the second system:

$$\boldsymbol{E}_n(x, y, z) = \boldsymbol{e}_n(x, y)e^{-jk_z},$$
$$\boldsymbol{H}_n(x, y, z) = \boldsymbol{h}_n(x, y)e^{-jk_z}.$$

We admit that the electromagnetic fields, $\boldsymbol{E}'(x, y, z)$ and $\boldsymbol{H}'(x, y, z)$, of the system (Σ') can be expressed as the superposition of normal modes of (Σ):

$$\boldsymbol{E}'_n(x, y, z) = \sum_m a_m(z)\boldsymbol{e}_m(x, y)e^{-jk_z},$$

$$\boldsymbol{H}_n(x, y, z) = \sum_m a_m(z)\boldsymbol{h}_n(x, y)e^{-jk_z}.$$

(12.A.4)

The summation should be made over all the modes of (Σ), including the radiated modes, which will be ignored in most cases.

Coupled Wave Guides

We consider the case of Figure 12.A.6(b); the arrangement (Σ) is defined by a permittivity repartition $\varepsilon(x, y, z)$, while (Σ') is defined by $\varepsilon'(x, y, z)$. Making use of the Lorentz reciprocity relation and of the modes orthogonality, we obtain, after some vector algebra manipulations, a set of differential equations

$$\frac{da_n}{dz} = -j\sum_m a_m e^{\Delta_{nm}z} C_{mn}. \tag{12.A.5}$$

$[\Delta_{nm} - (k_{zn} - k_{zm})]$ is called the phase mismatch between the modes m and n,

$$C_{nm}(z) = \text{sgn}(n)\frac{\omega}{2} \iint\limits_{S(z)} (\varepsilon' - \varepsilon)e_m e_n \, dS,$$

where $S(z)$ is the cross section of the guides at abscissa z.

The coefficients a_{mn} thus satisfy a set of coupled linear differential equations.

Forward Coupling of Two Monomode Guides

We consider the case of the two guides of Figure 12.A.6(a) and, furthermore, we suppose that they are monomode. The set of coupled equations then reduces to only two equations. An accurate description of the repartitions of the dielectric permittivity is needed and is given in the table below.

Initial system	System under study	Perturbation
Guide A is alone, guide B being rejected at infinity: ε_A inside guide A, ε outside.	Guides A and B are in close vicinity: ε_A inside guide A, ε outside, ε_B inside guide B.	$(\varepsilon' - \varepsilon) = (\varepsilon_B - \varepsilon)$ at the place of guide B. $(\varepsilon' - \varepsilon) = 0$ anywhere else.

Let $\{e_A(x, y); h_A(x, y)\}$ and $\{e_B(x, y); h_B(x, y)\}$, respectively, be the normalized modes of the two guides when they are considered independently; the electromagnetic field, $\{E'(x, y); H'(x, y)\}$, of the system made by the two coupled guides may be represented as a combination of the normalized modes,

$$\begin{Bmatrix} E'(x, y, z) \\ H'(x, y, z) \end{Bmatrix} = a_A(z)\begin{Bmatrix} e_A(x, y) \\ h_A(x, y) \end{Bmatrix}e^{-jk_{z,A}z} + a_B(z)\begin{Bmatrix} e_B(x, y) \\ h_B(x, y) \end{Bmatrix}e^{-jk_{z,B}z},$$

where $k_{z,A}$ and $k_{z,B}$ are the propagation constants of the normalized modes. The coefficients a_A and a_B follow a set of two differential equations,

$$\frac{da_A}{dz} = -j\big[C_{AA}a_A(z) + C_{BA}a_B(z)e^{j(k_{z,A}-k_{z,B})z}\big], \qquad (12.A.6.a)$$

$$\frac{da_B}{dz} = -j\big[C_{AB}a_A(z)e^{-j(k_{zB}-k_{zA})z} + C_{AB}a_B(z)\big], \qquad (12.A.6.b)$$

$$C_{AA} = \frac{\omega}{2}\iint_{S_B}(\varepsilon_B - \varepsilon)\mathbf{e}_A \mathbf{e}_A^* \, dS,$$

S_B = cross section of the guide B,

$$C_{BA} = \frac{\omega}{2}\iint_{S_B}(\varepsilon_B - \varepsilon)\mathbf{e}_B \mathbf{e}_A^* \, dS,$$

$$C_{BB} = \frac{\omega}{2}\iint_{S_A}(\varepsilon_A - \varepsilon)\mathbf{e}_B \mathbf{e}_B^* \, dS,$$

S_A = cross section of the guide A.

$$C_{AB} = \frac{\omega}{2}\iint_{S_A}(\varepsilon_B - \varepsilon)\mathbf{e}_A \mathbf{e}_B^* \, dS,$$

Inside the cross section of the guide B, $|\mathbf{e}_A|$ is much smaller than $|\mathbf{e}_B|$, so that C_{AA} is much smaller than C_{BA}; and, in the same way, $C_{BB} \ll C_{AB}$. In the end (12.A.6.b) can be simplified as

$$\frac{da_A}{dz} = -jC_{BA}a_B(z)e^{j\Delta kz},$$
$$\Delta k = (k_{z,A} - k_{z,B}). \qquad (12.A.6.c)$$
$$\frac{da_B}{dz} = -jC_{AB}a_B(z)e^{-j\Delta kz},$$

The set of differential equations (12.A.6.c) is made a set of algebraic equations by assuming that $a_A(z) = Ae^{-j\gamma_A z}$ and $a_B(z) = Be^{-j\gamma_B z}$; A, B, γ_A, and γ_B are unknown and don't depend on z,

$$\gamma_A A - C_{BA}Be^{j(\Delta k+\gamma_A-\gamma_B)z} = 0,$$
$$C_{AB}Ae^{-j(\Delta k+\gamma_A-\gamma_B)z} - \gamma_B B = 0. \qquad (12.A.7.a)$$

Since A and B don't depend on z, it is required that $\Delta k = (\gamma_B - \gamma_A)$,

$$\gamma_A A - C_{BA}B = 0,$$
$$C_{BA}A - \gamma_B B = 0. \qquad (12.A.7.b)$$

The only solution of (12.A.7.b) is the trivial one $A = B = 0$; except if the determinant of the system is equal to zero,

$$\left\| \begin{matrix} \gamma_A & -C_{BA} \\ C_{AB} & -\gamma_B \end{matrix} \right\| = C_{AB}C_{BA} - \gamma_A\gamma_B = 0,$$

$$C_{AB}C_{BA} = \gamma_A(\gamma_A + \Delta k) = \gamma_B(\gamma_B - \Delta k),$$

$$\gamma_A = -\frac{\Delta k}{2} \pm \Omega,$$
$$\text{with} \quad \Omega = \sqrt{\frac{\Delta k^2}{4} + C_{AB}C_{BA}}.$$
$$\gamma_B = +\frac{\Delta k}{2} \pm \Omega,$$

Two values are obtained for γ_A, and two values for γ_B. $a_A(z)$ and $a_B(z)$ are linear combinations of the complex exponential functions $e^{\pm j\gamma_A z}$ and $e^{\pm j\gamma_B z}$. By a proper choice of the coefficients, $a_A(z)$ can always be written as

$$a(z) = e^{j\frac{\Delta k}{2}z}(A_1 \cos \Omega z + A_2 \sin \Omega z), \qquad (12.A.8.a)$$

where $a_B(z)$ is obtained from $a_B(z) = j(1/C_{BA})e^{-j\Delta k z}(da_A(z)/dz)$; the two constants A_1 and A_2 are fixed by the input conditions. Let us suppose, for example, that at $z = 0$, all the energy is in the guide B, $a_A(0) = 0 \rightarrow A_2 = 0$:

$$a_A(z) = A_1 e^{j\frac{\Delta k z}{2}} \sin \Omega z,$$

$$a_B(z) = \frac{A_1}{C_{BA}} e^{-j\frac{\Delta k z}{2}}\left[-\frac{\Delta k}{2}\sin \Omega z + j\Omega \cos \Omega z\right].$$

The power that is carried by the modes is proportional to the square of the modulus $|a_A(z)|^2$ and $|a_A(z)|^2$, if P_0 is the power in the guide (B) at $z = 0$, we have

$$P_A(z) = P_0 \frac{|C_{AB}|^2}{\Omega^2}\sin^2 \Omega z,$$

$$P_B(z) = P_0\left[\frac{\Delta k^2}{\Omega^2}\sin^2 \Omega z + \cos^2 \Omega z\right].$$

In the case when the two guides are identical:
- The phase matching condition is satisfied, $\Delta k = 0$.
- The two coupling coefficients are equal, $C_{AB} = C_{BA} = \Omega$,

$$P_A(z) = P_0 \sin^2 \Omega z,$$
$$P_B(z) = P_0 \cos^2 \Omega z.$$

The coupling length, such as defined in Figure 12.A.3, is equal to $L_c = \pi/2\Omega$.

Annex 12.B

Attenuation of Silica Fibers

The attenuation coefficient of amorphous quartz Is $0.16 \, \text{db/km}$ *at* $\lambda = 1.55 \, \mu\text{m}$

The fascinating properties of silica fibers are a direct consequence of the fact that silicon and oxygen atoms easily combine to make a very stable and pure compound, SiO_2, which is called quartz and is an abundant constituent of the Earth's surface. Quartz may exist in the form of crystals, or as a vitreous and amorphous material also called fused quartz. Silica is one of the most transparent materials, at least for the case of very long samples. The reason for this comes from the fact that oxygen and silicon atoms have a great chemical affinity; they make very resistant covalent bonds, that don't allow the presence of free electrons inside the material. Furthermore, in the case of amorphous silica there are no dislocations and no grains on which the light could diffuse. In spite of all these good reasons, the remarkable transparency of SiO_2 was not really noticed before 1970, when it was suggested using silica fibers for the transportation of light waves over very long distances.

The losses have two main origins:

- Rayleigh diffusion by the electrons of the different atoms. The corresponding attenuation varies as $1/\lambda^4$, and decreases as the wavelength increases.
- Mechanical vibrations of the molecules (mostly SiO_2) and ions (mostly OH).
  - SiO_2 has two main absorption bands:
    - One is due to the vibration of the electrons with respect to the nuclei; it falls in the ultraviolet part of the spectrum, and doesn't play an important role in our case.
    - The other one is associated to the relative motion of the silicon and oxygen atoms, with a resonance frequency that corresponds to a wavelength of $\lambda = 2.4 \, \mu\text{m}$. Its effect is already felt in the near infrared ($\lambda \approx 1$–$2 \, \mu\text{m}$).

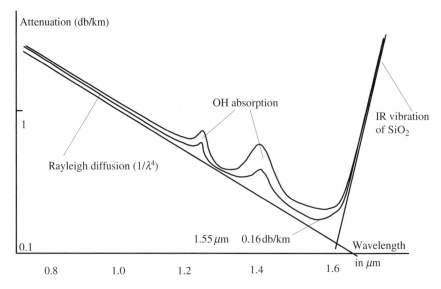

Figure 12.B.1. Dependence of the attenuation coefficient of silica on the wavelength. The lower curve corresponds to a purer silica (less OH ions). There is a local minimum at $1.3\,\mu$m and an absolute minimum at $1.55\,\mu$m.

○ OH: The presence of OH ions in silica leads to an absorption peak centered at $2.73\,\mu$m, which presents overtones and combinations with silica at $0.95\,\mu$m, $1.24\,\mu$m, and $1.39\,\mu$m. For a concentration of 1 ppm, the corresponding attenuations are about 1 db/km@$0.95\,\mu$m, 3 db/km@ $1.24\,\mu$m, and 40 db/km@$1.39\,\mu$m. State of the art fibers keep the OH concentration at levels below 0.1 ppm.

○ Si—O—H has several mechanical resonances between $l \approx 1.2\,\mu$m and $\lambda \approx 1.4\,\mu$m.

The Rayleigh diffusion and excitation of the $2.4\,\mu$m oscillation of SiO_2 have opposite effects, and the experiment shows that the attenuation curve has a minimum at $\lambda \approx 1.55\,\mu$m. The attenuation at minimum is as low as 0.16 db/km.

The presence of OH ions comes from the water vapor that is dissolved in the silica; recent progress in the technology of fabricating silica fibers has considerably reduced the concentration of OH ions which, in some cases, can be completely eliminated.

Annex 12.C

Elaboration of Optical Guides

It is out of the question to give an exhaustive description of the various methods for making optical waveguides; we will only describe the principle of the original methods that have then been the occasion of many tricks.

12.C.1. Optical Fibers

Optical fibers are very long pieces of material (hundreds of kilometers), with a very small diameter (hundreds of micrometers). The index profile is accurately controlled, especially in the core region (tens of micrometers). They are pulled out of a preform, a rod that is much shorter and thicker, and that has an index profile, which is about homothetic of the profile of the fiber. If D and d are the respective diameters of the preform and of the fiber, the conservation of the total mass and of the density implies that a preform with a length L will give a length l of fiber equal to $l = L(D/d)^2$. If $D = 5\,cm$ and $d = 125\,\mu m$, 160 km of fiber are obtained from 1 m of preform.

12.C.1.1. *Elaboration of the Preform*

Optical fibers may be made of silica or plastic; we will only describe the fabrication of silica fibers. Fibers are, of course, made of vitreous quartz. Silica can be mixed in any proportion with other oxides, such as germanium oxide, GeO_2, or phosphor oxide, P_2O_5, or boron oxide, B_2O_3; the index of refraction depends on the composition of the mixture.

The chemistry of the elaboration of the preform is largely inspired by the technology of silicon-based microelectronic devices. The key point is that many chloride compounds ($SiCl_4$, $GeCl_4$, $POCl_3$, B_2Cl_3) are gases; a mixture of gases is very homogeneous and has a chemical composition that can easily be controlled, using valves and taps. At high temperatures, these chlorides can be oxidized and give solid compounds that have the shape of soot and deposit either inside a pipe or around a bait rod. The oxides have two main

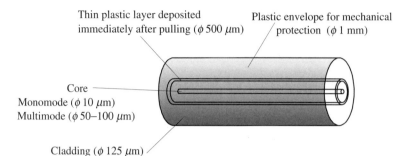

Thin plastic layer deposited
immediately after pulling (ϕ 500 μm)

Plastic envelope for mechanical
protection (ϕ 1 mm)

Core
Monomode (ϕ 10 μm)
Multimode (ϕ 50–100 μm)

Cladding (ϕ 125 μm)

Figure 12.C.1. Typical size of the main parts of an optical fiber.

effects; they modify the refractive index and they lower the melting point relative to pure silica. The main chemical reactions that are involved are summarized below:

$$SiCl_4 + O_2 \quad \rightarrow \quad SiO_2 \; + Cl_2 \qquad GeCl_4 + O_2 \quad \rightarrow \quad GeO_2 \; + Cl_2 \; .$$

gas   gas   solid soot gas   gas   gas   solid soot   gas

An excess of Cl_2, which is added at the various stages of the process, purges the OH ions to give HCl and O_2, which are rejected in the exhaust fumes. The presence of OH ions induces losses in the wavelength region of interest for telecommunications (1.3–1.6 μm).

In the case of the Inside Vapor Deposition (IVD) method, the chemical reactions are made inside a hollow tube, the soot deposit on the inner walls. At the end of the first operation, which is described in Figure 12.C.2(a), a suc-

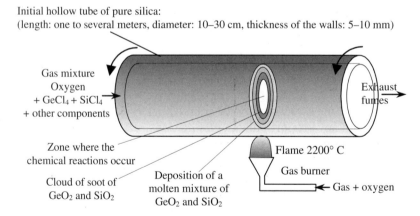

Initial hollow tube of pure silica:
(length: one to several meters, diameter: 10–30 cm, thickness of the walls: 5–10 mm)

Gas mixture
Oxygen
+ GeCl$_4$ + SiCl$_4$
+ other components

Exhaust
fumes

Zone where the
chemical reactions occur

Cloud of soot of
GeO$_2$ and SiO$_2$

Deposition of a
molten mixture of
GeO$_2$ and SiO$_2$

Flame 2200° C

Gas burner

Gas + oxygen

Figure 12.C.2(a). Inside Vapor Deposition (IVD) method. The gas burner periodically goes back and forth along the pipe, while the chemical composition is modified to obtain the desired chemical composition of the layers that are successively deposited.

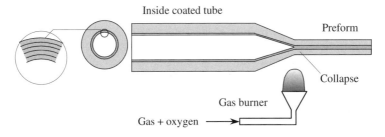

Figure 12.C.2(b). Elaboration of a preform in the case of the IVD method. After being coated with many layers of different compositions, the tube is heated and elongated: the internal hole collapses and the tube becomes a full rod.

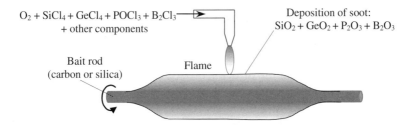

Figure 12.C.2(c). Outside Vapor Deposition (OVD) method. The successive layers are deposited on a bait rod that is finally removed; leaving along the center a small hole that disappears during the drawing of the fiber.

cession of concentric layers are deposited inside the tube; the next operation, which is described in Figure 12.C.2(b), is called *collapse*: the tube is heated very near to the melting point and pulled along a direction parallel to its axis. The surface energy, corresponding to capillary forces, is smaller when the central hole is filled: the hollow pipe spontaneously becomes a cylindrical rod, the core of which is made of the concentric layers and has a gradient of chemical composition. This rod is called a *preform*. The final fiber will be drawn out of the preform as shown in Figure 12.C.3.

In the case of the Outside Vapor Deposition (OVD) method, which is described in Figure 12.C.2(c), the preform is obtained directly by deposition of successive layers of soot directly on a bait rod made of graphite or of fused silica. This rod is removed from the preform after completion of the soot deposition. A porous preform is obtained, which is then transformed into a bubble-free clear preform after heating in a zone furnace to about $1500°$ C. At the end of the previous operation, which is called *preform sintering*, a small hole appears in the center, it will disappear during the drawing of the fiber. OVD preforms can have very large diameters and give very long optical fibers.

The last operation consists of drawing the fibers; this is achieved inside a specially designed tower (see Figure 12.C.3). The diameter of the cladding is

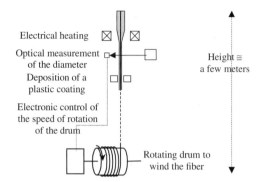

Figure 12.C.3. Fiber drawing. The preform is raised to a temperature close to the softening point of silica. Immediately after drawing, the fiber is protected against intrusion of OH ions. Thanks to an optical measurement and a feedback control of the drawing speed, the diameter is accurately controlled.

optically measured during the drawing process, and a feedback control of the speed of rotation of the drum allows an excellent control of the diameter (better than $\pm 1\,\mu m$); the core occupies the center of the fiber with the same accuracy. Immediately after pulling, the fiber is coated with a thin layer of plastic, which makes an efficient protection against humidity (avoiding reintrusion of OH ions).

12.C.1.2. *Cutting and Connecting Optical Fibers*

In spite of their very small diameters, optical fibers are not difficult to cut and connect. The connection losses are usually lower than 1 db. Cutting a fiber is a routine operation, which consists of creating a local stress on the fiber surface and pulling the fiber with a proper tension. The principle is illustrated in Figure 12.C.4, a blade (diamond or tungsten carbide) scores the fiber and then tension is applied. Because of the cylindrical symmetry and the small

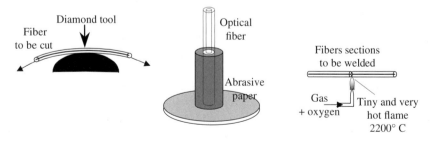

Figure 12.C.4. Principle of the main operations for cutting and connecting fibers. The gas burner is often replaced by an electric arc.

size of the diameter, the cross section is usually a mirror-like surface. The cleaved surface is easily and rapidly polished: after insertion inside a capillary tube, the fiber is gently wiped with a fine-graded abrasive paper ($0.1\,\mu$m).

The area inside which the electromagnetic energy is confined is very small, a few tenths of square micrometers in the case of monomode fibers. The connection of two fibers is thus a delicate operation. Not so delicate, in fact, because of the excellent calibration of the external diameter of the cladding and also because of the fact that the core is placed accurately at the center of the cross section. When two sections of fibers are to be connected, they are inserted inside hollow cylinders having an external diameter equal to the diameter of the cladding. Their tips can then be prepositioned and brought into close vicinity (micrometer). One fiber is excited and thanks to micromanipulators the energy transferred to the other fiber is optimized. At this stage there are two possibilities:

- A drop of glue is poured with a syringe and UV polymerized. The value of the index of refraction of the glue is important for eliminating Fresnel reflections on the interfaces.
- The two sections are pushed into contact and heated up to 2200° C with a miniblowtorch, which can be replaced by an electric arc or a CO_2 laser. No extra material must be added. Since they are more doped, the cores become liquid before the claddings, thanks to capillarity an accurate self-alignment is obtained.

Coupling Efficiency

The electromagnetic field of a guided wave cannot have any shape but should be a combination of specific functions, which are the modes of the guiding structure and are usually described by the normalized expressions $e_n(x,\ y)$ and $h_n(x,\ y)$ (see formulas (12.A.1)). Let us consider the excitation of a guide by an external source that illuminates the input face; only a fraction of the energy coming from the source is coupled to the guided modes, the remaining part is reflected (Fresnel reflection) or transferred to the cladding. The Fresnel reflection comes from the index difference between the core of the fiber and the material in which the source is embedded, antireflection coatings may be useful.

Let $Ef(x,\ y)$ be a vector describing the electric field that is created by the source on the surface of the input face, the nth mode will be all the more efficiently excited as $f(x,\ y)$ is similar to $e_n(x,\ y)$. If the function $f(x,\ y)$ is normalized, it can be shown that the coupling efficiency η is given by the following integral, which is called an *overlap integral*:

$$\eta = \int\limits_{-\infty}^{\infty} \int\limits_{-\infty}^{\infty} e_n(x, y) f(x, y)\, dx\, dy.$$

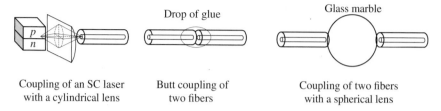

Coupling of an SC laser   Butt coupling of   Coupling of two fibers
with a cylindrical lens   two fibers   with a spherical lens

Figure 12.C.5. Some ways of exciting fibers. The cylindrical lens compensates for the important astigmatism of the beam emitted by a semiconductor laser. Glass marbles are cheap and efficient lenses.

When the two functions $e_n(x, y)$ and $f(x, y)$ are identical, the overlap integral is equal to one; which is the case when two identical fibers are put in contact (butt-coupling). If the two functions $e_n(x, y)$ and $f(x, y)$ are quite different, it is necessary to use a lens, see Figure 12.C.5. If the contact is not perfect, the two main parameters that control the efficiency are: the distance d between the two endfaces and the angle of tilt θ between the normal to the faces.

Of course $\eta \to 0$, if $d \to \infty$ or if $\theta \to \pi/2$. The coupling efficiency is significantly different from zero as long as the distance remains small as compared to the diameter of the core and the tilt is much smaller than the numerical aperture of the fiber.

12.C.2. Integrated Optics

The orders of magnitude concerning the guides that are used in integrated optics are quite different from those of optical fibers:

- Integrated optical wave guides are always made on a substrate.
- They are much shorter. The length is usually of a few millimeters, one-tenth of a centimeter being possible but exceptional.
- The losses remain low, but are quite high, 0.1 to 1 db/m, instead of db/km in the case of fibers.
- The index difference between the guide and the substrate or superstrate may be as small as the index difference between the core and the cladding of an optical fiber, ≈ 0.001 to 0.05. It can also be higher, ≈ 0.1 to 0.5.

12.C.2.1. *Classification of the Different Technologies of Integrated Optics*

The substrates are always planar and made of a very transparent material. The most common material substrates are: ordinary glass, amorphous silica, lithium niobate, and some semiconductors, such as gallium arsenide and indium phosphide and their derived ternary compounds.

Guides of practical interest are monomode, or weakly multimode, which corresponds to a thickness of $0.1\,\mu$m to $10\,\mu$m according to the index difference Δn between the substrate and the guide. An order of magnitude of the number of modes that can be supported by a given guide is obtained by dividing the optical thickness $e\Delta n$ by the wavelength.

The first problem is to make, on top of a given substrate, a layer with a higher index of refraction. The different technologies to do that can be divided into two main categories:

- Deposition of a layer of a well-chosen material.
- Modification of the chemical composition of the surface by the introduction of impurities across the surface.

12.C.2.2. *Deposition of Thin Layers*

Any method for depositing a thin layer of material on a substrate can, in principle, be used, especially vacuum evaporation and epitaxy. In many cases, a planar layer is first deposited, the stripe guides being etched, using conventional methods of photolithography.

A good example is given in Figure 12.C.6. Silicon cannot be directly used as a substrate, since it has a high absorption coefficient in the infrared and a very large index of refraction. Nevertheless, because of the possibility of having cheap and large size single crystals of high quality, silicon is very important for integrated optics. The chips are of the same kind as those that are used in microelectronics. At first, a thick buffer layer of oxide, SiO_2, is thermally grown on top of the surface before making optical guides; this layer of pure silica constitutes the substrate on which the guiding layer is deposited. There are two possibilities for making the guide:

- *Deposition of a layer of* Si_3N_4: This technology is derived from microelectronics where the silicon nitride is used to passivate the integrated circuits. Because of the high value of the index ($n \approx 2$), the layer must be very thin, if a monomode guide is desired ($\approx 0.5\,\mu$m).
- *Deposition of a layer of doped silica*: This method is a transposition of the technology of elaboration of silica fibers. The thickness of a monomode guide is typically equal to a few micrometers.

12.C.2.3. *Diffusion of Impurities Inside a Substrate*

12.C.2.3.1. *Wave Guides Made by Ion Exchange in Glass*

Glass is certainly one of the most popular materials in Optics and it should not be a surprise that one of the most convenient ways of making integrated optical devices takes advantage of the properties of this material. Common

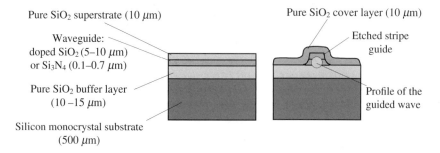

Pure SiO$_2$ superstrate (10 μm)

Waveguide:
doped SiO$_2$ (5–10 μm)
or Si$_3$N$_4$ (0.1–0.7 μm)

Pure SiO$_2$ buffer layer
(10 –15 μm)

Silicon monocrystal substrate
(500 μm)

Pure SiO$_2$ cover layer (10 μm)

Etched stripe
guide

Profile of the
guided wave

Figure 12.C.6. Elaboration of a planar or a stripe guide on top of a silicon chip. A first thick layer is first grown by oxidation of the silicon substrate. A guiding layer and a superstrate are then deposited. Stripe guides are etched by photolithography. A protection cover layer is often deposited.

glass is mainly a mixture of sand, which is made of silica, SiO$_2$, and of various oxides. Among those oxides the most important is sodium oxide, Na$_2$O. Historically, glass was invented by the Egyptians who had noticed that, although they could not reach the melting point of sand, a mixture of sand and sodium carbonate became plastic and then liquid below $800°$ C. After cooling, a new and very hard material, a piece of glass, was formed.

When the temperature is increased, the sodium carbonate transforms in carbon dioxide, CO$_2$, which escapes and in a solid phase, in principle made of sodium oxide, Na$_2$O. From a thermodynamic point of view the mixture becomes more stable if the atoms of sodium are ionized to give Na$^+$ ions while the remaining oxygen atoms join the network of silicon and oxygen atoms of silica, which now has a nonstoichiometric formula SiO$_{2+x}$. All the silicon atoms have four covalent bonds. Most of the oxygen atoms have two covalent bonds, but some of them have only one covalent bond, the second being a "pending bond"; those mono-coordinated atoms have an extra electron. A piece of glass can thus be considered as a fixed network of negative sites, inside which the positively charged sodium ions are relatively free to move. Low at room temperature, the mobility of the Na$^+$ ions becomes important when the temperature is increased and approaches the softening point.

We now consider a molten bath of some salt, M$^+$A$^-$, being in a liquid phase the M$^+$ and A$^-$ ions can move rapidly. If a glass slide is immersed in such a bath, because of the gradient of chemical concentration that exists in the immediate vicinity of the surface, the M$^+$ ions will exchange with the Na$^+$ ions of the glass sample: as a result, the chemical composition of the glass is modified, starting from the surface. After they have penetrated the glass, the M$^+$ ions continue their diffusion into the glass by further exchange with other sodium ions: a gradient of the concentration of M$^+$ ions is created inside the glass.

When the slide is taken out of the bath and goes back to room temperature, the diffusion stops. Because of the necessary conservation of the electric charge, only single-charged ions can be introduced using this technique. Silver ions are the easiest ions that can be introduced into a glass matrix; in the vicinity of the surface all the Na^+ ions can be replaced by Ag^+ ions.

To the modification of the chemical composition corresponds a correlative variation of the index of refraction. If the index of refraction is increased, the ion exchange process generates an optical wave guide on top of the glass slide. In the case of silver, the variation of the index can be as high as $\Delta n = 0.1$ for a total replacement. There are not many single-charged ions that can be used; beside silver, thallium and potassium can also be used. The variation of chemical composition is not the only reason for the modification of the index of refraction, mechanical stresses should be considered if the size of the introduced ions is different from that of the sodium ions. This is the case for potassium ions: the guides are often birefringent.

The experiment shows that the law of variation of the index, versus the depth z inside the glass, can be described by the following formula:

$$n(z) = n_0 + \Delta n f(z/a),$$

where

n_0 is the index of the glass prior to the ion exchange.

Δn depends on the glass material and on the ion M^+, $0.001 < \Delta n < 0.1$.

a is a parameter that is homogeneous to a length and that gives an order of magnitude of the depth of penetration of the ions. a is determined by the glass material, the temperature, and duration of the exchange.

$f(z/a)$ is a function of the dimensionless variable z/a, it goes to zero when $z \rightarrow \infty$. $f(z/a)$ usually has a maximum on the surface ($z = 0$), but the maximum can also occur below the surface (buried guide).

Ion exchange has proved to be one of the cheapest and most efficient ways to make passive optical waveguides. With silver or thallium ions Δn is of the order of 0.1, the index being increased from 1.5 up to 1.6. Such a value is often too high, lower values are obtained by diluting the M^+A^- salt with sodium nitrate. The depth of penetration, as in many diffusion processes, varies as the square root of the duration. Typically a is of the order of $1\,\mu m$ after a few

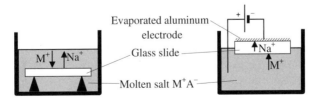

Figure 12.C.7. Principle of the fabrication of optical guides by ion exchange in glass.

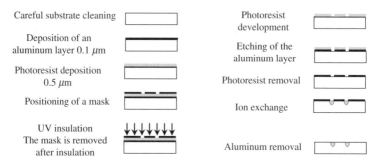

Figure 12.C.8. Typical procedure for making stripe guides by ion exchange.

minutes at 300° C. A depth, as large as 100 μm, can be reached after several hours.

The maximum of the function $f(z/a)$ can be obtained for an abscissa corresponding to a point that is located below the surface; to do so, the glass slide is first plunged into a molten bath of M^+A^- and then into a second bath of pure sodium nitrate. During the first immersion, M^+ ions are introduced inside the glass; during the second immersion some of them will leave the slide and be replaced by sodium ions. The guide is said to be buried; the main advantage is that the guided light becomes less sensitive to the imperfections of the surface, scratches or dust; this is the way to reach attenuation as low as 0.1 db/cm.

The ion exchange method is compatible with the usual photolithographic techniques as shown in Figure 12.C.7. An aluminum mask is transferred to the glass slide before immersion in the molten salt.

The ions can also be put in motion by a voltage applied across the glass sample: ions are introduced on one side of the slide, while other ions are extracted from the other side. This kind of electrical control of the exchange allows an easy control of the shape of the index profile. With a slide having a thickness of 1 mm, a temperature of 300° C and a voltage of 100 V, the current is of the order of 100 μA.

12.C.2.3.2. *Integrated Optical Amplifiers*

A severe drawback of glass-based integrated optical circuits comes from the passivity of the glass material which shows no electrooptic effects and has no direct amplification properties. This last inconvenience has recently been circumvented, and integrated laser amplifiers as well as integrated laser oscillators can be made using the ion exchange technique. The trick is to ask the glassmaker to introduce suitable ions inside a glass matrix; slides are then cut and polished before performing the ion exchange. Very interesting results are obtained with neodymium and erbium ions.

Figure 12.C.9 shows an example of integrated optical amplifiers made by Teemphotonics, this is a good illustration of the possibilities of integrated

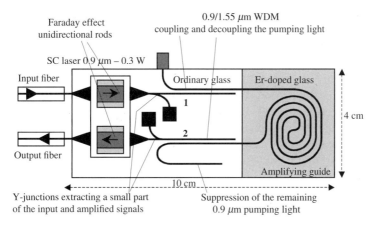

Figure 12.C.9. Integrated amplifier made by Teemphotonics.

optics. As Er-doped lasers have a three-level pumping scheme, the unpumped parts are absorbing and all the passive components are thus made in ordinary glass. To increase the length of the amplifying guide (40 cm) a special double spiral has been designed. The pumping light comes from a semiconductor laser and is coupled by a Wavelength Division Multiplexer (WDM 1). The remaining part of the pumping light should not reach the output fiber and is decoupled by a second WDM; a Bragg reflector, designed to reflect the pumping light, is added to increase the efficiency. Two Y-junctions extract a small part of the input and amplified signals are sent to two photodetectors; the detected signals are used to control the pumping power and keep the output signal at a given level. An important advantage of integrated amplifiers is their compactness and, most of all, the possibility of integrating many amplifiers on the same chip.

12.C.2.3.3. Wave Guides Made in Lithium Niobate

Lithium niobate, because of its nonlinear optical properties, is a very attractive material for integrated optics; especially for making modulators and electrically activated switches. It can also be doped with erbium and acquire amplification properties. The two main techniques for making guides on a lithium niobate substrate are the thermal diffusion of titanium atoms and the exchange of some lithium atoms with protons.

As for titanium diffusion, a thin layer of titanium (50–120 nm) is first deposited on the surface of the $LiNbO_3$ sample, which is then heated to 1000°C for about ten hours. The titanium atoms penetrate inside the crystal and induce an augmentation of the ordinary and extraordinary indices. Unfortunately, their presence makes the material more sensitive to optical damaging by intense light beams. However, in the telecommunications spectral band

(1300–1600 nm) the damage problem is not so severe and can be ignored as long as the power density remains below 0.1 W inside a $5 \times 5 \, \mu m^2$ monomode guide.

An easy technique for exchanging lithium ions with protons is to immerse the lithium niobate crystal in a molten bath of benzoic acid (C_6H_5COOH), which provides the protons H^+ that exchange with lithium ions. The exchange temperature is about 200° C and lasts for about two hours. The exchange induces mechanical strains that damage the crystal, which must be annealed at 350° C over several hours. The extraordinary index is mostly concerned by the exchange; good guides are obtained with losses of 0.15 db/cm in the 800–1500 nm band.

13

Fourier Analysis and Fourier Transform

Fourier analysis is a good example of a mathematical tool that has been exhibited by physicists and for which mathematicians have developed a very powerful and elegant theory, for the highest mutual benefit. We will consider Fourier analysis as a useful tool and will not raise any problem about the definition of real or complex functions or about their expression by means of Fourier integrals. For the reader who is familiar with this theory we hope that it will be an opportunity of a new visit, and that it will help others to become familiar with it.

13.1. Fourier Series

A periodic function $f(t)$, of angular frequency ω, can always be considered as the sum of harmonic functions whose frequencies are multiples of ω, which is called the *fundamental frequency*. Using obvious notations we write

$$f(t) = \sum_{n=0}^{\infty} (a_n \cos n\omega t + b_n \sin n\omega t) = \sum_{n=0}^{\infty} (c_n \cos(n\omega t + \phi_n)),$$

$$f(t) = \sum_{n=0}^{\infty} \frac{c_n}{2} \left(e^{j\phi_n} e^{jn\omega t} + e^{-j\phi_n} e^{-jn\omega t} \right) = \sum_{n=0}^{\infty} (\alpha_n e^{jn\omega t} + \alpha_n^* e^{-jn\omega t}),$$

$$f(t) = \sum_{n=-\infty}^{+\infty} (\beta_n e^{jn\omega t}). \tag{13.1}$$

Formula (13.1) is the *Fourier series* representation of the periodic function $f(t)$, it's also called the *Fourier development* of $f(t)$. It is strictly equivalent to know $f(t)$ or to know the set of complex numbers $\{\beta_n\}$.

The variable t will always be considered as *real*. If the function $f(t)$ is also real we have $\beta_n = \beta_{-n}^*$.

13.2. Fourier Integrals and Fourier Transform

For more general functions $f(t)$, i.e., nonperiodic functions, Fourier series are replaced by Fourier integrals. The function $f(t)$ is then the sum of infinity of sinusoidal functions, as indicated by the following formula:

$$f(t) = \int_{-\infty}^{+\infty} F(v)e^{j2\pi vt}\, dv. \tag{13.2.a}$$

Recall that, following Riemann's definition, an integral is actually the limit of a sum, as the number of elements goes to infinity while their size approaches zero,

$$\int_{-\infty}^{+\infty} F(v)e^{j(2\pi vt)}\, dv = \lim\left[\sum_{-\infty}^{+\infty} F(v_i)e^{j2\pi v_i t}\, \Delta v_i\right] \quad \text{if } \Delta v_i \to 0. \tag{13.2.b}$$

$F(v_i)$ is a complex number, which has a phase; using language that is familiar to opticians, for the time variation $f(t)$ that is represented in Figure 13.1, formula (13.2.b) can be understood as follows:

- At time $t = 0$, all the components have the same phase and add (interfere) constructively, giving a maximum of amplitude.
- Since the frequency is not the same for the different components, the phase concordance doesn't last long and the amplitude decreases.

From a mathematical point of view, formula (13.2) indicates a correspondence between two different spaces: the space of the functions of time t and the space of the frequency v. It can be established that the inverse transformation, that's to say, the transformation going from $F(v)$ to $f(t)$, is described by a very similar formula:

$$F(v) = \int_{-\infty}^{+\infty} f(t)e^{-j2\pi vt}\, dt, \tag{13.3}$$

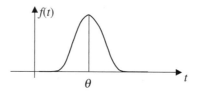

Figure 13.1. The function $f(t)$ is the result of the interference of all the Fourier components. At time $t = 0$, all the components have the same phase and interfere constructively to give a maximum. As time progresses the phase coincidence vanishes and so does the signal.

where $F(v)$ is said to be the Fourier transform (FT) of $f(t)$; conversely, $f(t)$ is the inverse Fourier transform of $F(v)$, (FT^{-1}). Formulas (13.2) and (13.3) can be summarized as

$$F(v) = \text{FT}[f(t)] \quad \text{and} \quad f(t) = \text{FT}^{-1}[F(v)]. \tag{13.4}$$

Some authors make use of the angular frequency ω, instead of the frequency v; the definition is almost the same, except that a coefficient $(2\pi)^{-1/2}$ should be introduced in the definition:

$$F(\omega) = (2\pi)^{-1/2} \int_{-\infty}^{+\infty} f(t) e^{-j2\pi\omega t}\, dt = \text{FT}[f(t)],$$

$$\tag{13.5}$$

$$f(t) = (2\pi)^{-1/2} \int_{-\infty}^{+\infty} F(\omega) e^{+j2\pi\omega t}\, dt = \text{FT}^{-1}[F(\omega)].$$

Bandwidth of a Signal

It is not so easy, and perhaps not so important, to give an exact definition of the bandwidth of a signal. For most functions used in practical problems in Physics the amplitude of the Fourier components goes to zero as the frequency is indefinitely increased.

- In the case of a periodic signal, the module of the coefficients β_n goes to zero if n becomes infinite.
- In the case of a nonperiodic function, $F(v) \to 0$ if $v \to \infty$.

The bandwidth of a signal is the frequency band, inside which the Fourier components significantly differ from zero.

13.3. Some Important Properties of the Fourier Transform

Fourier Transform of the Fourier Transform

$$\text{FT}[\text{FT}[f(t)]] = f(t).$$

Linearity

$$\text{FT}[f_1(t) + f_2(t)] = \text{FT}[f_1(t)] + \text{FT}[f_2(t)].$$

It's because of its linear character that the Fourier transform is so important in Physics and Signal Processing.

Scaling

If the variable t is replaced by t/τ (τ being real), the Fourier transform of $f(t/\tau)$ is proportional to $F(\nu\tau)$:

$$\text{FT}[f(t)] = F(\nu) \quad \leftrightarrow \quad \text{FT}\left[f\left(\frac{t}{\tau}\right)\right] = |\tau| F(\nu\tau).$$

According to this property, we see that the narrower the spectrum in one space, the broader it is in the other space. For example, the Fourier transform of a Dirac pulse, which is infinitely narrow, is a function that is constant, whatever the frequency.

Translation Theorem

Changing the origin of coordinates in one space produces a phase shift in the other space,

$$\text{FT}[f(t-\tau)] = e^{-j2\pi\nu\tau}\, \text{FT}[f(t)],$$

$$\text{FT}[F(\nu-\nu_0)] = e^{j2\pi\nu_0 t}\, \text{FT}[F(\nu)].$$

Fourier Transform of a Derivative

$$\text{FT}\left[\frac{d}{dt}(f(t))\right] = j2\pi\nu\, \text{FT}[f(t)] = j2\pi\nu F(\nu).$$

Fourier Transform of tf(t)

$$\text{FT}[tf(t)] = \frac{j}{2\pi}\frac{d}{d\nu}(\text{FT}[f(t)]) = \frac{j}{2\pi}\frac{d}{d\nu}(F(\nu)).$$

Fourier Transform of a Convolution Product

The convolution product of two functions $f_1(t)$ and $f_2(t)$ is defined by

$$f_1(t) \otimes f_2(t) = \int_{-\infty}^{\infty} f_1(t) f_2(t-\tau)\, d\tau.$$

A convolution in one space is equivalent to a multiplication in the other space,

$$\text{FT}[f_1(t) \otimes f_2(t)] = \text{FT}[f_1(t)]\text{FT}[f_2(t)] = F_1(\nu)F_2(\nu),$$

$$\text{FT}[f_1(t)f_2(t)] = \text{FT}[f_1(t)] \otimes \text{FT}[f_2(t)] = F_1(\nu) \otimes F_2(\nu).$$

Parseval's Theorem—Energy of a Signal

The instant power of a signal is proportional to its squared modulus $|f(t)|^2$, the total energy is obtained by the integral of the squared modulus. The following formula, known as Parseval's theorem, can be established:

$$\int_{-\infty}^{+\infty} |f(t)|^2 \, dt = \int_{-\infty}^{+\infty} |F(t)|^2 \, dv.$$

13.4. Two-Dimensional Fourier Transform

The Fourier transform is generalized to functions of several variables. We will write the formulas in the case of two variables. The two spaces that are connected now have two dimensions; the coordinates in the two spaces will, respectively, be (x, y) and (v_x, v_y). The Fourier transform of $f(x, y)$ is $F(v_x, v_y)$,

$$F(v_x, v_y) = \text{FT}[f(x, y)] = \int_{-\infty}^{+\infty} \int_{-\infty}^{+\infty} f(x, y) e^{j2\pi(xv_x + yv_y)} \, dx \, dy, \qquad (13.6.a)$$

$$f(x, y) = \text{FT}[F(v_x, v_y)] = \int_{-\infty}^{+\infty} \int_{-\infty}^{+\infty} F(v_x, v_y) e^{-j2\pi(xv_x + yv_y)} \, dv_x \, dv_y. \qquad (13.6.b)$$

Formula (13.6.b) is nothing other than the *planar wave decomposition* of $f(x, y)$.

13.5. Some Famous Fourier Transforms

13.5.1. *Gaussian Function*

The Fourier transform of a Gaussian function is another Gaussian function; the Gaussian function is an *eigenfunction* of the Fourier transform

$$\text{FT}\left[e^{-\pi t^2}\right] = e^{-\pi v^2}.$$

13.5.2. *Dirac Pulse $\delta(t)$*

The Dirac pulse was invented as a convenient mathematical representation of a very short signal of finite total energy. It can be considered as the limit of a rectangular signal of constant area, as its duration goes to zero while its amplitude becomes infinite,

Dirac pulse occurring at $t = 0$, $\begin{cases} \delta(t) = 0 & \text{if } t < 0 \text{ or } t > 0, \\ \delta(t) \to \infty & \text{as } t \to 0, \\ \int_{-\infty}^{+\infty} \delta(t) \, dt = 1, \end{cases}$

$$\text{FT}(\delta(t)) = 1.$$

The Fourier transform of a Dirac pulse is a constant function. In the same way the Fourier transform of a constant function is a Dirac pulse.

It is interesting to notice that the existence of discontinuities is not a serious handicap in taking the Fourier transform of a function. When a function has one or several discontinuities, it's often more convenient to work with its Fourier transform.

13.5.3. Rectangular Function

$$\text{Rect}(t) \quad \begin{cases} =1 & \text{if} \quad -\tfrac{1}{2} \le t \le +\tfrac{1}{2}, \\ =0 & \text{if} \quad |t| > +\tfrac{1}{2}, \end{cases}$$

$$\text{FT}[\text{Rect}(t)] = \frac{\sin \pi v}{\pi v} = \text{sinc}(\pi v).$$

13.5.4. Rectangle of Duration θ

Using the scaling theorem we obtain

$$\text{Rect}\!\left(\frac{t}{\tau}\right) \quad \begin{cases} =1 & \text{if} \quad -\tfrac{1}{2} \le \dfrac{t}{\tau} \le +\tfrac{1}{2}, \\[2mm] =0 & \text{if} \quad \left|\dfrac{t}{\tau}\right| > +\tfrac{1}{2}, \end{cases}$$

$$\text{FT}\!\left[\text{Rect}\!\left(\frac{t}{\tau}\right)\right] = \frac{\sin \pi \tau v}{\pi \tau v} = \text{sinc}(\pi \tau v).$$

13.5.5. Sinusoidal Signals with a Limited Duration

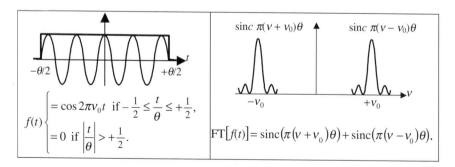

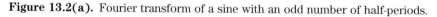

Figure 13.2(a). Fourier transform of a sine with an odd number of half-periods.

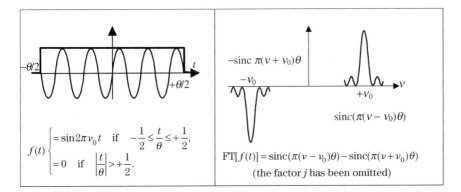

Figure 13.2(b). Fourier transform of a sine with an even number of half-periods.

13.5.6. *Sinusoidal Signals Lasting Forever*

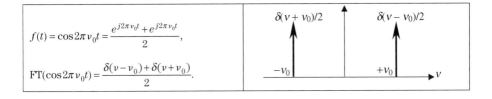

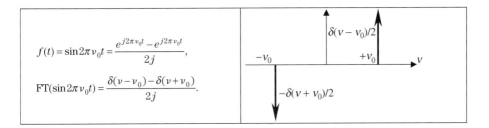

13.5.7. *Dirac Comb*

A Dirac comb is a succession of Dirac pulses that are periodically repeated. The Fourier transform is another comb,

$$\text{Comb}(t) = \sum_{n=-\infty}^{n=+\infty} \delta(t-n),$$

$$\text{FT}[\text{Comb}(t)] = \sum_{n=-\infty}^{n=+\infty} \delta(v-n) = \text{Comb}(v).$$

13.5.8. *Comb of Thin Periodic Teeth*

Infinite Comb of Identical Teeth

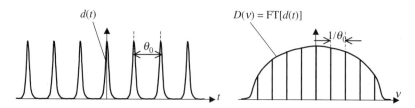

Figure 13.3. Fourier transform of an infinite comb.

Comb with a Finite Number of Identical Teeth

We now consider a comb made of a finite number of identical teeth. Such a signal is the product of a rectangle of duration θ by an infinite comb of identical teeth of duration τ. The Fourier transform is the convolution product of the Fourier transform of a rectangle (i.e., a *sinc* function) by the Fourier transform of an individual tooth.

Figure 13.5 represents a comb of short rectangular pulses of duration τ and periodicity δ, modulated by a rectangular pulse of duration θ. The assumption that τ is much shorter than θ makes the Fourier transform quite simple: it is a comb of a narrow sinc of periodicity $1/\delta$, modulated by a broader sinc having its first zero for $\nu = 1/\theta$.

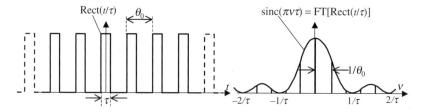

Figure 13.4. The Fourier transform of an infinite comb of identical teeth is a comb of Dirac pulses modulated by the Fourier transform of an individual tooth.

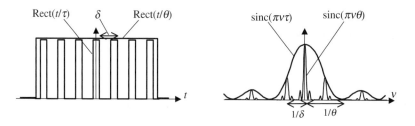

Figure 13.5. Fourier transform of a comb made of a finite number of identical teeth.

13.6. Wave Packet

We consider a sinusoidal wave (frequency v, wave vector k) modulated by a bell-shaped function $f(t)$:

$$e(t, z) = f(t)\cos(2\pi v_c t - kz).$$

- v_c is a frequency that is defined by the source (radio or optical frequency). The subscript c has been chosen to describe a carrier wave.
- $e(t)$ is called a *wave packet*. The variation of $f(t)$ is slow in comparison with the variation of the carrier, which means that v_c is higher than the frequency components of $f(t)$ and well outside the bandwidth. $k = n(2\pi v_c/c)$ is the wave vector, n is the index of refraction and varies with the frequency (dispersion).

The linear property of the Fourier transform makes it easy to obtain the spectrum of a wave packet by a simple application of the following trigonometric identity to each of the Fourier components of $f(t)$:

$$\cos(2\pi v_c t - kz)\cos 2\pi vt = \tfrac{1}{2}[\cos(2\pi(v_c + v)t - kz)] + [\cos(2\pi(v_c - v)t - kz)].$$

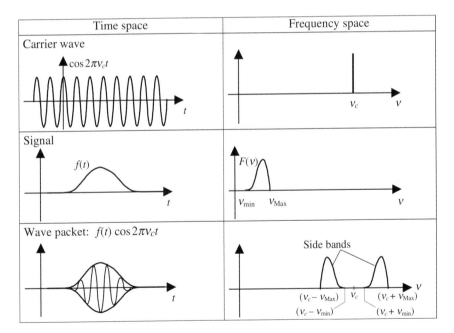

Figure 13.6. Respective representation of a carrier wave, a signal, and a modulated wave, and of their spectra.

Each Fourier component of $f(t)$ generates two components in the Fourier development of the modulated signal. If, as is usually the case, the Fourier transform $F(v)$ of $f(t)$ is a bell-shaped curve starting at v_{min} and ending at v_{Max}, the Fourier transform of the wave packet is made of two bell-shaped curves. Located on both sides of the carrier frequency they are easily deduced from $F(v)$, since their equations are, respectively, $F(v_c \pm v)$. They are called the side bands.

Phase Velocity—Group Velocity

A wave packet is the superposition of an infinite number of sinusoidal components that, having different frequencies, don't propagate at the same speed. If we want to study the propagation of a wave packet, we must study the propagation of its various components and calculate the Fourier integrals at each point. Let us write the equation of the wave packet at the origin of the coordinates and at a point of abscissa z.

- At the origin $z = 0$:

$$e(t, z = 0) = f(t)\cos 2\pi v_c t = \tfrac{1}{2} f(t)(e^{j2\pi v_c t} + e^{-j2\pi v_c t}),$$

$$E(v) = \frac{1}{2} \int_{-\infty}^{+\infty} f(t)e^{j2\pi(v_c + v)t}\, dv + \frac{1}{2} \int_{-\infty}^{+\infty} f(t)e^{j2\pi(v - v_c)t}\, dv.$$

By application of the translation theorem we again find the two side bands,

$$E(v) = \tfrac{1}{2}(F(v - v_c) + F(v + v_c)).$$

- At point M:

$$e(t, z) = \int_{-\infty}^{+\infty} E(v)e^{j(2\pi v t - k(v)z)}\, dv. \tag{13.7}$$

If the wave packet propagates in vacuum, the module of the wave vector is independent of the frequency and is equal to $k_0 = 2\pi(v/c)$.

$$e(t, z) = \int_{-\infty}^{+\infty} E(v)e^{j2\pi v\left(t - \frac{z}{c}\right)}\, dv. \tag{13.8.a}$$

By application of the translation theorem of Section 13.3, equation (13.8.a) becomes

$$e(t, z) = e((t - z/c), z = 0). \tag{13.8.b}$$

Formula (13.7.b) is a clear consequence of the fact that, on their arrival at the abscissa z, the Fourier components have kept the same phase repartition that they had at $z = 0$, since all of them travel at the same speed c. The signal that is seen, at time t by an observer sitting at abscissa z, is identical to the signal that was observed earlier at $(t - z/c)$ by an observer sitting at the origin.

Let us come back to a dispersive material for which the module $k(v)$ of the wave vector is a function of the frequency. The bandwidth of a side band is much smaller than the carrier frequency, which allows us to replace the wave vector by the approximate development,

$$k_{(v+v_c)} = k_{(v_c)} + (v+v_c)\frac{dk}{dv} + \frac{(v+v_c)^2}{2}\frac{d^2k}{dv^2}.$$

The contribution of the side band $(v + v_c)$ to the signal at the abscissa z is given by

$$F(v)e^{j[2\pi(v+v_c)t-k_{(vc)}z]} = F(v)e^{j[2\pi v_c t-k_{(vc)}z]}e^{j2\pi v\left(t-\frac{1}{2\pi}\frac{dk}{dv}z\right)},$$

$$f(t,z) = \frac{1}{2}e^{j[2\pi v_c t-k_{(vc)}z]}\int_{-\infty}^{+\infty} F(v)e^{j2\pi v\left(t-\frac{1}{2\pi}\frac{dk}{dv}z\right)}dv,$$

$$f(t,z) = \frac{1}{2}e^{j[2\pi v_c t-k_{(vc)}z]}f\left(t-\frac{z}{V_G}\right) \quad \text{with} \quad V_G = 2\pi\frac{dv}{dk}. \tag{13.9}$$

In formula (13.9) we recognize the carrier wave which propagates at the usual phase velocity $v_c/k_{(vc)}$. The second term, $f(t - z/V_G)$, which represents the speed of propagation of the maximum of the wave packet, is called the *group velocity*. It is often convenient to introduce the index of refraction n:

$$k_{(v)} = n2\pi\frac{v}{c} \quad \rightarrow \quad dk = \frac{2\pi}{c}\left[n+v\frac{dn}{dv}\right]dv,$$

$$V_G = \frac{c}{\left[n+v\dfrac{dn}{dv}\right]} \approx \frac{c}{n}\left[1+\frac{v}{n}\frac{dn}{dv}\right] = 1+\frac{\lambda}{n}\frac{dn}{d\lambda}.$$

While propagating, the shape of the wave packet is also modified and the pulse duration changes. Our first-order approximation is not accurate enough to evaluate this modification.

Index